PROCEEDINGS

OF THE

1993 INTERNATIONAL CONFERENCE

ON

PARALLEL PROCESSING

August 16 – 20, 1993

Vol. III Algorithms & Applications
Salim Hariri and P. Bruce Berra, Editors
Syracuse University

Sponsored by

THE PENNSYLVANIA STATE UNIVERSITY

CRC Press
Boca Raton Ann Arbor Tokyo London

ISSN 0190-3918
ISBN 0-8493-8983-6 (set)
ISBN 0-8493-8984-4 (vol. I)
ISBN 0-8493-8985-2 (vol. II)
ISBN 0-8493-8986-0 (vol. III)
IEEE Computer Society Order Number 4500-22

Additional copies may be obtained from:

CRC Press, Inc.
2000 Corporate Blvd., N.W.
Boca Raton, Florida 33431

Preface

Interest in parallel processing continues to be high; this year we received 477 papers from 28 different countries, representing a multiplicity of disciplines. In order to accommodate as many presentations as possible and yet maintain the high quality of the conference, we have accepted papers as regular and concise. Regular papers offer well-conceived ideas, good results, and clear presentations. While concise papers also present well-conceived ideas and good results, they do so in fewer pages. The table below summarizes the number of submissions by area:

Area	Submitted	Accepted	
		Regular	Concise
Architecture	166	16 (9.6%)	42 (25.3%)
Software	154	16 (10.4%)	39 (25.3%)
Algorithms/Applications	157	21 (13.4%)	31 (19.7%)
Total	477	53 (11.1%)	112 (23.5%)

This year the program activity was structured slightly differently than in past years since we had a program chair and three co-program chairs. With this organization we worked as a team and assigned papers to the three primary areas in somewhat of a balanced mode. Then each of the co-program chairs was largely responsible for their area. Each submitted paper was reviewed by at least three external reviewers. To avoid conflicts of interest, we did not process the papers from our colleagues at Syracuse University. These were handled by Professor Chita Das of Pennsylvania State University. We are very grateful for his professional and prompt coordination of the reviewing process.

We can truly state that the refereeing process ran much more smoothly than we had originally expected. We attribute this to two main reasons. The first is that we sent one copy of each paper outside of the United States for review; these reviews were thorough and timely. The second reason is that we used e-mail almost exclusively. We would like to express our sincere appreciation to the referees for their part in making the selection process a success. The quality of the conference can be maintained only through such strong support.

This year's keynote speaker is Dr. Ken Kennedy, an internationally known researcher in parallel processing. In addition, we have two super panels chaired by Dr. H.J. Siegel and Dr. Kai Hwang who are also internationally known in parallel processing. We are honored to have Dr.'s Kennedy, Siegel and Hwang sharing their visions on parallel processing with us.

Finally, the success of a conference rests with many people. We would like to thank Dr. Tse-yun Feng for his valuable guidance in the preparation of the program.In addition many graduate and undergraduate students aided us in the process and we gratefully acknowledge their help. We would like to thank Faraz Kohari for his help with the database management functions and Emily Yakawiak for putting the proceedings together. We gratefully acknowledge the New York State Center for Advanced Technology in Computer Applications and Software Engineering (CASE) and the Department of Electrical and Computer Engineering for their support.

P. Bruce Berra, Program Chair   The Case Center
C.Y. Roger Chen, Co-Program Chair-Architecture   Electrical and Computer Engineering
Alok Choudhary, Co-Program Chair-Software   Syracuse University
Salim Hariri, Co-Program Chair-Algorithms/Applications   Syracuse, NY 13244

LIST OF REFEREES - FULL PROCEEDINGS

Abdennadher, N.
Aboelaze, M. A.
Abu-Ghazaleh, N.
AbuAyyash, S.
Agarwal, A.
Agarwal, M.
Agha, G.
Agrawal, D. P.
Agrawal, P.
Agrawal, V.
Alijani, G.
Amano, H.
Andre, F.
Annaratone, M.
Antonio, J. K.
Anupindi, K.
Applebe, B.
Arif, G.
Ariyawansa, K. A.
Arthur, R. M.
Arunachalam, M.
Atiquzzaman, M.
Audet, D.
Avalani, B.
Ayani, R.
Ayguade, E.
Babb, J.
Babbar, D.
Bagherzadeh, N.
Baglietto, P.
Baldwin, R.
Banerjee, C.
Banerjee, P.
Banerjee, S.
Barada, H.
Barriga, L.
Basak, D.
Baskiyar, S.
Bastani, F.
Bayoumi, M. A.
Beauvais, J.
Beckmann, C.
Bellur, U.
Benson, G. D.
Berkovich, S.
Berson, D.

Bhagavathi, D.
Bhuyan, L.
Bhuyan, L. N.
Bianchini, R.
Bic, L.
Bitzaros, S.
Blackston, D.
Boku, T.
Bordawekar, R.
Boura, Y.
Bove Jr., V. M.
Brent, R. P.
Breznay, P.
Brown, J. C.
Burago, A.
Burkhardt, W. H.
Butler, M.
Cam, H.
Capel, M.
Casavant, T. L.
Casulleras, J.
Cavallaro, J.
Celenk, M.
Chalasani, S.
Chamberlain, R.
Chandra, A.
Chandy, J.
Chang, M. F.
Chang, Y.
Chao, D.
Chao, J.
Chao, L.
Chase, C.
Chase, C. M.
Chen, C.
Chen, C. L.
Chen, D.
Chen, K.
Chen, S.
Chen, Y.
Cheng, J.
Cheong, H.
Chiang, C.
Chien, C.
Chiueh, T.
Choi, J. H.

Christopher, T.
Chu, D.
Chung, S. M.
Clary, J. S.
Clifton, C.
Cloonan, T. J.
Cook, G.
Copty, N.
Cortadella, J.
Costicoglou, S.
Crovella, M.
Cukic, B.
Cytron, R.
Dahlgren, F.
Dandamudi, S.
Das, A.
Das, C.
Das, C. R.
Das, S.
Das, S. K.
Datta, A. K.
Davis, M. H.
Debashis, B.
Defu, Z.
Delgado-Frias, J. G.
Demuynck, M.
Deshmukh, R. G.
Dimopoulos, N. J.
Dimpsey, R.
Douglass, B.
Dowd, P. W.
Drach, N.
Du, D. H. C.
Duesterwald, E.
Duke, D. W.
Durand, D.
Dutt, N.
Dutt, S.
Dwarkadas, S.
Edirisooriya, S.
Efe, K.
El-Amawy, A.
Elmohamed, S.
Enbody, R. J.
Erdogan, S.
Erdogan, S. S.

Eswar, K.
Etiemble, D.
Evans, J.
Exman, I.
Farkas, K.
Felderman, R. E.
Feng, C.
Feng, W.
Fernandes, R.
Field, A. J.
Fienup, M. A.
Fraser, D.
Fu, J.
Ganger, G.
Garg, S.
Genjiang, Z.
Gessesse, G.
Gewali, L. P.
Ghafoor, A.
Ghonaimy, M. A. R.
Ghosh, K.
Ghozati, S. A.
Gibson, G. A.
Gokhale, M.
Goldberg
Gong, C.
Gonzalez, M.
Gornish, E.
Granston, E.
Greenwood, G. W.
Grimshaw, A. S.
Gupta, A. K.
Gupta, S.
Gupta, S. K.
Gurla, H.
Gursoy, A.
Haddad, E.
Hamdi, M.
Hameurlain, A.
Han, Y.
Hanawa, T.
Hao, Y.
Harathi, K.
Harper, M.
Hauck, S.
Heiss, H.

Helary, J.
Helzerman, R. A.
Hemkumar, N. D.
Hensgen, D.
Ho, C.
Holm, J.
Hong, M.
Horng, S. J.
Hossain
Hsu, W.
Huang, C.
Huang, G.
Huang, S.
Hughey, R.
Hui-Li
Hulina, P. T.
Hurley, S.
Hwang, D. J.
Hwang, K.
Hwang, S.
Hyder, S.
Ibarra, O. H.
Iqbal, A.
Ito, M.
Iwasaki, K.
Jadav, D.
Jaffar, I.
Jain, R.
Jaja, J.
Jamil, S.
Jazayeri, M.
Jha, N. K.
Jiang, H.
Jin, G.
Jing, W.
Johnson, T.
Jorba, A.
Jou, T.
Jun, Y.
Jung, S.
Kaeli, D. R.
Kain, R. Y.
Kale, L.
Kanevsky, A.
Kannan, R.
Karl, W.

Katti, R.
Kaushik, S. D.
Kavianpour, A.
Kaxiras, S.
Kee, K.
Kee, K. K.
Keleher, P.
Kelly, P. A. J.
Keryell, R.
Kessler, R.
Khan, J.
Khokhar, A. A.
Kim, G.
Kim, H.
Kim, H. J.
Kim, K.
King, C.
Kinoshita, S.
Klaiber, A.
Klaiber, K.
Kobel, C.
Koelbel, C.
Koester, D.
Kong, X.
Koppelman, D. M.
Kornkven, E.
Kothari, S.
Kothari, S. C.
Kravets, D.
Kremer, U.
Kreuger, P.
Krieger, O.
Krishnamoorty, S.
Krishnamurthy, E. V.
Ku, H.
Kudoh, T.
Kuhl, J.
Kumar, B.
Kumar, V.
Kuo, S.
Kurian, L.
Kyo, A.
Kyo, S.
LaRowe, R.
Ladan, M.
Lai, H.

Laszewski, G. v.
Lauria, A. T.
Lavery
Lee, D.
Lee, E.
Lee, G.
Lee, H. J.
Lee, K.
Lee, M.
Lee, O.
Lee, P.
Lee, S.
Lee, S. Y.
Leung, A.
Levine, G.
Li, A.
Li, J.
Li, Q.
Li, Z.
Liang, D.
Lin, A.
Lin, C.
Lin, F.
Lin, R.
Lin, W.
Lin, W. M.
Lin, X.
Lin, Y.
Lisper, B.
Liszka, K.
Liu, G.
Liu, J.
Liu, Y.
Llosa, J.
Lombardi, F.
Lopez, M. A.
Louri, A.
Lu, B.
Lu, P.
Ludwig, T.
Lyuu, Y.
Mackenzie, K.
Maggs, B.
Mahgoub, I.
Mahjoub, Z.
Makki, K.

Mannava, P. K.
Mao, A.
Mao, S.
Marek, T. C.
Maresca, M.
Marinescu, D.
Martel, C.
Masuyama, H.
Matias, Y.
Matloff, N.
Matsumoto, Y.
Mawnava, P. K.
Mayer, H.
Mazuera, O. L.
McDowell, C.
McKinley, P.
Melhem, R.
Menasce, D.
Mendelson, B.
Meyer, D.
Michael, W.
Mitchell, C. D.
Mohamed, A. G.
Mohapatra, P.
Moreira, J. E.
Mouhamed, M. A.
Mounes-Toussi, F.
Moyer, S.
Mrsic-Flogel, J.
Mukai, H.
Mukherjee, B.
Mullin, L.
Mullin, S.
Mutka, M.
Mutka, M. K.
Myer, H.
Nair, V. S. S.
Nakamura, T.
Nang, C. M.
Nassimi, D.
Natarajan, C.
Natarajan, V.
Ni, L.
Ni, L. M.
Nico, P.
Noh, S.

Novack, S.	Raatikainen, P.	Sheu, J.	Tout, W.
Nutt, G.	Radia, N.	Sheu, T.	Traff, J. L.
Obeng, M. S.	Rafieymehr, A.	Shi, H.	Trahan, J.
Oehring, S.	Raghavan, R.	Shin, K. G.	Tsai, W.
Oh, H.	Raghavendra, C. S.	Shing, H.	Tseng, W.
Okabayashi, I.	Raghunath, M. T.	Shiokawa, S.	Tseng, Y.
Okawa, Y.	Rajagopalan, U.	Shirazi, B.	Tzeng, N.
Olariu, S.	Ramachandran, U.	Shoari, S.	Ulusoy, O.
Olive, A.	Ramany, S.	Shu, W.	Unrau, R.
Omer, J.	Ravikumar, C.	Sibai, F. N.	Vaidya, N.
Omiecinski, E.	Ray, S.	Siegel, H. J.	Vaidyanathan, R.
Oruc, A. Y.	Reese, D.	Simms, D.	Valero-Garcia, M.
Ouyang, P.	Reichmeyer, F.	Sinclair, J. B.	Varma, G.
Paden, R.	Rigoutsos, I.	Singh, S.	Varma, G. S. D.
Panda, D.	Ripoli, A.	Singh, U.	Varman, P. J.
Panda, D. K.	Robertazzi, T.	Sinha, A.	deVel, O.
Pao, D.	Rokusawa, K.	Sinha, A. B.	Verma, R. M.
Parashar, M.	Rowland, M.	Sinha, B. P.	Vuppala, V.
Park, B. S.	Rowley, R.	Slimani, Y.	Wagh, M.
Park, C.	Roysam, B.	Snelick, R.	Wahab, A.
Park, H.	Ruighaver, A. B.	So, K.	Wakatani, A.
Park, J.	Ryan, C.	Soffa, M.	Wallace, D.
Park, K. H.	Saghi, G.	Son, S.	Wang, C.
Park, S.	Saha, A.	Song, J.	Wang, D.
Park, Y.	Sajeev, A.	Song, Q. W.	Wang, D. T.
Parsons, I.	Salamon, A.	Srimani, P. K.	Wang, H.
Patnaik, L. M.	Salinas, J.	Su, C.	Wang, H. C.
Pears, A. N.	Sarikaya, B.	Suguri, T.	Wang, M.
Pei-Yung-Hsiao	Sass, R.	Sun, X.	Watts, T.
Perkins, S.	Schaeffer, J.	Sundareswaran, P.	Weissman, J. B.
Peterson, G.	Schall, M.	Sunderam, V.	Wen, Z.
Petkov, N.	Schoinas, I.	Sussman, A.	Wills, S.
Pfeiffer, P.	Schwederski, T.	Sy, Y. K.	Wilson, A.
Phanindra, M.	Schwiebert, L.	Szafron, D.	Wilson, D.
Picano, S.	Seigel, H. J.	Szymanski, T.	Wilton, S. F.
Pinto, A. D.	Seo, K.	Takizawa, M.	Wilton, S. J. E.
Pissinou, N.	Sha, E.	Tan, K.	Wittie, L.
Podlubny, I.	Shah, G.	Tandri, S.	Wong, W.
Poulsen, D.	Shang, W.	Taylor, V. E.	Wu, C.
Pourzandi, M.	Shankar, R.	Tayyab, A.	Wu, C. E.
Pradhan, D. K.	Sharp, D.	Temam, O.	Wu, H.
Pramanick, I.	Sheffler, T. J.	Thakur, R.	Wu, J.
Prasanna, V. K.	Sheikh, S.	Thapar, M.	Wu, M.
Pravin, D.	Shekhar	Thayalan, K.	Wu, M. Y.
Puthukattukaran, J. J.	Shen, X.	Thekkath, R.	Xu, C.
Qiao, C.	Sheng, M. J.	Torrellas, J.	Xu, H.

Xu, Z.
Yacoob, Y.
Yalamanchili, S.
Yamashita, H.
Yang, C.
Yang, C. S.
Yang, Y.
Yen, I.
Yeung, D.
Youn, H. Y.
Young, C.
Young, H.
Young, H. C.
Youngseun, K.
Yousif, M.
Yu, C.
Yu, C. S.
Yu, S.
Yuan, S.
Yuan, S. M.
Yum, T. K.
Zaafrani
ZeinElDine, O.
Zeng, N.
Zhang, J.
Zhang, X.
Zhen, S. Q.
Zheng, S. Q.
Zhou, B. B.
Zhu, H.
Ziavras, S. G.
Zievers, W. C.
Ziv, A.
Zubair, M.

AUTHOR INDEX - FULL PROCEEDINGS

Volume I = Architecture
Volume II = Software
Volume III = Algorithms & Applications

TABLE OF CONTENTS
VOLUME III - ALGORITHMS & APPLICATIONS

SESSION 1C

NUMERICAL ALGORITHMS

SPACE – TIME REPRESENTATION OF ITERATIVE ALGORITHMS AND THE DESIGN OF REGULAR PROCESSOR ARRAYS

E.D. Kyriakis - Bitzaros, O.G. Koufopavlou[*], and C.E. Goutis
VLSI Design Laboratory
Dept. of Electrical Engineering
University of Patras, GREECE
e-mail: bitzaros@grpatvx1 (bitnet)

[*]IBM T.J. Watson Research Center
30 Saw Mill River Road
Hawthorne, NY 10532
e-mail: koufop@yktvmh (bitnet)

Abstract -- *A novel space-time representation of iterative algorithms, which can be expressed in nested loop form and may include non-constant dependencies is proposed and a systematic methodology for their mapping onto regular processor arrays is presented. In contrast to previous design methodologies, the execution time of any variable instance is explicitly expressed in the Dependence Graph, by the construction of the Space-Time Dependence Graph (STDG). This approach avoids the uniformization step of the algorithm and the requirement for fully indexing the variables. Also, in the STDG dependence vectors having opposite directions do not exist and therefore, a linear mapping of the STDG onto the processor array can always be derived. Efficient 2-D and 1-D regular processor arrays are produced by applying the method to the Warshall-Floyd algorithm.*

INTRODUCTION

A lot of efforts have been made for the exploitation of the parallelism of iterative algorithms, written in the form of nested loops, either by designing special purpose processor arrays [1-3,8,9,11-13,16-20,23] or by optimizing compilers [4,7,14,15,22] for shared memory multiprocessors. Various design concepts and methods can be found in the literature for the design of systolic, regular and piecewise regular processor arrays. Most of them are based on the pioneering work of Karp *et al* [6] and Lamport's Coordinate or Hyperplane method [10].

Many categories of algorithms have been defined, with respect to the form of the dependencies between the variables. Algorithms characterized by constant dependence vectors are known as Uniform Recurrence Equations (UREs) [6,17] or Regular Iterative Algorithms (RIAs) [18] and they are the most elaborated category. Affine Recurrence Equations (AREs) [16,20,23], Single Assignment Codes (SACs) and Weak SACs [1,19] are a wider class of algorithms, characterized by dependencies which are linear functions of the loop indices and by propagation of variable instances to multiple points of the index space [19,21]. A common restriction to these

descriptions is that, in order to derive the dependence vectors, all variables should be fully indexed, i.e., they should have dimensionality equal to the number of the loop indices.

The design of a special purpose processor array is accomplished in two steps. The derivation of the Dependence Graph, which is a representation of the *n-D* nested loop in the *n-D* integer Euclidean space, irrespectively of the dimensionality of the variables in the loop body and the mapping of the DG to the processor array space and time using linear [8,11,12,18] or quasi-linear [17,20] allocation and scheduling functions. An algorithm given in the form of Fortran-like nested loops should be modified in order to construct an equivalent single assignment form [4,13,19]. This modification is known as uniformization and it is a crucial point, since a systematic technique does not exist, the transformation is not unique, the use of additional intermediate variables which impose computational overhead, is inevitable, and an amount of potential parallelism may be lost, due to restrictions of the final description [19]. In many cases, even after the algorithm modification and the derivation of a regular DG it is not possible to obtain a processor array using linear mapping, due to existence of dependence vectors having opposite directions. A well known example of this category is the transitive closure and the shortest path algorithms, where heuristic transformations have been used to derive a DG without conversing dependencies [8,11,13,16].

The use of an additional index representing time was reported by Cappello and Steiglitz [3], in an attempt to unify the linear transformations for the design of processor arrays. According to their approach the additional index, adjoined to the feedback variables only, is initially set to zero, and then, a linear transformation is applied to interchange the time index with a space index and to define a correct sequence for the execution of the algorithm. This technique can be applied to UREs only and it is equivalent to the linear transformations which are widely used in the literature.

In this paper, a novel approach for the representation of iterative algorithms, is introduced. The key idea is to explicitly express the execution ordering of the variable instances in the DG and to construct a Space-Time DG (STDG), where both time and space are encountered, in contrast to previous approaches, where only the positions of the variables in the index space are considered. The nodes of the STDG are located to the integer points of a $(q+1)$-dimensional integer Euclidean space, where the first q coordinates correspond to space and the $(q+1)$th coordinate corresponds to time. The space coordinates of the position vector are the indices of the variable instance. The $(q+1)$-th order coordinate corresponds to the execution time of the variable instance, according to a preliminary timing, which is computed taking into account the ordering imposed by the algorithm and possibly an estimation of the delay due to hardware restrictions of the desired architecture.

The aforementioned Space-Time representation offers two significant advantages. It is not required to derive a fully indexed version of the variables and therefore, the uniformization step is avoided. Also, dependence vectors with opposite directions do not exist in the STDG. This is due to the fact that a dependence implies delay in the execution and thus the Space-Time Dependence Vectors (STDVs) have always positive entry in the time coordinate. As a consequence, a linear mapping of the STDG onto an array processor always exists. The constraints for the derivation of the linear transformation are equivalent to those of the conventional approaches, properly modified to be tailored to the needs of the Space-Time representation. A little computational overhead is imposed in the design phase, because loop unrolling techniques should be applied in order to derive the preliminary timing.

The paper is organised as follows. In section 2 the methodology for the construction of the STDG and the derivation of the linear transform for its mapping to the desired processor array is described. An application of the method to the Warshall-Floyd algorithm for the transitive closure problem is presented in section 3 and 2-D and 1-D architectures are systematically derived and compared to those known from the literature. Finally, the conclusions are discussed.

THE SPACE-TIME MAPPING METHOD

The general form of a nested loop algorithm is

$$
\begin{aligned}
&\text{for } i_1 = \ell_1 \text{ to } u_1, \ step_1 \\
&\qquad \vdots \\
&\qquad \text{for } i_n = \ell_n \text{ to } u_n, \ step_n \\
&\qquad loop\ body \qquad\qquad\qquad (1) \\
&\qquad \text{next } i_n \\
&\qquad \vdots \\
&\text{next } i_1
\end{aligned}
$$

In the loop body there are one or more statements, where arrays of variables are evaluated recursively. Each entry of the array is called *variable instance* and may be addressed multiple times during the algorithm execution. Without loss of generality it is assumed that $\ell_j = 1$ and $step_j = 1$ for all $j = 1, ..., n$. The upper bound of any index may be a function of the outer ones. The set

$$J = \{ i^n : i^n = (i_1, i_2, ..., i_n)^{\mathrm{T}}, \ \ell_j \leq i_j \leq u_j, j = 1,2,...,n \ \},$$

is called *index space* or *iteration space* of the loop. The n-D column vector is denoted by i^n. Let a variable instance computed at point i_1^n use the value of a variable instance generated at point i_2^n. The vector $d = i_1^n - i_2^\kappa$ is called *Dependence Vector (DV)*. Consequently, for the determination of the DVs it is necessary to obtain a fully indexed expression of all variable instances. The problem of such index transformations, when the dimensionality of the variables differs from that of the index space is still open [16].

In the loop body two types of variables may exist: *input variables* and *feedback variables*. The variables which are used only in the computations and they do not change their values during the loop execution, are called input variables. The variables which appear in both the left and the right hand side of the statements, are called feedback variables.

Space-Time Representation

In order to simplify the description let us, first, consider a nested loop of the form of (1) having one statement in the loop body and let q, $q \leq n$, be the dimensionality of the array of variables evaluated in the statement. Also, the execution time of the loop body is assumed to be unitary. The concept of the Space-Time representation is to map the q-D variable instances to the integer points of a $(q+1)$-D space, where the $(q+1)$th order coordinate corresponds to time. This is achieved by adding to all variable instances one more index. For the input variables this index can take any value, because they do not affect the execution ordering of the loop. These variables can be placed to the points where they are used a priori. The dependencies that affect the parallelism of the loop are those between the feedback variables. For any instance of a feedback variable the time index should be calculated. It should be noted that, the inherent lexicographical ordering of the indices in the execution of the loop by a serial computer contains no timing information, but only the sequence of the evaluation of the variable instances, which of course may not be altered in a parallel implementation.

An estimation of the degree of parallelism in an algorithm and a preliminary timing for the evaluation of each variable instance is given by the free schedule. The free schedule of an algorithm, expressed in URE form, is given by the formula [6]:

$$
t(i^n) = \begin{cases} 1 + \max_k \left\{ t(i^n - d_k^n) \right\}, & \text{if } (i^n - d_k^n) \in J \\ 1, & \text{otherwise} \end{cases} \quad (2)
$$

where k is the number of the dependence vectors.

The free schedule expresses the minimum possible execution time of an algorithm, since no hardware restrictions are taken into consideration and every variable instance is evaluated when all the operands on which it depends are available. All variable instances that have equal execution time can be executed simultaneously by a parallel processor without affecting the results of the algorithm. To obtain a valid free schedule eq.(2) should be computed using the *loop unrolling* technique [15]. Complete loop unrolling is a time consuming technique if the range of the indices is large. However, for a given algorithm, due to the regularity of the dependencies, it is possible to obtain a formula that gives the execution time as a function of the indices, after loop unrolling for small values of their limits.

The generalization of eq.(2) for algorithms including non-fully indexed variables is achieved by substituting the index point i^n with the variable instance $a(i_1, i_2, ..., i_q)$. Therefore, the free schedule is given by:

$$
t(a(i^q)) = \begin{cases} 1 + \max_k \left\{ t(a(i_k^q)) \right\}, & \text{if } a(i^q) \leftarrow a(i_k^q) \\ 1, & \text{otherwise} \end{cases} \quad (3)
$$

where k is the number of variable instances on which the instance $a(i^q)$ depends and $\leftarrow$ denotes the dependence between the instances.

A more general relation can be derived if the hardware restrictions are also taken into consideration

and the associated delay, t_k^{hd}, is included in eq.(3). The communication delay, may be estimated by the distance between the point of the generation and the point of the usage of the variable instance in the q-D space. Also architectural features of the Processing Elements (PEs), e.g., the time required for one operation, may be incorporated in t_k^{hd}. Therefore, eq.(3) is converted to:

$$
t(a(i^q)) = \begin{cases} \max_k \left\{ t(a(i_k^q)) + t_k^{hd} \right\}, & \text{if } a(i^q) \leftarrow a(i_k^q) \\ 1, & \text{otherwise} \end{cases} \quad (4)
$$

It should be noted that the *max* function is calculated over the sum $\left\{ t(a(i_k^q)) + t_k^{hd} \right\}$ due to the fact that $\left\{ t(a(i_{k_1}^q)) + t_{k_1}^{hd} \right\}$ may be greater than $\left\{ t(a(i_{k_2}^q)) + t_{k_2}^{hd} \right\}$ even when $t(a(i_{k_1}^q)) \leq t(a(i_{k_2}^q))$ if $t_{k_1}^{hd} > t(a(i_{k_2}^q)) - t(a(i_{k_1}^q)) + t_{k_2}^{hd}$. The minimum value of t_k^{hd} is equal to unity, in order to preserve the execution ordering, and eq.(4) is simplified to eq.(3).

After the preliminary timing for the execution of the loop has been obtained, one more index is adjoined to every variable instance according to the following scheme:

$$
a(i^q) \longrightarrow a(i^q, t(a(i^q))) \quad (5)
$$

Thus, the q-D array of variables is transformed to a $(q+1)$-D array, where the $(q+1)$-th order coordinate represents the time. The graphical representation (V, E) of this array in the $(q+1)$-D integer space is defined as the *Space-Time Dependence Graph* (*STDG*). The nodes (V_k) of the STDG are located to the integer points of the space, where a variable instance is defined according to relation (5). An edge $(E_{k_1 k_2})$ belongs to E if the variable instance mapped on V_{k_1} depends on the one mapped on V_{k_2}. The vector d^{q+1} joining the corresponding nodes is called *Space-Time Dependence Vector* (*STDV*). Let $d^{q+1} \equiv (d_s^q, d_t)$, where d_s^q and d_t denote the space and the time part of the STDV respectively, and D denotes the set of the STDVs. From the definition, it holds that the time coordinate of all STDVs is always positive.

Up to now the method for the construction of the STDG for a single statement loop has been analyzed. From this description it is evident that the variable indices are interpreted as space indices of the STDG. This is a major difference between the space time representation and conventional methods, where the loop indices are used as space indices of the DG. Using this basic concept we proceed to the construction of the STDG for multiple statement loops, where the variables may have different number of indices.

Initially, consider that all variables in the loop body have the same dimensionality. In this case all variable instances with equal indices are mapped to points of the STDG which are characterized by equal coordinates of the space part. The execution time of each variable can be computed again according to eqs (3) or (4) but the *max* function should be evaluated over all variables, irrespectively of their names. According to this technique simultaneously executable operations may be mapped to a point of the STDG, since no additional constraints have been taken into consideration, and may lead to increase of the complexity of the PE. This approach is similar to that of the well known methods of [8,12], where no distinction between different variables is made. More sophisticated methods for the separation variables with different names analogous to those of [9,17,18] may be applied after appropriate modifications.

In the more general case, where the dimensionalities of the variables are different to each other the space part of the STDG should have dimensionality equal to the dimensionality of the variable with the maximum number of indices. Therefore, the variables with smaller dimensionality, which lie in a subspace of this space should be placed in a specific position inside the larger space. This is achieved by setting the missing variable indices equal to the lower limit of the corresponding loop index. For example, let $a(i_1, i_3, i_2)$ and $b(i_1)$ be two variables in the body of a loop. The space part of the STDG is 3-*D* and it is specified by i_1, i_3 and i_2; consequently, the space part of b should be (i_1, ℓ_3, ℓ_2). This selection is in accordance with the basic interpretation of the indices of the proposed approach and preserves the regularity of the dependencies.

The construction of the STDG is illustrated using the example of Fig.1. The code of the nested loop is shown in Fig.1a. Every instance of both variables is evaluated four times, for the different values of i and j. Since the variables have only one index the STDG of the algorithm is 2-*D* and it has one space and one time axis, as it is depicted in Fig.1b. The time coordinate has been calculated using eq.(3). In Fig.1b the white circles correspond to instances of variable a, the hatched circles to those of b and finally black circles are the points of the STDG where instances of both variables are located. It can be observed that there are no dependencies having opposite directions, since the STDVs belong in a pointed cone defined by the vectors $(1,1)$ and $(-2,1)$.

Mapping of the STDG

The next step of the proposed method is the derivation of the desired array processor architecture, from the STDG. The target architecture is an *m-D* regular array of processors, where the PEs are located to the integer points of an *m-D* space, the *processor space*. By adding, as previously, one more index corresponding to time, the processor array can be represented by an $(m+1)$-*D* hyperparallelepiped graph. The time coordinate can take any positive integer value in the interval $[1, +\infty)$. The communication links are expressed by integer vectors that connect the corresponding points of the graph. Their time coordinate express the delay associated to the link. The mapping of the STDG into the $(m+1)$-*D* space-time of the processor array is achieved by using a linear transformation. The constraints for the derivation of a valid array structure from the DG of an algorithm, are known from the literature [8,12,16]. Similar restrictions hold for the mapping of the STDG, but they must be adapted to the space-time representation.

Let Φ be a linear transformation such that:

$$(i_1, i_2, ..., i_q, t)^T \xrightarrow{\Phi} (p_1, p_2, ..., p_m, t')^T$$

where p_i, $i = 1, ..., m$ are the space coordinates of the processor array and t' is the activation time of the PE computing variable instance $a(i^q, t)$. In matrix formulation the transform can be written as:

$$\Phi : \begin{bmatrix} S \\ T \end{bmatrix} = \begin{bmatrix} s_{i_1 1} & s_{i_2 1} & \cdots & s_{i_q 1} & s_{t1} \\ \vdots & \vdots & \ddots & \vdots & \vdots \\ s_{i_1 m} & s_{i_2 m} & \cdots & s_{i_q m} & s_{tm} \\ t_{i_1} & t_{i_2} & \cdots & t_{i_q} & t_t \end{bmatrix} \quad (6)$$

where S is a $(q+1) \times m$ submatrix and T is a $(q+1)$ row vector representing the space and the time part of the transform respectively. From eq.(6) it is clear that the time coordinate of the nodes of the STDG affects the allocation of the variable instances.

A transform is valid if the correct ordering of the algorithm is preserved, and if every operation is executed when all of its operands are available [8,12]. To obtain the desired transformation these constraints should be expressed by using the dependence vectors. By letting $d' \equiv (d'_s, d'_t)^T = \Phi \cdot d^q$, the required movement of the data within the array is expressed by d'_s while the associated delay by d'_t. In order to simplify the notation, the dimensionality of the transformed vectors is omitted; obviously d' is $(m+1)$-*D*, while d'_s is *m-D*.

From the construction of the STDG the time coordinate of a STDV gives the minimum possible delay between the operations connected by this vector. Therefore the execution ordering is preserved if this delay after the transformation is greater than or equal to the initial delay. This is expressed by the inequality:

$$\forall d \in D, \quad T \cdot (d_s, d_t)^T \geq d_t \qquad (7)$$

To ensure that all the operands are available before the execution of the operation in which they are involved the delay of the transformed STDVs must be greater than or equal to the time required for the data transfer along the transformed space part of the STDVs. This restriction is satisfied if

$$\forall d \in D, \quad t(d_s') \leq d_t' \qquad (8)$$

where $t(d_s')$ is the time required for the data movement between two PEs connected by d_s', and it depends on the available interconnections in the processor array. In general, the vector d_s' can be written as a linear combination of the existing links in the array with integer coefficients, $\{c_l\}$, $l = 1, 2, ..., L$ Then,

$$t(d_s') = \sum_{l=1}^{L} |c_l| \cdot t_l \qquad (9)$$

where $|\alpha|$ denotes the absolute value of α and t_l is the delay associated to each link.

A widely used and easily implemented architecture is the *m-D* mesh connected array. In this architecture every PE is directly connected to its near neighbours in the *m-D* space. It is assumed that a data transfer between neighbour PEs can be performed in one time unit. Consequently, the time required for a data movement across d_s' is equal to the sum of the absolute values of its coordinates. Therefore, eq.(9) is simplified to:

$$t(d_s') = \sum_{l=1}^{m} |d_l'| \qquad (10)$$

The solution of the integer programming problem composed by eqs (7) and (8) and (9) or (10) gives the valid transformations. As it has already been mentioned, this problem always has a solution because, the STDVs belong to a pointed cone. The number of the inequalities can be reduced drastically as it suffices to use only the rays of the cone for the solution of the system [5,17]. Additional constraints, e.g., the total execution time and/or the number of the PEs, can be imposed to reduce the set of feasible solutions.

PROCESSOR ARRAYS FOR THE ALGEBRAIC PATH PROBLEM

Array Architectures for the Warshall-Floyd Algorithm

Algorithms for the Algebraic Path Problem (APP) can be used to compute the transitive closure and all shortest paths in a connected graph [8,11]. It has also been proved that the non-singular matrix inversion problem, using Gauss-Jordan elimination, is a special case of the APP. Processor arrays designed for the transitive closure can be easily modified to compute the APP [8,16]. A well known algorithm for the solution of these problems is the Warshall-Floyd algorithm [8,11]. The original form of the algorithm is:

```
for k = 1 to N
  for j = 1 to N
    for i = 1 to N
      C(i,j) = C(i,j) ⊕ [C(i,k) ⊗ C(k,j)]     (11)
    next i
  next j
next k
```

For the transitive closure problem C is the connectivity matrix of the graph and $\oplus$ and $\otimes$ represent the logical "OR" and the logical "AND" operations, respectively. In the output matrix $C(i,j) = 1$ if nodes i and j are directly or indirectly connected. For the all shortest paths problem, initially $C(i,j)$ contains the length of the connection between nodes i and j and $\oplus$ represents the *min* function while $\otimes$ the addition. In the output, $C(i,j)$ is equal to the shortest path from i to j.

It can be easily seen that the algorithm specified in eq.(11) is not in single assignment form. Using the properties of the specific problem instances, heuristic transformations were applied to generate single assignment forms [8,11]. The resulting algorithm has five dependence vectors. Many conditional statements are also used to define the region of the index space where each dependence is valid. Spiral, hexagonal, and orthogonal 2-D arrays [8,13,16] as well as linear arrays [11] have been derived by using linear projections of the DG. A simpler algorithm, in the form of Affine Recurrence Equations was produced by Quinton *et al* [16] but dependence vectors with opposite directions arise after the uniformization procedure. An ad-hoc method was used to overcome the problem by splitting the algorithm in two steps, where the input of the second step are the results of the first one. The orthogonal array architecture proposed in [8,16], is not completely regular, due to the existence of diagonal links in some PEs. Moreno and Lang derived a 2-*D* array for the APP problem using the Multimesh Graph method [13]. The bidirectional propagation of the data in the fully parallel DG of the algorithm is heuristically eliminated and the resulting DG is projected along the axes' directions. Although the resulting architecture is mesh connected, it is not orthogonal and delay elements are used in the boundaries for the transmition of data. These two architectures are almost identical except from the delay elements and the arrangement of the output, as shown in Fig. 2.

Application of the proposed methodology

The proposed Space-Time representation is applied to the Warshall-Floyd algorithm for the derivation of 2-D and 1-D regular processor arrays from the original description of the algorithm. As the array of the variables involved is 2-D, the STDG of the algorithm should be 3-D. The values of the loop indices, the variable instance evaluated in each point and its execution time according to eq.(4), are shown in Table 1, for $N=3$. In eq.(4) only the communication overhead has been included. The delay is estimated by the sum of the absolute values of the coordinates of the vector connecting the depended variable instances in the 2-D space.

The formula for the determination of the execution time as a piecewise linear function of the loop indices, is produced by observation of the unrolled algorithm for some values of N. The parts of the index space, are bounded by the lines where the sign of the space coordinates of the STDVs changes.

$$t(a(i,j)) = \begin{cases} 3(k-1)+|i-(k-1)|+|j-(k-1)|-1, \\ \quad \text{if } (i>k-1 \text{ and } j>k-1) \\ \quad \text{or } (i<k-1 \text{ and } j<k-1) \\ 3(k-1)+|i-(k-1)|+|j-(k-1)|, \quad (12) \\ \quad \text{if } (i=k-1 \text{ or } j=k-1) \\ \quad \text{or } (i>k-1 \text{ and } j<k-1) \\ \quad \text{or } (i<k-1 \text{ and } j>k-1) \end{cases}$$

The total execution time of the algorithm according to eq.(12) is $5N$-8 for $N>3$, while the free schedule, computed using eq.(3), is $3N$-2.

Returning to our example, for $N=3$, there are ten STDVs:

$d_1 = (0,0,1),$ $d_2 = (0,0,2),$ $d_3 = (1,0,1)$
$d_4 = (2,0,2),$ $d_5 = (0,1,1),$ $d_6 = (0,2,2)$
$d_7 = (-1,0,1),$ $d_8 = (-2,0,2),$ $d_9 = (0,-1,1)$
$d_{10} = (0,-2,2)$

These vectors belong in a pointed cone defined by d_4, d_6, d_8, and d_{10}. It can be easily seen that in the general case the STDVs belong to the cone defined by the vectors

$(N-1,0,N-1),$ $(0,N-1,N-1),$ $(-(N-1),0,N-1),$ and $(0,-(N-1),N-1)$.

2-D array. The 3×3 identity matrix is a valid transform, which satisfies eqs. (7), (8) and (10), for the derivation of a 2-D mesh connected processor array. Thus,

$$\Phi = \begin{bmatrix} 1 & 0 & 0 \\ 0 & 1 & 0 \\ 0 & 0 & 1 \end{bmatrix}$$

The resulting array processor has 9 PEs and each PE evaluates all variable instances having indices equal to its position coordinates, as shown in Fig.3. The architecture is completely regular and each PE is connected only with its four near neighbours in the 2-D space by bi-directional links. In general, the Warshall-Floyd algorithm for a $N \times N$ input matrix, using the proposed method, can be mapped on a $N \times N$ array processor and the total execution time is $5N$-8, for $N>3$. In [8], the minimum possible execution time was $5N$-4 units, due to the overhead added by the transformation to the single assignment form, which is avoided here. Equivalent architectures can be derived by changing the 1s of the space part of the transform to -1 or by interchanging the two first rows of Φ. The resulting arrays differ in the direction of the data movement, and the allocation of the variable instances to the PEs.

1-D array. For the derivation of a linear array, similar to that presented in [11], an acceptable solution of eqs. (7), (8) and (10) is:

$$\Phi = \begin{bmatrix} 0 & 0 & 0 \\ 1 & 1 & 0 \\ 0 & -1 & 2 \end{bmatrix}$$

In Fig.4 the space time mapping of the STDG onto an 1-D array for $N=3$ is shown. The position of the instances in the STDG, according to table 1, are given in the parentheses. For example, in PE1 the three instances of $C(1,1)$ have been allocated and their execution time is 1, 5 and 13 respectively, instead of 1, 3 and 7 according to the preliminary timing. The total execution time is 15 time units, while in [11] it is 17 units. Also the PE and the interconnection complexity is not increased.

CONCLUSIONS

The concept of the space time representation of nested loop algorithms for the design of processor arrays has been developed. According to the proposed method the execution time of each variable is explicitly expressed in the DG, thus, resulting in a space time DG. The STDG can be constructed without the need for the derivation of a fully indexed version of all variables and therefore the uniformization step of the algorithm is avoided. The STDVs, due to the insertion of the time coordinate, which is always positive, belong to a cone and therefore, the linear programming problem used for the derivation of the mapping transform has a solution. The methodology is applied to the Warshall-Floyd algorithm, for which a lot of heuristic transformations were applied in the past in order to avoid the opposite dependence vectors, and leads to efficient 2-D and linear arrays, both in terms of hardware and execution time.

ACKNOWLEDGMENT

This work was partially supported by the Commission of the EC under contract BRA 3281 (ASCIS) in context of the ESPRIT II programme.

REFERENCES

[1] M.K. Birbas, D.J. Soudris, C.E. Goutis, "Design Methodology for Mapping Iterative Algorithms on Array Architectures", *ISCAS 1991*, pp. 3058-3061

[2] J. Bu, and E.F. Deprettere, "Analysis and Modelling of Sequential Iterative Algorithms for Parallel and Pipeline Implementations", *ISCAS 1988*, pp. 1961-1965.

[3] P.R. Cappello, and K. Steiglitz, "Unifying VLSI Array Design with Linear Transformations of Space - Time", in *Advances in Computer Research*, Vol. 2, JAI Press Inc., 1984, pp 23-65.

[4] R. Cytron, and J. Ferrante, "What's in a Name? or The Value of Renaming for Parallelism Detection and Storage Allocation" *ICPP*, 1987, pp. 19-27.

[5] G. Hadley, *Linear Programming*, Addison Wiley, Reading Massachusets, 1962.

[6] R. Karp, R.E. Miller, and S. Winograd, "The Organization of Computations for Uniform Recurrence Equations", *J. of the ACM*, Vol. 14, No.3, pp. 563-590, July 1967.

[7] D.J. Kuck, *The structure of Computers and Computations*, John Willey & Sons, 1978.

[8] S.Y. Kung, *VLSI Array Processors*, Prentice Hall, New Jersey, 1988.

[9] E.D. Kyriakis-Bitzaros, and C.E. Goutis, "An Efficient Decomposition Technique for Mapping Nested Loops with Constant Dependencies into Regular Processor Arrays", *J. of Parallel and Distributed Computing*, Vol. 14, Apr. 1992.

[10] L. Lamport, "The Parallel Execution of DO Loops", *Communications of the ACM*, Vol. 17, No. 2, pp. 83-93, Feb. 1974.

[11] P. Lee, and Z.M. Kedem, "Synthesizing Linear Array Algorithms from Nested For Loop Algorithms", *IEEE Trans. on Computers*, Vol. 37, No 12, pp. 1578-1598, Dec. 1988.

[12] D.I. Moldovan, J.A.B. Fortes, "Partitioning and Mapping Algorithms into Fixed Size Systolic Arrays", *IEEE Trans. on Computers*, Vol C-35, No.1, Jan. 1986, pp. 1-12.

[13] J.H. Moreno and T. Lang, "Matrix Computations on Systolic Type Meshes", *IEEE Computer*, Vol. 23, pp. 32-51, Apr. 1990.

[14] J.K. Peir, and R. Cytron, "Minimum Distance: A Method for Partitioning Recurrences for Multiprocessors", *IEEE Trans. on Computers*, Vol. 38, No 8, pp. 1203-1211, Aug. 1989.

[15] C.D. Polychronopoulos, *Parallel Programming and Compilers*, Kluwer Academic Publishers, 1988.

[16] P. Quinton, and V.V. Dongen, "The mapping of Linear Recurrence Equations on Regular Arrays", *J. of VLSI Signal Processing*, 1, 1989, pp. 95-113.

[17] P. Quinton, "The Systematic Design of Systolic Arrays", in *Automata Networks in Computer Science*, F. Fogelman, Y. Robert, M. Tchuente, (eds), 1987, Ch.9, pp. 229-260.

[18] S.K. Rao, and T. Kailath, "Regular Iterative Algorithms and their Implementation on Processor Arrays", *Proc. of the IEEE*, Vol. 76, No.3, Mar 1988, pp. 259-269.

[19] V.P. Roychowdhury, S.K. Rao, L. Thiele, and T. Kailath, "On the Localization of Algorithms for VLSI Processor Arrays", *VLSI Signal Processing III*, R.W. Brodersen, H.S.Moscovitz, (eds), IEEE Press, 1988, pp 459-470.

[20] V. Van Dongen, "Systolic Design of Parameterized Recurrences", *W.D. 042*, Philips Research Lab., Brussels, Jan. 1987.

[21] Y. Wong, and J.M. Delosme, "Broadcast Removal in Systolic Algorithms", *Proc. of the Int. Conf. on Systolic Arrays*, 1989, pp. 403-412.

[22] M.E. Wolf, and M.S. Lam, "A Loop Transformation Theory and an Algorithm to Maximize Parallelism", *IEEE Trans. on Parallel and Distributed Systems*, Vol. 2, No 4, pp. 452-471, Oct. 1991.

[23] Y. Yaacoby and R. Cappello, "Scheduling a System of Affine Recurrence Equations onto a Systolic Array", *Proc. of the Int. Conf. on Systolic Arrays*, 1988, pp. 373-381.

```
for i = 1 to 4
    for j = 1 to 4
        a(j) = a(i-1) * a(j-1)
        b(j) = a(j) * b(j-2)
    next j
next i
```

(a)

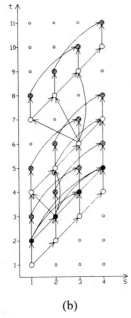

(b)

Fig. 1. Example: (a) A nested loop, and (b) the corresponding Space-Time Dependence Graph

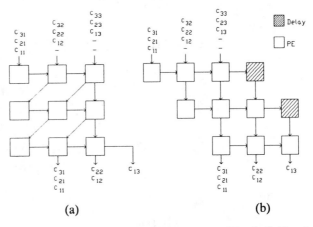

(a)   (b)

Fig. 2. The 2-D arrays for the Warshall-Floyd algorithm proposed in [9] (a) and in [15] (b)

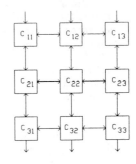

Fig. 3. The proposed regular orthogonal array for the Warshall-Floyd algorithm.

TABLE 1. PARALLEL EXECUTION OF THE WARSHALL-FLOYD ALGORITHM.

Loop index k,i,j	Var inst	Exec time	Loop index k,i,j	Var inst	Exec time	Loop index k,i,j	Var inst	Exec time
1,1,1	C(1,1)	1	2,1,1	C(1,1)	3	3,1,1	C(1,1)	7
1,1,2	C(1,2)	2	2,1,2	C(1,2)	4	3,1,2	C(1,2)	7
1,1,3	C(1,3)	3	2,1,3	C(1,3)	5	3,1,3	C(1,3)	8
1,2,1	C(2,1)	2	2,2,1	C(2,1)	4	3,2,1	C(2,1)	7
1,2,2	C(2,2)	3	2,2,2	C(2,2)	4	3,2,2	C(2,2)	6
1,2,3	C(2,3)	4	2,2,3	C(2,3)	5	3,2,3	C(2,3)	7
1,3,1	C(3,1)	3	2,3,1	C(3,1)	5	3,3,1	C(3,1)	8
1,3,2	C(3,2)	4	2,3,2	C(3,2)	5	3,3,2	C(3,2)	7
1,3,3	C(3,3)	5	2,3,3	C(3,3)	6	3,3,3	C(3,3)	7

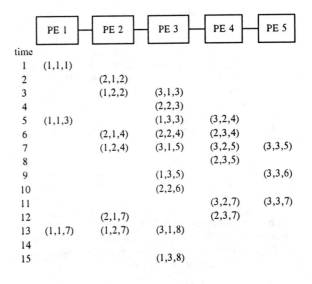

Fig. 4. Processor allocation and timing for the proposed linear processor array computing the Warshall-Floyd algorithm.

On the Parallel Diagonal Dominant Algorithm *

Xian-He Sun

ICASE

Mail Stop 132C

NASA Langley Research Center

Hampton, VA 23681-0001

sun@icase.edu

Abstract

The Parallel Diagonal Dominant (PDD) algorithm is a highly efficient, ideally scalable tridiagonal solver. In this paper, a detailed study of the PDD algorithm is given. First the PDD algorithm is introduced. Then the algorithm is extended to solve periodic tridiagonal systems. A variant, the reduced PDD algorithm, is also proposed. Accuracy analysis is provided for a class of tridiagonal systems, the symmetric and skew-symmetric Toeplitz tridiagonal systems. Implementation results show that the analysis gives a good bound on the relative error, and the algorithm is a good candidate for the emerging massively parallel machines.

1 Introduction

Solving tridiagonal systems is one of the key issues in computational fluid dynamics (CFD) and many other scientific applications [1]. The alternating direction implicit (ADI) method solves partial differential equations (PDEs) by solving tridiagonal systems alternately in each coordinate direction. Discretization of partial differential equations by compact difference schemes also leads to a sequence of tridiagonal systems. However, tridiagonal systems are difficult to solve efficiently on parallel computers. Intensive research has been done on the development of efficient parallel tridiagonal solvers. Many algorithms have been proposed [2, 3]. Most of these parallel tridiagonal solvers increase parallelism by adding additional computation. Recently, Sun, Zhang, and Ni [1] have proposed three parallel algorithms, the parallel partition LU (PPT) algorithm, the parallel hybrid (PPH) algorithm, and the parallel diagonal dominant (PDD) algorithm, for solving tridiagonal systems. Compared with other tridiagonal solvers, which all have at least $O(\log p)$ communication cost, the PDD algorithm has only a small fixed communication cost and a small amount of additional computation. In fact, the PDD algorithm is perfectly scalable, in the the sense that the communication cost and the computation overhead do not increase with the problem size or with the number of processors available.

*This research was supported by the National Aeronautics and Space Administration under NASA contract NAS1-19480 while the author was in residence at the Institute for Computer Applications in Science and Engineering (ICASE), NASA Langley Research Center, Hampton, VA 23681-0001.

Modern technological advances have made it possible to build computers containing more and more processors. The emerging parallel computers, such as the Intel Paragon, Thinking Machine Corporations's CM-5, and Cray's MPP, are noted for their scalable architecture and massively parallel processing. Scalability has become an important metric of parallel algorithms [4, 5]. Its perfect scalability and high efficiency make the PDD algorithm, when applicable, an ideal choice on these new machines. In this paper we give a detailed study of the PDD algorithm. The PDD algorithm proposed in this paper is slightly different from the algorithm proposed in [1]. Extended study is provided for different applications, such as periodic systems, and systems with multiple right-sides. The reduced PDD algorithm is also proposed. Simple formulas are provided for accuracy checking for symmetric and skew-symmetric Toeplitz tridiagonal systems.

This paper is organized as follows. Section 2 will introduce the sequential and parallel PDD algorithms. The applications of the PDD algorithm will be discussed in Section 3. This section will also give the variant of the PDD algorithm for periodic systems and the reduced PDD algorithm. Section 4 will give the accuracy study for the PDD algorithm and the reduced PDD algorithm. Experimental results on an Intel/860 multicomputer will be presented in Section 5. Finally, Section 6 gives the final remarks.

2 The PDD Algorithm

We are interested in solving a tridiagonal linear system of equations

$$Ax = d \qquad (1)$$

where A is a tridiagonal matrix of order n

$$A = \begin{bmatrix} b_0 & c_0 & & & & \\ a_1 & b_1 & c_1 & & & \\ & \cdot & \cdot & \cdot & & \\ & & \cdot & \cdot & \cdot & \\ & & & a_{n-2} & b_{n-2} & c_{n-2} \\ & & & & a_{n-1} & b_{n-1} \end{bmatrix} \qquad (2)$$

$x = (x_0, \cdots, x_{n-1})^T$ and $d = (d_0, \cdots, d_{n-1})^T$. We assume that A, x, and d have real coefficients. Extension to the complex case is straightforward.

2.1 The Matrix Partition Technique

The matrix A in Eq. (2) can be written as

$$A = \tilde{A} + \Delta A, \qquad (3)$$

where

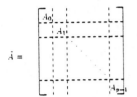 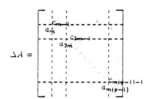

The submatrices $A_i (i = 0, \cdots, p-1)$ are $m \times m$ tridiagonal matrices. For convenience we assume that $n = pm$. Let e_i be a column vector with its ith ($0 \le i \le n-1$) element being one and all the other entries being zero. We have

$$\Delta A = [a_m e_m, c_{m-1} e_{m-1} a_{2m} e_{2m}, c_{2m-1} e_{2m-1}, \cdots,$$

$$c_{(p-1)m-1} e_{(p-1)m-1}] \begin{bmatrix} e_{m-1}^T \\ e_m^T \\ \cdot \\ \cdot \\ e_{(p-1)m}^T \end{bmatrix} = V E^T.$$

Thus, we have

$$A = \tilde{A} + V E^T.$$

Based on the matrix modification formula originally defined by Sherman and Morrison [6] for rank-one changes and assuming that all A_i's are invertible, Eq. (1) can be solved by

$$x = \tilde{A}^{-1} d - \tilde{A}^{-1} V (I + E^T \tilde{A}^{-1} V)^{-1} E^T \tilde{A}^{-1} d. \qquad (4)$$

Let

$$\tilde{A}\tilde{x} = d \qquad (5)$$
$$\tilde{A}Y = V \qquad (6)$$
$$h = E^T \tilde{x} \qquad (7)$$
$$Z = I + E^T Y \qquad (8)$$
$$Zy = h \qquad (9)$$
$$\Delta x = Yy. \qquad (10)$$

Equation (4) becomes

$$x = \tilde{x} - \Delta x. \qquad (11)$$

In Eq.s (5) and (6), $\tilde{x}$ and Y are solved by the LU decomposition method. By the structure of $\tilde{A}$ and V,

this is equivalent to solve

$$A_i[\tilde{x}^{(i)}, v^{(i)}, w^{(i)}] = [d^{(i)}, a_{im}e_0, c_{(i+1)m-1}e_{m-1}], \qquad (12)$$

$i = 0, \cdots, p-1$. Here $\tilde{x}^{(i)}$ and $d^{(i)}$ are the ith block of $\tilde{x}$ and d, respectively, and $v^{(i)}, w^{(i)}$ are possible nonzero column vectors of the ith row block of Y.

2.2 The Algorithm

Solving Eq. (9) is the major computation involved in the conquer part of our algorithms. The matrix Z in Eq. (9) has the form

$$Z = \begin{bmatrix} 1 & w_{m-1}^{(0)} & 0 & & & & \\ v_0^1 & 1 & 0 & w_0^{(1)} & & & \\ v_{m-1}^{(1)} & 0 & 1 & w_{m-1}^{(1)} & & & \\ & & \cdot & \cdot & \cdot & & \\ & & & \cdot & \cdot & \cdot & \\ & & & & 1 & 0 & w_0^{(p-2)} \\ & & & & & 1 & w_{m-1}^{(p-2)} \\ & & & & & v_0^{(p-1)} & 1 \end{bmatrix}$$

where $v^{(i)}, w^{(i)}$ for $i = 0, \cdots, p-1$ are solutions of Eq. (12) and the 1's come from the identity matrix I. In practice, especially for a diagonal dominant tridiagonal system, the magnitude of the last component of $v^{(i)}, v_{m-1}^{(i)}$, and the first component of $w^{(i)}, w_0^{(i)}$, may be smaller than machine accuracy when $p \ll n$. In this case, $w_0^{(i)}$ and $v_{m-1}^{(i)}$ can be dropped, and Z becomes a diagonal block system consisting of $(p-1)$ 2×2 independent blocks. Thus, Eq.(9) can be solved efficiently on parallel computers, which leads to the highly efficient *parallel diagonal dominant* (PDD) algorithm.

Using p processors, the PDD algorithm consists of the following steps:

Step 1. Allocate $A_i, d^{(i)}$, and elements $a_{im}, c_{(i+1)m-1}$ to the ith node, where $0 \le i \le p-1$.

Step 2. Solve (12). All computations can be executed in parallel on p processors.

Step 3. Send $\tilde{x}_0^{(i)}, v_0^{(i)}$ from the ith node to the $(i-1)$th node, for $i = 1, \cdots, p-1$.

Step 4. Solve

$$\begin{bmatrix} 1 & w_{m-1}^{(i)} \\ v_0^{(i+1)} & 1 \end{bmatrix} \begin{pmatrix} y_{2i} \\ y_{2i+1} \end{pmatrix} = \begin{pmatrix} \tilde{x}_{m-1}^{(i)} \\ \tilde{x}_0^{(i+1)} \end{pmatrix}$$

in parallel on the ith node for $0 \le i \le p-2$. Then send y_{2i} from the ith node to the $(i+1)$th node, for $i = 0, \cdots, p-2$.

Step 5. Compute (10) and (11). We have

$$\Delta x^{(i)} = [v^{(i)}, w^{(i)}] \begin{pmatrix} y_{2(i-1)} \\ y_{2i} \end{pmatrix}$$

$$x^{(i)} = \tilde{x}^{(i)} - \Delta x^{(i)}$$

In all of these, one has only two neighboring communications.

Empirically, for most distributed-memory computers, the communication time for a neighboring communication is a linear function of the problem size [1]. Let S be the number of bytes to be transferred. Then the transfer time of a neighboring communication can be expressed as $\alpha + S\beta$, where α represents a fixed startup overhead and β is the incremental transmission time per byte. Assuming 4 bytes are used for each real number, Step 3 and Step 4 take $\alpha + 8\beta$ and $\alpha + 4\beta$ communication respectively. Notice that Y has at most two nonzero entries in every row, and Z is a diagonal block matrix with 1's as diagonal elements, we conclude that the parallel PDD algorithm needs $17\frac{n}{p} - 4$ parallel computation and $2(\alpha + 6\beta)$ communication.

2.3 Scalability Analysis

As parallel machines have been built with more and more processors, the performance metric *scalability* becomes more and more important. Thus, the question is how an algorithm will perform when the problem size is scaled up linearly with the number of processors. Let $T(p, W)$ be the execution time for solving a system with W work (problem size) on p processors. The ideal situation would be when both the number of processors and the amount of work are scaled up N times, the execution time remains unchanged:

$$T(N \times p, N \times W) = T(p, W) \qquad (13)$$

How one should define problem size, in general, is a style under debate. However, it is commonly agreed that the floating point (flop) operation count is a good estimate of problem size for scientific computations. To eliminate the effect of numerical inefficiencies in parallel algorithms, in practice the flop count is based upon some practical optimal sequential algorithm. In our case, the LU decomposition has chosen as the sequential algorithm. It takes $8n - 7$ floating point operations, where 7 is a negligible constant number when n is large. As the problem size W increases N times to W', we have

$$\begin{aligned} W' &= N \times 8n = 8n' \\ n' &= N \cdot n. \end{aligned} \qquad (14)$$

Let τ_{comp} represent the unit of a computation operation normalized to the communication time. The time required to solve (1) by the PDD algorithm with p processors is

$$T(p, W) = (17\frac{n}{p} - 4)\tau_{comp} + 2(\alpha + 6\beta), \qquad (15)$$

and

$$T(N \cdot p, N \cdot W) = (17\frac{n'}{N \cdot p} - 4)\tau_{comp} + 2(\alpha + 6\beta)$$

$$= (17\frac{N \cdot n}{N \cdot p} - 4)\tau_{comp} + 2(\alpha + 6\beta)$$

$$= (17\frac{n}{p} - 4)\tau_{comp} + 2(\alpha + 6\beta)$$

$$= T(p, W).$$

The PDD algorithm has the ideal scalability. Similar arguments could be applied to periodic systems (see Section 3) and the same result would be obtained.

Using the isospeed approach, scalability has been formally defined in [4]. The average unit speed is defined as the quotient of the achieved speed of the given computing system and the number of processors. Since Eq.(13) is true if and only if the average unit speed of the given computing system is a constant, the scalability is defined as the ability to maintain the average unit speed [4]. Let W be the amount of work of an algorithm when p processors are employed in a machine, and let W' be the amount of work of the algorithm when $p' = N \cdot p$ processors are employed to maintain the average speed, then the scalability from system size p to system size $N \cdot p$ of the algorithm-machine combination is defined as

$$\psi(p, N \times p) = \frac{N \cdot p \cdot W}{p \cdot W'} = \frac{N \cdot W}{W'}. \qquad (16)$$

The average unit speed can be represented as

$$A\_S(p, W) = \frac{W}{p \cdot T(p, W)}, \qquad (17)$$

where W is the problem size, p is the number of processors, and $T(p, W)$ is the corresponding execution time. From our early discussion, for the PDD algorithm, when $W' = N \cdot W$, we have $T(N \times p, W') = T(p, W)$. Therefore

$$\begin{aligned} A\_S(N \times p, W') &= \frac{W'}{N \cdot T(N \times p, W')} \\ &= \frac{N \cdot W}{N \cdot T(p, W)} = \frac{W}{T(p, W)}. \end{aligned}$$

That is $W' = N \cdot W$ has maintained the average unit speed, and the scalability is

$$\psi(p, N \times p) = \frac{N \cdot W}{W'} = \frac{N \cdot W}{N \cdot W} = 1. \qquad (18)$$

It is the ideal scalability.

3 Special Applications

In this section, we first discuss some tridiagonal systems arising in CFD applications, the *symmetric* and *skew-symmetric Toeplitz tridiagonal systems*. Then two variants of the PDD algorithm, the *reduced PDD algorithm* and the PDD algorithm for periodic systems, will be presented.

3.1 Toeplitz Tridiagonal Systems

A Toeplitz tridiagonal matrix has the form

$$A = \begin{bmatrix} b & c & & & \\ a & b & c & & \\ & \cdot & \cdot & \cdot & \\ & & \cdot & \cdot & c \\ & & & a & b \end{bmatrix} = [a, b, c]. \qquad (19)$$

Symmetric Toeplitz tridiagonal systems often arise in solving partial differential equations and in other scientific applications. Compact finite-difference scheme [7] is a discretization scheme for solving PDE's. Using the compact scheme, the general approximation of a first derivative has the form:

$$\beta f'_{i-2} + \alpha f'_{i-1} + f'_i + \alpha f'_{i+1} + \beta f'_{i+2}$$
$$= c\frac{f_{i+3} - f_{i-3}}{6h} + b\frac{f_{i+2} - f_{i-2}}{4h} + a\frac{f_{i+1} - f_{i-1}}{2h}$$

Letting

$$\alpha = \frac{1}{3}, \beta = 0, a = \frac{14}{9}, b = \frac{1}{9}, c = 0, \qquad (20)$$

the scheme becomes formally sixth order accurate and the resulting system is $[\frac{1}{3}, 1, \frac{1}{3}]$, a symmetric Toeplitz tridiagonal system. Similarly, a resulting matrix of a sixth order difference scheme of a second derivative is $[\frac{2}{11}, 1, \frac{2}{11}]$. It is symmetric and Toeplitz. Discretized in time, the one dimensional wave equation $u_t = a \cdot u_x$ and the heat equation $u_t = a \cdot u_{xx}$ can be represented as

$$u^{n+1} = u^n + \Delta t \cdot a \cdot u^n_x, \qquad (21)$$

and

$$u^{n+1} = u^n + \Delta t \cdot a \cdot u^n_{xx} \qquad (22)$$

respectively. Using ADI methods [8], parabolic and hyperbolic systems can be solved by solving a sequence of symmetric Toeplitz tridiagonal systems.

Skew-symmetric Toeplitz tridiagonal systems also arise in solving PDEs [8]. For instance, to solve the wave equation $u_t + a \cdot u_x = f$, we have the approximation

$$\frac{a\lambda}{4}v^{n+1}_{m+1} + v^{n+1}_m - \frac{a\lambda}{4}v^{n+1}_{m-1} =$$
$$-\frac{a\lambda}{4}v^n_{m+1} + v^n_m + \frac{a\lambda}{4}v^n_{m-1} + \frac{k}{2}(f^{n+1}_m + f^n_m).$$

The left side is an *skew-symmetric Toeplitz tridiagonal matrix*, $A = [\frac{a\lambda}{4}, 1, \frac{-a\lambda}{4}]$.

3.2 Periodic Tridiagonal Systems

Many PDE's arisen in real applications have periodic boundary conditions. The corresponding resulting linear systems have the form of

$$A = \begin{bmatrix} b_0 & c_0 & & & & & a_0 \\ a_1 & b_1 & c_1 & & & & \\ \cdot & \cdot & \cdot & & & & \\ & \cdot & \cdot & \cdot & & & \\ & & & & \cdot & & \\ & & & & a_{n-2} & b_{n-2} & c_{n-2} \\ c_{n-1} & & & & & a_{n-1} & b_{n-1} \end{bmatrix},$$

and are called *periodic tridiagonal systems*. On sequential machine, periodic tridiagonal systems are solved by combining the solutions of two different right-sides [9], which increases the operation count from $8n - 7$ to $14n - 16$.

The PDD algorithm can be extended to periodic tridiagonal systems. The difference is that, after dropping $w^{(i)}_0$, and $v^{(i)}_{m-1}$, the matrix Z becomes a periodic system of order $2p$:

$$Z = \begin{bmatrix} 1 & w^{(0)}_{(m-1)} & & & v^{(0)}_0 \\ v^{(1)}_0 & 1 & 0 & & \\ & 0 & 1 & \cdot & \\ & & \cdot & \cdot & \cdot \\ & & & \cdot & \cdot & w^{(p-2)}_{m-1} \\ w^{(p-1)}_{m-1} & & v^{(p-1)}_0 & & 1 \end{bmatrix}$$

The dimension of Z is slightly higher than in the non-periodic case, which slightly makes the load on the 0th and (p-1)th processor identical to load on all of the other processors. The parallel computation time remains the same. For periodic systems, the communication at step 3 and 4 changes from one dimensional array communication to ring communication. The communication time is also unchanged. Figure 1 shows the communication pattern of the PDD algorithm for periodic systems.

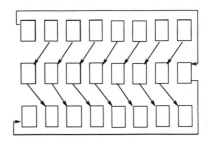

Figure 1. Communication Pattern for Solving Periodic Systems.

3.3 The Reduced PDD Algorithm

In the last step, Step 5, of the PDD algorithm, the final solution, x, is computed by combining the intermediate results concurrently on each processor,

$$x^{(i)} = \tilde{x}^{(i)} - y_{2(i-1)}v^{(i)} - y_{2i}w^{(i)}, \qquad (23)$$

which requires $4(n - 1)$ operations in total and $4m$ parallel operations, if $p = n/m$ processors are used. The PDD algorithm drops off the the first element of

w, w_0, and the last element of v, v_{m-1}, in solving Eq. (9). In [9] we have shown that, for symmetric and skew-symmetric Toeplitz tridiagonal systems,

$$v = \frac{1}{\lambda(a + b\sum_{i=0}^{m-1} b^{2i})}$$

$$\times (\sum_{i=0}^{m-1} b^{2i}, \sum_{i=0}^{m-1} b^{2i}/(-b), \cdots, (-b)^{m-1})^T.$$

So, when m is large enough, we may drop off v_i, $i = \frac{m}{2}, \cdots, m-1$, and $w_i, i = 0, 1, \cdots, \frac{m}{2}-1$, while maintaining the same accuracy. If we replace v_i by $\tilde{v}_i$, where $\tilde{v}_i = v_i$ for $i = 0, 1, \cdots, \frac{m}{2}-1$, $\tilde{v}_i = 0$, for $i = \frac{m}{2}, \cdots, m-1$; and replace w by $\tilde{w}$, where $\tilde{w}_i = w_i$ for $i = \frac{m}{2}, \cdots, m-1$, and $\tilde{w}_i = 0$, for $i = 0, 1, \cdots, \frac{m}{2}-1$; and use $\tilde{v}, \tilde{w}$ in Step 5, we have

Step 5'

$$\Delta x^{(i)} = [\tilde{v}, \tilde{w}] \begin{pmatrix} y_{2(i-1)} \\ y_{2i} \end{pmatrix} \qquad (24)$$

$$x^{(i)} = \tilde{x}^{(i)} - \Delta x^{(i)}. \qquad (25)$$

It requires $2\frac{n}{p}$ parallel operation. Replacing Step 5 of the PDD algorithm by Step 5', we get the reduced PDD algorithm which requires $15\frac{p}{n} - 4$ parallel computations.

4 Accuracy Analysis

In this section we give an accuracy study for a particular class of tridiagonal systems, *symmetric and skew-symmetric Toeplitz tridiagonal systems*. To meet the paging limitation, only final results and outline of proofs are presented here. The reader may refer [9] for more information.

4.1 The Decay Rate of v_{m-1} and w_0

Symmetric Toeplitz tridiagonal systems have the form $A = [\lambda, \beta, \lambda] = \lambda[1, c, 1]$, where $c = \beta/\lambda$. We assume the matrix A is diagonal dominant. That is $|c| > 2$. To study the accuracy of the solution of $Ax = b$, we first study the matrix

$$\tilde{B} = \begin{pmatrix} a & 1 & & & \\ 1 & c & 1 & & \\ & 1 & \cdot & \cdot & \\ & & \cdot & \cdot & 1 \\ & & & 1 & c \end{pmatrix} = [b, 1, 0] \cdot [0, a, 1]$$

where a and b are the real solutions of

$$b + a = c, \quad b \cdot a = 1. \qquad (26)$$

Since $a \cdot b = 1$ and $|c| > 2$, we may further assume that $|a| > 1$, and $|b| < 1$.

Let

$$\Delta B = \begin{pmatrix} b \\ 0 \\ \cdot \\ \cdot \\ 0 \end{pmatrix} (1, 0, \cdots, 0) = \tilde{V}\tilde{E}^T, \qquad (27)$$

and

$$B = \tilde{B} + \Delta B = [1, c, 1]. \qquad (28)$$

Then, by the matrix modification formula (4), the solution of $By = d$ is

$$y = \tilde{B}^{-1}d - \tilde{B}^{-1}\tilde{V}(I + \tilde{E}^T\tilde{B}^{-1}\tilde{V})^{-1}\tilde{E}^T\tilde{B}^{-1}d \quad (29)$$

For $d = (1, 0, \cdots, 0)^T$, by direct calculation, we find

$$y = \frac{b}{a} \cdot \frac{\sum_{i=0}^{n-1} b^{2i}}{a + b\sum_{i=0}^{n-1} b^{2i}} \begin{pmatrix} \sum_{i=0}^{n-1} b^{2i} \\ \cdot \\ (-b)^{n-1} \end{pmatrix}.$$

Thus, for the last element of y, y_{n-1}, and the first element of y, y_0, we have

$$|y_{n-1}| \leq \frac{|b|^{n-1}}{|a|} = |b|^n$$

and

$$|y_0| = \left| \frac{b(1 - b^{2n})}{1 - b^{2(n+1)}} \right| < |b|.$$

For the original system $Ax = d, A = \lambda[1, c, 1]$, the first and last element of x is

$$x_0 = \frac{y_0}{\lambda}, \qquad x_{n-1} = \frac{y_{n-1}}{\lambda}.$$

With similar arguments, we can prove that for $d = (0, \cdots 0, 1)^T$, $Ax = d$ has solution

$$x_i = \frac{y_{n-(i+1)}}{\lambda}. \qquad (30)$$

Since for Toeplitz tridiagonal systems, each submatrix $A_i, i = 0, \cdots, p-1$, has the same structure as A, we have the following theorem:

Theorem 1 *If $\frac{b^{m-1}}{\lambda a}$, $m = n/p$, is less than machine accuracy, then the PDD algorithm gives an approximation to the true solution within machine accuracy.*

4.2 Accuracy of the PDD Algorithm

Theorem 1 says that if v_{m-1}, w_0 are less than machine accuracy, the PDD algorithm gives a satisfactory solution. In most scientific applications, the accuracy requirement is much weaker than machine accuracy. We now study how the decay rate of v_{m-1}, w_0 influences the accuracy of the final solution.

Let x be the solution of the matrix partition formula (4). Let x^* be the corresponding final solution of the PDD algorithm. Then

$$x - x^* = \tilde{A}^{-1}V(I + E^T\tilde{A}^{-1}V - D)^{-1}D \cdot E^Tx,$$

where D is the $2(p-1) \times 2(p-1)$ matrix which contains all the $v_{m-1}^{(i)}, w_0^{(i)}$ elements. Thus,

$$\frac{||x - x^*||}{||x||} \leq ||\tilde{A}^{-1}V(I + E^T\tilde{A}^{-1}V - D)^{-1}DE^T||.$$
(31)

The inequality (31) holds for general tridiagonal systems. For symmetric Toeplitz tridiagonal system, with direct calculation, we have

$$||\tilde{A}^{-1}V(I + E^T\tilde{A}^{-1}V - D)^{-1}DE^T|| \leq \frac{|\tilde{b}|}{|1 - |\tilde{a}||}||v||,$$
(32)

where $v_0^{(i)} = w_{m-1}^{(i)} = \tilde{a}$, $v_{m-1}^{(i)} = w_0^{(i)} = \tilde{b}$, for $i = 0, \cdots, p-1$, and $v^{(i)} = v$, for $i = 0, \cdots, p-1$ (see Eq. (12)). From our results given in Section 4.1,

$$|\tilde{a}| = \left|\frac{b(1 - b^{2m})}{\lambda(1 - b^{2(m+1)})}\right| \leq \left|\frac{b}{\lambda}\right|, \quad |\tilde{b}| \leq \left|\frac{b^m}{\lambda}\right|, \quad (33)$$

and, with direct calculation, we have

$$||v|| \leq \frac{1}{|\lambda|(|a| - 1)}$$
(34)

Combining the inequalities (31), (32), and (34) we obtain the final results

$$\frac{||x - x^*||}{||x||} \leq \frac{|\tilde{b}|}{|\lambda(1 - |\tilde{a}|)| \times (|a| - 1)}$$
(35)

$$\frac{||x - x^*||}{||x||} \leq \frac{|b|^m}{|\lambda|\left(|\lambda| - \left|\frac{b(1 - b^{2m})}{1 - b^{2(m+1)}}\right|\right)(|a| - 1)}$$
(36)

Inequality (35) shows how the values of v_{m-1} and w_0 influence the accuracy of the final results. Inequality (36) gives an error bound of the PDD algorithm. When $\left|\frac{b}{\lambda}\right| < 1$, inequality (36) can be simplified to

$$\frac{||x - x^*||}{||x||} \leq \frac{|b|^m}{|\lambda|(|\lambda| - |b|)(|a| - 1)}$$

4.3 Accuracy of the Reduced PDD Algorithm

For the sake of writing, in this and the next sections we assume $m = n/p$ is an even integer. Let x' be the solution of the reduced PDD algorithm. Notice that x^* is the solution of the PDD algorithm (see Section 4.2). Then, by Eq. (4) and the relation $(I + E^T\tilde{A}^{-1}V)E^T\tilde{A}^{-1}d = E^Tx$,

$$x' - x^* = (\tilde{A}^{-1}\tilde{V} - \tilde{A}^{-1}V)E^Tx.$$

Therefore,

$$\frac{||x' - x^*||}{||x||} \leq \frac{|b^{\frac{m}{2}}|}{|\lambda|(|a| - 1)}$$

Equation (37) gives the accuracy of the reduced PDD algorithm.

$$\frac{||x - x'||}{||x||} \leq \frac{||x - x^*||}{||x||} + \frac{||x^* - x'||}{||x||}. \quad (37)$$

4.4 Skew-Symmetric Toeplitz Tridiagonal Systems

A skew-symmetric Toeplitz tridiagonal matrix A has the form $A = [-\lambda, \beta, \lambda] = \lambda \cdot [-1, c, 1]$. Let $B = [-1, c, 1]$. Then, for the corresponding matrix $\tilde{B}$ (see Section 4.1)

$$\tilde{B} = [b, 1, 0] \times [0, a, 1] \times [0, 1, -b],$$

where a, b are the solutions of

$$b + a = c, \quad b \cdot a = -1. \quad (38)$$

Comparing with symmetric case, the only difference are $-b$ in matrix $[0, 1, -b]$ and $b \cdot a = -1$ in Eq. (38). Following the steps given in the study of symmetric systems, we can find for skew-symmetric Toeplitz tridiagonal systems $v_0^{(i)} = w_{m-1}^{(i)} = \tilde{a}$, $v_{m-1}^{(i)} = -w_0^{(i)} = \tilde{b}$, for $i = 0, \cdots, p-1$, and

$$|\tilde{b}| \leq \frac{|b^m(1 + b^2)|}{|\lambda|};$$

$$|\tilde{a}| = \left|\frac{-b \cdot (1 - b^{2m})}{\lambda(1 + b^{2(m+1)})}\right| \leq \frac{|b|}{\lambda}.$$

The corresponding relative error

$$\frac{||x - x^*||}{||x||} \leq \frac{|\tilde{b}|}{|\lambda(1 - |\tilde{a}|)(|a| - 1)|}$$
(39)

in terms of $\tilde{a}$ and $\tilde{b}$; and

$$\frac{||x - x^*||}{||x||} \leq \frac{|b|^m(1 + b^2)}{|\lambda(|\lambda| - \left|\frac{b(1 - b^{2m})}{1 + b^{2(m+1)}}\right|)(|a| - 1)|}$$
(40)

in terms of a and b. When $\frac{|b|}{|\lambda|} < 1$, we have

$$\frac{||x - x^*||}{||x||} \leq \frac{|b|^m(1 + b^2)}{|\lambda(|\lambda| - |b|)(|a| - 1)|}.$$

For the reduced PDD algorithm, when the system is skew-symmetric, we have

$$\frac{||x - x'||}{||x||} \leq \frac{||x - x^*||}{||x||} + \frac{|b|^{m/2}}{|\lambda|(|a| - 1)}.$$

5 Experimental Results

Table 1 gives the computation and communication count of the PDD algorithm. The computation and communication count of solving multiple right-side systems is also listed in Table 1, where the factor-

System	Matrix	Best sequential	the PDD	
			Computation	Communication
Single	Non-periodic	8n-7	$17\frac{n}{p} - 4$	$2\alpha + 12\beta$
System	Periodic	14n-16	$17\frac{n}{p} - 4$	$2\alpha + 12\beta$
Multiple	Non-periodic	$(5n - 3) * n1$	$(9\frac{n}{p} + 1) * n1$	$(2\alpha + 8\beta) * n1$
right-side	Periodic	$(7n - 1) * n1$	$(9\frac{n}{p} + 1) * n1$	$(2\alpha + 8\beta) * n1$

Table 1. Computation and Communication Counts of the PDD Algorithm

System	the Reduced PDD	
	Computation	Comm.
Single system	$15\frac{n}{p} - 4$	$2\alpha + 12\beta$
Multiple right-side	$(7\frac{n}{p} + 1)n1$	$(2\alpha + 8\beta)n1$

Table 2. Computation and Communication Counts of the Reduced PDD

ization of matrix A is not considered and $n1$ is the number of right-sides. Note for multiple right-side systems, the communication cost increases with the number of right-sides. Table 2 gives the computation and communication counts of the reduced PDD algorithm. As the PDD algorithm, it has the same parallel computation and communication counts for periodic and non-periodic systems.

A matrix A resulting from the compact scheme is chosen to illustrate and verify the algorithm and theoretical results given in previous sections.

$$A = [\frac{1}{3}, 1, \frac{1}{3}] = \frac{1}{3} \cdot [1, 3, 1] =$$

$$\frac{1}{3} \cdot ([b, 1, 0] \times [0, a, 0] \times [0, 1, b] - \Delta B),$$

where ΔB is given by Eq.(27), and

$$\lambda = \frac{1}{3}, c = 3, a = \frac{3 + \sqrt{5}}{2}, b = \frac{3 - \sqrt{5}}{2}. \qquad (41)$$

The algorithm was implemented on a 32-node Intel/860 to measure the speedup over Thomas algorithm [9], a commonly used practical sequential algorithm for periodic tridiagonal systems. Figure 2 depicts the decay rate of v_{m-1} of matrix A, where the x-coordinate is the order of the sub-system A_i and the y-coordinate is the value of v_{m-1}. Accuracy comparisons of the PDD and the reduced PDD algorithms are given in Fig. 3 and Fig. 4 respectively. The right-side vector, d, was randomly generated. The x-coordinate is the order of matrix A, and the y-coordinate is the relative error in the 1-norm. These figures show that our accuracy analysis provides a very good bound.

Figure 5 and 6 give the speedup of the PDD algorithm over Thomas algorithm. For single system, the order of matrix A is limited by the machine memory for $n = 6400$. For multiple right-sides, the system is limited for $n = 128$ and $n1 = 4096$. Factorization time

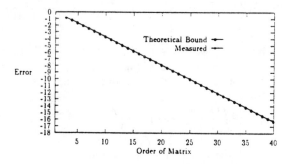

Figure 2. Measured and Predicted Decay Rate.

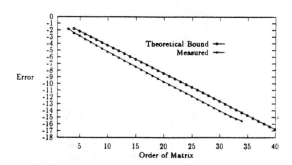

Figure 3. Measured and Predicted Accuracy of the PDD Algorithm.

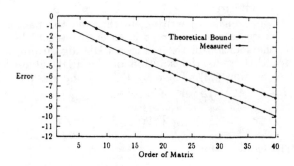

Figure 4. Measured and Predicted Accuracy of the Reduced PDD .

is not included in Fig. 6. From Fig. 5 we can see that the speedup of solving a single system increases linearly with the number of processors. Figure 6 shows that the linear increasing property does not hold for multiple right-side systems. The lower speedup is due to the reducing of the matrix size and the increase of the number of right sides. By Table 1, the communication cost increases linearly with the number of right sides. Since the Intel/860 has a very high (communication speed)/(computation speed) ratio, we can expect a better speedup on an Intel Paragon or even on an Intel/iPSC2 [10] multicomputer.

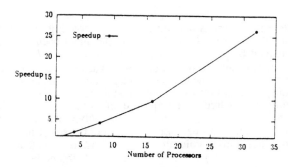

Figure 5. Measured Speedup Over Thomas Algorithm *Single System*.

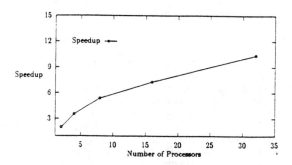

Figure 6. Measured Speedup Over Thomas Algorithm *Multiple Right-side*.

6 Conclusion

A detailed study has been given for the efficient tridiagonal solver, the Parallel Diagonal Dominant (PDD) algorithm. The PDD algorithm is extended to periodic systems. A variant, the reduced PDD algorithm, was also introduced. Accuracy analysis is provided for a class of tridiagonal systems, the symmetric and skew-symmetric Toeplitz tridiagonal systems. Implementation results were provided for both accuracy analysis and for the proposed algorithm. They showed that the accuracy analysis provides a very good theoretical bound and that the algorithm is highly efficient for both single and multiple right-side systems. The algorithm is a good candidate for large scale computing, where the number of processors and the problem size are large. The discussion is based on distributed-memory machines. The result can be easily applied to shared-memory machines as well.

The PDD algorithm and the reduced PDD algorithm proposed in this paper can be extended to band systems and block tridiagonal systems. The accuracy analysis, which gives a good, simple relative error bound, is for symmetric and skew-symmetric Toeplitz tridiagonal systems only. It is unlikely that the analysis can be extended for general case with the same technique.

Acknowledgements

I would like to thank my colleague John R. Van Rosendale at ICASE for his valuable suggestions and comments which improved the technical quality and presentation of this paper.

References

1. X.-H. Sun, H. Zhang, and L. Ni, "Efficient tridiagonal solvers on multicomputers," *IEEE Transactions on Computers*, vol. 41, no. 3, pp. 286–296, 1992.

2. J. Ortega and R. Voigt, "Solution of partial differential equations on vector and parallel computers," *SIAM Review*, June 1985.

3. C. Ho and S. Johnsson, "Optimizing tridiagonal solvers for alternating direction methods on boolean cube multiprocessors," *SIAM J. Sci. Stat. Comput.*, vol. 11, no. 3, pp. 563–592, 1990.

4. X.-H. Sun and D. Rover, "Scalability of parallel algorithm-machine combinations." Technical Report, IS-5057, UC-32, Ames Laboratory, U.S. Department of Energy, 1991. Presented at Supercomputing '92 workshop.

5. X.-H. Sun and L. Ni, "Another view on parallel speedup," in *Proc. of Supercomputing '90*, (NY, NY), pp. 324–333, 1990.

6. J. Sherman and W. Morrison, "Adjustment of an inverse matrix corresponding to changes in the elements of a given column or a given row of the original matrix," *Ann. Math. Stat.*, vol. 20, no. 621, 1949.

7. S. Lele, "Compact finite difference schemes with spectral-like resolution," *J. Comp. Phys.*, vol. 103, no. 1, pp. 16–42, 1992.

8. J. C. Strikwerda, *Finite Difference Schemes and Partial Differential Equations*. Wadsworth & Brooks/Cole, Mathematics Series, 1989.

9. X.-H. Sun, "Application and accuracy of the parallel diagonal dominant algorithm." ICASE Technical Report, 93-6, ICASE, NASA Langley Research Center, 1993.

10. X.-H. Sun and J. Gustafson, "Toward a better parallel performance metric," *Parallel Computing*, vol. 17, pp. 1093–1109, Dec 1991.

SUPERNODAL SPARSE CHOLESKY FACTORIZATION ON DISTRIBUTED-MEMORY MULTIPROCESSORS

Kalluri Eswar   P. Sadayappan   Chua-Huang Huang
Department of Computer and Information Science
The Ohio State University
Columbus Ohio 43210
{eswar,saday,chh}@cis.ohio-state.edu

V. Visvanathan
Indian Institute of Science
Bangalore 560012, India
vish@cad.iisc.ernet.in

Abstract – *The concept of supernodes has been widely used in the design of algorithms for the solution of sparse linear systems of equations. This paper discusses the use of supernodes in the design of algorithms for sparse Cholesky factorization on distributed-memory multiprocessors. A new algorithm that is communication efficient, has good load balance, and benefits significantly from supernodes is presented. A taxonomy of distributed sparse Cholesky factorization algorithms is proposed. Performance results on an Intel iPSC/860 multiprocessor are reported.*

1 Introduction

Several parallel algorithms for sparse Cholesky factorization on shared-memory and distributed-memory multiprocessors have been presented in the literature and a survey can be found in [7]. The recognition of supernodes (collections of columns with identical sparsity structure) in a sparse matrix has played an important role in the development of efficient sparse factorization algorithms on vector supercomputers uniprocessor workstations and shared-memory multiprocessors [7, 8]. This paper addresses the use of supernodes in distributed sparse factorization algorithms. A new algorithm for sparse Cholesky factorization is presented that is able to exploit supernodes while simultaneously achieving low communication and good load balance.

The paper is organized as follows. § 2 considers the use of supernodes in two of the existing algorithms for distributed sparse Cholesky factorization and includes performance results on an Intel iPSC/860 multiprocessor. § 3 contains a new algorithm for effective exploitation of supernodes and compares its performance results with those given in § 2. A taxonomy of distributed sparse Cholesky factorization algorithms is presented in § 4. § 5 provides concluding remarks.

2 Supernodes in Distributed Factorizers

This section considers the utility of exploiting supernodes in distributed sparse Cholesky factorization algorithms. The two algorithms that will be considered are the *fan-out* [6] algorithm and the *fan-in* [2] algorithm. It is assumed that the columns of the matrix have been distributed among the processors using a communication-reducing and load-balancing heuristic such as *recursive partitioning* [3]. § 2.1 explains the concept of supernodes. § 2.2 discusses the benefits expected from the use of supernodes in the fan-out and the fan-in algorithms. § 2.3 presents and analyzes performance results obtained on an Intel iPSC/860 multiprocessor.

2.1 Supernodes

Let A be an $n \times n$ sparse symmetric positive definite matrix and let L be its Cholesky factor. The *elimination tree* [5] of L has n nodes, one corresponding to each of the columns of the matrix. The parent of each node is defined by $parent(j) = \min \{i : i > j \text{ and } L_{ij} \neq 0\}$. It can be shown that a column updates only a subset of its ancestors in the elimination tree and in turn is updated only by a subset of its descendants [5]. A *chain* in an elimination tree is a sequence of nodes $j_1, j_2, \ldots, j_m$ such that, for $1 \leq l < m$, j_l is the only child of j_{l+1} in the elimination tree. Let $R(j)$ denote the set of nonzero row positions in column j of L. If it is also true that, for $1 \leq l < m$, $R(j_l) = R(j_{l+1}) \cup \{l+1\}$, the nodes in the chain are said to constitute a *supernode*. In other words, a supernode is a set of nodes in a chain whose columns have the same nonzero structure. This property has been exploited to reduce the amount of storage required to store the sparsity structure of a matrix. Supernodes have been used to improve the efficiency of several algorithms associated with sparse linear system solution. A few of those situations are discussed next.

Performing an update operation in sparse Cholesky factorization requires an extra level of indirection in accessing either the source column or the target column or both because their nonzero structures will, in general, be different and they are stored in compact form. Operations such as these are called *sparse* operations in contrast with *dense* operations where array subscripts do not involve indirection. Since columns in a supernode have the same nonzero structure, updates between them can be performed using dense operations. Moreover, the set of columns outside the supernode updated by each of the columns in the supernode is identical. The combined update of all the columns in the supernode can therefore be computed using dense operations, with sparse operations used to apply the combined update to the target column, thus replacing a number of sparse operations by dense operations, which are often less expensive than sparse operations, and also vectorize better.

2.2 Expected Benefits of Using Supernodes in Distributed Factorizers

The primary impediment to the utility of supernodes in distributed factorizers seems to be the fact that the columns of a supernode might be distributed among several processors [7] in order to improve load balance. It can, however, be expected that matrices for which using supernodes makes a significant difference in other situations, like on a uniprocessor, and whose factorization time without the use supernodes on a certain number of processors of a distributed-memory multiprocessor is reasonably high, will have many supernodes large enough so that splitting them among the same number of processors will still leave each processor with "mini-supernodes" big enough that significant benefit can be expected from their use.

Another minor problem caused by the algorithm used to map columns onto processors is that the columns of a supernode assigned to a processor may not be consecutively numbered. This makes the code running on each processor slightly more complicated. The mapping scheme used can be modified to use a *supernodal elimination tree* whose vertices are supernodes. Since parallelism within a supernode is important, the mapping algorithm should map supernodes onto *groups* of processors. The recursive partitioning algorithm performs such a mapping. The columns within the supernode can then be mapped in a block-wrap manner among the processors in the group, thus making the minisupernodes assigned to each processor to contain consecutively numbered columns. While load balance through time between processors may be negatively impacted by using such a mapping scheme, there is a tradeoff with reduced communication costs for intra-supernodal updates.

Algorithm *supernodal fan-out*
> **On each processor do**
>> $count :=$ number of owned columns
>> **for** each owned leaf k **do**
>>> normalize column k
>>> **if** k is the last column in its minisupernode **then**
>>>> send minisupernode of k to needy processors
>>> **else**
>>>> send column k to myself
>>> **end**
>>> $count := count - 1$
>> **end**
>> **while** $count \neq 0$ **do**
>>> receive a column k or a minisupernode ending in k
>>> **if** a column k is received **then**
>>>> **for** each column j after k in its minisupernode **do**
(*)>>>>> update column j using column k (dense)
>>>> **end**
>>> **else**
>>>> **for** each owned column j updated by column k **do**
>>>>> **for** each column k' in minisupernode **do**
(**)>>>>>> accumulate update of column k' to j (dense)
>>>>> **end**
(***)>>>>> update column j using accumulated values (sparse)
>>>>> **if** column j is completely updated **then**
>>>>>> normalize column j
>>>>>> **if** j is the last column in its minisupernode **then**
>>>>>>> send minisupernode of j to needy processors
>>>>>> **else**
>>>>>>> send column j to myself
>>>>>> **end**
>>>>>> $count := count - 1$
>>>>> **end**
>>>> **end**
>>> **end**
>> **end**
> **end**
end *supernodal fan-out*

Figure 1: Supernodal fan-out Cholesky factorization

A supernodal fan-out algorithm is shown in Figure 1. The basic structure is similar to that of the fan-out algorithm. When the last column of a minisupernode is normalized, the entire minisupernode is sent out to other processors. If an intermediate column in a minisupernode is normalized, it is sent by the processor to itself for use in updating subsequent columns within the minisupernode. As in the fan-out algorithm, in practice, columns and minisupernodes are not actually sent by a processor to itself, but a local queue is used.

Based on the algorithm given in Figure 1, the following potential benefits can be expected for the fan-out algorithm from using supernodes.

Dense operations: As discussed in § 2.1, the use of supernodes replaces sparse operations with dense operations. In the supernodal fan-out algorithm, the update of a column by another column in the same minisupernode can be performed using dense operations (line marked (*) in Figure 1). When a minisupernode updates a column, the updates of the individual columns in the minisupernode are accumulated using dense operations (line marked (**) in Figure 1).

Reduced searching overhead: The fan-out algorithm suffers from an element-level searching overhead when performing the update of a target column by a source column. More details about this can be found in [4]. In the supernodal fan-out algorithm, this searching need only be done while performing a sparse update of the target column using the accumulated update of an entire minisupernode (line marked (* * *) in Figure 1). The searching overhead is thus amortized over all the columns of a minisupernode.

Better cache usage: There are many distributed-memory multiprocessors in which each processor has a data cache. On such machines, it can be expected that a reduction in cache misses of the order of the average number of columns in a minisupernode can be expected in the supernodal fan-out algorithm over the nonsupernodal fan-out algorithm. This is because the supernodal algorithm uses all the columns of a minisupernode consecutively to update local columns, whereas the nonsupernodal algorithm uses the columns at different times. A more detailed analysis can be found in [4].

A supernodal fan-in algorithm is given in Figure 2. The basic structure is similar to that of the fan-in algorithm. When a column j needs to be updated or its combined contribution needs to be formed, the source columns are grouped by minisupernodes when performing the updates. This allows dense operations to be used to accumulate the contributions of the columns in the minisupernode, and sparse operations are used only for applying the accumulated values. This is a benefit that the supernodal fan-in algorithm derives from the use of supernodes. It can be expected that the supernodal fan-in algorithm will not have any significant improvement in its cache usage over the fan-in algorithm. The target-driven nature of the algorithms makes it likely that several minisupernodes will be used between two successive uses of a minisupernode. Therefore, it is likely that the minisupernode would have been displaced from the cache before it is used again. The cache usage behavior of the supernodal fan-in algorithm will hence not be very different from that of the nonsupernodal fan-in algorithm.

2.3 Performance of Supernodal Factorizers

The nonsupernodal and the supernodal versions of the fan-out and fan-in algorithms were implemented in Fortran77 and evaluated on a 32-node iPSC/860 hypercube multiprocessor. A test matrix derived from a 127x127 grid using a 9-point stencil is used. This matrix has 16129 columns and 518578 nonzeroes in L. The performance results for the four algorithms are summarized in Tables 1 and 2.

```
Algorithm supernodal fan-in
  On each processor do
    for j := 1 to n do
      count[j] := number of combined contributions to be received
                  for column j
    end
    for j := 1 to n do
      if owned(j) then
        for each owned minisupernode σ updating j do
          for each column k in the minisupernode σ do
            accumulate update of column k to column j (dense)
          end
          update column j using accumulated values (sparse)
        end
        for each column k before column j in its minisupernode do
          update column j using column k (dense)
        end
        while count[j] ≠ 0 do
          receive and apply a combined contribution for, say,
            column j' (dense)
          count[j'] := count[j'] − 1
        end
        normalize column j
      elsif there exists at least one owned k updating j then
        initialize combined contribution to zero
        for each owned minisupernode σ updating j do
          for each column k in the minisupernode σ do
            accumulate update of column k (dense)
          end
          update combined contribution using accumulated
            values (sparse)
        end
        send combined contribution to owner of j
      end
    end
  end
end supernodal fan-in
```

Figure 2: Supernodal fan-in Cholesky factorization

It can be seen in Table 1 that the supernodal fan-out algorithm has a factorization time that is much less than that for the fan-out algorithm. The other interesting observation that can be made is the reduction in the number of messages that need to be communicated in the supernodal case. This is because of columns being grouped together as minisupernodes before being sent out. This is yet another benefit of using supernodes for the fan-out algorithm. The message volume remains essentially the same because the same values are being sent (the slight difference is because the fan-out algorithm does not send out the diagonal element in each column, whereas the supernodal fan-out algorithm sends them because they are stored along with the rest of the column and a contiguous buffer is to be specified when sending out a minisupernode).

From Table 2, it can be seen that while there is significant improvement due to the use of supernodes for

Table 1: Fan-out versus Supernodal Fan-out

	Fan-out			Supernodal Fan-out		
p	Fact. Time	Mesg. Count	Mesg. Vol	Fact. Time	Mesg. Count	Mesg. Vol
1	19.82	0	0	8.20	0	0
2	11.14	1267	119337	4.37	202	120604
4	7.26	4020	378634	2.54	620	382654
8	4.83	10390	931108	1.62	1724	941498
16	4.01	21831	1979026	1.55	3836	2000857
32	4.47	43443	4038796	2.06	8947	4082239

Table 2: Fan-in versus Supernodal Fan-in

	Fan-in			Supernodal Fan-in		
p	Fact. Time	Mesg. Count	Mesg. Vol	Fact. Time	Mesg. Count	Mesg. Vol
1	11.89	0	0	8.72	0	0
2	6.20	127	8128	4.57	127	8128
4	3.45	507	44418	2.61	507	44418
8	2.70	1643	178766	2.22	1643	178766
16	2.19	4127	465502	1.95	4127	465502
32	1.59	9367	1065114	1.53	9367	1065114

smaller values of p, the amount of improvement is very low for higher values of p. This is because the fan-in algorithm is not able to take much advantage of supernodes. The communication required in the supernodal fan-in algorithm is identical to that in the fan-in algorithm. It is also interesting to note that the supernodal fan-out algorithm performs better than the supernodal fan-in algorithm until $p = 16$, but its performance deteriorates after that and in fact, does worse on 32 processors than on 16 processors. Although the supernodal fan-out algorithm is able to better exploit supernodes, its main limiting factor appears to be the large volume of communication it incurs. The next section considers how both the desired objectives of better exploitation of supernodes and low communication volume can be achieved.

3 A New Supernodal Algorithm

The results from § 2.3 show that although the supernodal fan-out algorithm has a low message count, it has essentially the same high message volume as the fan-out algorithm. The fan-in algorithms, nonsupernodal and supernodal, have a much lower message volume. Aggregating all updates from a processor to a target column before sending out the combined contribution is the basic idea of the fan-in algorithm. If this idea is used with the supernodal fan-out algorithm, an algorithm that uses supernodes well and has low communication like the fan-in algorithms can be expected. Figure 3 shows the resulting new algorithm, which will be given a formal name in § 4.

The outermost **while** loop in the algorithm encloses two **while** loops. These are, respectively, for handling columns in the local queue which are ready to be normalized, and for handling combined contributions arriving from other processors. After a column is normalized, it is used to update subsequent columns in its minisupernode if it is not the last column in the minisupernode. If it is the last one, the entire minisupernode is used to update targets. For each target column, dense operations are used to accumulate the update of the minisupernode. This is used to update the target column if it is locally owned, and the target is added to the queue if it is ready to be normalized. If the target column is not locally owned, the accumulated values are used to update the target column's combined contribution from this processor. If all local contributions to this contribution have been applied, it is sent to the processor which owns the target column.

The new algorithm has exactly the same communication as the fan-in algorithms. While this algorithm uses the idea of aggregating local updates to a target column, the updates are not aggregated together during a continuous period of time, but are added to the combined contribution at different times. This requires that each target's combined contribution have separate stor-

```
Algorithm new algorithm
  On each processor do
    initialize queue to contain owned leaves
    while queue not empty or messages to be received do
      while queue not empty do
        delete a column, say k, from queue
        normalize column k
        if k is not the last column in its minisupernode then
          for each subsequent column j in minisupernode do
            update column j using column k (dense)
            if column j is completely updated then
              add column j to queue
            end
          end
        else
          for each column j updated by column k do
            for each column k' in minisupernode do
              accumulate update of column k' to column j (dense)
            end
            if owned(j) then
              apply accumulated values to column j (sparse)
              if column j is completely updated then
                add column j to queue
              end
            else
              apply accumulated values to combined contribution
                for column j (sparse)
              if column j is completely updated then
                send combined contribution to owner of column j
              end
            end
          end
        end
      end
      while messages are available do
        receive and apply a combined contribution for, say,
          column j (dense)
        if column j is completely updated then
          add column j to queue
        end
      end
    end
  end
end new algorithm
```

Figure 3: New algorithm

age available. However, the fan-in algorithm can reuse the storage for the combined contributions.

The new algorithm retains the good cache usage behavior of the supernodal fan-out algorithm because all the targets of a minisupernode are updated during a continuous period of time. Dense operations once again replace sparse operations because of the use of supernodes. While there is a searching overhead during the sparse updates, it is amortized over all the columns in a minisupernode, just as it is in the supernodal fan-out algorithm.

Table 3 shows performance results obtained using the new algorithm on the test matrix used earlier. The factorization times of the supernodal fan-out and supernodal fan-in algorithms are reproduced from Tables 1 and 2 for ease of comparison. In addition, the measured total idle time on all processors in each case is included. It can be seen that the new algorithm consistently outperforms both the supernodal fan-in algorithm and the supernodal fan-out algorithm. The supernodal fan-in algorithm has a much higher idle time than the new algorithm. This problem is inherent in the fan-in algorithm. When a processor is at an iteration j of its outermost loop, it cannot proceed to perform useful work on later columns

Table 3: Supernodal algorithms performance comparison

p	Supernodal Fan-out		Supernodal Fan-in		New Algorithm	
	Fact. Time	Idle Time	Fact. Time	Idle Time	Fact. Time	Idle Time
1	8.20	0.00	8.72	0.00	7.98	0.00
2	4.37	0.13	4.57	0.10	4.07	0.10
4	2.54	0.68	2.61	0.49	2.11	0.25
8	1.62	1.20	2.22	5.69	1.20	0.69
16	1.50	7.18	1.95	14.84	0.77	1.04
32	2.06	37.30	1.53	23.88	0.57	1.57

until all the external updates for column j have been received. The *compute-ahead fan-in* algorithm [1] has been proposed to partially alleviate this problem. Consider the case of $p = 16$ in Table 3. The average idle time per processor for the supernodal fan-in algorithm is $14.84/16 = 0.93$. If all of this idle time is assumed to be eliminated by using the compute-ahead fan-in algorithm, the expected factorization will be $1.95 - 0.93 = 1.02$, which is still more than the factorization time for the new algorithm. Since both algorithms incur the same communication, the reason for the better performance of the new algorithm must be because the local operations on each processor are being performed more efficiently.

4 A Taxonomy of Distributed Sparse Cholesky Factorizers

Distributed sparse Cholesky factorization algorithms can be classified based on three characteristics:

Ordering of operations: In a distributed factorizer, the operations that need to be performed by each processor are determined by the mapping of the columns of the sparse matrix and the type of the updates (see below) that are performed. The local ordering of these operations is determined by whether the local computation is to be source-driven or target-driven. In a *source-driven* algorithm, a source column or a group of source columns is used to update targets successively. In a *target-driven* algorithm, source columns or groups of source columns are used successively to update a target column. The terms *kji* and *jki* will be used instead of "source-driven" and "target-driven" hereafter.

Type of updates: When a source column updates a target column owned by a different processor, the update can happen on the processor owning the target column. Such an update is referred to as a *direct update*. If all updates to a particular target column from columns owned by a processor are added together on the source processor and the result sent to the processor owning the target column where it is subtracted from the target column, it is referred to as an *aggregated update*. The terms *dir* and *agg* will be used instead of "direct update" and "aggregated update" hereafter.

Type of sources: When performing updates, a *nodal* algorithm considers each of the source columns of a target individually, whereas a *supernodal* algorithm considers source columns belonging to the same supernode together. The terms *nod* and *sup* will be used instead of "nodal" and "supernodal" hereafter.

Each of these three characteristics or dimensions has two possible choices, giving rise to a total of eight different possible combinations. These are shown in Figure 4

Table 4: Impact of algorithm characteristic on performance

Algorithm characteristic (direction of change)	Effect on performance measures			
	Message count	Message volume	Load balance	Local efficiency
Ordering of operations ($jki \longrightarrow kji$)	Unchanged	Unchanged	Improves	$\dfrac{nod \;\; \text{Worsens}}{sup \;\; \text{Improves}}$
Type of updates ($dir \longrightarrow agg$)	$\dfrac{nod \;\; \text{Decreases generally}}{sup \;\; \text{Indeterminate}}$	Decreases	$\dfrac{kji \;\; \text{Worsens}}{jki \;\; \text{Improves}}$	Unchanged
Type of sources ($nod \longrightarrow sup$)	$\dfrac{dir \;\; \text{Decreases}}{agg \;\; \text{Unchanged}}$	Unchanged	$\dfrac{dir \;\; \text{Worsens slightly}}{agg \;\; \text{Unchanged}}$	$\dfrac{kji \;\; \text{Improves significantly}}{jki \;\; \text{Improves slightly}}$

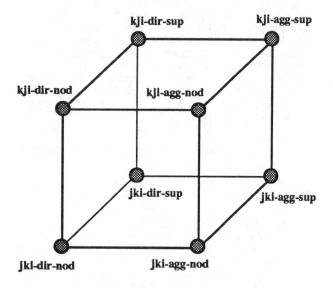

Figure 4: Taxonomy cube

where the label of each vertex indicates the choices made for each of the three characteristics. The algorithms discussed so far fit into this taxonomy. The fan-out algorithm is the *kji-dir-nod* algorithm. The fan-in algorithm is the *jki-agg-nod* algorithm. The supernodal fan-out and the supernodal fan-in algorithms are, respectively, the *kji-dir-sup* and *jki-agg-sup* algorithms. The new algorithm presented in § 3 is the *kji-agg-sup* algorithm. The *kji-agg-nod* algorithm is simply a nonsupernodal version of the new algorithm and can be easily constructed. The remaining two algorithms, *jki-dir-nod* and *jki-dir-sup*, can be expected to have relatively poor performance based on the analysis referred to below.

The impact of the choices made for each of the three dimensions on the various measures of the performance of the distributed factorization algorithm can be analyzed. Table 4 contains a summary of this analysis, the details of which can be found in [4]. Each of the dimensions is considered and the effect of going from one choice to the other along that dimension is given. These transitions are represented by the edges of the taxonomy cube. The measures considered are message count, message volume, load balance, and local efficiency, where local efficiency refers to the efficiency of execution of local operations on each processor. It is assumed that the same mapping of columns to processors is used in all cases.

Based on the analysis, a supernodal aggregated source-driven algorithm (the new algorithm) can be expected to have good local efficiency, low communication, and good load balance. The performance results from § 3 confirm these expectations.

5 Conclusion

The use of supernodes in distributed sparse Cholesky factorization algorithms has been considered. Supernodes have been shown both by analysis and actual performance results to benefit the fan-out algorithm more than the fan-in algorithm. A new supernodal algorithm that has the communication behavior of the fan-in algorithm and that exploits supernodes as well as the supernodal fan-out algorithm was presented. The new algorithm's performance was found to be superior to that of the supernodal fan-out and the supernodal fan-in algorithms. A taxonomy of distributed sparse Cholesky factorization algorithms was proposed and the effect of the various dimensions in the taxonomy summarized.

References

[1] C. Ashcraft, S. C. Eisenstat, J. W. H. Liu, B. W. Peyton, and A. H. Sherman, *A compute-ahead implementation of the fan-in sparse distributed factorization scheme*, Dept. of Computer Science, York University, Tech. Report No. CS-90-03, 1990.

[2] C. Ashcraft, S. Eisenstat, and J. W.-H. Liu, "A fan-in algorithm for distributed sparse numerical factorization," *SIAM J. Sci. Statist. Comput.*, Vol. 11, pp. 593-599, 1990.

[3] K. Eswar, P. Sadayappan, and V. Visvanathan, "Multifrontal factorization of sparse matrices on shared-memory multiprocessors," *Proceedings of the Twentieth International Conference on Parallel Processing*, St. Charles, IL, Vol. III, pp. 159-166, 1991.

[4] K. Eswar, P. Sadayappan, and C.-H. Huang, *Supernodal sparse Cholesky factorization on distributed-memory multiprocessors*, Department of Computer and Information Science, The Ohio State University, Tech. Report No. OSU-CISRC-4/93-TR17,16 pp.,1993.

[5] A. George and J. W.-H. Liu, *Computer Solution of Large Sparse Positive Definite Systems*, Prentice-Hall, Englewood Cliffs, NJ, 1981.

[6] A. George, M. Heath, J. W.-H. Liu, and E. G.-Y. Ng, "Sparse Cholesky factorization on a local-memory multiprocessor," *SIAM J. Sci. Statist. Comput.*, Vol. 9, pp. 327-340, 1988.

[7] M. T. Heath, E. Ng, and B. W. Peyton, "Parallel algorithms for sparse linear systems," *SIAM Review*, Vol. 33, pp. 420-460, 1991.

[8] E. Rothberg and A. Gupta, "Techniques for improving the performance of sparse matrix factorization on multiprocessor workstations," *Proceedings of Supercomputing '90*, pp. 232-241, 1990.

Parallel FFT Algorithms for Cache Based Shared Memory Multiprocessors*

Akhilesh Kumar and Laxmi N. Bhuyan

Department of Computer Science
Texas A&M University
College Station, TX 77843-3112.

Abstract – Shared memory multiprocessors with cache require careful consideration of cache parameters while implementing an algorithm to obtain optimal performance. In this paper, we study the implementation of some existing FFT algorithms and analyze the number of cache misses based on the problem size, number of processors, cache size, and block size. We also propose a new FFT algorithm which minimizes the number of cache misses.

1 Introduction

Discrete Fourier Transform (DFT) is widely used in many scientific applications. Because of its wide area of application, a good amount of research effort has been directed towards making the computation of DFT as fast as possible. The Fast Fourier Transform (FFT) introduced by Cooley and Tukey [2] has been a principal tool for computing DFT. After the introduction of FFT, all the research in DFT has been aimed towards optimizing the performance of FFT on various computer architectures [4, 7, 8, 9].

In this paper, we shall focus our discussion on the general purpose shared memory MIMD computers. All shared memory multiprocessors have small cache near the processors to avoid frequent access to global memory modules. Usually, the cache memories are organized either as direct mapped or 2^l set-associative, where l is usually a small integer. We shall assume that the system follows write-back policy to update the main memory. Cache coherency is maintained by invalidating the stale copies in the system.

The performance of FFT algorithms have been studied on various multicomputer organizations [1, 6]. Jeon and Reeves [4] introduced an algorithm suitable for multicluster architectures. This algorithm was also studied by Norton and Silberger [6] for shared memory architectures. In their analysis Norton and Silberger assumed that the object code and private data are cacheable, but the shared data is not cacheable. Horiguchi and Nakada [3] also studied the performance of FFT on a shared memory multiprocessor with cache. However, they did not study the effect of different cache parameters, problem sizes, and number of processors on the performance. In this paper we concentrate on the effect of cache on the performance of two FFT algorithms. The analysis considers that the shared data are cacheable. We also develop a new FFT algorithm for cache based shared memory multiprocessors.

*This research is supported by NSF grant MIP-9002353 and Texas ATP grant 999903-165.

The next two sections discuss about the performance of Cooley-Tukey (C-T) FFT and Jeon-Reeves (J-R) FFT algorithms, respectively. Section 4 presents a new algorithm which is a modification of Jeon-Reeves' algorithm to optimize on the number of cache misses. In the last section we compare the number of cache misses due to these three algorithms depending on cache size, block size, problem size, and number of processors.

The notations used in this paper are as follows:

$N = 2^n$ = Number of FFT data points
$P = 2^p$ = Number of processors
C = Size of cache memory per processor
B = Cache block size
S = Size of each complex data point
S' = Size of array indices

All the logarithms used in this paper have base 2. To keep the analysis simple, we will not consider the effect of cache interference due to object code or other data such as twiddle factor table. Also, we will not consider the overhead due to initial bit-reversal since it is common to all the algorithms discussed here.

2 Cooley-Tukey FFT Algorithm

A data flow diagram of C-T FFT algorithm can be found in [5, 6]. It is mapped on multiprocessors by equally dividing the bit-reversed data among P processors. Each of the processors operate on N/P data elements during every stage. All the processors perform first $(n-p)$ stages of FFT on disjoint sets of data, concurrently. In each of the rest p stages, every processor needs the results obtained by some other processor to compute further. Barrier synchronization is used to order the sequence of computation.

The FFT data is organized as a one-dimensional array in the shared memory space. The data set is divided into 2^{n-s} blocks of size 2^s data elements, where $s = 0, 1, \ldots, n-1$, is the stage number in the FFT algorithm. At any stage s, a processor simultaneously operates on two of the adjacent blocks, reading data from them and modifying them at every stage. The algorithm works in-place during the first $(n-p)$ stages of computation, i.e., it overwrites the new values on the previous one.

Now, the number of cache misses due to the C-T FFT algorithm can be quantified as follows:

Case 1: $C \leq NS/P$.

(a) When $s = 0, 1, \cdots, \log(C/S) - 1$, i.e., $C \geq 2^{s+1}S$. In this case when a processor is operating on two adjacent blocks of data, the cache can accommodate both these blocks. However, if $C < NS/P$ the cache cannot accommodate all the data accessed by the processor during a stage. This causes the data to be replaced from the cache before it is accessed in the next stage. Even the set-associative mapping of memory blocks to cache is not of much help, since if the set size is 2^k blocks, the number of sets is reduced proportionately. Thus, for this case the number of cache misses during each stage of computation is $NS/(PB)$. So, the total number of cache misses during the first $\log(C/S)$ stages are $\frac{NS}{PB}\log(C/S)$.

(b) When $s = \log(C/S), \cdots, (n-p-1)$, i.e., $C < 2^{s+1}S$. In this case the cache cannot accommodate two adjacent blocks of data involved in computation. If the cache is direct mapped then both data units involved during one butterfly computation map to the same cache block. This generates large number of cache misses if these two data are accessed alternately during the computation. This can be mitigated to some extent by storing the data in processor registers and operating on these registers rather than from memory. Still, this computation causes at least $3\lceil S/B \rceil$ cache misses per butterfly computation, $\lceil S/B \rceil$ each to fetch the two data units and another $\lceil S/B \rceil$ to modify the data that was replaced in the cache. Since each processor does $N/2P$ butterfly computations in each stage, the total number of cache misses in this case is $[n - p - \log(C/S)]\frac{3N}{2P}\lceil \frac{S}{B} \rceil$.

If the cache is set-associative, the number of cache misses will be $[n - p - \log(C/S)]\frac{2SN}{2PB}$.

(c) When $s = (n-p), \cdots, n-1$. During these stages, each data is shared by two processors to compute the next value. Since these processors are working asynchronously, if we overwrite on the old data we cannot guarantee that the other processor used it before it was overwritten. Thus, in these stages the algorithm cannot work in-place and we need an auxiliary array, which along with the original array works alternately as source and result of computations at a stage.

Using two arrays for the main and auxiliary storage causes interference in a direct mapped cache. To avoid this interference, the data from the main and auxiliary arrays can be defined as a record or structure with two fields, and an array of such record can be used. Using such record as storage unit gives the effect of interleaving of main and auxiliary arrays. Such an interleaving results in mapping of the data from main and auxiliary component with the same index to two different cache blocks such that there is no interference between them.

Since $C \leq NS/P$, cache cannot accommodate all the data accessed by the processor during a stage. The two data units involved in a computation are $2^s S$ apart which is larger than the cache size. Thus both these data units are mapped to the same cache block in a direct mapped cache. Note that only one of the two points in a butterfly is computed by one processor during these stages. So, the data which is the result of computation of other part of the butterfly, need not be fetched in the cache.

Therefore, the number of cache misses per computation in a direct mapped cache is $(\lceil 2S/B \rceil + \lceil S/B \rceil)$. Each processor does N/P such computations in every stage (only half of each butterfly computation is done by a processor), and there are p such stages of computations. So, the total number of cache misses in a direct mapped cache in this case is $p\frac{N}{P}(\lceil 2S/B \rceil + \lceil S/B \rceil)$. If the cache is set-associative, the number of cache misses in the worst case will be $p\frac{N}{P}(2S/B + \lceil S/B \rceil)$, because of invalidations due to shared component of data that is modified by another processor in that stage.

Case 2: $C > NS/P$

(a) When $s = 0, 1, \cdots, (n-p-1)$. In this case all the data accessed during these stages can be accommodated in the cache during the first stage and will be available in the cache for $(n-p)$ stages. Thus the total number of cache misses during these stages is $\frac{NS}{PB}$.

(b) When $s = (n-p), \cdots, \min(n, \log(C/2S)) - 1$. During these stages the data organization used is as described in Case 1(c). The cache size at each processor is enough to accommodate all the data referenced during these stages. Each of the processors references N/P of the interleaved data units during all these stages and another of N/P data units from only one of the components of main or auxiliary array depending upon the source array at that stage. This 'source only' component have disjoint array indices during different stages in this case. Thus, in every stage new 'source only' data will be loaded in the cache.

So, the total number of cache misses during these stages for direct mapped as well as set-associative cache is $\frac{2NS}{PB} + \frac{N}{P}\lceil S/B \rceil[\min(n, \log(C/2S)) + p - n]$.

(c) When $s = \log(C/2S), \cdots, n-1$. Again the data organization is same as in Case 1(c). During these stages the difference between the indices used to reference the data accessed in one computation is larger than the cache size. This causes interference in direct mapped cache. The total number of cache misses during these stages in a direct mapped cache is $\frac{N}{P}(\lceil 2S/B \rceil + \lceil S/B \rceil)[n - \log(C/2S)]$.

In case of set-associative cache, such interference is avoided since there is enough space to accommodate all the data used during one stage. Thus the total number of cache misses during these stages in a set-associative cache is $\frac{2NS}{PB} + \frac{N}{P}\lceil S/B \rceil[n - \log(C/2S)]$.

3   Jeon-Reeves FFT Algorithm

Pease [7] proposed an FFT algorithm suitable for parallel processing on an SIMD machine. Pease's implementation is not very efficient for multiprocessors because of the large overhead due to data reorganization in every stage. Jeon and Reeves [4] proposed a generalization of the Pease FFT algorithm suitable for multiprocessors. A data flow diagram of this algorithm can be found in [4, 5].

This algorithm can also be mapped similar to C-T FFT algorithm on a multiprocessor by equally diving the bit-reversed data among all processors. Each processor computes $N/2P$ butterflies in every stage of FFT. All the processors perform first $(n-p)$ stages of FFT on disjoint sets of data, concurrently. After $(n-p)$ stages, the data is reorganized such that for another $(n-p)$ stages all the computation is done on disjoint sets of data. The reorganization pattern is a p shuffle or $(n-p)$ inverse-shuffle. This shuffling is repeated after every $(n-p)$ stages.

The shuffling of data is not symmetric and cross interprocessor boundaries, therefore, it cannot be done in-place. Thus we need two data arrays. Only one of these arrays is used between two shuffling stages because the computation can be done in-place. So, the data array used during computation alternates between the two arrays with every shuffling stage. It must be pointed out that it is not necessary to interleave these two arrays as in the case of C-T FFT algorithm since only one of them is used at a time. However, we need to add another component to the data unit which is initialized to the starting array index of the data. This keeps track of the original index of the data and helps to compute the twiddle factor. So, the unit data size is now $(S + S')$, where S' is the size of an integer array index in bytes.

Now, the number of cache misses due to the J-R FFT algorithm can be quantified as follows:

Case 1: When $C < N(S + S')/P$.

(a) When $s \bmod (n-p) = 0, 1, \cdots, \log(C/(S + S')) - 1$. This case is similar to Case 1(a) of C-T FFT algorithm. Since $C < N(S + S')/P$ the cache cannot accommodate all the data referenced by a processor during a stage. Thus, the number of cache misses during each stage of computation is $\frac{N(S+S')}{PB}$. So, the total number of cache misses for both direct mapped and set-associative cache during these stages will be

$$\frac{N(S + S')}{PB} \left[\lfloor n/(n-p) \rfloor \log\left(\frac{C}{S + S'}\right) + \min\left(n \bmod (n-p), \log\left(\frac{C}{S + S'}\right)\right) \right]$$

(b) When $s \bmod (n-p) = \log(C/(S + S')), \cdots, (n - p - 1)$. In this case the number of cache misses can be derived similar to Case 1(b) of the C-T FFT algorithm. The total number of cache misses for a direct mapped cache will be

$$\frac{3N}{2P} \left[\lfloor n/(n-p) \rfloor \left((n-p) - \log\left(\frac{C}{S + S'}\right)\right) + \left((n \bmod (n-p)) \dot{-} \log\left(\frac{C}{S + S'}\right)\right) \right] \left\lceil \frac{(S + S')}{B} \right\rceil$$

where $\dot{-}$ represents proper subtraction, *i.e.*, $m \dot{-} n$ is equal to $m - n$ for $m \geq n$, and zero for $m < n$.

The total number of cache misses for the set-associative cache will be

$$\left[\lfloor n/(n-p) \rfloor \left((n-p) - \log\left(\frac{C}{S + S'}\right)\right) + \left((n \bmod (n-p)) \dot{-} \log\left(\frac{C}{S + S'}\right)\right) \right] \frac{N(S + S')}{PB}$$

It should be noted that the number of cache misses accounted for in the above expressions are for the computation stages only. We still have to account for the

number of cache misses due to shuffling stages. Some of the processors are allocated completely new set of data after the shuffling. These processors fetch data that are not in its cache and write them to alternate array that is also not in the cache. So the number of cache misses for these processors will be equal to $\frac{2N(S+S')}{PB}$ for every shuffling stage. Since there are $\lceil n/(n-p) \rceil$ shuffling stages (taking into account the shuffling required at the end to bring data in order), the total number of cache misses due to shuffling stages will be $\frac{2N(S+S')}{PB} \lceil n/(n-p) \rceil$.

Case 2: When $C \geq N(S + S')/P$ and $C < 2N(S + S')$. In this case the cache can accommodate all the data referenced by the processor during the stages of computation. In fact the data loaded into the cache during shuffling stage can be used during next $(n-p)$ stages of computation. Thus the cache misses are only due to the shuffling stages. So, the total number of cache misses in this case for both direct mapped and set-associative cache will be $\lceil n/(n-p) \rceil \frac{2N(S+S')}{PB}$.

Case 3: When $C \geq 2N(S + S')$. In this case cache has enough space to accommodate all the data. Thus the cache misses will be only due to the first time reference of a data. The number of cache misses will only depend on the number of different data units referenced during the whole computation. So the number of cache misses in this case for both direct mapped and set-associative cache will be at most $\lceil n/(n-p) \rceil \frac{N(S+S')}{PB}$.

4 New FFT Algorithm

The cache size affects the performance of an algorithm if the data in the cache is not fully utilized before it is replaced. This may happen in case of FFT algorithms discussed earlier if the data size is large such that $C < N(S + S')/P$. The algorithm must be suitably modified to avoid such situations. Here we shall present a modification of J-R FFT algorithm which minimizes the number of cache misses.

The idea comes from the observation in Case 2 of J-R FFT algorithm where the data loaded in the cache is used in multiple computation stages before it is replaced due to the shuffling stage. This happens only when $C \geq N(S + S')/P$. In other cases where $C < N(S + S')/P$, the data is replaced in every stage and needs to be reloaded in the next stage. This is due to the fact that a processor completes all the computation in one stage before it starts the computation of the next stage. Now, if the data allocated to one processor is divided into smaller segments such that these segments can be accommodated in the cache and the processor computes FFT of these segments one by one and then shuffles the segments, again computes and shuffles, and so on till the FFT of all the data allocated to it is completed, then the number of cache misses can be reduced considerably. At the interprocessor level this algorithm works exactly like J-R FFT algorithm.

A data flow diagram of this algorithm is shown in Figure 1 for $N = 32$, $P = 2$, and $C = 4(S + S')$. Let $k = \log(C/(S + S'))$. The shuffling pattern after every k stages in a set of $(n-p)$ stages is a k inverse-shuffle on N/P data. At the end of every $(n-p)$

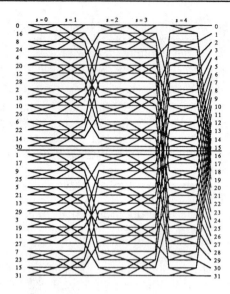

Figure 1: Data flow diagram for the new FFT for $N = 32$, $P = 2$, and $C = 4(S + S')$.

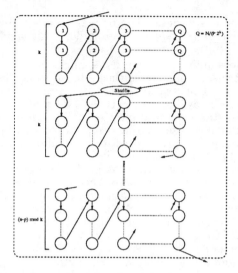

Figure 2: Sequence of computation inside each processor in the new FFT algorithm.

stages, the data allocated to each processor need to be k inverse-shuffled or $((n - p) \bmod k)$ inverse-shuffled depending upon whether $((n-p) \bmod k)$ is zero or not. After this shuffling, an $(n-p)$ inverse-shuffle need to be performed on all N data as in J-R FFT algorithm. To avoid some of the reorganization overhead and cache misses, these two shuffling stages can be combined.

The data organization for this algorithm is similar to J-R FFT algorithm. The N/P data allocated to each processor is divided into segments of 2^k data units. The processor operates on these segments one by one. The data in each segment is operated on for k stages before data in the next segment is operated. The data is shuffled after all the segments are computed. The sequence of computation by each processor during a set of $(n-p)$ computation stages is shown in Figure 2. The circles in the figure shows one stage of computation on segments of 2^k data. The numbers in the circle indicate the segment number.

Now, let us quantify the number of cache misses based on the data access pattern described above.
Case 1: When $C < N(S + S')/P$. In this case the algorithm sequentially works on segments of 2^k data units. When the processor starts working on a segment, it needs to fetch all the data in this segment from the memory and load it into cache. So, the number of cache misses for each of these segments is $2^k(S+S')/B$. Each of the processor works on $N/(P2^k)$ such segments. This process is repeated every k stages of computation. So, every k stage the number of cache misses is $N(S + S')/(PB)$. The across the processor shuffling stages are at every $(n - p)$ stages. It should be noted that the cache misses due to shuffling of intraprocessor segments can be reduced by doing it piecemeal before the beginning of next computation on a segment rather than doing the shuffle completely and then starting the computation. Also, one of the intraprocessor shuffling can be merged with across the

processor shuffling. Now, the total number of cache misses per processor for both direct mapped and set-associative cache can be expressed as

$$\left[\left\lfloor \frac{n}{n - p} \right\rfloor \left(\left\lceil \frac{n - p}{k} \right\rceil - 1\right) + \left\lceil \frac{n \bmod (n - p)}{k} \right\rceil - 1\right] \frac{N(S + S')}{PB}$$

The expressions presented above for the number of cache misses does not account for the cache misses due to across the processor shuffling stages. The number of cache misses due to these stages remains same as in the case of J-R FFT algorithm, *i.e.*, $\frac{2N(S+S')}{PB} \lceil n/(n - p) \rceil$ misses per processor.
Case 2: When $C \geq N(S+S')/P$ and $C < 2N(S+S')$. Same as Case 2 of J-R FFT algorithm.
Case 3: When $C \geq 2N(S + S')$. Same as Case 3 of J-R FFT algorithm.

5 Results and Conclusion

The number of cache misses per processor due to the three FFT algorithms discussed in this paper are shown in Figure 3 through Figure 6. The J-R FFT algorithm performs even worse than C-T FFT algorithm when $C < N(S + S')/P$ because of the cache misses due to shuffling stages. Also, the size of each data unit is bigger in J-R FFT algorithm than it is in C-T FFT algorithm. However, when $C \geq N(S + S')/P$, J-R FFT algorithm and the new algorithm perform equally well, as shown in Figure 3.

When $B > (S+S')$, all the data in a cache block will not be used before it is replaced during some stages in the first two algorithms. This is not the case in the new algorithm since all the data present in the cache is used for multiple stages before they are replaced. This is the reason why the new algorithm consistently performs much better than the other two algorithms for larger block sizes as shown in Figure 4.

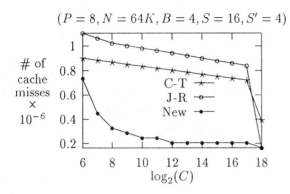

Figure 3: Comparison of FFT algorithms with respect to cache misses for different cache sizes.

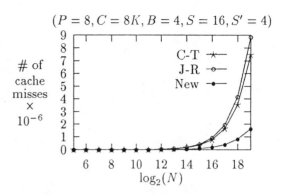

Figure 5: Comparison of FFT algorithms with respect to cache misses for different problem sizes.

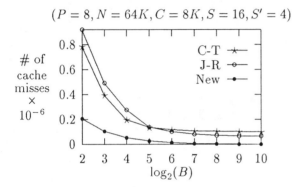

Figure 4: Comparison of FFT algorithms with respect to cache misses for different block sizes.

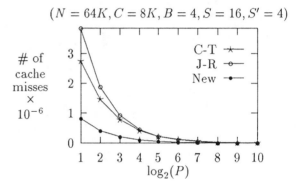

Figure 6: Comparison of FFT algorithms with respect to cache misses for different number of processors.

The plots in Figures 5 and 6 compare the number of cache misses for different problem sizes and for different number of processors, respectively. From these plots we can conclude that the new algorithm proposed in this paper performs much better than the other algorithms on a cache based shared memory multiprocessor. The improvement will be prominent in case of large problem sizes and limited number of processors. The next step in our research is to consider various cache coherence protocols [10] while analyzing these FFT algorithms.

References

[1] L.N. Bhuyan and D.P. Agrawal, "Performance analysis of FFT algorithm on multiprocessor systems," *IEEE Trans. Software Engg.*, vol. SE-9, no. 4, July 1983, pp. 512-521.

[2] J.W. Cooley and J.W. Tukey, "An algorithm for the machine calculation of complex Fourier series," *Math. of Computation*, vol. 19, Apr. 1965, pp. 297-301.

[3] S. Horiguchi and T. Nakada, "Performance evaluation of parallel fast Fourier transform on a multiprocessor workstation," *Journal of Parallel and Distributed Computing*, vol. 12, 1991, pp. 158-163.

[4] C.H. Jeon and A.P. Reeves, "The fast Fourier transform on a multicluster computer architecture," *Proc. 19th Annual Hawaii Intl. Conf. on System Sciences*, 1986, pp. 103-110.

[5] A. Kumar and L.N. Bhuyan, "Parallel FFT Algorithms for Cache Based Shared Memory Multiprocessors," Tech. Rep. TAMU 93-022, Dept. of Computer Science, Texas A&M University, April 1993.

[6] A. Norton and A.J. Silberger, "Parallelization and performance analysis of Cooley-Tukey FFT algorithm for shared-memory architectures," *IEEE Trans. Computer*, vol. C-36, no. 5, 1987, pp. 581-591.

[7] M.C. Pease, "An adaptation of the fast Fourier transform for parallel processing," *Journal of the ACM*, vol. 15, no. 2, 1968, pp. 252-264.

[8] P.N. Swarztrauber, "FFT algorithms for vector computers," *Parallel Computing*, vol. 1, 1984, pp. 45-63.

[9] P.N. Swarztrauber, "Multiprocessor FFTs," *Parallel Computing*, vol. 5, 1987, pp. 197-210.

[10] Q. Yang, L.N. Bhuyan, and B.C. Liu, "Analysis and comparison of cache coherence protocols for a packet switched multiprocessor," *IEEE Trans. Computer*, Aug. 1989, pp. 1143-1153.

SESSION 2C

PARALLEL ALGORITHMS

An Analysis of Hashing on Parallel and Vector Computers

Thomas J. Sheffler
Carnegie Mellon University
Pittsburgh, PA 15213
sheffler@ece.cmu.edu

Randal E. Bryant
Carnegie Mellon University
Pittsburgh, PA 15213
bryant@cs.cmu.edu

Abstract — Parallel hashing is well-known in the folklore of parallel computing, but there has been a remarkable dearth of literature describing it in detail. Because many parallel algorithms such as histogramming, set intersection and dictionary lookup can make use of hashing as a core step, hashing is a fundamental parallel operation. This paper sheds light on the performance that may be achieved using parallel hashing algorithms and should lend credibility to their use.

Parallel hashing is described as the solution to a more general problem called the "naming problem," which may be solved by either sorting or hashing techniques. An analysis of the hashing algorithm shows that it performs $S = O(\lg n)$ expected parallel steps and $W = O(n)$ expected work. Since the algorithm performs no more work than its serial counterpart, parallel hashing is work efficient. Performance data collected from implementations of the algorithms on the Connection Machine CM-2 and the CRAY Y-MP show that hashing is faster than sorting by up to a factor of four on each machine.

INTRODUCTION

Hashing techniques have long been used to perform table lookup on serial machines. While some hashing techniques have also been extended to parallel and vector computers [8], it is only recently that the algorithmic complexity of parallel hashing algorithms has been studied [11, 1]. In [8], a vectorized algorithm was described that is similar to the one presented here, however no bounds were placed on the expected performance of the algorithm. While [11] provided the framework for understanding the total number of parallel steps required, and [1] provided an empirical study of parallel step counts, neither recognized the importance of work complexity. In this paper we place expected case bounds on the parallel step count of our algorithm as well as its work complexity. We show that an important indicator of actual performance is the total amount of *work* performed, and that to achieve high performance a parallel algorithm must be carefully designed to perform no more total work than necessary.

We also extend parallel hashing to the solution of a more general problem called the *naming* problem. Algorithms that solve the naming problem may be based on either sorting or hashing. When a total order of the elements to be named is not required, the sorting algorithm performs more work than necessary. This is borne out by the algorithmic complexities of the two methods in which general parallel sorting performs $\Omega(n \log n)$ work for arbitrary sized keys but the hashing algorithm performs only $O(n)$ expected work.

The remainder of this paper is organized as follows. First, the naming problem is defined as a generalization of hash table insertion and a number of fundamental parallel algorithms are outlined that use the naming algorithm as a core step. The basic parallel hash table insertion algorithm is then presented along with some variants that attempt to reduce the total number of steps required by the algorithm. The analysis of the expected case behavior of the parallel algorithm shows that the parallel hashing algorithm performs no more work than its serial counterpart, thus parallel hashing is *work efficient*. Finally, we present performance data collected from implementations of our algorithms on the Connection Machine CM-2 and CRAY Y-MP and discuss the implications of work efficiency with regard to parallel step count.

THE NAMING PROBLEM

When building a table using hashing, each insertion of a new data item involves finding either a previous instance of the same key, or an empty site in which to put the key. A parallel algorithm for key insertion may be constructed in which a vector of keys are inserted together. The value returned by such a parallel algorithm is a vector of table indices identifying the final hash sites of each key.

This problem attempts to find a mapping of keys from a potentially large universe to a small hash table. This is an instance of the *naming* problem which is as follows: a multiset X of n elements selected from some universe U are to be assigned names in the range $\{1..m\}$, $m \geq n$, such that two elements are given the same name if and only if they are the same element from U.

The naming problem is one that is fundamental to many algorithms that appear in scientific supercomputing applications. In many of these examples, the keys are tuples composed of a number of quantities. For example, the keys may encode the velocity and position of a particle in particle-in-cell simulations, or quantities associated with an event in high-energy physics experiments. Naming serves as a canonicalization step that gives each unique key a small integer identifier. The integer identifiers assigned can then be used as simple array indices to complete the algorithm. Some specific examples follow.

Removal of Duplicates: This operation is fundamental to almost all data analysis and report generation. After assigning names to each of the keys, each key attempts to write its value to the index of its name in a temporary array. The array will be left with one of each unique key.

Histogramming: The frequency of events in high-energy physics experiments are typically reported through histogramming. Each event is a tuple and unique tuples must be recognized before the frequency of each unique even may be calculated [2]. The naming step reduces the problem of histogramming the tuples to that of counting the frequency of each integer, for which there are known algorithms for parallel and vector computers.

Dictionary Lookup: In the diagonalization of quantum-spin models, a spin configuration is encoded in a key for which an action on the spin must be retrieved from a table [10]. After the naming step, table lookup becomes a simple array access by the integer identifier.

Algorithms

In many applications, there is a total order defined for the keys and the naming operation may be implemented by using a sort procedure. In this method, after the keys are placed in sorted order, they are divided into groups such that a new group begins where a key differs from its left neighbor. The first element of each group is called its "leader." The available names are distributed to the leaders, which then distribute the names to all members of their groups.

The sorting step dominates the time spent in the procedure outlined above, and may not be applicable if there is not a total order on the keys. Even when applicable, because sorting requires $\Omega(n \log n)$ operations it is not work efficient for the naming problem.

In this paper we present a family of efficient parallel algorithms that solve the naming problem by using parallel hashing. The first step of the algorithm uses a hash function, h, that uniformly distributes the n keys across the name space. At each step of the iterative procedure, a large number of keys claim names. These keys are removed from the working set by using a "packing" step so that the working set size decreases as the algorithm proceeds. Many keys find names in just the first few parallel steps, after which, the number of keys remaining in the working set decreases quickly.

THE PARALLEL VECTOR MODEL OF COMPUTING

The parallel vector model of computation builds on the standard RAM model by adding a separate vector memory and vector processing unit [4]. Operations on vectors have the distinction of being performed in parallel. For example, the addition of two vectors can be performed for all elements of the result vector simultaneously. A standard RAM would iterate through each of the elements to calculate the same result.

The parallel vector model imposes no restrictions on the length of a vector. Two complexity measures are introduced for algorithms in the vector model. The first is the *step* complexity, which measures the total number of vector instructions issued. Another complexity measure, the *work* complexity, measures the total number of vector elements operated on over all instructions. The work complexity of an algorithm is the same as the time complexity of the algorithm on a single processor since it measures the total number of single element operations performed. A parallel algorithm is work efficient if it has a work complexity no greater than that of an equivalent serial algorithm that solves the same problem.

The *step* and *work* complexity measures are related to the PRAM model through the following relation. For an algorithm with step complexity S and work complexity W, the PRAM complexity of the algorithm is

$$t = O(W/p + S)$$

for a p-processor PRAM. This result is due to a simulation argument in which the vector algorithm is simulated on an EREW PRAM that provides unit-step parallel-prefix operations [4]. Given an unbounded number of processors, the complexity is dominated by the step complexity. Conversely, the work complexity indicates the total number of operations performed when only one processor is available. All real parallel machines have a processor count somewhere between these two extremes; when designing algorithms for the vector model it is important to keep both measures in mind.

Analysis of Algorithms

All algorithms presented in this paper will be analyzed for their step and work complexities in the vector model. In the parallel hashing algorithms the step complexity is related to the number of iterations required to complete the insertion of all n keys, while the work complexity is the sum of the number of active elements over all steps.

PARALLEL HASHING

Traditional serial hashing algorithms are concerned with the efficient creation of a hash table (inserting all n elements) as well as being able to locate keys already in the table [9]. The naming problem only concerns itself with the insertion phase of hash table use; its final result is the set of key destinations.

Closed table hash algorithms begin with a function h that is designed to map each of the keys uniformly to some number in the range $\{1..m\}$ for the given m. As each key is inserted, its hash location is probed to examine whether another key has already been inserted there. If so, an iterative search procedure is used to find an empty site. The sequence of hash sites examined by each key is called a "path." Ideally, each key examines a different path so as to avoid a situation where many keys collide repeatedly along the same path. This phenomenon is called "clustering."

In the parallel version of hashing, the insertion of many keys is attempted at each step. Those that fail all examine the next site in their path in parallel. The process continues until all keys find a hash site.

Pseudo-code for parallel hash-table insertion is presented below. The algorithm accepts a vector of n *keys* and returns a vector of *names*. A hash table of length m is allocated and initialized so that all locations have the special value EMPTY. Upon completion, the name associated with *key[i]* is located at *name[i]*.

In the initialization step (Lines 1–4), the keys are copied to a vector called *active* and the initial hash sites are calculated by **hash-fn**. The *home* vector records the starting location of each key. After each iteration of the algorithm, these temporary vectors are compressed so that they contain only the information relevant to those keys that have yet to find a hash site.

The main step of the algorithm is described by the parallel-do loop of Line 6. In parallel, all keys check to

```
input:       keys[n], table[m];
output:      names[n];
temporary:   hash[n], get[n], active[n],
             home[n], notdone[n];
     procedure parallel-insert {
(1)      pardo (0 ≤ i < n){
(2)          active[i] = keys[i];
(3)          home[i] = i;
(4)          hash[i] = hash-fn(keys[i], m);
         }
(5)      while (length(active) ≠ 0) {
(6)          pardo (0 ≤ i < length(active)) {
(7)              get[i] = table[hash[i]];
(8)              If (get[i] == EMPTY) {
(9)                  table[hash[i]] = keys[i];
(10)                 get = table[hash[i]];
                 }
(11)             notdone[i] = (get[i] ≠ keys[i]);
(12)             If (not(notdone[i]))
(13)                 names[home[i]] = hash[i];
             }
(14)         active = pack(active, notdone);
(15)         home = pack(home, notdone);
(16)         hash = pack(hash, notdone);
(17)         pardo (0 ≤ i < length(hash)) {
(18)             hash[i] = next-fn(hash[i], active[i]);
             }
         }
     }
```

see if their hash site is empty. Those that find an empty site try to claim it by sending their key there. When multiple keys are sent to the same site, one overwrites all others ensuring that at least one key claims the site. With one more table reference, the hash site of each key is examined to determine which key claimed it. Keys that find their own value are marked done and their hash site is sent *home* to the *names* vector. All other keys continue with further probe steps by calculating a new hash site in Line 18.

The **pack** step on Lines 14–16 produces a new vector containing only the values of a first vector that are flagged **true** in a second. This is the familiar *compress* operation on vector computers. On parallel computers it is implemented with a prefix-sum instruction and a permutation of the data values. In either case, the **pack** operation is a unit-step operation of the parallel vector model. Packing ensures that as the algorithm progresses, the length of the active vector decreases.

Probe Strategies

The function **next-fn** may be designed to implement a number of different probe strategies. The path that each key follows may be defined by either "Linear

Probing" or "Double Hashing" probe strategies. Linear probing simply searches the hash sites in sequence. Double hashing attempts to reduce "clustering" by using a probe strategy where the path each key follows is a function of its value. To implement this algorithm, a second hash function, h_2, is employed and the ith site examined for the key is $((h(A) + ih_2(A)) \mod m)$.

Another variation also adopted in many serial hashing algorithms is the use of an "Overflow" hash table. In this approach, two hash tables (designated T_1 and T_2) are used such that each key makes its first probe into T_1 and makes all other probes in T_2. This arrangement takes advantage of the fact that many keys require only one probe, leaving table T_2 nearly empty.

In the next section, we show that for Linear Probing, our algorithm performs a total of $S = O(\log n)$ expected steps, and $W = O(n)$ expected work. Analyses of double hashing algorithms show that on the average, they reduce the total number of probe sites examined [9]. We implemented a version of double hashing and compare our results to linear probing on the CM-2. On the CRAY we implemented both probe strategies with and without an overflow table. The performance of our implementations is discussed after an analysis of the algorithmic complexity of the algorithm.

ANALYSIS OF THE ALGORITHM

The behavior of the parallel hashing algorithm for linear probing will be analyzed by comparing it to its serial counterpart. Whereas the analysis of the running time of the serial algorithm is concerned with the *average* time of insertion for each of the n keys, the parallel algorithm must consider two measures of performance.

The first is the step complexity, S, which measures the total number of primitive vector instructions issued. Because every line of the algorithm is a unit step operation in the vector model, this measure is the same as the total iteration count of the outer loop. The number of iterations is governed by the *maximum* number of probes required to insert any of the keys since the algorithm cannot terminate until all keys find a hash site.

The second measure is the total work performed, W, and is the sum of the lengths of the active vector over all iterations. Since each key remains in the active vector for only as many iterations as it requires probe steps, the total work is the sum of the number of probes required to insert all n keys. When the hash function and hash table size are the same for the serial and parallel algorithms, the total number of probes required by each of the algorithms is shown to be the same.

The number of iterations required to insert all of the keys could conceivably be n for a particularly bad arrangement of keys, however this is almost never the case. An expected bound on the number of iterations required will be made by examining the expected number of probes required to insert the worst-case key.

The total work required to insert N keys

A surprising property of the serial algorithm for Linear Probing was first pointed out by W. W. Peterson [12]:

Theorem 1 *The total number of probes required to insert N keys remains unchanged regardless of the order in which the keys are inserted in the table.*

Proof: (due to Knuth, [9]) The keys are presented in some order $A_1, A_2, \ldots A_N$. It suffices to show that the total number of probes needed to insert the keys is the same as the total number needed for $A_1 \ldots A_{i-1} A_{i+1} A_i A_{i+2} \ldots A_N, 1 \leq i < N$. There is clearly no difference unless the $(i+1)$st key in the second ordering falls into the position occupied by the ith in the first ordering. But then the ith and the $(i+1)$st merely exchange places, so the total number of probes for the $(i+1)$st is decreased by the same amount the number for the ith is increased. $\square$

A similar theorem holds for inserting keys in parallel.

Theorem 2 *The total number of hash sites examined remains unchanged if two or more keys are inserted in parallel.*

Proof: Consider two keys presented in order A_1, A_2. If, in parallel, the two keys begin their probes at two different locations, then because they each move only one site forward each iteration, the two keys will find the same empty hash sites as they would had they been inserted in sequence. If the keys happen to begin at the same site, then the first two empty sites past their starting point are the two sites that the keys will find. The two possible arrangements of these two keys into the two sites are exactly the two arrangements dictated by the serial presentation order A_1, A_2 and A_2, A_1. Thus, parallel insertion has the same effect as a random exchange of the order in which keys are inserted by a serial algorithm, which by the previous theorem does not affect the total number of probe steps made. $\square$

These two theorems state that the parallel linear probing algorithm performs exactly the same number of probes as its serial counterpart. The analysis of the numbers of probes required to insert n elements into a hash table using linear probing is well understood. From [9] we have the following: The average number of probes required to insert the ith key is

$C_i' = \sum_{1 \le r \le m} r P_r(i)$ where $P_r(i)$ is the probability that the ith key requires r probes to find an unoccupied site. The average number of probes to insert all n keys is $C_n = \frac{1}{n} \sum_{0 \le i \le n} C_i'$. The total work is simply $W = nC_n$ for n keys. Knuth gives the value of C_n for linear probing as $C_n \approx \frac{1}{\alpha} \log \frac{1}{1-\alpha}$ where $\alpha = \frac{n}{m}$. Thus, when the hash table is sized so that $m = \Theta(n)$, the expected work performed by the parallel algorithm for linear probing is

$$W = nC_n \approx n\frac{1}{\alpha} \log \frac{1}{1-\alpha} = O(n).$$

The total number of iterations required to insert N keys

The Longest Length Probe Sequence (LLPS) quantity associated with a hashing algorithm measures the longest sequence of probes needed to locate any of the n keys inserted in the table. The LLPS measure is a random variable whose maximum value is obviously n for linear probing. However its average value over all possible arrangements of keys in the hash table, the average LLPS, is also of interest.

Since the parallel linear probing algorithm requires as many iterations as the longest length probe sequence for any key, the average LLPS figure of the serial algorithm measures the average number of iterations that our parallel hashing algorithm requires. Gaston Gonnet [7] has derived bounds on the average LLPS for linear probing and shows that it is $O(\log n)$, when $m > n$. Thus, the parallel step complexity for n keys is

$$S = O(\log n).$$

The same expected bound was given for parallel uniform hashing in [11]. Our experience shows that the average number of iterations required for the hashing algorithm to complete grows slowly with the number of keys. Data presented later shows that the insertion of one million keys is achieved in less than 20 iterations, on average. However, the work figure is a more accurate measure of the total amount of time required by the algorithm.

IMPLEMENTATION ON THE CONNECTION MACHINE CM-2

We implemented our algorithms on the Connection Machine model CM-2. Through the virtual processor (VP) mechanism of the Paris assembly language we were able to support the varying sized vectors that our **pack** procedure produced. As stated earlier, this was necessary to ensure work efficiency.

Our choice of a hash function was guided by the target architecture for implementation. One particularly good hash function is based on a result from coding theory [6]. It treats an n-bit key as a polynomial in $GF(2^n)$ and computes an m-bit hash value as the remainder of division by a primitive irreducible polynomial of the field $GF(2^m)$. These hash functions are often not used on conventional architectures because of the bit manipulations that must be performed. Because the CM-2 is a bit-serial machine, these hash functions are extremely fast. Also, because these hash functions require a table whose size is a power of two, they are perfectly suited to the size of VP sets and guarantee that keys are evenly distributed over all processors of the CM-2.

A minor modification was made to the insertion procedure. Because communications operations can require as much as 1000 times as much time as elementwise operations on the CM-2, the procedure **parallel-insert** was modified to reduce the total number of communications operations performed by omitting the first **get** step. Instead, as hash sites are claimed, a flag is set. During the probe step, all keys send their values regardless of the state of their hash site. Then, non-empty hash sites overwrite the value sent there with the key that previously claimed the site. Even with large hash tables, these elementwise operations take virtually no time, and are outweighed by the savings from eliminating the extra communications operations. Theoretically, this increases the work complexity of the algorithm by $O(m \log n)$, but does not have an effect when $\log n < 1000$, which is a reasonable assumption for almost any problem size hoped to be solved on a real machine.

Results

A series of trials were taken for n ranging from 50,000 to 2,000,000 elements on a 16k processor CM-2. The key size in this trial was 32 bits but the universe of possible keys was restricted to be a random set of size n. In this manner, the keys represent a random mapping of a set into itself. Each trial point was repeated a total of five times and the average time reported was measured in CM seconds, the amount of time the Connection Machine required to complete all iterations. Both Linear Probing and Double Hashing schemes were compared to a naming algorithm that sorted the keys using the RANK function of the system software and enumerated the leaders to assign names. Figure 1 summarizes our results.

Through all trials, the hash based algorithms outperformed the sorting based algorithm by a factor of 3 to 4. The sharp jumps in times for the sort based algorithm occurred where the vector size increased past a VP ratio boundary. Our data spans VP ratios of 2 to

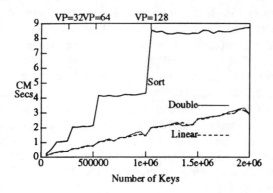

Figure 1: A comparison of the hashing algorithms to a naming algorithm that uses sorting on the CM. Through all trials the hashing algorithms performed better by a factor of 3 to 4.

128 and shows that the hashing algorithms have linear characteristics even over VP ratio boundaries.

It is interesting to notice that the two variants of hashing require almost the same time for all key set sizes. Even though the total number of iterations required by the Linear Probing variant *always* exceeded those required by Double Hashing, the extra iterations were performed on such small vectors that the additional time is negligible. This point deserves further examination.

Figure 2 shows the average number of iterations required for each algorithm for different input set sizes. The sharp drops in the curves are due to the fact that the hash table and overflow table sizes were rounded to the next power of two when the key set size exceeded a VP ratio boundary. In those situations when the hash table was very large with respect to the number of keys, the number of iterations became quite small. While we see that double hashing keeps the number of iterations performed smaller than by using the linear probing method, the total times spent by the algorithms are roughly the same, and are not affected very much by the occasional large jumps in the number of iterations.

Figure 3 provides further insight into the impact of the total number of iterations on the completion time. The trials for the insertion of 1,900,000 keys using double hashing required a wide range of iterations to complete. Even as the iterations increased by almost a factor of 2, the total time grew by only 7%. The time is dominated by the first few probe iterations with long vectors; all other iterations are performed with very short vectors and require a small percentage of the total time.

The decrease in the number of active keys is shown

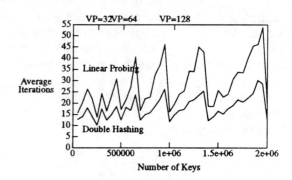

Figure 2: The average number of iterations required by the two hashing algorithms. The data show that double hashing keeps the total number of iterations lower than the number required by linear probing. The unusual drops in the curves occur when the hash table increased to the next VP set size on the CM, resulting in a hash table with a very low load factor.

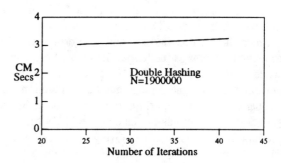

Figure 3: A comparison of the execution times when the number of iterations varied greatly. Even though the number of iterations spanned a wide range, the time increased by only a small amount when the number of iterations grew large. All of the extra iterations were performed with very small vectors and required very little time.

in Figure 4. When inserting one million keys using the double hashing algorithm, by the eighth iteration the total number of active keys fell below 100. Probe iterations performed with small vector lengths are completed quickly.

IMPLEMENTATION ON THE CRAY Y-MP

The implementation of the hashing algorithm on the CRAY Y-MP was straightforward because all operations completely vectorize. We used the CVL (C Vector Library) developed by Guy Blelloch at Carnegie Mellon University for most of the vector operations [3], but developed our own hash function in CRAY vectorized C.

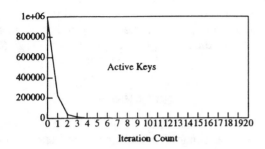

Figure 4: The number of active keys at each step in the hashing algorithm. The hashing algorithm eliminates many keys from the active set in just a few iterations. In this graph, the total number of active keys is less than 100 after eight iterations. Most of the iterations are performed with very short vectors.

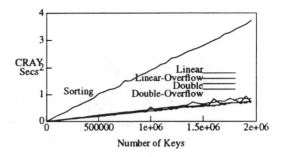

Figure 5: A comparison of the times of the hashing algorithms to the time to sort the keys on the CRAY Y-MP. Through all trials, the hashing algorithms performed better by a factor of 4 to 5. While the more complicated probe strategies required fewer total iterations, the total time required by each was nearly equal.

As with the implementation for the CM-2, the hash function was highly optimized for the target architecture. We chose to use a standard remainder operator and sized the hash tables to be the least prime number greater than the number of keys. However, because the standard modulo operator ('%') of CRAY C calls a subroutine that computes the remainder between two 64-bit integers, it was much too slow. We wrote a highly optimized modulo function that took advantage of the facts that the same modulus was applied to all keys, and that it had far fewer than 64-bits of significance. Using these observations, we divided the computation into two parts that each used an integer division and a subtraction operation, yielding a very fast hash function.

The next-site function, **next-fn**, was written to allow the choice of Linear or Double probing, with or without an overflow table. Double probing was described earlier. To implement an overflow table, we conceptually divided the hash table into two parts: it's lower half acting as T_1, and the upper half acting as T_2. The next-site function was written to supply a first probe site in the lower half, with all other probes indexing sites in the upper half. In this manner, a single site address could specify one of the two tables and a site within that table.

Results

We measured the performance of our hashing algorithm relative to sorting. This time, the sorting time measured represents *only* that time spent using the ORDERS subroutine of the system software [5]. The key size in this trial was 64 bits and the keys once again represented the random mapping of a set into itself. Figure 5 summarizes the performance for each of the

four combinations of probe strategies relative to sorting. Throughout all trials, the hashing algorithms outperformed the CRAY ORDERS subroutine by up to a factor of 5.

It is interesting to notice that the total times required by each of the four approaches to hashing are nearly equal. In general, the fastest approach was the simple Linear Probing search, while the slowest was Double Hashing with an Overflow table. These results are somewhat surprising in light of the total number of iterations required by each approach.

Figure 6 displays the average number of iterations per key required by each of the variations. The effect observed here is that the extra complexity of the next-site function (**next-fn**) when using Double Hashing and an Overflow table far outweighs any benefits achieved by a reduction in the number of iterations. Thus, the simple and straightforward Linear Probing approach is the fastest even with the cost of additional iterations. While the more sophisticated probing algorithms did reduce the total number of probe steps required, their extra complexity caused a loss in actual performance.

This effect is due to the fact that computation and communication are so nearly balanced on the CRAY Y-MP. While some machines perform local arithmetic operations much faster than they can permute a vector, on the CRAY Y-MP these operations require nearly the same amount of time per element. Because of this, it is faster to perform the simple Linear Probing hashing algorithm than to use a more clever (and complicated) probing function. On a machine with slower communication, it might be wise to use the more sophisticated probing schemes.

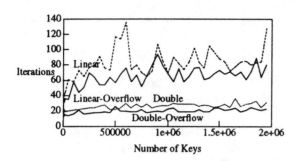

Figure 6: The average number of iterations per key on the CRAY Y-MP for each of the four approaches. In general, the more complicated probing schemes reduced the total number of probe steps, as expected. However, as the complexity of the probe function increased, so did the execution time, making the simplest probe scheme preferable.

CONCLUSIONS

The naming problem is one that is fundamental to many parallel algorithms. While the most straightforward implementation of this parallel kernel uses sorting to organize the keys, we have presented an efficient randomized hashing-based algorithm that is more efficient both in theory and practice.

The main idea that should transfer from this work is the importance of reducing the problem size as the algorithm progresses. Many Connection Machine algorithms have been designed with a fixed number of processors in mind. This design influence reflects an earlier view of the CM-2 when the number of virtual processors was fixed throughout the execution of a program. Because virtual processor sets may be allocated dynamically, the view of the CM-2 is one of a dynamically reconfigured vector processor, whose length adjusts to fit the problem size. We have shown the importance of reducing the problem size and using smaller VP sets when possible. On vector computers such the CRAY Y-MP, the benefit of *compressing* vectors has long been recognized.

Work efficiency for parallel algorithms has often been sacrificed in favor of reducing the total number of parallel steps an algorithm requires. This paper shows that for some algorithms, variations in the total number of parallel steps can have a small impact on the actual performance and that work complexity is an important measure.

ACKNOWLEDGMENTS

This research was supported by the Defense Advanced Research Project Agency, ARPA Order 7597. Connection Machine time was provided by the Northeast Parallel Architectures Center, and by the Pittsburgh Supercomputing Center, grant number ESC910004P. Cray Y-MP time was provided by the Pittsburgh Supercomputing Center (Grant ASC890018P).

REFERENCES

[1] Paul B. Anderson. Parallel hashed key access on the connection machine. In Ronnie Mills, editor, *Second Symposium On The Frontiers of Massively Parallel Computations*, pages 643–645. IEEE, IEEE Computer Society Press, 1988.

[2] A. Bale, E. Gerelle, J. Jessersmith, R. Warren, and J. Hoek. Event analysis using a massively parallel processor. In *8th Conference on Computing in High Energy Physics*, pages 436–441, 1990.

[3] Guy Blelloch, Siddhartha Chatterjee, Jay Sippelstein, and Marco Zagha. CVL: a C Vector Library. School of Computer Science, Carnegie Mellon University, 1991.

[4] Guy E. Blelloch. *Vector models for data-parallel computing*. MIT Press, 1990.

[5] Cray Research Inc., Mendota Heights, Minnesota. *ORDERS(3SCI) Manual Page SR-2081 5.1*, 1988.

[6] R. G. Gallagher. *Information Theory and Reliable Communication*. John Wiley and Sons, Inc., 1968.

[7] Gaston H. Gonnet. *Handbook of Algorithms and Data Structures*. Addison Wesley, 1990.

[8] Yasusi Kanada. A vectorization technique of hashing and its application to several sorting algorithms. In *PARBASE-90, International Conference on Databases, Parallel Architectures, and Their Applications*, 1990.

[9] Donald Knuth. *The Art of Computer Programming; Volume 3: Sorting and Searching*. Computer Science and Information Processing. Addison-Wesley, 1968.

[10] H. Q. Lin. Exact diagonalization of quantum-spin models. *Physical Review B (Condensed Matter)*, 42(10):6561–6567, Oct. 1990.

[11] Fabrizio Luccio, Andrea Pietracaprina, and Geppino Pucci. Analysis of parallel uniform hashing. *Information Processing Letters*, 37:67–69, January 1991.

[12] W. W. Peterson. *IBM J. Research and Development*, 1:130–146, 1957.

MULTIPLE QUADRATIC FORMS: A CASE STUDY
IN THE DESIGN OF SCALABLE ALGORITHMS

Mu-Cheng Wang[†*] Wayne G. Nation[†*] James B. Armstrong[†*] Howard Jay Siegel[†*]
Shin-Dug Kim[†*] Mark A. Nichols[†*] Michael Gherrity[‡]

[†]Parallel Processing Laboratory
School of Electrical Engineering
Purdue University
West Lafayette, IN 47907-1285 USA

[‡]Superconcurrency Research Team
Naval Ocean Systems Center, Code 421
San Diego, CA 92152-5000 USA

Abstract -- *Parallel implementations of the computationally intensive task of solving multiple quadratic forms (MQFs) have been examined. Coupled and uncoupled parallel methods are investigated, where coupling relates to the degree of interaction among the processors. Also, the impact of partitioning a large MQF problem into smaller non-interacting subtasks is studied. Trade-offs among the implementations for various data-size/machine-size ratios are categorized in terms of complex arithmetic operation counts, communication overhead, and memory storage requirements. From the complexity analyses, it is shown that none of the algorithms presented in this paper is best for all data-size/machine-size ratios. Thus, to achieve scalability (i.e., good performance as the number of processors available in a machine increases [4]), instead of using a single algorithm, the approach proposed is to have a set of algorithms from which the most appropriate algorithm or combination of algorithms is selected based on the ratio calculated from the scaled machine size. The analytical results have been verified from experiments on the MasPar MP-1, nCUBE 2, and PASM prototype.*

1. INTRODUCTION

Several parallel algorithms are developed and analyzed in this research for computing the multiple quadratic forms (MQFs) that are part of an adaptive beamformer calculation with minimum variance distortionless response (MVDR) [10]. The implementations of the MQF problem for various data-size/machine-size ratios are evaluated in terms of the number of complex arithmetic operations, communication overhead, and memory storage requirements. Furthermore, both coupled and uncoupled parallel methods are investigated, where coupling relates to the degree of interaction of the processors. Then, the impact of partitioning a large MQF problem into smaller non-interacting subtasks is studied. The analytical results show that none of the algorithms presented in this paper is best for all data-size/machine-size ratios. Thus, to achieve scalability (i.e., good performance as the number of processors available in a machine increases), instead of using a single algorithm, the approach proposed is to have a set of algorithms from which the most appropriate algorithm or combination of algorithms is selected based on the ratio calculated from the scaled machine

size. In this sense, the term scalable is applied to a set of algorithms. The importance of using a set of algorithms also has been recognized by other researchers (e.g., [12]). Experimental results from the MasPar MP-1, nCUBE 2, and PASM parallel processing systems are shown to support the theoretical results derived herein.

To date, no related work has been found that examines the problem of determining parallel implementations for computing multiple quadratic forms for the same matrix (the type of computation studied here). However, many publications exist that examine the more general problem of parallel implementations of matrix multiplication (e.g., [1, 3, 5, 8, 11]).

An overview of parallel matrix multiplication algorithms based on different processor interconnection topologies, such as the hypercube, mesh, and perfect shuffle networks, is given in [5, 11]. Concerning the hypercube topology [5], that paper presents some parallel matrix multiplication algorithms for $p = n^3$ and $p = n^2$ cases. The computational complexity of the parallel algorithm is $O(\log_2 n)$ when $p = n^3$ and $O(n)$ when $p = n^2$.

Several papers (e.g., [1, 3, 8]) have been published that discuss performing matrix multiplication on hypercube machines by using a logical mesh of processors. These papers differ from [5], [11], and the discussion in this paper in that the analyses are specifically for hypercube machines. [8] shows that the optimal performance of matrix multiplication on the Cosmic Cube [Sei85] is achieved (in terms of communication overhead and load balancing) by decomposing the problem into square sub-blocks. Similarly, [3] discusses partitioning the matrix and the impact of communication overhead when performing matrix multiplication problem on an nCUBE [9]. The work presented in [1] is an extension of [8] in that it considers the restrictive condition of only having nearest-neighbor one-to-one communication.

Unlike previous work, this paper deals with the computation of *multiple* quadratic forms and discusses the impact of the two primary aspects of system configuration on the computation: interprocessor communication and partitionability. Consequently, the results set forth in the above publications are not directly applicable to the MQF problem discussed here.

In Section 2, the MQF problem is defined. The parallel processing system model is overviewed in Section 3. Section 4 describes the uncoupled algorithm. Several coupled parallel implementations and the impact of problem partitioning are presented in Section 5. A generalized method, of which the uncoupled and coupled methods are special cases, is defined in Section 6. In Section 7, the theoretical results of Sections 4, 5, and 6 that are useful for choosing an optimal algorithm are reviewed. Also in Section 7, a combined coupled/uncoupled approach is presented. Experimental results are discussed in Section 8.

*This research was supported by Rome Laboratory under contract number F30602-92-C-0150, by the Naval Ocean Systems Center under the High Performance Computing Block, ONT, by the Office of Naval Research under grant number N00014-90-J-1937, by the National Science Foundation under grant number CDA-9015696, and by National Aeronautical and Space Administration under grant number NGT-50961.

Wayne G. Nation, Shin-Dug Kim, and Mark A. Nichols are currently with IBM Corporation, Rochester, MN, KwangWoon University, Seoul, Korea, and NCR Corporation, San Diego, CA, respectively.

2. THE MQF PROBLEM

Let s-vector be an $n{\times}1$ vector of complex numbers and $\underline{v}$ be the total number of s-vectors. Define $\underline{M}$ to be an $n{\times}n$ matrix of complex numbers and $M(i,j)$ to be the element of M in row i and column j, for $0 \le i,\ j < n$. The q-th s-vector is denoted by $\underline{s_q}$, for $0 \le q < v$. Element m of the s-vector q is denoted $s_q(m)$, for $0 \le m < n$. Let H denote the Hermitian transposition, i.e., the complex conjugate transposition of the s-vector (where the complex conjugate of $a + bi$ is $a - bi$). Then, the MQF calculation can be formally defined as:
$$w_q = s_q^H M s_q \quad \text{for } 0 \le q < v .$$

In addition to the MVDR problem, this type of computation also appears in other problems. For example, if M is a positive definite matrix, the quadratic $P(x) = (1/2)x^T M x - x^T b$ is minimized at the point where $Mx = b$ and the minimum value is $P(M^{-1}b) = -(1/2)b^T M^{-1}b$. The parallel algorithms proposed here can be easily generalized to compute the equation $x^T M y$ that appears in the fundamental variational principles of physics [18].

If the computation of v quadratic forms is performed on a serial machine, $v(n^2+n) = vn^2 + vn$ complex multiplications and $v(n(n-1)+(n-1)) = vn^2 - v$ complex additions are required. Also, the serial machine must have enough memory space to store the entire M matrix and v s-vectors, with n^2 and vn complex numbers, respectively. It is assumed that each complex number is represented by two floating point numbers.

3. PARALLEL SYSTEM MODEL

The model of a parallel system that is assumed here includes p processing elements (PEs) of the same computing power, where each PE is a processor and memory module pair. Such a configuration is often referred to as a physically distributed memory machine, and is used in most current parallel systems with 64 or more processors, e.g., MasPar [2] and nCUBE 2 [9], as well as in the PASM prototype [6, 16]. No particular interconnection network is assumed for the proposed algorithms, but the multistage cube and hypercube (single-stage cube) networks [13] are shown to be flexible interconnection networks that can perform the PE-to-PE permutations required by the MQF implementations without any conflicts.

The proposed parallel algorithms can be implemented on a SIMD, MIMD, or mixed-mode machine, where a mixed-mode machine can execute in either SIMD or MIMD mode and can dynamically switch between the modes at instruction-level granularity with negligible overhead [6].

4. THE UNCOUPLED PARALLEL METHOD

The uncoupled method uses the obvious approach to parallelism. Assume that p divides v, i.e., $v = c{\times}p$ where c is an integer. By distributing v s-vectors evenly among p PEs, each PE can compute c quadratic forms in parallel without having to communicate with any other PE. If p does not divide v, i.e., $v \ne c{\times}p$, $\lceil v/p \rceil$ s-vectors will be assigned to each of $v \bmod p$ PEs, and $\lfloor v/p \rfloor$ s-vectors will be assigned to each of the remaining PEs. However, in this case some PEs will do more work than others. If $v < p$, only v PEs will actually be utilized. Thus, for the uncoupled method, utilization is good when either p divides v or $v \gg p$.

Solving a quadratic form can be decomposed into two phases: (1) the calculation of $r_q = s_q^H M$ and (2) the calculation of $w_q = r_q s_q$. The number of complex multiplications and additions

required for each s-vector during phase (1) is n^2 and $n(n-1)$, respectively. For phase (2), n complex multiplications and $(n-1)$ complex additions are performed for each s-vector. Thus, the total number of complex multiplications and additions performed for each s-vector is $n^2 + n$ and $n^2 - 1$, respectively. Because some PEs are assigned $\lceil v/p \rceil$ s-vectors, the maximum number of parallel complex multiplications and additions is $\lceil v/p \rceil(n^2+n)$ and $\lceil v/p \rceil(n^2-1)$, respectively.

For the uncoupled method, the maximum storage requirement per PE is n^2 complex elements for the M matrix and $\lceil v/p \rceil n$ complex elements for the s-vectors. This assumes that as each element of r_q is computed, it is used to calculate w_q and is not stored (i.e., phases (1) and (2) are temporally interleaved).

5. THE COUPLED PARALLEL METHOD

5.1 Overview

This section describes how to map the MQF problem onto an $a{\times}b\ (=p)$ logical grid of p PEs (not necessarily a physical grid or mesh), where $1 \le a, b \le n$, and a, b, and p are powers of two. Furthermore, the advantages of partitioning the problem are discussed. The interconnection network assumed is either the multistage cube network or the hypercube network [13]. The special cases of $p = n\ (a = 1,\ b = n)$ and $p = n^2\ (a = b = n)$ were studied first and used to develop this general case [19].

For ease of presentation, n is assumed to be a power of two initially. Based on the description of the coupled implementation, the proposed method can be readily extended to cover the case when n is not a power of two. This case will be briefly discussed in Subsection 5.4.

5.2 Algorithm Description and Evaluation

Let $PE(i,j)$ denote the PE in row i and column j in an $a \times b$ logical PE grid, where $0 \le i < a$ and $0 \le j < b$. Fig. 1 illustrates how M, s_q^H, and s_q are distributed across the PEs of the $a \times b$ grid. Initially, the s-vectors are loaded into the PE memories such that an (n/a)-element part of the Hermitian of each s-vector, s_q^H, and an (n/b)-element part of at most $\lceil v/a \rceil$ s-vectors, s_q, are stored in each PE memory. Each PE also holds an $(n/a) \times (n/b)$ portion of M. The exact elements stored in each PE are defined for general $PE(i,j)$ in Fig. 1. For example, if $v = 6$, $n = 4$, $a = 2$, and $b = 4$, $PE(1,2)$ contains $M(2,2)$, $M(3,2)$, $s_q^H(2)$ and $s_q^H(3)$ for $0 \le q < 6$, $s_1(2)$, $s_3(2)$, and $s_5(2)$. The data layout for the s-vectors for this example is shown in Fig. 2.

There are two common data layouts of the M matrix and s vectors in memory, referred to in [7] as the block and scattered decompositions. For the MQF problem, the block decomposition method has contiguous entries of the M matrix and s vectors assigned to PEs in blocks. For the ease of presentation, only the block decomposition is presented here.

As in the uncoupled method, the calculation can be decomposed into two phases. Letting $*$ denote scalar multiplication, the Phase (1) calculation is $r_q(y) = \sum_{x=0}^{n-1} s_q^H(x) * M(x,y)$, and the Phase (2) calculation is $w_q = \sum_{y=0}^{n-1} \sigma_q^y$, where $\sigma_q^y = r_q(y) * s_q(y)$. For each s-vector s_q^H, each PE calculates (n/a) products for each of (n/b) $r_q(y)$'s. $PE(i,j)$ calculates $\sum_{x=i(n/a)}^{(i+1)(n/a)-1} s_q^H(x) * M(x,y)$, for $j(n/b) \le y < (j+1)(n/b)$ and $0 \le q < v$. Thus, each PE performs $v(n/b)(n/a) = vn^2/p$ complex multiplications followed by $v(n/b)(n/a-1) = (vn/p)(n-a)$ complex additions. All p PEs do these calculations simultaneously.

PE(0,0) memory
$M(x,y)$
$0 \le x < n/a, \; 0 \le y < n/b$
$s_q^H(m)$
$0 \le q < v, \; 0 \le m < n/a$
$s_t(z)$
$t = fa, \; 0 \le f < \lceil v/a \rceil$
$0 \le z < n/b$

PE(0,b−1) memory
$M(x,y)$
$0 \le x < n/a, \; (b-1)(n/b) \le y < n$
$s_q^H(m)$
$0 \le q < v, \; 0 \le m < n/a$
$s_t(z)$
$t = fa, \; 0 \le f < \lceil v/a \rceil$
$(b-1)(n/b) \le z < n$

PE(i,j) memory
$M(x,y)$
$i(n/a) \le x < (i+1)(n/a), \; j(n/b) \le y < (j+1)(n/b)$
$s_q^H(m)$
$0 \le q < v, \; i(n/a) \le m < (i+1)(n/a)$
$s_t(z), \; t = fa + i$
$0 \le f < \lceil v/a \rceil, \; j(n/b) \le z < (j+1)(n/b)$

PE(a−1,0) memory
$M(x,y)$
$(a-1)(n/a) \le x < n, \; 0 \le y < n/b$
$s_q^H(m)$
$0 \le q < v, \; (a-1)(n/a) \le m < n$
$s_t(z)$
$t = fa + (a-1), \; 0 \le f < \lceil v/a \rceil$
$0 \le z < n/b$

PE(a−1,b−1) memory
$M(x,y)$
$(a-1)(n/a) \le x < n, \; (b-1)(n/b) \le y < n$
$s_q^H(m)$
$0 \le q < v, \; (a-1)(n/a) \le m < n$
$s_t(z)$
$t = fa + (a-1), \; 0 \le f < \lceil v/a \rceil$
$(b-1)(n/b) \le z < n$

Fig. 1: Distribution of M and v steering vectors onto $a \times b$ PEs. (The maximum value t can have is $v - 1$.)

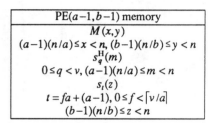

PE(0,0)	PE(0,1)	PE(0,2)	PE(0,3)
$s_0(0)$	$s_0(1)$	$s_0(2)$	$s_0(3)$
$s_2(0)$	$s_2(1)$	$s_2(2)$	$s_2(3)$
$s_4(0)$	$s_4(1)$	$s_4(2)$	$s_4(3)$
$s_1(0)$	$s_1(1)$	$s_1(2)$	$s_1(3)$
$s_3(0)$	$s_3(1)$	$s_3(2)$	$s_3(3)$
$s_5(0)$	$s_5(1)$	$s_5(2)$	$s_5(3)$
PE(1,0)	PE(1,1)	PE(1,2)	PE(1,3)

Fig. 2: Distribution of s-vectors over the PEs for $v = 6$, $n = 4$, $a = 2$, and $b = 4$.

Next, the summation in phase (1) is performed. The elements to be totaled to calculate $r_q(y)$ are resident in the a PEs of a given column. Recursive doubling can be used to compute the sums [17]. Performing interleaved recursive doubling procedures [14, 15] to find simultaneously the sum of the elements of ρ distinct vectors is referred to here as a ρ-recursive doubling.

To help clarify the operations involved in the ρ-recursive doubling technique, Fig. 3 illustrates how $N = 4$ PEs would sum the elements for an arbitrary set of vectors. In this example, there are $\rho = 6$ vectors, α_η for $0 \le \eta < 6$, each having N elements. If α_η^j is the j-th element of vector η with α_η^j initially stored in PE(j), then the sum that is sought is $\sum_{j=0}^{3} \alpha_\eta^j$ for each η, $0 \le \eta < 6$. In Fig. 3, the arrows denote the transfer of data from one PE to another followed by an addition; together these form a transfer-add operation. Five transfer-adds are required: three for the first step and two for the second step.

In general, the final sums for all ρ vectors, each with N elements, are computed in $\log_2 N$ steps, where $\lceil \rho/2^i \rceil$ transfer-add operations occur at step i for $1 \le i \le \log_2 N$. Thus, the total number of transfer-adds required in the ρ-recursive doubling procedure is $\sum_{i=1}^{\log_2 N} \lceil \rho/2^i \rceil$. Fig. 4 gives an algorithm to perform the ρ-recursive doubling procedure.

In the coupled algorithm, because each of the a PEs in a given column contributes to n/b different $r_q(y)$'s for each of the v s-vectors, a (vn/b)-recursive doubling procedure, utilizing $\sum_{i=1}^{\log_2 a} \lceil (vn/b)/2^i \rceil$ transfer-add operations, on the a PEs of each column can be used. All b logical columns of PEs simultaneously perform the b (vn/b)-recursive doublings (one per column). Both the multistage cube and hypercube networks can perform the needed inter-PE communications for all PEs with no conflicts. As a result, PE(i,j) will hold $r_q(y)$, where $j(n/b) \le y < (j+1)(n/b)$ and for all q, $0 \le q < v$, where q mod $a = i$. As an example for $v = 6$, $n = 4$, $a = 2$, and $b = 4$, the calculation of $r_q(j)$ (for all q,j) is illustrated in Fig. 5.

Phase (2) of the coupled algorithm involves the formation of $w_q = \sum_{y=0}^{n-1} \sigma_q^y$, where $\sigma_q^y = r_q(y) * s_q(y)$. After the first phase of

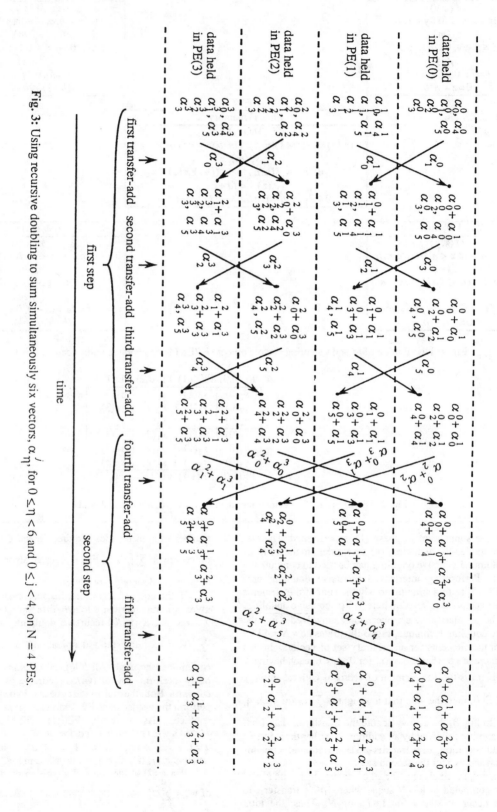

Fig. 3: Using recursive doubling to sum simultaneously six vectors, α_η^j, for $0 \le \eta < 6$ and $0 \le j < 4$, on $N = 4$ PEs.

```
/* This procedure will be performed on PE(j), 0 ≤ j < N. */

for i = 1 to log₂N do {   /* step i */
    k = j ⊕ 2^(i-1);      /* the destination PE number */
    for β = 0 to (⌊ρ/2^i⌋ - 1) do {
        η = (k mod 2^i) + β2^i;
        transfer local partial sum of vector η to PE(k);
        perform the complex addition with the received
        data and the corresponding local partial sum
        in PE(j);
    }
    if (ρ/2^i ≠ ⌊ρ/2^i⌋) {
        β = ⌊ρ/2^i⌋;
        η = (k mod 2^i) + β2^i;
        if (j & 2^(i-1) ≠ 0) {
            transfer local partial sum of vector η to PE(k);
        } else {
            perform the complex addition with the received
            data and the corresponding local partial sum
            in PE(j);
        }
    }
}
```

Fig. 4: An algorithm to perform the ρ-recursive doubling procedure.

computation, each PE holds at most $\lceil v/a \rceil (n/b)$ distinct $r_q(y)$ components and the corresponding elements of the s-vectors, $s_q(y)$, required to form σ_q^y. PE(i,j) forms the local partial sums of the products $\sum_{y=j(n/b)}^{(j+1)(n/b)-1} \sigma_q^y$, for all q, $0 \le q < v$, where q mod $a = i$. This is shown in Fig. 6 for $v=6$, $n=4$, $a=2$, and $b=4$. To do this, $\lceil v/a \rceil (n/b)$ complex multiplications and $\lceil v/a \rceil (n/b-1)$ complex additions are needed. Then, similar to that of the first phase, a $\lceil v/a \rceil$-recursive doubling procedure is employed in each row of b PEs to combine the partial sums above to form w_q for all q. All PEs in row i, i.e., PE(i,j) for $0 \le j < b$ and i fixed, participate in the same $\lceil v/a \rceil$-recursive doubling procedure to form at most $\lceil v/a \rceil$ w_q values, for all q, $0 \le q < v$, where q mod $a = i$. To compute all v values of w_q in parallel, a independent $\lceil v/a \rceil$-recursive doubling procedures are performed simultaneously (one per row). Both the multistage cube and hypercube networks can perform the required inter-PE transfers with no conflicts.

After the $\lceil v/a \rceil$-recursive doubling procedures, each PE stores at most $\lceil v/p \rceil$ elements of w_q. Specifically, the results placed in PE(i,j) are w_q, for all q, $0 \le q < v$, where $q = ja + i + fab$, for $0 \le f < \lceil v/p \rceil$. This is illustrated in Fig. 7 for $v=6$, $n=4$, $a=2$, and $b=4$. To form w_q, $\sum_{i=1}^{\log_2 b} \lceil v/(2^i a) \rceil$ transfer-add operations are performed in this phase.

In summary, the first phase requires vn^2/p complex multiplications followed by $(vn/p)(n-a)$ complex additions and $\sum_{i=1}^{\log_2 a} \lceil vn/(2^i b) \rceil$ transfer-adds. The second phase requires $\lceil v/a \rceil (n/b)$ complex multiplications followed by $\lceil v/a \rceil (n/b-1)$ complex additions and $\sum_{i=1}^{\log_2 b} \lceil v/(2^i a) \rceil$ transfer-adds.

By regarding a transfer-add as an inter-PE transfer and a complex addition, the complex additions required in the coupled scheme becomes $(vn/p)(n-a) + \lceil v/a \rceil (n/b-1)$

$+ \sum_{i=1}^{\log_2 a} \lceil vn/(2^i b) \rceil + \sum_{i=1}^{\log_2 b} \lceil v/(2^i a) \rceil$. Fixing v, p, and n and varying a and b, the lower bounds for the number of complex operations can be derived to be $(v/p)(n^2+n)$ multiplications, $(v/p)(n^2-1)$ additions, and $(vn/p)(a-1) + (v/p)(b-1)$ inter-PE transfers.

Recall that with the coupled method, $p=a \times b$, for $0 < a, b \le n$, and a, b, and n are assumed to be powers of two. The lower bounds for the number of complex multiplications and additions are independent of a and b, and there is a factor of p speedup on the required number of serial arithmetic operations. However, to minimize the number of inter-PE transfers, a should be set as small as possible while remaining a power of two (and b should not be greater than n due to the structure of the algorithm). Thus, if $p \le n$, a $1 \times p$ logical grid of PEs should be chosen to compute the problem and consequently, the recursive doubling procedure is involved only in the second phase of computation. If $n < p \le n^2$, a logical $(p/n) \times n$ grid of PEs is the optimal choice and the recursive doubling procedure is required to total the partial sums stored on different PEs during both the first and second phases of computation. If $p > n^2$, according to the computational characteristics of the coupled method, only n^2 PEs can be utilized and the remaining PEs are idle.

The data memory storage requirements for this method are $n^2/p + vn/a + \lceil v/a \rceil (n/b)$ complex numbers. The n^2/p term is for the $(n/a) \times (n/b)$ subarray of M, the vn/a term is for the (n/a)-element part of each of the v s-vector Hermitians, s_q^H, and the $\lceil v/a \rceil (n/b)$ term is for the (n/b)-element part of $\lceil v/a \rceil$ s-vectors, s_q, stored in each PE memory.

5.3 Comparison of the Single and Multiple Groups Coupled Methods

When employing the coupled method, it is assumed that all active PEs are working together as a single group at execution time. However, some machines (e.g., nCUBE, PASM) allow another possibility. For the $n < p \le n^2$ case, if p PEs are partitioned into $a = p/n$ independent groups of $1 \times n$ PEs, $\lceil v/a \rceil$ s-vectors are assigned to each of v mod a groups, and $\lfloor v/a \rfloor$ s-vectors are assigned to each of the remaining groups. Then the recursive doubling becomes unnecessary during the first phase of computation. This may reduce overall execution time. Assuming $n < p \le n^2$ and $a = p/n$, Table 1 shows the computational complexity of the single group coupled method (i.e., one group of $a \times n$ PEs) and the multiple groups coupled method (i.e., "a" groups of $1 \times n$ PEs). More storage is required for the multiple groups coupled method; however, computational complexity is the performance metric used here.

As shown in Table 1, if a divides v (i.e., $v = ca$ and c is an integer), the multiple groups coupled method is always better than the single group coupled method. When $v \ne ca$, although the number of inter-PE transfers is reduced by partitioning all PEs into "a" groups of $1 \times n$ PEs (consequently, the communication overhead is reduced), the number of complex additions and multiplications is increased. Thus, if the reduction in communication time is greater than the increase in computation time, the multiple groups coupled method will outperform the single group coupled method; otherwise the single group coupled method will be at least as good.

5.4 Coupled Method When n is Not a Power of 2

In this subsection, the coupled method is extended to include the case when n is not a power of 2. The data allocation among the PEs is slightly modified, but the algorithm for the

$r_q(y)$

PE(0,0)	PE(0,1)	PE(0,2)	PE(0,3)
$r_0(0)$	$r_0(1)$	$r_0(2)$	$r_0(3)$
$r_2(0)$	$r_2(1)$	$r_2(2)$	$r_2(3)$
$r_4(0)$	$r_4(1)$	$r_4(2)$	$r_4(3)$
$r_1(0)$	$r_1(1)$	$r_1(2)$	$r_1(3)$
$r_3(0)$	$r_3(1)$	$r_3(2)$	$r_3(3)$
$r_5(0)$	$r_5(1)$	$r_5(2)$	$r_5(3)$
PE(1,0)	PE(1,1)	PE(1,2)	PE(1,3)

$=$

$s_q^H(y)$

$s_0^H(0)$ $s_0^H(1)$	$s_0^H(2)$ $s_0^H(3)$
$s_1^H(0)$ $s_1^H(1)$	$s_1^H(2)$ $s_1^H(3)$
$s_2^H(0)$ $s_2^H(1)$	$s_2^H(2)$ $s_2^H(3)$
$s_3^H(0)$ $s_3^H(1)$	$s_3^H(2)$ $s_3^H(3)$
$s_4^H(0)$ $s_4^H(1)$	$s_4^H(2)$ $s_4^H(3)$
$s_5^H(0)$ $s_5^H(1)$	$s_5^H(2)$ $s_5^H(3)$
PE(0,0)	PE(1,0)
PE(0,1)	PE(1,1)
PE(0,2)	PE(1,2)
PE(0,3)	PE(1,3)

$s_q^H(2i)$ and $s_q^H(2i+1)$ are in PE(i,j)
$0 \le j < 4$, $0 \le q < 6$

$\times$

M

PE(0,0)	PE(0,1)	PE(0,2)	PE(0,3)
$M(0,0)$	$M(0,1)$	$M(0,2)$	$M(0,3)$
$M(1,0)$	$M(1,1)$	$M(1,2)$	$M(1,3)$
$M(2,0)$	$M(2,1)$	$M(2,2)$	$M(2,3)$
$M(3,0)$	$M(3,1)$	$M(3,2)$	$M(3,3)$
PE(1,0)	PE(1,1)	PE(1,2)	PE(1,3)

Fig. 5: Distribution of data over the PEs for $v = 6$, $n = 4$, $a = 2$, and $b = 4$, where $r_q(y) = \sum_{x=0}^{3} s_q^H(x) * M(x,y)$.

$\sum_{y=j(n/b)}^{(j+1)(n/b)-1} r_q(y) * s_q(y)$

PE(0,0)	PE(0,1)	PE(0,2)	PE(0,3)
$r_0(0)*s_0(0)$	$r_0(1)*s_0(1)$	$r_0(2)*s_0(2)$	$r_0(3)*s_0(3)$
$r_2(0)*s_2(0)$	$r_2(1)*s_2(1)$	$r_2(2)*s_2(2)$	$r_2(3)*s_2(3)$
$r_4(0)*s_4(0)$	$r_4(1)*s_4(1)$	$r_4(2)*s_4(2)$	$r_4(3)*s_4(3)$
$r_1(0)*s_1(0)$	$r_1(1)*s_1(1)$	$r_1(2)*s_1(2)$	$r_1(3)*s_1(3)$
$r_3(0)*s_3(0)$	$r_3(1)*s_3(1)$	$r_3(2)*s_3(2)$	$r_3(3)*s_3(3)$
$r_5(0)*s_5(0)$	$r_5(1)*s_5(1)$	$r_5(2)*s_5(2)$	$r_5(3)*s_5(3)$
PE(1,0)	PE(1,1)	PE(1,2)	PE(1,3)

$=$

$r_q(y)$

PE(0,0)	PE(0,1)	PE(0,2)	PE(0,3)
$r_0(0)$	$r_0(1)$	$r_0(2)$	$r_0(3)$
$r_2(0)$	$r_2(1)$	$r_2(2)$	$r_2(3)$
$r_4(0)$	$r_4(1)$	$r_4(2)$	$r_4(3)$
$r_1(0)$	$r_1(1)$	$r_1(2)$	$r_1(3)$
$r_3(0)$	$r_3(1)$	$r_3(2)$	$r_3(3)$
$r_5(0)$	$r_5(1)$	$r_5(2)$	$r_5(3)$
PE(1,0)	PE(1,1)	PE(1,2)	PE(1,3)

$*$

$s_q(y)$

PE(0,0)	PE(0,1)	PE(0,2)	PE(0,3)
$s_0(0)$	$s_0(1)$	$s_0(2)$	$s_0(3)$
$s_2(0)$	$s_2(1)$	$s_2(2)$	$s_2(3)$
$s_4(0)$	$s_4(1)$	$s_4(2)$	$s_4(3)$
$s_1(0)$	$s_1(1)$	$s_1(2)$	$s_1(3)$
$s_3(0)$	$s_3(1)$	$s_3(2)$	$s_3(3)$
$s_5(0)$	$s_5(1)$	$s_5(2)$	$s_5(3)$
PE(1,0)	PE(1,1)	PE(1,2)	PE(1,3)

Fig. 6: Distribution of data over the PEs for $v = 6$, $n = 4$, $a = 2$, and $b = 4$ (* represents element-wise scalar multiplication).

PE(0,0)	PE(0,1)	PE(0,2)	PE(0,3)
w_0	w_2	w_4	—
w_1	w_3	w_5	—
PE(1,0)	PE(1,1)	PE(1,2)	PE(1,3)

Fig. 7: Distribution of w_q's at conclusion of procedure for $v = 6$, $n = 4$, $a = 2$, and $b = 4$.

coupled method described earlier is still valid. Consequently, it will not be discussed in great detail.

The distribution of the M matrix and the Hermitians of v s-vectors across $a \times b$ PEs are described below. Initially, a $\lceil v/a \rceil$-element part of the Hermitian of each s-vector, s_q^H, a $(\lceil n/b \rceil)$-element part of $\lceil v/a \rceil$ s-vectors, s_q, and a $(\lceil n/a \rceil) \times (\lceil n/b \rceil)$ portion of M are stored in each PE's local memory. Stated precisely, the memory of PE(i,j) is loaded with $s_q^H(m)$ for $i(\lceil n/a \rceil) \le m < (i+1)(\lceil n/a \rceil)$, $0 \le m < n$, and $0 \le q < v$ and $s_t(z)$, where $t = fa + i$ for $j(\lceil n/b \rceil) \le z < (j+1)(\lceil n/b \rceil)$, $0 \le z < n$, and $0 \le f < \lceil v/a \rceil$. PE$(i,j)$ holds $M(x,y)$ for $i(\lceil n/a \rceil) \le x < (i+1)(\lceil n/a \rceil)$, $j(\lceil n/b \rceil) \le y < (j+1)(\lceil n/b \rceil)$, and $0 \le x,y < n$.

Similar to the previous case, the calculation of MQF is decomposed into two phases, i.e., to compute (1) $r_q(y) = \sum_{x=0}^{n-1} s_q^H(x)*M(x,y)$, and (2) $w_q = \sum_{y=0}^{n-1} \sigma_q^y$, where $\sigma_q^y = r_q(y)*s_q(y)$. Because of the data distribution described

above, phase (1) requires at most: $v\lceil n/b \rceil \lceil n/a \rceil$ complex multiplications followed by $v\lceil n/b \rceil(\lceil n/a \rceil - 1)$ complex additions and $\sum_{i=1}^{\log_2 a} \lceil (v\lceil n/b \rceil)/2^i \rceil$ transfer-adds. After the first phase, each PE holds at most $\lceil v/a \rceil \lceil n/b \rceil$ distinct $r_q(y)$ components and the corresponding elements of the s-vectors $s_q(y)$ required to form σ_q^y. Thus, in phase (2), $\lceil v/a \rceil \lceil n/b \rceil$ complex multiplications followed by $\lceil v/a \rceil(\lceil n/b \rceil - 1)$ complex additions and $\sum_{i=1}^{\log_2 b} \lceil v/(2^i a) \rceil$ transfer-adds are performed.

The data memory storage requirements of the coupled method for this case is $\lceil n/a \rceil \lceil n/b \rceil + v\lceil n/a \rceil + \lceil v/a \rceil \lceil n/b \rceil$ complex numbers, i.e., $\lceil n/a \rceil \lceil n/b \rceil$ for the $\lceil n/a \rceil \times \lceil n/b \rceil$ subarray of M, $v\lceil n/a \rceil$ for the $\lceil n/a \rceil$-element part of each of the v s_q^H vectors, and $\lceil v/a \rceil \lceil n/b \rceil$ for the $\lceil n/b \rceil$-element part of $\lceil v/a \rceil$ s_q vectors stored in each PE memory. The analyses in Subsections 5.2 and 5.3 can be used here. Depending on the trade-off between the increase in communication time and the decrease in computation time, the optimal implementation can be either the (single or multiple groups) coupled or uncoupled method.

6. GENERALIZING THE MQF ALGORITHM

Given p PEs where $p = a \times n \le n^2$, the multiple groups coupled method described in the previous subsection partitions all p PEs into a independent groups of $1 \times n$ PEs. This idea can be generalized to allow different methods of solving the MQF problem; three of which are the uncoupled method (Section 4), the single group coupled method (Subsection 5.2), and the multiple group coupled method (Subsection 5.3). Let $p = \gamma \times \alpha \times \beta$, where α, β, and γ are powers of two and $1 \le \gamma \le p$ and $1 \le \alpha$, $\beta \le n$. Given p PEs where $p \le n^2$, if all p PEs can be partitioned into γ groups of $\alpha \times \beta$ PEs, then the problem is to compute the MQF for $\lceil v/\gamma \rceil$ s-vectors on $\alpha \times \beta$ PEs. It can be seen that the single group coupled and uncoupled methods represent two extreme cases of the generalized method, i.e., for $\gamma = 1$, $\alpha = a$, $\beta = b$, and $\gamma = p$, $\alpha = \beta = 1$, respectively.

	L: one group of $p = a \times n$ PEs, where $a = p/n$	R: "a" groups of $(1 \times n)$ PEs	$v = ca$	$v \neq ca$
M	$vn/a + \lceil v/a \rceil$	$\lceil v/a \rceil (n+1)$	L=R	L≤R
A	$(v/a)(n-a)$ $+ \sum_{i=1}^{\log_2 a} \lceil v/2^i \rceil$ $+ \sum_{i=1}^{\log_2 n} \lceil v/(2^i a) \rceil$	$\lceil v/a \rceil (n-1)$ $+ \sum_{i=1}^{\log_2 n} \lceil v/(2^i a) \rceil$	L=R	L≤R
T	$\sum_{i=1}^{\log_2 a} \lceil v/2^i \rceil$ $+ \sum_{i=1}^{\log_2 n} \lceil v/(2^i a) \rceil$	$\sum_{i=1}^{\log_2 n} \lceil v/(2^i a) \rceil$	L>R	L>R

Table 1: Computational complexity for the one group $a \times n$ PEs and "a" groups of $1 \times n$ PEs coupled methods, where $n < p \le n^2$ and M, A, and T represent multiplications, additions, and inter-PE transfers, respectively.

Let v, a, and b in Table 1 be replaced by $\lceil v/\gamma \rceil$, α, and β, respectively. Then, the computational complexity of the generalized method is: $\lceil v/\gamma \rceil (n^2/(\alpha\beta)) + \lceil v/(\gamma\alpha) \rceil (n/\beta)$ complex multiplications, $\lceil v/\gamma \rceil (n/(\alpha\beta))(n-\alpha) + \lceil v/(\gamma\alpha) \rceil (n/\beta - 1)$ $+ \sum_{i=1}^{\log_2 \alpha} \lceil vn/(\gamma\beta 2^i) \rceil + \sum_{i=1}^{\log_2 \beta} \lceil v/(\gamma\alpha 2^i) \rceil$ complex additions, and $\sum_{i=1}^{\log_2 \alpha} \lceil vn/(\gamma\beta 2^i) \rceil + \sum_{i=1}^{\log_2 \beta} \lceil v/(\gamma\alpha 2^i) \rceil$ inter-PE transfers.

Similar to the single group coupled method, given v, p, and n values, the actual number of complex operations and inter-PE transfers are dependent on the values of γ, α, and β. By fixing v, p, and n, the lower bounds for those operations are derived to be $(vn/p)(\alpha - 1) + (v/p)(\beta - 1)$ inter-PE transfers, $(v/p)(n^2 + n)$ multiplications, and $(v/p)(n^2 - 1)$ additions. The lower bounds are attainable when v/p is an integer.

The lower bounds for multiplications and additions are independent of α and β. When the number of PE groups (γ) is decided, to minimize the number of inter-PE transfers, α should be set as small as possible while remaining a power of two and β should not be greater than n, which is identical to the results

obtained for the single group coupled method.

When $p > n^2$, not all of the PEs can be utilized by the single group coupled method. However, by partitioning p PEs into γ groups, where γ is a power of two and $1 \le p/\gamma \le n^2$, the resources can be fully utilized. This is done by assigning $\lceil v/\gamma \rceil$ s-vectors to each of v mod γ groups, and $\lfloor v/\gamma \rfloor$ s-vectors to each of the remaining groups. Within each group, whether the single group or multiple groups optimal coupled method should be used can be determined based on their relative performance. Thus, the computational complexity required for this problem is equal to the complexity of computing the MQF problem for $\lceil v/\gamma \rceil$ s-vectors on p/γ PEs. The optimal value of γ is dependent on v, n, and the communication overhead. Table 2 summarizes the computational complexity of the uncoupled, single group, and generalized methods.

	Coupled Method where $p = a \times b = \gamma \alpha \beta$		Uncoupled Method
	One group of $(a \times b)$ PEs	γ groups of $(\alpha \times \beta)$ PEs	
M	$\dfrac{vn^2}{p}$ $+ \lceil \dfrac{v}{a} \rceil \dfrac{n}{b}$	$\lceil \dfrac{v}{\gamma} \rceil \dfrac{n^2}{\alpha\beta}$ $+ \lceil \dfrac{v}{\gamma\alpha} \rceil \dfrac{n}{\beta}$	$\lceil \dfrac{v}{p} \rceil (n^2 + n)$
A	$\dfrac{vn}{p}(n-a)$ $+ \lceil \dfrac{v}{a} \rceil (\dfrac{n}{b} - 1)$ $+ \sum_{i=1}^{\log_2 b} \lceil \dfrac{vn}{2^i b} \rceil$ $+ \sum_{i=1}^{\log_2 a} \lceil \dfrac{v}{2^i a} \rceil$	$\lceil \dfrac{v}{\gamma} \rceil (\dfrac{n}{\alpha\beta})(n-\alpha)$ $+ \lceil \dfrac{v}{\gamma\alpha} \rceil (\dfrac{n}{\beta} - 1)$ $+ \sum_{i=1}^{\log_2 \beta} \lceil \dfrac{vn}{\gamma\beta 2^i} \rceil$ $+ \sum_{i=1}^{\log_2 \alpha} \lceil \dfrac{v}{\gamma\alpha 2^i} \rceil$	$\lceil \dfrac{v}{p} \rceil (n^2 - 1)$
T	$\sum_{i=1}^{\log_2 a} \lceil \dfrac{vn}{2^i b} \rceil$ $+ \sum_{i=1}^{\log_2 b} \lceil \dfrac{v}{2^i a} \rceil$	$\sum_{i=1}^{\log_2 \alpha} \lceil \dfrac{vn}{\gamma\beta 2^i} \rceil$ $+ \sum_{i=1}^{\log_2 \beta} \lceil \dfrac{v}{\gamma\alpha 2^i} \rceil$	−

Table 2: Computational complexity for the coupled and uncoupled methods when $p \le n^2$ and M, A, and T represent the number of multiplications, additions, and inter-PE transfers, respectively.

Consider the two cases when p divides v and when p does not divide v. If p divides v, both uncoupled and single group coupled methods require the same number of complex additions and multiplications. However, because there are inter-PE transfers associated with the single group coupled method, the uncoupled method will outperform the single group coupled method in this case.

If p does not divide v, the single group coupled method takes less time to perform the needed complex additions and multiplications than the uncoupled method, but the single group coupled method has the overhead of inter-PE transfers. Thus, if the reduction in communication time is greater than the increase in computation time, the uncoupled method will outperform the single group coupled method; otherwise the single group coupled method will be at least as good.

It can be seen from Table 2 that, in general, if α and p are kept constant and γ is doubled (β is halved), the number of transfers decreases. However, if the number of vectors, v, is not a multiple of 2γ, then more computation per PE is required when γ is doubled; otherwise, the same amount of computation takes place. In the former case, the γ value that can provide the best performance is dependent on the trade-off between the increase in computation time and the reduction in communication time. In the latter case, it would be preferable to double the number of groups (i.e., double γ).

7. CHOOSING AN OPTIMAL ALGORITHM

Results from the previous sections that relate different algorithm approaches include the following. Section 6 presents the complexity of the algorithm proposed in this paper in terms of v, n, γ, α, and β. The parameters v and n are input data parameters and γ, α, and β are logical system configuration parameters used by the generalized method. Section 4 and Subsection 5.2 describe the structure of the algorithm when certain restrictions are placed on the γ, α, and β. The uncoupled method applies when $\alpha\beta=1$. When $\gamma=1$, $\alpha\beta=p$ and $\beta\le n$, the coupled method is used. Subsection 5.2 showed that, when using the coupled method and γ is fixed, it is best to make α as small as possible, yet keeping β less than or equal to n. Subsection 5.3 made a strong argument for partitioning the problem in certain situations. It demonstrated that if $\gamma=a$, $\alpha=1$, and $\beta=n$, when $v=ca$, it is better to use a groups of $1\times n$ PEs then 1 group of $a\times n$ PEs. Finally, Section 6 concludes that the logical configuration $2\gamma\times\alpha\times\beta/2$ (when $\beta/2$ is an integer) will outperform the logical configuration $\gamma\times\alpha\times\beta$ when 2γ divides v: another argument for partitioning the problem.

Now consider the many possible ways to solve the MQF problem by the algorithms presented in this paper, given v, n, and p. One can use any variant of the generalized method (e.g., the uncoupled method, the single group coupled method, the multiple group method), or a combined approach. A combined approach may use a different method to compute the MQF problem for each different subset of steering vectors. For example, if $v>p$, then the best algorithm *may* use the uncoupled method for $v-(v\bmod p)$ vectors and the single group coupled method for $v\bmod p$ vectors. The combined approach permits any combination of methods, each being used to compute the MQF problem for a different subset of steering vectors. The total number of complex multiplications, complex additions, and inter-PE transfers for any algorithm can be computed from Table 2. The relative cost of a complex multiplication, complex addition, and inter-PE transfer for a given system can be obtained by experimentation and used to determine the optimal algorithm. Trade-offs involved in changing the logical system configuration, such as those mentioned earlier in this section, can be used to limit the number of possible algorithm choices to solve the MQF problem for a given v, n, and p.

8. EXPERIMENTAL STUDIES

The goal of this section is to validate experimentally many of the theoretical results found earlier. Both the uncoupled and coupled methods for solving MQFs have been implemented on the 16-PE small-scale PASM prototype, the 64-PE nCUBE 2, and the 16K-PE MasPar MP-1. The implementations assume that a, b, and p are powers of 2.

Fig. 8 shows that for a logical $a\times b$ machine configuration when using the single group coupled method, the communication overhead for the PASM prototype (Fig. 8) decreases as b increases. The same is true for the MasPar and nCUBE (not shown), where communication time nearly doubled when the parameter a was doubled [19]. These experimental results confirm the conclusion presented in Subsection 5.2 that com-

munication overhead is minimized when a is chosen as small as possible (i.e., a and b being a power of 2 and $p=a\times b$). Consequently, all the following experiments using the coupled method chose a to be as small as possible.

Fig. 9 shows the results of executing the uncoupled and single group coupled methods on the nCUBE 2, using 16, 32, and 64 PEs for a varying number (v) of steering vectors of size $n=16$. The logical system configuration ($a\times b$) used to generate the graph for 16 PEs was 1×16, for 32 PEs was 2×16, and for 64 PEs was 4×16 (recall that b should be less than or equal to n). Similar results were generated using the PASM prototype [19].

The experimental results validate the theoretical conclusions summarized in Table 1 and Table 2. Table 2 says that whenever $v=cp$ the uncoupled method outperforms the single group coupled method. Fig. 9 verifies this for $p=16$, 32, and 64. Furthermore, when p does not divide v, the choice of method depends on the relationship between communication and computation costs. For example, as shown in Fig. 9, when $p=32$, the uncoupled method outperforms the single group coupled method for $v=24$, 32, 48, 56, 64, 80, 88, 96. The exact points of intersection depend on the communication and computation cost relationship.

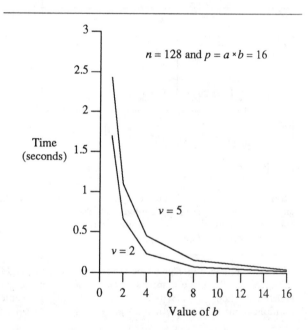

$n=128$ and $p=a\times b=16$

$v=5$

$v=2$

Fig. 8: Communication cost on the PASM prototype for various $a\times b$ logical configurations.

Now consider the computational complexities shown in Table 3. When $v=ca$, the table shows that "a" groups of $(1\times n)$ PEs will outperform one group of $p=a\times n$ PEs. Let $v=80$, $a=4$, and $n=16$. From Fig. 9, one group of $64=4\times16$ PEs takes .01 seconds. To compute $v=80$, it would take four groups of (1×16) PEs as long as it takes one group of (1×16) PEs to compute 20 steering vectors. By looking at the graph for the single group coupled method using $p=16$ PEs for $v=20$ vectors, one can deduce that four groups of (1×16) PEs take .0077 seconds for $v=80$, which is less than .01 seconds. Thus, in this case, the multiple group method outperforms the single group method, as predicted by Table 1.

The combined approach, discussed in Section 7, can be

"fine grain parallelism" that characterizes the coupled method.

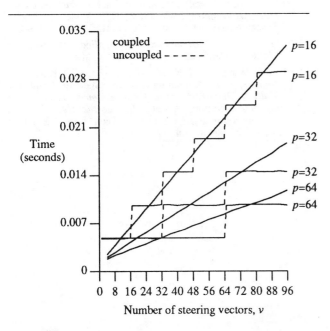

Fig. 9: Execution time of the uncoupled and coupled methods versus number of steering vectors on the nCUBE 2 for $n = 16$.

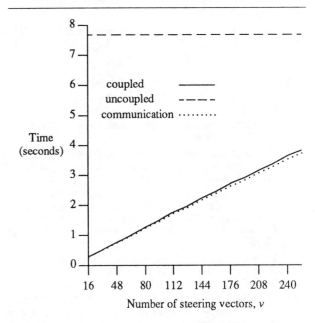

Fig. 10: Performance comparison for the uncoupled and coupled methods when 16K-PE MasPar MP-1 is employed.

used to compute the MQF problem for v steering vectors. For instance, the combined approach may consist of the following two steps: (1) use the uncoupled method for the first $(\lfloor v/p \rfloor \times p)$ vectors, and (2) compute the remaining $(v - \lfloor v/p \rfloor \times p)$ vectors by the single group coupled method. For example, if $v = 80$ and $p = 64$, then step 1 would compute 64 vectors by the uncoupled method (.0049 seconds), and step 2 would compute 16 vectors by the coupled method (.0032 seconds). By this combined method, the MQF problem for $v = 80$ takes .0081 seconds on the nCUBE 2. The coupled and uncoupled methods take .01 seconds and .0097 seconds, respectively. Therefore, the combined method outperforms the single group coupled and uncoupled methods. Consequently, when choosing an optimal algorithm for a particular set of v, n, and p values, different combined approaches may need to be considered.

The MasPar MP-1 results in Fig. 10 show the number of steering vectors versus execution time for the single group coupled and uncoupled methods. It also graphs the time spent doing communication for the coupled method. Because the data is distributed across a large number of PEs ($p = n^2 = 128 \times 128$), the computation per PE is negligible compared to the communication time. Despite this fact, the coupled method greatly outperforms the uncoupled method for the number of vectors tested. If the coupled method line (solid line) was extrapolated until it intersected the uncoupled method line (dashed line), the value of v would be approximately 516 at the point of intersection. For $v < 516$, the added computation cost of the uncoupled method outweighs the communication cost of the coupled method. This shows that when massive parallelism is available to implement the MQF problem ($p \gg v$), the more sophisticated coupled method is needed to exploit the available parallelism, despite incurring high communication overhead. In contrast, when $516 < v \le 16{,}384$, the uncoupled method using only v PEs will outperform the single group coupled method using 16,384 PEs. This is the communication overhead price that is paid for the

9. SUMMARY

Several parallel implementations of the computationally intensive task of computing the MQF problem have been examined in this paper. Trade-offs among the implementations for various data-size/machine-size ratios are categorized in terms of complex arithmetic operation counts, communication overhead, and memory storage requirements. The results showed that when p divides v, the uncoupled method is the optimal parallel implementation and a speedup of p on the number of complex multiplications and complex additions can be achieved. However, when v is not a multiple of p, a combined approach using both the uncoupled and coupled methods should be considered. In addition, trade-offs between the "single-group" and "multiple-group" decomposition for the coupled method were presented.

This research can be directly applied to SIMD, MIMD, and mixed-mode parallel machines interconnected with either the multistage cube or the hypercube interconnection network topology. This work can be extended further to mesh-connected topologies. The experiments performed on the PASM prototype, nCUBE 2, and MasPar MP-1 were used to validate some of the analytically derived relationships among input data and logical system parameters that apply across all machines.

Both analytical and experimental results demonstrated that a combined approach may be better than any one method. Choosing an optimal algorithm for an MQF problem with a given n and v is machine and p dependent. Therefore, by having a set of algorithms that perform the MQF problem efficiently for various input data parameters (e.g., n, v) and system parameters (e.g., p, communication time, complex addition time, complex multiplication time), an automatic algorithm or combined algorithm selection methodology can be implemented using the analysis established. This analysis of the MQF problem that has been presented and experimentally verified supports the viability of such an automatic method that can be developed for a variety of applications.

ACKNOWLEDGMENTS: The authors thank Professor Michael Zoltowski of Purdue University for his assistance.

REFERENCES

[1] J. Berntsen, "Communication efficient matrix multiplication on hypercubes," *Parallel Comput.,* v. 12, no. 3, 1989, pp. 335-342.

[2] T. Blank, "The MasPar MP-1 architecture," *IEEE Compcon,* Feb. 1990, pp. 20-24.

[3] V. Cherkassky and R. Smith, "Efficient mapping and implementation of matrix algorithms on a hypercube," *Supercomput. J.,* v. 2, 1988, pp. 7-27.

[4] J. Choi, J. J. Dongarra, R. Pozo, and D. W. Walker, "ScaLAPACK: a scalable linear algebra library for distributed memory concurrent computers," *Fourth Symp. Frontiers Massively Parallel Comput.,* Oct. 1992, pp. 120-127.

[5] E. Dekel, E. Nassimi, and S. Sahni, "Parallel matrix and graph algorithms," *SIAM J. Comput.,* v. 10, no. 4, 1981, pp. 657-675.

[6] S. A. Fineberg, T. L. Casavant, and H. J. Siegel, "Experimental analysis of a mixed-mode parallel architecture using bitonic sequence sorting," *J. Parallel Distrib. Comput.,* v. 11, Mar. 1991, pp. 239-251.

[7] G. C. Fox, M. A. Johnson, G. A. Lyzenga, S. W. Otto, J. K. Salmon, and D. W. Walker, *Solving Problems on Concurrent Processors, Volume 1,* Prentice Hall, Englewood Cliffs, NJ, 1988.

[8] G. C. Fox, S. W. Otto, and A. J. G. Hey, "Matrix algorithms on a hypercube I: matrix multiplication," *Parallel Comput.,* v. 4, no. 1, 1987, pp. 17-31.

[9] J. P. Hayes and T. Mudge, "Hypercube supercomputers," *Proc. IEEE,* v. 77, Dec. 1989, pp. 1829-1841.

[10] S. Haykin, *Adaptive Filter Theory,* Prentice-Hall, Englewood Cliffs, NJ, 1986.

[11] J. J. Modi, *Parallel Algorithms and Matrix Computations,* Oxford University Press, New York, NY, 1988.

[12] C. M. Pancake, "Software support for parallel computing: where are we headed?," *Comm. ACM,* v. 34, Nov. 1986, pp. 53-64.

[13] H. J. Siegel, *Interconnection Networks for Large-Scale Parallel Processing: Theory and Case Studies, 2nd Edition,* McGraw-Hill, New York, NY, 1990.

[14] H. J. Siegel, J. B. Armstrong, and D. W. Watson, "Mapping computer-vision-related tasks onto reconfigurable parallel-processing systems," *Computer,* v. 25, Feb. 1992, pp. 54-63.

[15] H. J. Siegel, L. J. Siegel, F. C. Kemmerer, P. T. Mueller, Jr., H. E. Smalley, Jr., and S. D. Smith, "PASM: a partitionable SIMD/MIMD system for image processing and pattern recognition," *IEEE Trans. Comput.* v. C-30, Dec. 1981, pp. 934-947.

[16] H. J. Siegel, T. Schwederski, J. T. Kuehn, and N. J. Davis IV, "An overview of the PASM parallel processing system," in *Computer Architecture,* edited by D. D. Gajski, V. M. Milutinovic, H. J. Siegel, and B. P. Furht, IEEE Computer Society Press, Washington, D. C., 1987, pp. 387-407.

[17] H. S. Stone, "Parallel computers," in *Introduction to Computer Architecture, 2nd Edition,* H. S. Stone, ed., Science Research Associates, Inc., Chicago, IL, 1980, pp. 363-425.

[18] G. Strang, *Introduction to Applied Mathematics,* Wellesley-Cambridge Press, Wellesley, MA, 1986.

[19] M. C. Wang, W. G. Nation, H. J. Siegel, S. D. Kim, M. A. Nichols, J. B. Armstrong, and M. Gherrity, *Computing Multiple Quadratic Forms for a Minimum Variance Distortionless Response Adaptive Beamformer Using Parallelism: Analyses and Experiments,* Tech. Rep., EE School, Purdue, in preparation.

DATA-PARALLEL R-TREE ALGORITHMS*

Erik G. Hoel†
Geography Division
Bureau of the Census
Washington, D.C. 20233

Hanan Samet
Computer Science Department
Center for Automation Research
Institute for Advanced Computer Sciences
University of Maryland
College Park, Maryland 20742

Abstract – *Data-parallel algorithms for R-trees, a common spatial data structure are presented, in the domain of planar line segment data (e.g., Bureau of the Census TIGER/Line files). Parallel algorithms for both building the data-parallel R-tree, as well as determining the closed polygons formed by the line segments, are described and implemented using the SAM (Scan-And-Monotonic-mapping) model of parallel computation on the hypercube architecture of the Connection Machine.*

INTRODUCTION

The SAM (Scan-And-Monotonic-mapping) model of parallel computation [1] may be defined by one or more linearly ordered sets of processors which allow element-wise and scan-wise operations to be performed. A *scan* operation [2] takes an associative operator $\bigoplus$, a vector $[a_0, a_1, \ldots, a_{n-1}]$, and returns the vector $[a_0, (a_0 \bigoplus a_1), \ldots, (a_0 \bigoplus a_1 \bigoplus \ldots \bigoplus a_{n-1})]$. Both within and between each linearly ordered set of processors, monotonic mappings may also be performed. Being a subset of the scan-model [2], the SAM model considers scan operations as taking unit time, thus allowing sorting operations to be performed in $O(\log n)$.

The R-tree [3] is a data structure for representing spatial data based upon spatial occupancy. Such methods decompose the space from which the data is drawn into regions called *buckets*. The R-tree buckets the data on the basis of minimal bounding rectangles (in the 2-d case). Objects are then grouped into hierarchies, and then stored in another structure such as a B-tree.

The R-tree's drawback is that it does not result in a disjoint decomposition of space. An object is associated with a single bounding rectangle, but the area spanned by the rectangle may be included in several other bounding rectangles. To determine which object is associated with a particular point in a 2-d space, we may have to search the entire database in the degenerate case. Alternatives such as the R+-tree and the PMR quadtree [4] which decompose the space into disjoint cells which are then mapped into buckets are not described here.

This paper is organized as follows; First, we present the data-parallel R-tree and give an R-tree building algorithm. Next, we describe the parallel implementation of a polygonization.

---

*This work was supported in part by the National Science Foundation under Grant IRI-90-17393.

†Also with the Center for Automation Research at the University of Maryland.

PARALLEL R-TREES

We limit ourselves to objects which are line segments although our techniques are applicable to other objects as well. Standard sequential R-trees are constructed in a manner whereby all leaf nodes appear at the same level in the tree. Each entry in a leaf node is a 2-tuple of the form (R, O) so that R is the smallest rectangle enclosing line segment O (where O points to the actual line segment). Each entry in a directory (non-leaf) node is a 2-tuple of the form (R, P), where R is the minimal rectangle enclosing the rectangles in the child node pointed at by P. An R-tree of order (m, M) means that each node in the tree, with the exception of the root, contains between $m \leq \lceil M/2 \rceil$ and M entries. The root node has at least two entries unless it itself is a leaf node.

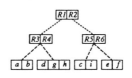
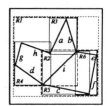

Figure 1: Example order (1,3) R-tree.

R-trees are built in the same way as B-trees. Line segments are inserted into leaf nodes. The appropriate leaf node is determined in a top-down fashion by traversing the R-tree starting at its root and at each step choosing the subtree whose corresponding bounding rectangle will have to be enlarged the least by the addition of line segment x. Once a leaf node is determined, a check is made to see if the insertion of x will cause the leaf node to overflow. If the leaf node is overflowing, it is then split and the $M + 1$ 2-tuples are redistributed among the two resulting nodes. Splits may propagate up the R-tree. Note that the tree's final shape depends on the insertion order of the line segments. In the data parallel environment, all line segments are inserted at the same time, and thus the final shape of the parallel R-tree will likely not be the same as its sequential one.

The parallel R-tree building algorithm proceeds as follows. Initially, one processor is assigned to each line of the data set, and one processor to the resultant R-tree (e.g., Fig. 2). Assume an order (1,3) R-tree. In the figure, label A denotes the line processor set, C_0 denotes the R-tree node processor set, with the associated square region containing the line processor identifier for the corresponding line segment group (i.e., the line processor segment group starting at processor 0 in A), Within set A, the square regions contain the line identifiers and indicate which processor is associated with which line. Each of the line processors is associated with the single R-tree node processor. A downward scan operation (i.e., a scan which returns the vector $[(a_0 \oplus a_1 \oplus \ldots \oplus a_{n-1}), (a_1 \oplus a_2 \oplus \ldots \oplus a_{n-1}), \ldots, a_{n-1}]$) is performed on the line processor set to determine the number of lines associated with the single R-tree processor (shown in Fig. 2 as the count field beneath the line processor set). The number of lines in the *segment* (a collection of line processors associated with a single R-tree node processor) is then passed to the single R-tree node processor. If the number of lines in the segment exceeds M, then split the R-tree root node into two leaf nodes and a root node (as is also done for the sequential R-tree). The two new leaf nodes are inserted into the R-tree processor set, with the former root node/processor updated to reflect the two new children.

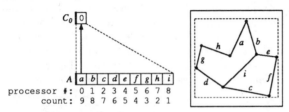

Figure 2: Initial processor assignments.

We use a node splitting algorithm that first sorts all lines in the segment according to the left edge of their bounding boxes. A parallel upward scan operation (i.e., a scan which returns the vector $[a_0, (a_0 \oplus a_1), \ldots, (a_0 \oplus a_1 \oplus \ldots \oplus a_{n-1})]$), is used to determine the extents of the bounding box formed by all lines preceding a line in the sorted segment. Similarly, a downward scan determines the bounding box for all following lines in the segment. For all legal splits (i.e., where each of the two resulting nodes receives at least m/M of the lines being redistributed), the amount of bounding box overlap is calculated, with the split corresponding to the minimal amount of overlap being selected as the x-axis candidate. An analogous procedure is employed for the y-axis in obtaining the y-axis candidate split coordinate value. Next, we choose the candidate split coordinate value corresponding to the minimal bounding box overlap. In case of a tie, use another metric such as the split with the minimal bounding box perimeter lengths. The complexity is $O(\log n)$ at each stage of the building operation as we employ two $O(\log n)$ sorts and a constant number of scan operations.

Once the splitting axis and the coordinate value are chosen, an *un-shuffle* operation [1] (where two intermixed types are rearranged into two disjoint groups via two monotonic mappings) is used to concentrate those line processors together into two new segments, each of which corresponds to one of the two R-tree leaf node

processors (Fig. 3). For example, all lines with a midpoint less than the split coordinate value are monotonically shifted toward the left, while those which are

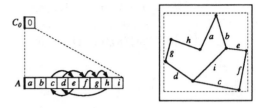

Figure 3: Un-shuffle operation.

greater than the split coordinate value are monotonically shifted toward the right among the line processors. Note that the root node of the R-tree is now associated with two segments in the line processor set A (i.e., (a, b, e, h) and (c, d, f, g, i)), and must itself be subdivided in an analogous manner. Fig. 4 shows the result which consists of two segments in the line processor set A, and two different R-tree processor sets C_0 and C_1 (each set corresponding to a node at a different height in the R-tree).

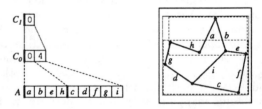

Figure 4: Completion of root node split operation.

The insertion algorithm proceeds iteratively as before, with each segment determining the number of lines it contains, and transmitting the count to the associated R-tree processor. If the number of lines in the segment is greater than M, then the segment (and corresponding R-tree node processor) are forced to subdivide (i.e., split). Note that this subdivision process may result in processors that correspond to internal nodes in the R-tree splitting themselves (with these splits possibly propagating up the R-tree). The building process terminates when all nodes in the R-tree processor set have at most M child processors. The R-tree root node corresponds to the single processor in set C_2, the leaf nodes are contained in processor set C_0, and all lines area grouped in segments of length ≤ 3 in processor set A (recall that our R-tree is order $(1, 3)$).

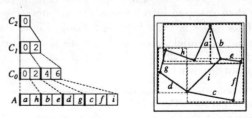

Figure 5: Completion of the R-tree building operation.

The data-parallel R-tree building operation is $O(\log^2 n)$. Each of the $O(\log n)$ stages is $O(\log n)$ (a constant number of scans and two bounding box sorts).

POLYGONIZATION

Polygonization is the process of determining all closed polygons formed by a collection of planar line segments. We identify each polygon uniquely by the bordering line with the lexicographically minimum identifier (i.e., line number) and the side on which the polygon borders the line. Polygonization can be done without a data-parallel R-tree. Basically, the lines could be sorted based upon their identifier in $O(\log n)$ time, then each line in sorted sequence would transmit its endpoint coordinates, line identifier, and current left and right polygon identifiers to all following lines via a sequence of $O(n)$ scan operations. Each line can independently determine the identifiers of the left and right polygons. The drawback is that it is an $O(n)$ operation with a large amount of scans. The R-tree decomposition can reduce the amount of global scan operations (i.e., a scan across the entire processor set) by instead relying upon segmented scans executed in parallel, thus speeding the computation.

Given a data-parallel R-tree, start by constructing a partial winged-edge representation (see, e.g., [4]) (an association between the incident line segments forming the minimal and maximal angles at each endpoint of each segment). This representation enables us to determine all edges that comprise a face (i.e., polygon) and all edges that meet at a vertex in time proportional to the number of edges. It consists of face, vertex, and edge tables. The face table has an entry for each face which is one of the face's constituent edges. The vertex table has an entry for each vertex which is one of the edges that meets at the vertex. The edge table has an entry for each edge which consists of the two vertices defining the edge, the two adjacent faces, and the preceding and following edges for each of these faces.

We implicitly construct the entire data structure although our example only illustrates how we determine the adjacent faces of each edge. We proceed by broadcasting the endpoints of each line in a segment group to all other lines in the segment group through a series of scans. By *broadcast* we mean the process of transmitting a constant value from a single processor to all other processors in the same segment group via a scan operation (i.e., the vector $[a_0, a_0, \ldots, a_0]$). Locally, each line processor maintains the minimal and maximal angles formed at each endpoint as well as the identities of the corresponding lines. Once the broadcasts are done, each line processor locally assigns an initial polygon identifier for the bordering polygon on the left and right side (moving from source to destination endpoint).

In Fig. 6, the left polygon identifier for line segment z is selected from the minimum identifiers of the source endpoint minimal angle (w_R, where w is the line identifier and R denotes the right side of w), the destination endpoint maximal angle (y_R), and the line identifier itself (z_L). Similarly, for the right side polygon identifier, the minimum identifier among the source endpoint

maximum angle (x_R), the destination endpoint minimal angle (v_R), and the line identifier (z_R) is selected.

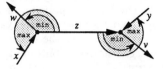

Figure 6: Selecting the initial polygon identifiers.

In Fig. 6, line z is assigned w_R as the initial left polygon identifier, and v_R as the right polygon identifier. Fig. 7 shows the initial polygon assignment for our example where the left and right polygon identifiers are contained in processor sets L_{ID} and R_{ID}, respectively. Since we are restricted to considering only lines that share the same R-tree node (e.g., in Fig. 5 in line processor set A, lines a, h compose the first segment group; lines b, e the second segment group; lines d, g the third segment group; and lines c, f, i the final segment group) when constructing the initial winged-edge representation, line i in Fig. 7 is assigned identifiers c_L and c_R rather than b_R and c_R as would be the case had line b also shared the same R-tree node.

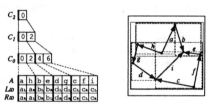

Figure 7: Initial polygon assignments.

Next, starting at the leaf level, merge all sibling lines together into the parent nodes. Mark all lines that intersect any of the overlapping regions formed by the bounding boxes of the nodes that have been merged for rebroadcasting among the lines in the merged nodes. This enables propagating the equivalence between the different identifiers in the merged nodes representing the same polygon (e.g., Fig. 8a where two R-tree nodes A and B are to be merged). In Fig. 8, A contains lines (a, c, g, h), and B contains lines (b, d, e, f). Lines (a, b, d) must be rebroadcast to the merged set of lines (i.e., lines (a, b, c, d, e, f, g, h)) as they intersect the overlapping region formed by the bounding boxes of A and B. This operation updates the winged-edge representations of any necessary lines (i.e., lines a and b in Fig. 8a). During the update, we note any polygon identifiers that must also be updated (e.g., line b has both its left and right polygon identifiers updated; b_L in Fig. 8a becomes a_L in Fig. 8b, and similarly, b_R becomes a_R). Neither of line a's polygon identifiers are updated because they are lexicographically minimal.

Broadcast the updates to all other lines in the merged node via scan operations (e.g., b_L to a_L and b_R to a_R in Fig. 8). Locally, if the transmitted polygon update matches either the left or right polygon identifiers of the local line, the local polygon identifier is updated to reflect the polygon identifiers that have been broadcast. In Fig. 8a, the right polygon identifier of line e is updated to show that polygon identifier b_R becomes a_R. Similarly, the left side polygon identifiers of lines d and f are updated to show that polygon identifiers b_L becomes a_L. Fig. 8b shows the resulting polygon identifiers and merged nodes. Continue the process up the entire R-tree until all lines are contained in a single node and all necessary broadcasts have been made. Fig. 9 is the final configuration of our example. The identifiers assigned to the three polygons are circled in the figure.

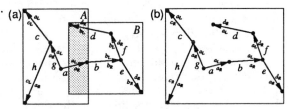

Figure 8: (a) Example of two nodes merging; and (b), the result of the merge operation.

The R-tree's spatial sort greatly limits the amount of inter-segment communication necessary as compared with a non-spatially sorted dataset where all lines would have to communicate their endpoints and polygon identifiers to all others. However, the non-disjoint decomposition of the R-tree causes more work in the local broadcasting phase of the sibling merge operation in comparison to an analogous disjoint decomposition spatial data structure such as the PMR quadtree and the R+-tree [4] because often many lines fall in the intersecting areas when the R-tree nodes are merged. Representations based on a disjoint decomposition of space mean that only those lines intersecting the splitting lines would need to be locally broadcast during the sibling merge operation.

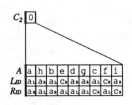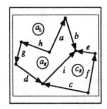

Figure 9: Completion of the polygonization operation.

Now, let us estimate the number of broadcasts necessary during the polygon identification process due to the lines intersecting overlapping regions. Assume that each R-tree node has an average fanout of M. Let c (where $0 \le c \le 1$) be the fraction of the lines in each node intersecting one or more of the overlapping regions formed by bounding boxes of nodes that have been merged. Let h be the height of the R-tree (without loss of generality, $h = \log_M n$, n = number of lines in the tree). As $M^h = n$, it can be shown that the number of local broadcasts B that must be made during merging phases due to the intersection of lines with overlapping regions is $B = \sum_{i=2}^{h} cM^i \le n(M/(M-1))$, which is $O(n)$.

bucket	build		polygonization	
size	time	scans	time	scans
25	105.6	1607	1042.1	92795
35	86.6	1315	997.7	81124
45	76.0	1147	1047.5	92496
55	73.0	1107	1046.6	92149
65	65.5	980	1741.8	115152
75	62.9	940	1695.9	111443
85	60.1	898	1479.4	104366
95	59.9	898	1625.6	108281

Table 1: Performance statistics for Montgomery Co., MD.

However, on the average it is expected to be lower. In particular, the average complexity of the line broadcasting step depends on the ability of the node splitting algorithm to partition the buckets as much as possible (thereby lowering the fraction c of lines intersecting overlapping regions).

CONCLUDING REMARKS

The data-parallel R-tree construction algorithm has been implemented on a Thinking Machines CM-2 with 16K processors and benchmarked against a sequential implementation on a Sun SPARCStation 1+. Data-parallel R-trees were observed to be approximately 10× faster during the build operation than their sequential counterparts for large datasets (i.e., 250K line segments). Table 1 shows some of the results for the Montgomery County, MD dataset (90K line segments). Note that build times decrease with increasing bucket sizes; this is due to larger buckets requiring fewer subdivisions, and consequently fewer scan and sorting operations. Polygonization times behave differently, with larger bucket sizes implying that more buckets are being merged at each level. This results in more lines present in the intersecting bucket regions, thereby causing increased numbers of line rebroadcasts and winged-edge updates.

REFERENCES

[1] T. Bestul, *Parallel Paradigms and Practices for Spatial Data*, Comp. Sci. Dept., Univ. of Maryland, Ph.D. diss., CS-TR-2897, College Park, MD, 1992.

[2] G. E. Blelloch and J. J. Little, "Parallel Solutions to Geometric Problems on the Scan Model of Computation", *Proc. of the 17th Intl. Conf. on Parallel Processing*, 1988, St. Charles, IL, 218–222.

[3] A. Guttman, "R-trees: A Dynamic Index Structure for Spatial Searching", *Proceedings of the SIGMOD Conference*, June 1984, San Diego, 47–57.

[4] H. Samet, *The Design and Analysis of Spatial Data Structures*, Addison-Wesley, Reading, MA, 1990.

Time Parallel Algorithms for Solution of Linear Parabolic PDEs

Amir Fijany

Jet Propulsion Laboratory, California Institute of Technology
Pasadena, CA 91109

Abstract- In this paper, fast time- and space-parallel algorithms for solution of parabolic PDEs are developed. It is shown that the seemingly strictly serial time-stepping procedures for solution of problem can be completely decoupled. This is achieved by developing time-parallel algorithms which allow the solution for all time steps to be computed in parallel. The time-parallel algorithms have a highly decoupled structure and hence can be efficiently implemented on emerging massively parallel MIMD architectures with minimum communication and synchronization overhead.

I. Introduction

In this paper, we consider the linear parabolic equation in a bounded domain Ω (which can be one-two-, or three-dimensional) with boundary Ω' as

$$\frac{\partial U}{\partial t} = \alpha^2 \mathcal{L} U \qquad \text{in } \Omega \text{ and } T>t>0 \qquad (1)$$

with boundary and initial conditions as

$$U = g \qquad \text{on } \Omega' \text{ and } T>t>0$$
$$U = f \qquad \text{in } \Omega \text{ and } t = 0$$

where $\mathcal{L}$ is the Laplace operator and α is constant. For two- and three- dimensional problems, Ω is assumed to be regular, i.e., a square or a cube. Also, for the sake of clarity, we first consider the problem with zero boundary conditions, i.e., $g = 0$. The case of nonzero boundary conditions will be discussed later.

The discretization of Eq. (1) by superimposing a uniform grid on Ω and using the usual finite-difference schemes leads to a family of iterative methods as

$$(I + 2\gamma\delta M_L)U^{(i)} = (I - 2\delta(1 - \gamma)M_L)U^{(i-1)} \qquad (2)$$

for i = 1 to M, where I is the unit matrix of appropriate size, M_L is the matrix arising from discretization of $\mathcal{L}$, δ is constant, $\Delta t = k$ is the time step size, and M = T/k.

The three methods for the problem are characterized by the parameter γ as

$\gamma = 0$: Explicit method and Eq. (2) becomes

$$U^{(i)} = (I - 2\delta M_L)U^{(i-1)} \qquad i = 1 \text{ to } M \qquad (3)$$

$\gamma = 1$: Implicit method and Eq. (2) becomes

$$(I + 2\delta M_L)U^{(i)} = U^{(i-1)} \qquad i = 1 \text{ to } M \qquad (4)$$

$\gamma = 1/2$: Crank-Nicholson (C-N) method and Eq. (2) becomes

$$(I + \delta M_L)U^{(i)} = (I - \delta M_L)U^{(i-1)} \qquad i = 1 \text{ to } M \qquad (5)$$

The C-N method is usually preferred over Implicit and Explicit methods since it is second-order accurate in both time and space.

The iterations in Eqs. (3)-(5) represent the time-stepping, or marching in time, procedure for solution of the problem. Computationally, the problem is both time (i.e., number of time steps, M) and space (i.e, size of grid, N) dependent. In this paper, the term space-parallel is used for algorithms that exploit parallelism only in each solution of Eqs. (3)-(5) while the term time-parallel refers to algorithms that exploit full or partial parallelism in computation of all vectors $U^{(i)}$. A both time- and space-parallel algorithm is then the one that exploits parallelism in computation of all vectors $U^{(i)}$ as well as parallelism in computation of each vector.

The coefficient matrices in Eqs. (4)-(5) have a structure similar to those arising in solution of Poisson equation. In fact, both Eqs. (4) and (5) represent a sequence of Poisson equations and hence fast Poisson Solvers can be used for their serial and parallel direct solution. However, the time-stepping procedure in Eqs. (3)-(5) seems to imply a strict sequentiality of the computation in time. This has motivated the development of new approaches to increase parallelism in time [1-4]. However, these time-parallel algorithms achieve a rather limited parallelism in time. In fact, Womble [3] supports the assessment of [5] wherein simultaneous solution for all time steps is not considered practical.

In this paper we develop time-parallel algorithms that, for the class of problems defined by Eq. (1), allow the iterations in Eqs. (3)-(5) to be fully decoupled. However, the main emphasis is on time-parallel computation of the C-N method. Our results suggest that, unlike the general assumption, the iteration in Eq. (5) can be more efficiently parallelized in time than in space. In addition to fully exploiting parallelism in time, the time-parallel algorithms allow the use of more processors to exploit parallelism in space. With a sufficient number of processors, the time- and space- parallel algorithms achieve the time lower bound of $O(\text{Log } N) + O(\text{Log } M)$. However, for most cases, the complexity of time- and space-parallel algorithms is of $O(\text{Log } N)$ and hence independent of M.

This paper is organized as follows. In Sec. II-III the time-parallel for two- and three-dimensional problems are developed. In Sec. IV practical implementation of the time-parallel algorithms is discussed. Finally, some concluding remarks are made in Section V.

II. Two-Dimensional (2D) Parabolic Equation
A. Problem Statement and Crank-Nicholson Method

For 2D case, Ω is considered to be a unit square. Superimposing a uniform grid on Ω ($\Delta x = \Delta y = h$) and using the 5-point finite difference scheme, the C-N method is given by

$$(I_{N2} + \delta \mathcal{M}_{L2})U^{(1)} = (I_{N2} - \delta \mathcal{M}_{L2})U^{(1-1)} \qquad (6)$$

where $h = 1/N+1$ and $\delta = \alpha^2 k/2h^2$. I_N and I_{N2} are the $N{\times}N$ and $N^2{\times}N^2$ unit matrices. $\mathcal{M}_{L2}$ is an $N^2{\times}N^2$ block tridiagonal matrix as
$\mathcal{M}_{L2} = \text{Tridiag}[-I_N, A, -I_N]$ and $A = \text{Tridiag}[-1, 4, -1]$.

The coefficient matrix in Eq. (6) has a structure similar to $\mathcal{M}_{L2}$ which is the matrix arising in solution of Poisson Equation. Hence, the fast Poisson Solvers, i.e., Cyclic Reduction (CR) algorithm [6], Matrix-Decomposition (MD) algorithm [6], and Fourier Analysis (FA) algorithm [7], can be used for fast serial direct solution of Eq. (6). This leads to a complexity of $O(N^2 \text{Log } N)$ for each solution of Eq. (6) and an overall serial computational complexity of $O(MN^2 \text{Log } N)$ for the problem.

Both the MD and FA algorithms can be computed in $O(\text{Log } N)$ steps with $O(N^2)$ processors [8]. This leads to a computational complexity of $O(M\text{Log } N)$ for space-parallel solution of the problem which indicates that
a) The computation is fully parallelized in space, i.e., computation of each step is fully parallel and the time lower bound of $O(\text{Log } N)$ is achieved.
b) The computation is strictly serial in time.

B. Time-Parallel Algorithm

Following theorem is used in derivation of time-parallel algorithms.

Theorem 1. The eigenvalue-eigenvector decomposition of an $N{\times}N$ symmetric tridiagonal Toeplitz matrix $T = \text{Tridiag}[\beta, \alpha, \beta]$ is given as

$$T = Q\lambda Q \qquad (7)$$

The matrix $Q = (2/N+1)^{1/2}\{\sin (ij\pi/N+1)\}$, i and $j = 1$ to N, is the set of normalized eigenvectors of T and the diagonal matrix $\lambda = \text{Diag}\{\lambda_i\}$ with $\lambda_i = \alpha + 2\beta\cos(i\pi/N+1)$ is the set of eigenvalues of T.

Proof. See for example [6]. □

Note that, Q is a symmetric orthonormal matrix, and hence $Q = Q^t = Q^{-1}$.

The derivation of time-parallel algorithm is based on the eigenvalue-eigenvector decomposition of matrix $\mathcal{M}_{L2}$. Let us first consider an $N^2{\times}N^2$ matrix $\mathcal{Q} \triangleq \text{Diag}[Q, Q, \ldots, Q, Q]$. It follows that $\mathcal{Q}$ is a symmetric orthonormal matrix, i.e., $\mathcal{Q} = \mathcal{Q}^t = \mathcal{Q}^{-1}$. Also, consider an $N^2{\times}N^2$ symmetric permutation matrix P. If two vectors W and V are defined as $W = \text{col}\{w_{ij}\}$ and $V = \text{col}\{v_{ij}\}$,

for i and $j = 1$ to N, then $W = PV$ implies that $w_{ij} = v_{ji}$. Since P is a symmetric permutation matrix it follows that $P = P^t$ and $PP = I_{N2}$.

Theorem 2. The matrix $\mathcal{M}_{L2}$ has an eigenvalue-eigenvector decomposition as

$$\mathcal{M}_{L2} = \mathcal{Q}P\mathcal{Q}\lambda_2 \mathcal{Q}P\mathcal{Q} \qquad (8)$$

where $\lambda_2 = \text{Diag}\{\lambda_{2ij}\}$, i and $j = 1$ to N, and
$\lambda_{2ij} = 4 - 2\cos(i\pi/N+1) - 2\cos(j\pi/N+1)$

Proof. See [13]. □

Note that, from the definition of $\mathcal{Q}$ and P, it follows that the matrix $\theta = \mathcal{Q}P\mathcal{Q}$ is orthonormal (see also below). The time-parallel algorithm is derived by replacing Eq. (8) into Eq. (6) as

$$\mathcal{Q}P\mathcal{Q}(I_{N2} + \delta\lambda_2)\mathcal{Q}P\mathcal{Q}U^{(1)} = \mathcal{Q}P\mathcal{Q}(I_{N2} - \delta\lambda_2)\mathcal{Q}P\mathcal{Q}U^{(1-1)}$$

Let $D_2 \triangleq (I_{N2} + \delta\lambda_2)^{-1}(I_{N2} - \delta\lambda_2)$ and $\tilde{U}^{(1)} \triangleq \mathcal{Q}P\mathcal{Q}U^{(1)}$. Since the matrix $\theta = \mathcal{Q}P\mathcal{Q}$ is orthonormal and hence is nonsingular, it then follows that

$$\tilde{U}^{(1)} = D_2\tilde{U}^{(1-1)} \qquad i = 1 \text{ to } M \qquad (9)$$

which implies that

$$\tilde{U}^{(1)} = (D_2)^1\tilde{U}^{(0)} \qquad i = 1 \text{ to } M \qquad (10)$$

The diagonal matrix D_2 is only a function of problem size and the time and space discretization parameters. If a same problem is solved many times for different boundary and/or initial conditions then all the matrices $(D_2)^1$ can be precomputed (see also Sec. IV). In this case, starting with $\tilde{U}^{(0)}$, all $\tilde{U}^1$ can be computed in parallel from Eq. (10). The time-parallel algorithm is then given as

Preprocessing: $(D_2)^1 = D_2(D_2)^{1-1}$, $i = 1$ to M (11)

Step I: $\tilde{U}^{(0)} = \mathcal{Q}P\mathcal{Q}U^{(0)}$

Step II: $\tilde{U}^{(1)} = (D_2)^1\tilde{U}^{(0)}$ $i = 1$ to M

Step III: $U^{(1)} = \mathcal{Q}P\mathcal{Q}\tilde{U}^{(1)}$ $i = 1$ to M

The computations in Steps II-III are completely decoupled. If these computations are performed in parallel then the computational complexity of time-parallel algorithm is independent of M.

C. Comparison of Serial, Space-Parallel, and Time-Parallel Algorithms

The matrix Q is the 1D Discrete Sine Transform (DST) operator [11]. It follows that the matrix $\theta = \mathcal{Q}P\mathcal{Q}$ is the 2D DST operator. Therefore, the multiplication by the matrix θ in Steps I and III is tantamount to performing a 2D DST which, by using fast techniques, can be performed in $O(N^2 \text{Log } N)$. This leads to an $O(MN^2 \text{Log } N)$ complexity for serial implementation of the time-parallel algorithm. Hence, the time-parallel algorithm is, asymptotically, as fast as the best serial algorithm for the problem.

By using $O(M)$ processors, that corresponds to a coarse grain parallel implementation, the computational cost of time-parallel algorithm is of $O(N^2 \text{Log } N)$. This represents a linear speedup of $O(M)$ over the best serial algorithms. By using $O(MN^2)$ processors, parallelism in both time and space can be fully exploited. In this case, the computations in Steps I and III can be performed in $O(\text{Log } N)$- since a 2D DST consists of two steps wherein at each step N decoupled 1D DST's are performed and each DST can be computed in $O(\text{Log } N)$ with $O(N)$ processors- and the computation in Step II can be performed in $O(1)$. This leads to a computational cost of $O(\text{Log } N)$ for both time- and space-parallel algorithm.

IV. Three-Dimensional Parabolic Equation

A. Problem Statement and Crank-Nicholson Method

For 3D case, Ω is considered to be a unit cube. Superimposing a uniform grid on Ω ($\Delta x = \Delta y = \Delta z = h$) and using the 7-point finite difference scheme, the C-N method is given by

$$(I_{N3} + \delta \mathcal{M}_{L3})U^{(1)} = (I_{N3} - \delta \mathcal{M}_{L3})U^{(1-1)} \qquad (12)$$

where h and δ are defined as before, I_{N3} is the $N^3 \times N^3$ unit matrix and $\mathcal{M}_{L3}$ is an $N^3 \times N^3$ block tri-diagonal matrix as $\mathcal{M}_{L3} = \text{Tridiag}[-I_{N2}, B, -I_{N2}]$. B is an $N^2 \times N^2$ block tridiagonal matrix as $B = \text{Tridiag}[-I_N, A', I_N]$ and $A' = \text{Tridiag}[-1, 6, -1]$.

The coefficient matrix in Eq. (12) has a structure similar to $\mathcal{M}_{L3}$ which is the matrix arising in solution of 3D Poisson equation. The extension of MD and FA algorithm for solution of 3D Poisson equation is studied in [9,10,13]. Both algorithms can compute each solution of Eq. (11) in $O(N^3 \text{Log } N)$, leading to an $O(MN^3 \text{Log } N)$ complexity for serial solution of the problem. Based on the analysis in [9,10], it can be shown that [13], by using $O(N^3)$ processors, both MD and FA algorithm can be computed in $O(\text{Log } N)$. This leads to a computational complexity of $O(M\text{Log } N)$ for space-parallel solution of the problem which indicates that the computation is fully parallel in space but is strictly serial in time.

B. Time-Parallel Algorithm

For 3D case, the derivation of time-parallel algorithm is also based on the eigenvalue-eigenvector decomposition of matrix $\mathcal{M}_{L3}$. Let us first define two $N^3 \times N^3$ matrices Q and Θ as $Q = \text{Diag}[Q, Q, \ldots, Q, Q]$ and $\Theta = \text{Diag}[\theta, \theta \ldots, \theta, \theta]$. Q and Θ are symmetric orthonormal matrices, i.e, $Q = Q^t = Q^{-1}$ and $\theta = \theta^t = \theta^{-1}$. Also, an $N^3 \times N^3$ permutation matrix, similar to that arising in 3D FFT, is defined as $\mathcal{P}$ (see also [13]). $\mathcal{P}$ is not symmetric but we have $\mathcal{P}\mathcal{P}^t = \mathcal{P}^t\mathcal{P} = I_{N3}$ since $\mathcal{P}$ is a permutation matrix.

Theorem 3. The eigenvector-eigenvalue decomposition of matrix $\mathcal{M}_{L3}$ is given by

$$\mathcal{M}_{L3} = \Theta\mathcal{P}^t Q\lambda_3 Q\mathcal{P}\Theta \qquad (13)$$

where $\lambda_3 = \text{Diag}\{\lambda_{3ijk}\}$, i, j, and k = 1 to N, and $\lambda_{3ijk} = 6 - 2\cos(i\pi/N+1) - 2\cos(j\pi/N+1) - 2\cos(k\pi/N+1)$

Proof. See [13]. □

The matrix $\Phi = Q\mathcal{P}\Theta$ is not symmetric but it is orthogonal since $\Phi\Phi^t = (Q\mathcal{P}\Theta)(\Theta\mathcal{P}^t Q) = I_{N3}$. The time-parallel algorithm is derived by replacing Eq. (13) into Eq. (12) as

$$\Theta\mathcal{P}^t Q(I_{N3} + \delta\lambda_3)Q\mathcal{P}\Theta U^{(1)} = \Theta\mathcal{P}^t Q(I_{N3} - \delta\lambda_3)Q\mathcal{P}\Theta U^{(1-1)}$$

Let $\tilde{U}^{(1)} \triangleq Q\mathcal{P}\Theta U^{(1)}$ and $D_3 \triangleq (I_{N3}+\delta\lambda_3)^{-1}(I_{N3}-\delta\lambda_3)$. Since Φ is nonsingular, it follows that

$$\tilde{U}^{(1)} = D_3 \tilde{U}^{(1-1)}$$

which suggests that

$$\tilde{U}^{(1)} = (D_3)^1 \tilde{U}^{(0)}$$

Again, assuming that the diagonal matrices $(D_3)^1$ are precomputed, the time-parallel algorithm is given as

Step I: $\tilde{U}^{(0)} = Q\mathcal{P}\Theta U^{(0)} = \Phi U^{(0)}$

Step II: $\tilde{U}^{(1)} = (D_3)^1 \tilde{U}^{(0)}$ i = 1 to M

Step III: $U^{(1)} = \Theta\mathcal{P}^t Q\tilde{U}^{(1)} = \Phi^t \tilde{U}^{(1)}$ i = 1 to M

Again, the computations in Steps II-III are completely decoupled. If these computations are performed in parallel then the computation complexity of time-parallel algorithm is independent of M.

C. Comparison of Serial, Space-Parallel, and Time-Parallel Algorithms

The matrices Φ and Φ^t are the operators for 3D direct and inverse DST. By using fast techniques, multiplication by Φ and Φ^t in Steps I and III is performed in $O(N^3 \text{Log } N)$, leading to an $O(N^3 \text{Log } N)$ complexity for serial implementation of time-parallel algorithm. Hence, asymptotically, the algorithm is as fast as best serial algorithm for the problem. Interestingly, in terms of actual number of operations, the time-parallel algorithm is more efficient than both MD and FA algorithms for serial solution of problem [13].

With $O(M)$ processors, the computational cost of time-parallel is of $O(N^3 \text{Log } N)$ which represents a linear speedup of $O(M)$ over the best serial algorithm. By using $O(MN^3)$ processors, maximum parallelism in both time and space can be exploited. In this case, the 3D DST's in Steps I and III can be performed in $O(\text{Log } N)$ and the computation in Step II in $O(1)$. This leads to a computational cost of $O(\text{Log } N)$ for both time- and space-parallel algorithm.

IV. Some Issues in Practical Implementation of Time-Parallel Algorithms

A. On-line Computation of Matrices $(D_j)^1$

In previous sections, it was assumed that the diagonal matrices $(D_j)^1$, i = 1 and M and j = 2 or 3 (j indicates the dimension of problem) can be precomputed. Here, we consider the case wherein the problem is solved only once and hence the cost of computing $(D_j)^1$ needs to be included in overall computational cost of algorithms. Only 2D problem is considered below since the extension to 3D case is straightforward.

Note that, the computation of $(D_2)^1$ and $\tilde{U}^{(0)}$ in Step I are decoupled and can be performed in parallel. By using $O(MN^2)$ processors, parallel computation of $(D_2)^1$ from the linear recurrence in Eq. (11) can be performed in $O(\text{Log } M)$ steps [12] while computation of $\tilde{U}^{(0)}$ takes $O(\text{Log } N)$ steps. If these two computations are performed in parallel then the overall cost of computing $\tilde{U}^{(0)}$ and $(D_2)^1$ is $\max(O(\text{Log } N), O(\text{Log } M))$. Since the computation of Steps II and III are performed in $O(\text{Log } N)$, it then follows that the complexity of the time-parallel algorithms is of $\max(O(\text{Log } N) + O(\text{Log } M)) + O(\text{Log } N)$.

However, the computation of $(D_2)^1$ can be performed in $O(\text{Log } N)$ with no communication so that the highly decoupled structure of the algorithms is preserved. Consider parallel implementation of the algorithm by using M groups of processors. Any group, say Group i, needs only to compute $(D_2)^1$ and not all intermediate powers of D_2. In this case, each group of processors performs following computations.

a) For $i = 2^n$, the computation is performed as

$$(D_2)^{2^k} = \left((D_2)^{2^{k-1}}\right)^2 \qquad k = 1 \text{ to } \text{Log}_2 i \qquad (14)$$

with a serial computational cost of $O(\text{Log}_2 i)$.

b) For $i \ne 2^n$ but $2^n > i > 2^{n-1}$, we can write i as

$$i = \sum_{k=0}^{n-1} a_k 2^k \qquad (15)$$

where a_k = 0 or 1. Here, first all $(D_2)^{2^k}$, k = 1 to (n-1), are computed from Eq. (14) in $O(n-1)$ steps and then $(D_2)^1$ is computed as

$$(D_2)^1 = \Pi_{k=0}^{(n-1)} (D_2)^{2^k} \qquad (16)$$

In the above product only those $(D_j)^{2^k}$ for which a_k = 1 need to be included. For the worst case, i.e., where all a_k = 1, Eq. (15) involves the multiplication of n terms and hence its serial computation can be done in $O(n-1)$ or $O(\lfloor \text{Log}_2 i \rfloor)$ steps. ($\lfloor x \rfloor$ denotes the greatest integer smaller than or equal to x.) The overall computation cost is determined by that of Group M which for the worst case, i.e., $M = 2^m - 1$, is of $O(2\lfloor \text{Log}_2 M \rfloor)$ where $\lfloor \text{Log}_2 M \rfloor = m$.

B. Nonzero Boundary Condition

We now consider the case wherein nonzero boundary condition is given by a space-dependent function g(x,y) on Ω'. In this case, Eq. (6) becomes

$$(I_{N2} + \delta \mathcal{M}_{L2})U^{(i)} = (I_{N2} - \delta \mathcal{M}_{L2})U^{(i-1)} + b \qquad (17)$$

where b is a constant (i.e., time-independent) sparse vector resulting from the discretization of g(x,y) on Ω'. Eq. (9) is now written as

$$\tilde{U}^{(i)} = D_2 \tilde{U}^{(i-1)} + \tilde{b} \qquad i = 1 \text{ to } M \qquad (18)$$

where $\tilde{b} = QPQb$. Eq. (18) represents a linear recurrence and hence can be computed in $O(\text{Log } M)$. However, Eq. (18) can be modified so that it can be computed in $O(1)$ and hence the highly decoupled structure of the algorithm is preserved. To this end, note that, from Eq. (18) it follows

$$\tilde{U}^{(i)} = (D_2)^1 \tilde{U}^{(0)} + ((D_2)^{i-1} + (D_2)^{i-2} + \ldots + I_{N2})\tilde{b}$$

but

$$(D_2)^{i-1} + (D_2)^{i-2} + \ldots + I_{N2} = ((D_2)^1 - I_{N2})(D_2 - I_{N2})^{-1}$$

Let us define a vector $\tilde{b}^{(i)}$ given by

$$\tilde{b}^{(i)} = ((D_2)^1 - I_{N2})(D_2 - I_{N2})^{-1}\tilde{b} \qquad i = 1 \text{ to } M \quad (19)$$

Eq. (18) can be written as

$$\tilde{U}^{(i)} = (D_2)^1 \tilde{U}^{(0)} + \tilde{b}^{(i)} \qquad i = 1 \text{ to } M \qquad (20)$$

If $(D_2)^1$ are computed from Eq. (14) and (16), then all vectors $\tilde{b}^{(i)}$ can be computed in parallel from Eq. (19) and all vectors $\tilde{U}^{(i)}$ can also be computed in parallel from Eq. (20). The time-parallel algorithm is given as

Step I: Compute in parallel $(D_2)^1$, i = 1 to M, from Eqs. (12) and (14).

Step II: Compute in parallel $\tilde{U}^{(0)} = QPQU^{(0)}$ and $\tilde{b} = QPQb$.

Step III: Compute in parallel $\tilde{b}^{(i)}$, i = 1 to M, from Eq. (19). Compute in parallel $\tilde{U}^{(i)}$, i = 1 to M, from Eq. (20).

Step IV: Compute in parallel $U^{(i)} = QPQ\tilde{U}^{(i)}$, i = 1 to M

Note that, as stated before, the computation of Step I and Step II are decoupled and can be performed in parallel. The time-parallel algorithm for 3D case can be also modified in a similar fashion. If parallelism in both time- and space is fully exploited, then the computational complexity of the algorithms is of $\max(O(\text{Log } N) + O(\text{Log } M)) + O(\text{Log } N)$.

V. Discussion and Conclusion

We presented time-parallel algorithms for solution of a class of parabolic PDEs given by Eq. (1). Our results prove that, contrary to the general assumption, the time-stepping procedure can be fully parallelized in time. In fact, our results prove that parallelization of the computation can be performed much more efficiently in time than in space.

The performance of coarse grain implementation of time-parallel algorithms on MIMD parallel architectures and by using M processors is analyzed in detail in [13]. It has been shown that under very realistic assumptions, a speedup very close to the linear one can be achieved with minimum communication and synchronization overhead. A more detailed description of the time-parallel algorithms along with their extension to Neumann boundary condition is presented in [13].

It should be emphasized that our time-parallel computing approach can be also applied to a wider class of problems. The extension of time-parallel algorithm for solution of linear inhomogeneous parabolic PDEs with constant and variable coefficients is presented in [14]. The extension of time-parallel computing approach for solution of a more general class of evolutionary PDEs, including both parabolic and hyperbolic on regular and irregular domain in presented in [15].

ACKNOWLEDGMENT

The research described in this paper was performed at the Jet Propulsion Laboratory, California Institute of Technology, under contract with the National Aeronautics and Space Administration (NASA). I am highly indebted to my colleagues Drs. J. Barhen and N. Toomarian for insightful discussions and suggestions. I also thank an anonymous reviewer for constructive suggestions.

References

[1] E. Lelarasmee *et al*, "The Waveform relaxation method for the time domain analysis of large scale integrated circuits," IEEE Trans. Computer-Aided Design, Vol. (1), 1982.

[2] J. Saltz and V. Nail, "Towards Developing Robust Algorithms for Solving Partial Differential Equations on MIMD Machines," Parallel Computing, Vol. 6, 1988.

[3] D. Womble, "A time-stepping algorithm for parallel computers," SIAM J. Sci. Stat. Comput., Vol. 11(5), pp. 824-837, 1990.

[5] W. Hackbusch, "Parabolic multi-grid methods," Proc. 6th Int. Symp. on Computing Methods in Applied Sciences and Engineering, Dec. 1983.

[5] G. Strang and G. Fix, *An Analysis of the Finite Element Method*, Prentice-Hall, 1973.

[6] B. Buzbee, G. Golub, and C. Nielson, "On Direct Methods for Solving Poisson's Equations," SIAM J. Numer. Anal., Vol. 7, 1970.

[7] R. Hockney, "A Fast Direct Solution of Poisson's Equation Using Fourier Analysis," J. ACM, Vol. 12, pp. 95-113, 1965.

[8] P. Swarztrauber and R. Sweet, "Vector and Parallel methods for the Direct Solution of Poisson's Equation," J. Computional & Applied Math., Vol. 27, pp. 241-263, 1989.

[9] A.H. Sameh, "A fast Poisson solver for multiprocessors," *Elliptic Problem Solvers II*, Academic Press, 1984.

[10] R. Sweet *et al* "FFTs and Three-Dimensional Poisson Solvers for Hypercube," Parallel Computing, Vol. 17, pp. 121-131, 1991.

[11] C. Van Loan, *Computational frameworks for the Fast Fourier Transform*, SIAM, Philadelphia 1992.

[12] R. Hockney and C. Jesshope, *Parallel Computers*, Adam Hilger Ltd., 1981.

[13] A. Fijany, "Fast Time- and Space-Parallel Algorithms for Solution of Linear Parabolic Partial Differential Equations," Submitted to IEEE Trans. Parallel and Dist. Syst..

[14] A. Fijany, "Fully Time-Parallel Algorithms for Solution of Linear Inhomogeneous Heat Equation with constant and variable Coefficients," Submitted to SIAM J. Sci. Comput.

[15] A. Fijany, "The Structure of Time-Parallel Algorithms for Solution of Linear Evolutionary PDEs," Submitted to IEEE Trans. on Comp.

SESSION 3C

GRAPH ALGORITHMS (I)

COMPUTING CONNECTED COMPONENTS AND SOME RELATED APPLICATIONS ON A RAP

Tzong-Wann Kao, Shi-Jinn Horng,
Horng-Ren Tsai
Department of Electrical Engineering, National Taiwan Institute of Technology
Taipei, Taiwan
horng@twnntit.bitnet

Abstract

A *reconfigurable array of processors* (RAP) is a parallel processing system which has the ability to change dynamically the supported interconnection scheme during the execution of an algorithm. Assume the data bus is n-bit wide. In this paper, two basic operations are introduced first: It takes $O(1)$ time to embed an undirected graph in a 2-D $N \times N$ RAP; the time for finding the minimum number of $N \log N$-bit unsigned integers on a RAP using N processors is $O(T)$, where $1 \leq T \leq \frac{\log N}{\log \log N}$ and $\log N \leq n \leq N$. Assume $n = N$, then these basic operations can be used to design several $O(1)$ time parallel algorithms for graph problems on a RAP using $N \times N$ processors. These problems include the connected components problem, the transitive closure problem and the dominators problem, respectively.

1   Introduction

Finding the connected components of an undirected graph with N vertices and M edges can be found in the literatures [3, 5, 7, 12, 17, 18, 21, 23, 25, 28]. Most of the well-known algorithms for the connected component problem are proposed on the PRAM (parallel random access machine) models. On the CREW (concurrent read, excluded write) model, Hirschberg, Chandra, and Sarwate [7] proposed an $O(\log^2 N)$ time parallel connectivity algorithm for a graph with N vertices and M edges using $N^2 / \log N$ processors. Later Chin et al. [3] proposed an $O(N^2/P + \log^2 N)$ time algorithm using P processors for dense graphs. Han et al. [5] proposed an $O(M/P + (N \log N)/P + \log^2 N)$ time algorithm using P processors. On the CRCW model, Shiloach and Vishkin [25] proposed an $O(\log N)$ time algorithm for this problem using $O(M + N)$ processors. On the hypercube computer,

Ryu et al. [21] proposed an $O((N + M)/P \log \log N)$ time algorithm for this problem using P processors with all-port communication, where $M + N = \Omega(P^{1+\epsilon})$, for any constant $\epsilon > 0$. On a processor array with a reconfigurable bus system, Wang et al. [28] proposed an $O(1)$ time algorithm for this problem using either a 3-D $N \times N \times N$ processors or a 2-D $N^2 \times N^2$ processors.

Most of well-known parallel algorithms to find the connected components of an undirected graph are based on the breadth-first search approach [17, 18], transitive closure approach [12, 17, 28], and vertex collapse approach [3, 7, 17, 23, 25]. In this paper, we propose a novel technique called the *component isolation* approach to find the connected components of an undirected graph. Our approach consists of two phases. First, we isolate the connected components. For each connected component, we then select a minimum identity index to label it, respectively. As a result, this approach also can be used to solve the image connected component labeling problem [10].

A reconfigurable array of processors (abbreviated to RAP) is defined to be an array of processors connected to a reconfigurable bus system whose configuration can be dynamically changed by setting up the local switch of each processor. There are many reconfigurable parallel processing systems such as the bus automaton [19, 20], the reconfigurable mesh [15], the polymorphic-torus network [13] and the processor arrays with reconfigurable bus system (abbreviated to PARBS) [28] which are functionally equivalent to the RAP.

In this paper, first we introduce two basic operations for embedding an undirected graph in a 2-D RAP and finding the minimum number of $N \log N$-bit unsigned integers, respectively. Then, these basic operations can be used to design several efficient algorithms for graph problems such as the connected component problem, the transitive closure problem and the dominators problem, respectively.

The rest of this paper is organized as follows. We first describe the reconfigurable array of processors on which our algorithms are based in Section 2. Section 3 introduces two basic operations. Section 4 deals with several efficient algorithms which include the connected component problem, the transitive closure problem and the dominators problem, respectively. Finally, some concluding remarks are included in the last section.

2 The Computation Model and Basic Notations

A *k-dimensional* (*k*-D) RAP of size H contains H processors arranged in a *k*-D grid. That is, the bus system can be thought of as logically arranged as in a *k*-D array $A(n_{k-1}, n_{k-2}, \ldots, n_0)$, where n_j, $0 \le j \le k - 1$, is the size of the jth dimension and $H = n_{k-1} \times n_{k-2} \times \ldots \times n_0$. Each processor is identified by a k-tuples unique index $(i_{k-1}, i_{k-2}, \ldots, i_0), 0 \le j \le k-1, 0 \le i_j \le n_j - 1$, and denoted by $P_{i_{k-1}, i_{k-2}, \ldots, i_0}$. Each processor has $2k$ ports and each port has n bits data bus width denoted by $-S_j, +S_j$, $0 \le j \le k-1$ and $-S_j(i), +S_j(i), 0 \le i \le n-1$, respectively. Processor $P_{i_{k-1}, i_{k-2}, \ldots, i_j, \ldots, i_0}$ connects its port $+S_j$ to the i_j-dimension bus and processor $P_{i_{k-1}, i_{k-2}, \ldots, i_{j+1}, \ldots, i_0}$ also connects its port $-S_j$ to the i_j-dimension bus, for $0 \le j \le k-1$, and $0 \le i_j \le n_j - 1$. Assume each port has 4 bits data bus. We show an example for a 2-D 4×4 RAP in Figure 1.

For a unit time, assume each processor can either perform arithmetic and logic operation or communicate with others by broadcasting data on a bus. It allows multiple processors to broadcast data on the different buses or to broadcast the same data on the same bus simultaneously at a time unit, if there is no collision.

Any configuration of the bus system that is derivable by properly establishing the local connection among the data bus of each port within each processor. To represent the local connection within a processor, we use the notations $\{g_0\}, \{g_1\}, \ldots, \{g_{t-1}\}$, where g_i, $0 \le i \le t - 1$, denotes a group of buses that are connected together [27, 28]. That is, each processor has t connections and each connection is described by $g_i, 0 \le i < t$. For example, in an 1-D RAP, if the local connection of a processor is $\{-S_0(i), +S_0(i + 1), 0 \le i < n - 1\}$, then the n bits

data are shift left (or right) one bit after passing through this processor. By properly establishing the local connection within a processor, Figure 2 also shows five interesting configurations derivable from a processor in a 2-D RAP.

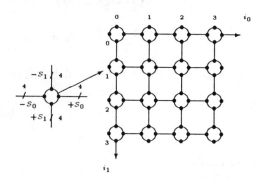

Figure 1: A 2-D 4×4 RAP, with $n = 4$ bits.

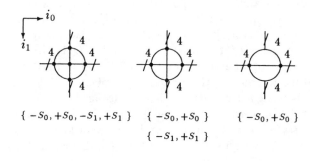

$\{-S_0, +S_0, -S_1, +S_1\}$ $\{-S_0, +S_0\}$ $\{-S_0, +S_0\}$

$\{-S_1, +S_1\}$

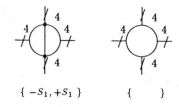

$\{-S_1, +S_1\}$ $\{\ \ \}$

Figure 2: Four switch configurations in a 2-D RAP with $n = 4$ bits.

Let $var(k)$ denote the local variable var (memory or register) in a processor with index k. For example, $sum(0, 0, 1)$ is a local variable sum of processor $P_{0, 0, 1}$.

A RAP is operated in an SIMD (single instruction stream, multiple data streams) model. The bus bandwidth is not unlimited between processors. We assume the bus bandwidth is bounded by n bits wide, where n is an integer. A constant time can be achieved for transferring an n bits data between processors under such an assumption. The I/O loading (up load or down load) time is fully dependent on how complex the I/O interface between processors and peripherals will be. Ben-Asher et al. [1] used an *initializing network* to deliver the input to the switches of a 2-D reconfigurable mesh. Therefore, the complexity of an algorithm is assumed to be the sum of the computation time of processors and the communication time among processors. This assumption was also used by many researchers [1, 13, 15, 27, 28].

3 Basic Operations

Some data operations will be described in this section. These data operations are used for deriving several efficient algorithms in Section 4.

3.1 Embedding an Undirected Graph

For an undirected graph $G = (V, E)$, we can embed it in a 2-D $N \times N$ RAP, where $N = |V|$.

Embedding rules : Processor $P_{i_1, i_0}, 0 \leq i_1, i_0 < N$, establishes the local connection $\{-S_0, +S_0, -S_1, +S_1\}$ if $(i_1, i_0) \in E$ or $i_1 = i_0$; $\{-S_0, +S_0\}, \{-S_1, +S_1\}$ otherwise.

Figure 3 shows an undirected graph that is embedded in a 2-D RAP. As Horng [9] and Wang [28], the correctness of embedding an undirected graph in a 2-D RAP can be straightforwardly proved. This leads to the following lemma.

Lemma 1 *For an undirected graph $G = (V, E)$, it can be directly embedded in an $N \times N$ RAP, where $N = |V|$.*

Note that a 2-D $N \times N$ RAP with one bit bus width is enough to embed an undirected graph $G = (V, E)$ with N vertices. Hence, a 2-D $N \times N$ RAP with N bits data bus width can be used to embed N undirected graphs $G_0, G_1, \ldots, G_{N-1}$ each with at most N vertices, respectively.

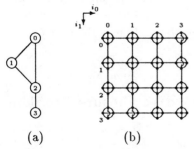

(a)   (b)

Figure 3: (a). An undirected graph G.
(b). Embed (a) with a 4×4 RAP.

3.2 Finding the Minimum of Integer Numbers

Given N integer numbers $D = \{D_0, D_1, \ldots, D_{N-1}\}$, where $0 \leq D_i \leq N - 1$, for $0 \leq i \leq N - 1$, the minimum operation of these N numbers is to determine if there is a number which is not greater than the others. Without loss of generality, usually assume that all numbers are distinct and each D_i is represented by the base-2 system, where

$$D_i = \sum_{j=0}^{R-1} b_{i, j} 2^j, \qquad (1)$$

for $1 \leq R \leq \lfloor \log_2 N \rfloor + 1$, $b_{i, j} \in \{0, 1\}$, $0 \leq D_i \leq N - 1$, $0 \leq i \leq N - 1$, and $0 \leq j \leq R - 1$. Instead of using the base-2 system to represent D_i, D_i can be represented by the base-n system as follows:

$$D_i = \sum_{k=0}^{T-1} d_{i, k} n^k, \qquad (2)$$

for $1 \leq T \leq \lfloor \log_n N \rfloor + 1$, $0 \leq i \leq N - 1$, $0 \leq k \leq T - 1$, and $0 \leq d_{i, k} \leq n - 1$. Assume D_i is allocated in processor i. In general, $d_{i, k}$ can be obtained from D_i by the division operation in $O(T)$ time on processor i. Without loss of generality, assume $n = 2^R$, $1 \leq R \leq \lfloor \log_2 N \rfloor + 1$. The relationship between $d_{i, k}$ and $b_{i, j}$ is $d_{i, k} = \{b_{i, j} \mid Rk \leq j \leq R(k+1) - 1\}$ and $d_{i, k}$ can be obtained from $b_{i, j}$ by the shift operation in $O(T)$ time also.

From Eq. (2), each D_i is represented by T digits and each digit is bounded within the interval $[0, n-1]$. Based on the prune-and-search technique, Kao et al. [10] first designed an $O(T)$ time *MINA algorithm* to find the minimum number of these N unsigned integers, where $1 \leq T \leq \frac{\log N}{\log \log N}$. The detailed algorithm can be referred to Kao et al. [10].

Lemma 2 *[10] The MINA procedure can be performed in $O(T)$ time, $1 \leq T \leq \frac{\log N}{\log \log N}$, on a linear N RAP, where n is the data bus width and $\log N \leq n \leq N$. Initially, D_i is stored in processor P_i, for all $0 \leq i \leq N - 1$.*

Instead of using an 1-D RAP, note that it is easy to modify the MINA procedure for processing the nondistinct data case under any configuration.

4 Application to Graph Problems

Efficient parallel algorithms for several graph problems may be obtained using the basic operations described in Section 3. We discuss some of them in the following sections.

4.1 The Connected Component Problem

Let $V = \{0, 1, \ldots, N-1\}$, and let $G = (V, E)$ be an undirected graph with vertex set V and edge set E. We represent G by its adjacency matrix $A = [a_{i, j}]$, which is an $N \times N$ matrix, where $a_{i, j} = 1$ if $(i, j) \in E$ and $a_{i, j} = 0$, otherwise. A subgraph of G is a graph whose vertices and edges are in G. A graph is connected if there is a path between any pair of vertices in G. A maximal connected subgraph of G is called its connected component. Each vertex belongs to exactly one connected component, and the identifier of each connected component in G is set to be the minimum index of vertices belonging to it. The connected component problem is to label each vertex by its corresponding connected component identifier.

Based on the transitive closure approach, Wang et al. [28] proposed an $O(1)$ time algorithm for the connected component problem using a 3-D $N \times N \times N$ PARBS. Note that only one bit data bus is enough to solve the connected component problem as that proposed by Wang et al. [28]. If the data bus between processors is n–bit wide and $\log N \leq n \leq N$, then the connected component problem can be solved more efficiently as follows. Using the technique developed by Wang et al. [28], first the transitive closure of any vertex can be computed in $O(1)$ time on a 2-D $N \times N$ PARBS with one bit data bus, and then the transitive closure of N vertices, A^*, can be straightforwardly computed in $O(N/n)$ time on a 2-D $N \times N$ PARBS or in $O(1)$ time on a 3-D $N \times N \times N/n$ PARBS. Finally, based on the algorithm proposed by Wang et al. [28], the connected component problem can be solved in the same time and processor complexities as those proposed for the transitive closure problem. When the n-bit data bus is considered, the processor complexity is improved by N times if $n = N$. We are not satisfied by this straightforward result. Hence, we propose another approach in the following.

Let RAPG be an $N \times N$ RAP which embeds an undirected graph according to the embedding rules. We can use a RAPG to compute the connected components of an undirected graph more efficiently. Instead of using the transitive closure technique, we propose another approach called the component isolation approach. Our approach consists of two major steps. First, we isolate the connected components. For each connected component, we then select a minimum identity index to label it, respec-

tively. Assume that n is the data bus width and $\log N \leq n \leq N$. Let vid denote the index of vertex i, where $vid = i$. Initially, $vid(i_0, i_0)$ is loaded with i_0 for each vertex i_0 and the adjacency matrix a_{i_1, i_0} of an undirected graph is also allocated in processor $P_{i_1, i_0}, 0 \leq i_1, i_0 \leq N-1$. Finally, the minimum identity index of each connected component is stored in processor $P_{i_0, i_0}, 0 \leq i_0 \leq N - 1$. The detailed algorithm CONNECT is described as follows. A snapshot of procedure CONNECT is shown in Figure 4.

procedure CONNECT(A, vid);

Step 1: // Isolate each component. //

Using the embedding rules to embed an undirected graph $G = (V, E)$ in a 2-D $N \times N$ RAP. Then, information can be only carried on those processors belonged to the same component. See Figure 4(a) and 4(b) for an example.

Step 2: // Label each component with the minimum index. //

(1). The processor with the minimum index $vid(i_0, i_0)$ for each component can be identified by applying the minimum basic operation, which is described in Section 3.2, to each component simultaneously. See Figure 4(c).

(2). After the minimum index $vid(i_0, i_0)$ is determined for each component, the processor with the minimum index $vid(i_0, i_0)$ broadcasts its $vid(i_0, i_0)$ to the correspondent component by the established bus from Step 1. See Figure 4(d).

Theorem 1 *Given an undirected graph $G = (V, E)$ with $N = |V|$, the connected component problem can be solved in $O(T)$ time, where $1 \leq T \leq \frac{\log N}{\log \log N}$, on a 2-D $N \times N$ RAP with n-bit data bus width, and $\log N \leq n \leq N$. Initially, the adjacency matrix a_{i_1, i_0} is allocated in processor $P_{i_1, i_0}, 0 \leq i_1, i_0 \leq N - 1$, respectively.*

Proof : Clearly, the algorithm is correct. Since each connected component is connected together to the same bus by using the embedding rules, there always exists a unique index to label each component. The time complexity is analyzed as follows. Step 1 takes $O(1)$ time and Step 2 takes $O(T)$ time by Lemma 2, respectively. Hence, the time complexity is $O(T)$.

Q. E. D.

4.2 The Transitive Closure Problem

Given an undirected graph $G = (V, E)$ with $N = |V|$, the adjacency matrix $A = [a_{i, j}], 0 \leq i, j \leq N-1$, of G is the $N \times N$ matrix for which $a_{i, j} = 1$ if $(i, j) \in E$ or $i = j$; $a_{i, j} = 0$, otherwise. The transitive closure

of G, denoted by G^*, is the undirected graph with vertex set V and edge set E^*, where
$E^* = \{(i, j) \mid$ there is an undirected path between vertex i and vertex j in $G\}$.
Hence, the transitive closure of A, denoted by $A^* = [a^*_{i, j}]$, is the adjacency matrix of G^*.

There are many ways to compute the transitive closure of an undirected graph G. We solve this problem first by computing the connected components of an undirected graph G, and then setting $a^*_{i_1, i_0} = 1$ if and only if i_1 and i_0 are in the same component. From Section 4.1, the component identifier is the minimum index belonging to it. After each connected component is determined, each diagonal processor holds the correspondent component identifier by Theorem 1. $a^*_{i_1, i_0}$ can be determined in a constant time by letting processor $P_{i_0, i_0}, 0 \leq i_0 \leq N-1$, broadcast its component identifier on both i_0-dimension bus and i_1-dimension bus (all processor establish the local connection $\{-S_0, +S_0\}, \{-S_1, +S_1\}$). Then, processor $P_{i_1, i_0}, 0 \leq i_1, i_0 \leq N-1$, sets $a^*_{i_1, i_0} = 1$ if it receives the same component identifiers both from the i_0-dimension bus and the i_1-dimension bus or $i_1 = i_0$; sets $a^*_{i_1, i_0} = 0$, otherwise. This leads to the following theorem.

Theorem 2 *Given an undirected graph $G = (V, E)$ with $N = |V|$, the transitive closure problem can be solved in $O(T)$ time, where $1 \leq T \leq \frac{\log N}{\log \log N}$, on a 2-D $N \times N$ RAP with n-bit data bus width and $\log N \leq n \leq N$. Initially, the adjacency matrix a_{i_1, i_0} is allocated in processor $P_{i_1, i_0}, 0 \leq i_1, i_0 \leq N-1$, respectively.*

Proof : The correctness is directly followed from Theorem 1. The time complexity can be easily verified to be $O(T)$ time, where $1 \leq T \leq \frac{\log N}{\log \log N}$.
$$Q. E. D.$$

In the ensuing discussion, for simplicity, it is assumed that the data bus is N bits wide.

4.3 Counting the Connected Components

Lemma 3 *[11] Given a binary sequence $b_i, 0 \leq i \leq N-1$, then the summation of these N binary data can be computed in $O(1)$ time on a linear N RAP with N-bit data bus width. Initially, b_i is stored in processor P_i, for all $0 \leq i \leq N-1$.*

Wang et al. [27] first proposed an $O(1)$ time algorithm for this problem using $N \times N$ processors. Then Kao et al. [11] reduced the number of processor to N processors but with the same time complexity. An

$O(1)$ time algorithm can be obtained using N processors in the following. Assume bit b_i is allocated at processor P_i, $0 \leq i \leq N-1$. Establish the local connection $\{-S_0(j), +S_0(j+1), 0 \leq j \leq N-1\}$ of processor P_i, $0 \leq i \leq N-1$, if b_i is 1; $\{-S_0, +S_0\}$, otherwise. Then processor P_0 broadcasts a signal on the established bus through the bit 0 bus of port $-S_0 (i.e. -S_0(0))$. If processor P_{N-1} receives that signal at bit m bus of port $-S_0 (i.e. -S_0(m))$ then the binary sequence sum is obtained by adding bit b_{N-1} to m, where $0 \leq m \leq N-1$.

The number of components of an undirected graph can be determined easily by Theorem 1 and Lemma 3. First, we can easily obtain all connected components of an undirected graph using Theorem 1. Then, the leader processor P_{i_0, i_0} with the minimum index of each component sets $b(i_0, i_0) = 1$, and other processors set $b(i_0, i_0) = 0$. Finally, the number of connected component can be computed by using Lemma 3. This leads to the following lemma.

Lemma 4 *Given an undirected graph $G = (V, E)$ with $N = |V|$, the number of connected components problem can be solved in $O(1)$ time on a 2-D $N \times N$ RAP with N-bit data bus width. Initially, the adjacency matrix a_{i_1, i_0} is allocated in processor P_{i_1, i_0}, $0 \leq i_1, i_0 \leq N-1$, respectively.*

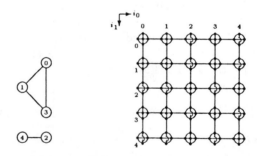

(a). An undirected graph G. (b). Embed (a).

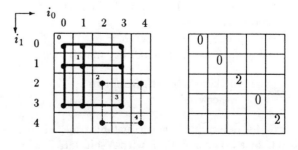

(c) After Step 2.1. (d) After Step 2.2.

Figure 4: A snapshot for procedure CONNECT.

4.4 The Dominators Problems

Most of the well-known algorithms for computing dominators in parallel are proposed under the CREW PRAM model. Savage [22] proposed an $O(\log^2 N)$ time algorithm using $O(N\ P_t(N))$ processors, where $P_t(N)$ is the processor complexity of computing the transitive closure. Hirschberg [6] and Chandra [2] proposed two algorithms for computing the transitive closure using $O(N^3/\log N)$ and $O(N^{2.81}/\log N)$ processors, respectively. Like Savage [22], Pawagi et al. [16] developed an $O(\log^2 N)$ time algorithm but reduced the number of processors to $O(P_t(N))$ for a directed acyclic graph. Unfortunately, their algorithms are incorrect [see [8], for detail]. On the hypercube machine, Horng [8] proposed two $O(\log^2 N)$ time algorithm for this problem using N^3 processors. Horng [9] also proposed an $O(1)$ time algorithm under the RAP machine using N^4 processors. Compared to the algorithms proposed by Horng [9], our algorithms reduce the number of processors by N^2 times but achieve the same time complexity.

Let $G = (V,\ E)$ be an undirected graph or a directed graph with $N = |V|$ and rooted at t. Assume the vertices are numbered from 0 to $N - 1$. We say that vertex j is a dominator of vertex k if j is on every path from t to k. If j is a dominator of k and every other dominator j' of k also dominates j, then j is called an immediate dominator of k. The dominator relation can be represented by a dominator tree rooted at t. If i is the immediate dominator of j then $<i,\ j>$ is an edge of the dominator tree. The dominator matrix is defined as

$$dom[j,k] = \begin{cases} 1 & \text{if } j \text{ is a dominator of } k; \\ 0 & \text{otherwise,} \end{cases} \quad (3)$$

where $0 \leq j,\ k \leq N - 1$. Like Savage [22], we use the reachability matrix approach to compute the dominator matrix.

Let G be represented by an adjacency matrix $a_{i,\ j}$, $0 \leq i,\ j \leq N - 1$. $a_{i,\ j} = 1$ if $(i,\ j) \in E$ or $i = j$; 0, otherwise. Assume $a_{i,\ j}$, $0 \leq i,\ j \leq N - 1$, is allocated in processor $P_{i,\ j}$. Let $R = [r_{t,\ j}]$, $0 \leq j \leq N - 1$, be defined as the reachability matrix, where

$$r_{t,\ j} = \begin{cases} 1 & \text{if there is a path from } t \text{ to } j; \\ 0 & \text{otherwise,} \end{cases} \quad (4)$$

R can be determined in a constant time. First, embed G by a 2-D $N \times N$ RAP with N-bit data bus by the embedding rules as described in Section 3.1. After embedding, there are N same established buses to embed G. Only bit t of the established bus is enough to compute R. That is, processor $P_{t,\ t}$ broadcasts

a signal '1' on the bit t of the established bus. If processor $P_{j,\ j}$ receives a signal '1' from the bit t of the established bus, then processor $P_{j,\ j}$ sets $r_{t,\ j} = 1$; 0, otherwise. This leads to the following lemma.

Lemma 5 *Given an undirected graph $G = (V,\ E)$ with $N = |V|$ and rooted at t, the reachability matrix R can be computed in $O(1)$ time, on a 2-D $N \times N$ RAP with one bit data bus. Initially, the adjacency matrix $a_{i_1,\ i_0}$ is allocated in processor $P_{i_1,\ i_0}$, $0 \leq i_1,\ i_0 \leq N - 1$, respectively.*

Let G^i, $0 \leq i \leq N - 1$ and $i \neq t$, be a subgraph of G, where $G^i = (V - \{i\},\ E)$ and $R^i = [r_{t,\ j}^i]$, $0 \leq j \leq N - 1$, represent the reachability matrix of G^i. G^i can be embedded as that of embedding G except reestablishing the local connection $\{-S_0(i),\ +S_0(i)\}$ and $\{-S_1(i),\ +S_1(i)\}$ for the bit i of each processor of the ith row and the ith column of a 2-D RAP. R^i, $0 \leq i \leq N - 1$ and $i \neq t$, can be determined simultaneously in a constant time similarly as that of R. That is, processor $P_{t,\ t}$ broadcasts a signal '1' on all N bits of the established bus. If processor $P_{j,\ j}$ receives a signal '1' on bit i of the established bus, then processor $P_{j,\ j}$ sets $r_{t,\ j}^i = 1$; 0, otherwise.

Theorem 3 *Given an undirected graph $G = (V,\ E)$ with $N = |V|$ and rooted at t, the reachability matrix R^i, $i = 0,\ 1,\ \ldots,\ N - 1$ and $i \neq t$ can be computed in $O(1)$ time on a 2-D $N \times N$ RAP with N bits data bus. Initially, the adjacency matrix $a_{i_1,\ i_0}$ is allocated in processor $P_{i_1,\ i_0}$, $0 \leq i_1,\ i_0 \leq N - 1$, respectively.*

After the computation, $R = [r_{t,\ j}]$, $0 \leq j \leq N - 1$, is stored at bit t of the processor $P_{j,\ j}$. Similarly, $R^i = [r_{t,\ j}^i]$, $0 \leq i,\ j \leq N - 1$ and $i \neq t$ is also stored at bit i of the processor $P_{j,\ j}$. Vertex i is a dominator of vertex j, if j is reachable in R but not in R^i. For $0 \leq i,\ j \leq N - 1$, Eq. (4) can be reformulated as

$$dom[i,\ j] = \begin{cases} r_{t,\ j} & \text{if } i = t; \\ 1 & \text{if } i = j; \\ r_{t,j} \oplus r_{t,j}^i & \text{otherwise,} \end{cases} \quad (5)$$

By establishing the local connection $\{-S_1,\ +S_1\}$ for each processor $P_{j,\ j}$, $0 \leq j \leq N - 1$, broadcasts R and R^i along each established column bus. Then each processor $P_{i,\ j}$ has both $r_{t,\ j}$ and $r_{t,\ j}^i$ at bit t and bit i, respectively. Finally, all processors compute Eq. (5) simultaneously and $dom[i,\ j]$ is stored at processor $P_{i,\ j}$.

Theorem 4 *Given an undirected graph $G = (V,\ E)$ with $N = |V|$ and rooted at t, the dominators problem can be solved in $O(1)$ time on a 2-D $N \times N$ RAP with*

N bits data bus. Initially, the adjacency matrix a_{i_1, i_0} is allocated in processor P_{i_1, i_0}, $0 \leq i_1, i_0 \leq N-1$, respectively.

The dominator tree can be created straightforwardly by determining the immediate dominator for each vertex. Recall that the dominator tree is unique. The immediate dominator is unique also and it is the closest dominator of that vertex. The number of ancestor of any vertex in the dominator tree is equal to its dominators. Therefore, if vertex i has d dominators then the immediate dominator of vertex i is a dominator of i and has $d-1$ dominators. Let $nd[i_0]$, $0 \leq i_0 \leq N-1$, be the number of dominators of vertex i_0. Then the dominator tree is defined as

$$domt[i_1, i_0] = \begin{cases} 1 & \text{if } nd[i_1] - nd[i_0] = -1 \\ & \text{and } dom[i_1, i_0] = 1; \\ 0 & \text{otherwise,} \end{cases} \qquad (6)$$

where $0 \leq i_1, i_0 \leq N-1$. The $nd[i_0]$, $0 \leq i_0 \leq N-1$, can be computed in a constant time by using Lemma 3. After the number of dominators of each vertex is computed, Eq. (6) can be computed in a constant time straightforwardly.

Theorem 5 *Given an undirected graph $G = (V, E)$ with $N = |V|$ and rooted at t, the dominator tree problem can be solved in $O(1)$ time on a 2-D $N \times N$ RAP with N bits data bus. Initially, the adjacency matrix a_{i_1, i_0} is allocated in processor P_{i_1, i_0}, $0 \leq i_1, i_0 \leq N-1$, respectively.*

If the bus width is restricted to n bits, then the algorithm can be easily modified to run in $O(1)$ time but using $N \times N \times N/n$ processors. Note that if the bus of a RAP is directional as described by Horng [9] and the data bus width is N bits then Theorem 3, 4 and 5 are also applicable for any directed graph using the embedding rules proposed by Horng [9]. As a result, for those applications based on the dominators and dominator tree problems under either an undirected graph or an arbitrary directed graph such as finding back edges in a flow graph, recognizing acyclic flow graph and recognizing reducible flow graph all can be solved in a constant time on a 2-D $N \times N$ RAP with N bits data bus, respectively. [See Horng[9], for detail.]

5 Concluding Remarks

The size of a data item and the bus capacity determine the transmission time of a data item between processors. Their sizes are not unlimited. In this paper, assume the data bus is $n-$bit wide and n is an integer. First we introduce two basic operations: one is to embed an undirected graph with N vertices in a 2-D $N \times N$ RAP with one bit data bus and the other is to find the minimum number of $N \log N$-bit unsigned integers, respectively. The former operation takes $O(1)$ time but that of the latter requires $O(T)$ time, where $1 \leq T \leq \frac{\log N}{\log \log N}$.

Assume that $n = N$, then based on these basic operations and using $N \times N$ processors, we derive several $O(1)$ time algorithms for undirected graph problems. These problems include the connected components problem, the transitive closure problem and the dominators problem, respectively.

Note that the complexity of a switch in a RAP is fully dependent on how many dimensions of a RAP and the bandwidth between processors will be. There are at least two projects to demonstrate the feasibility and benefits of a 2-D RAP; one is the YUPPIE (Yorktown Ultra-Parallel Polymorphic Image Engine) chip [14] and the other is the GCN (Gated-Connection Network) chip [26]. We agree the connection delay will depend on the problem size so that the constant time for broadcasting delay is not true. However, although it is not true, the broadcasting delay is very small. For a 10^6 processors YUPPIE, only 16 machine cycles are enough for broadcasting. GCN has further reduced the delay by using the pre-charged circuits. Ideally, it has been shown in [24] that the $O(1)$ time claim may be made true if the reconfigurable bus system is implemented using Optical Fibers [4] as the underlying global bus system and Electrically Controlled Coupler Switches (ECS) [4] for connecting or disconnecting two fibers. With the advance of the VLSI technology, we believe that a practical RAP machine will be built in the near future.

References

[1] Y. Ben-Asher, D. Peleg, R. Ramaswami and A. Schuster, "The Power of Reconfiguration," *Journal of Parallel and Distributed Computing*, Vol. 13, (1991), pp. 139–153.

[2] A. K. Chandra, Maximal Parallelism in Matrix Multiplication, *RC 6193*, IBM Rept. (1976).

[3] F. Y. Chin, J. Lam and I. N. Chen, "Efficient Parallel Algorithms for Some Graph Problems," *Commum. ACM*, Vol. 25, No. 9, (Sept. 1982), pp. 659–665.

[4] D. G. Feitelson, Optical Computing, MIT press, (1988).

[5] Y. Han and R. A. Wagner, "An Efficient and Fast Parallel-Connected Component Algorithm," *Journal of the Association for Computing Machinery*, Vol. 37, No. 3, (July 1990), pp. 626–642.

[6] D. S. Hirschberg, "Parallel Algorithms for the Transitive Closure and the Connected Component Problem," *Proc. Eighth STOC*, (1976), pp. 55–57.

[7] D. S. Hirschberg, A. K. Chandra and D. V. Sarwate, "Computing Connected Components on Parallel Computers," *Commum. ACM*, Vol. 22, No. 8, (Aug. 1979), pp. 461–464.

[8] S. J. Horng, "Computing Dominators on a Cube-Connected Machine," *Tech. Rept.*, Dept. of Elec. Eng. Nati. Taiwan Inst. of Tech., (1992).

[9] S. J. Horng, "Computing Dominators on a RAP," *Tech. Rept.* Dept. of Elec. Eng. Nati. Taiwan Inst. of Tech. (1991).

[10] T. W. Kao and S. J. Horng, "Efficient Parallel Algorithms for Image Processing on RAP," submitted for publication.

[11] T. W. Kao, S. J. Horng and Y. L. Wang, "Efficient Parallel Algorithms for Summation and Sorting on RAP," Technical Report, Department of Electrical Engineering, National Taiwan Institute of Technology, (1991).

[12] L. Kucera, "Parallel Computation and Conflicts in Memory Access," *Information Processing Letters*, Vol. 14, No. 2, (Apr. 1982), pp. 93–96.

[13] H. Li and M. Maresca, "Polymorphic-Torus Network," *IEEE Transactions on Computers*, Vol. 38, No. 9, (Sep. 1989), pp. 1345–1351.

[14] M. Maresca and H. Li, "Connection Autonomy in SIMD Computers: A VLSI Implementation," *Journal of Parallel and Distributed Computing*, Vol. 7, No. 2, (1989), pp. 302–320.

[15] R. Miller, V. K. P. Kumar, D. Reisis and Q.F. Stout, "Meshes with Reconfigurable Buses," *Proceedings of the MIT Conference on Advanced Research in VLSI*, (Mar. 1988), pp. 163–178.

[16] S. Pawagi, P. S. Gopalakrishnan, and I. V. Ramakrishnan, A parallel algorithm for dominators, *Proc. ICPP*, (1986) 877-879.

[17] M. J. Qunn and N. Deo, "Parallel Graph Algorithms", *Computing Surveys*, Vol. 16, No. 3, (Sep. 1984), pp. 319-348.

[18] E. Reghbati (Arjomandi) and D. G. Corneil, "Parallel Computations in Graph Theory," *SIAM J. Computing*, Vol. 7, No. 2, (1978), pp. 230–237.

[19] J. Rothstein, "On the Ultimate Limitations of Parallel Processing," *Proc. ICPP*, (1976), pp. 206–212.

[20] J. Rothstein, "Bus Automata, Brains, and Mental Models," *IEEE Transactions on Systems, Man, and Cybernetics*, Vol. 18, No. 4, (Apr. 1988), pp. 522–531.

[21] K. W. Ryu and J. JáJá, "Efficient Algorithms for List Ranking and for Solving Graph Problems on the Hypercube," *IEEE Transactions on Parallel and Distributed Systems*, Vol. 1, No. 1, (Jan. 1990), pp. 83–90.

[22] C. Savage, Parallel algorithms for some graph problems, *TR 784*, Dept. of Mathematics, Univ. of Illinois, Urbana, (1977).

[23] C. D. Savage and J. JáJá, "Fast Efficient Parallel Algorithms for Some Graph Problems," *SIAM J. Computing*, Vol. 10, No. 4, (1981), pp. 682–691.

[24] A. Schuster and Y. Ben-Asher, "Algorithms and Optic Implementation for Reconfigurable Networks," *Proc. of the 5th Jerusalem Conf. on Info. Tech.*, (Oct. 1990).

[25] Y. Shiloach and U. Vishkin, "An $O(\log n)$ Parallel Connectivity Algorithm," *J. Algorithms*, Vol. 3, No. 1, (Mar. 1982), pp. 57–67.

[26] D. B. Shu and J. G. Nash, "The Gated Interconnection Network for Dynamic Programming," S. K. Tewsburg et al. (ed), Concurrent Computing, Plenum, New York, (1988).

[27] B. F. Wang, G. H. Chen and F. C. Lin, "Constant Time Sorting on a Processor Array with a Reconfigurable Bus System," *Information Processing Letters*, Vol. 34, No. 4, Apr. 1990, pp. 187–192.

[28] B. F. Wang and G. H. Chen, "Constant Time Algorithms for The Transitive Closure and Some Related Graph Problems on Processor Arrays with a Reconfigurable Bus System," *IEEE Transactions on Parallel and Distributed Systems*, Vol. 1, No. 4, (Oct. 1990), pp. 500–507.

EMBEDDING GRIDS, HYPERCUBES, AND TREES IN ARRANGEMENT GRAPHS

Khaled Day
Department of Computer Science
University of Bahrain
P.O. Box: 32038
Isa Town, State of Bahrain
kday@BHUOB00.BITNET

Anand Tripathi
Department of Computer Science
University of Minnesota
Minneapolis, MN, 55455, USA
tripathi@cs.umn.edu

Abstract-- *The use of a new topology, called the arrangement graph, as a viable interconnection scheme for parallel and distributed systems, has been recently proposed and examined. The arrangement graphs represent a rich class of generalized star graphs with many attractive properties. In this paper we obtain (1) unit expansion embeddings of a variety of multi-dimensional grids, (2) dilation 1, 2, 3, and 4 embeddings of hypercubes and (3) hierarchical and greedy spanning trees to support broadcasting and personalized communication in arrangement graphs.*

INTRODUCTION

In designing large multi-processor systems, networks of processors are often organized into various configurations. Some of the important interconnection networks which have been studied so far are: hypercubes, grids, trees, shuffle exchange, cube-connected cycles, and star graphs [1], [5], [19], [20], [23], [25], [26]. Among these topologies the star graph has received the most recent attention [1], [2], [11], [15], [17], [18], [21], [22]. It has been proposed as an attractive alternative to the hypercube with many superior characteristics [1, 11, 17].

A class of generalized star graphs -the arrangement graphs- has been recently proposed as a possible interconnection network for parallel and distributed computing systems [10]. The arrangement graph has been shown [10] to be vertex and edge symmetric, strongly hierarchical, maximally fault tolerant, and strongly resilient. A simple shortest path routing algorithm has also been obtained [10]. This topology exhibits graceful degradation as well as other fault tolerance properties including complete families of node-disjoint paths between any two nodes [9]. Furthermore, it has been shown in [8] that an arrangement graph embeds cycles of arbitrary lengths ranging between 3 and the number of nodes in the graph (provided that it is not a star graph in which case it embeds only even length cycles).

Over the years many parallel algorithms have been designed to solve different problems on various famous interconnection networks such as grids, hypercubes, and tree configured networks. It would be of interest to be able to execute these algorithms on interconnection networks based on the arrangement graph topology. To do this, we need to map the algorithms' underlying network structure into the arrangement graph.

In the first part of this paper we study the mapping of grids and hypercubes in arrangement graphs. In other words, we study the one-to-one association of the processors of a grid or a hypercube with the nodes of an arrangement graph and we evaluate the costs of these mappings. Similar graph embeddings have been widely used in the literature to simulate well studied topologies, such as cycles, trees and grids, on other important network topologies, such as hypercubes and star graphs [6], [7], [16], [18], [21] - [24], [26]. We finally describe some spanning trees for these graphs. Spanning trees are essential to solve many computation and communication problems such as broadcasting and personalized communication [13]. Embedded trees can also be totally or partially used to simulate parallel algorithms which are originally designed for tree based networks.

BACKGROUND

Graph definition and basic properties

Let n and k be two integers satisfying $1 \leq k \leq n-1$. Denote $<n> = \{1, 2, ... , n\}$ and $<k> = \{1, 2, ... , k\}$. Let P_k^n be the set of permutations of the n elements of $<n>$ taken k at a time, i.e., the set of arrangements of k elements out of the n elements of $<n>$. The k elements of an arrangement p are denoted $p_1, p_2, \ldots, p_k$; we write $p = p_1 p_2 \ldots p_k$, and we refer to p_i as the i'th element of p.

Definition 1. The *(n,k)-arrangement graph $A_{n,k}$* is an undirected graph (V,E) given by:

$V = \{ p_1 p_2 \ldots p_k \mid p_i$ in $<n>$ and $p_i \neq p_j$ for $i \neq j \} = P_k^n$,

$E = \{(p,q) \mid p$ and q in V and for some i in $<k>$, $p_i \neq q_i$ and $p_j = q_j$ for $j \neq i \}$.

. That is, the vertices (or nodes) of $A_{n,k}$ are the arrangements of k elements out of the n elements of $<n>$, and the edges of $A_{n,k}$ connect arrangements which differ in exactly one of their k positions. For example in $A_{5,3}$ the node $p = 513$ is connected to the nodes 213, 413, 523, 543, 512, and 514. An edge of $A_{n,k}$ connecting two arrangements p and q which differ only in position i, is called an *i-edge*. For all values of n and k, $A_{n,k}$ is a regular graph with a number of nodes $\frac{n!}{(n-k)!}$, a degree $k(n-k)$, and a diameter $\lfloor \frac{3}{2} k \rfloor$ [10]. Let $p = p_1 p_2 \ldots p_k$ be an arrangement in P_k^n; we define the set EXT(p) = $<n> - \{p_1, p_2, \ldots, p_k\}$ of the n-k elements of $<n>$ not appearing in the arrangement p. Similarly we define the set INT(p) = $\{p_1, p_2, \ldots, p_k\} = <n> -$ EXT(p) of the k elements of $<n>$ appearing in the arrangement p. We also refer to the elements of EXT(p) as the *external elements* of p, and we refer to the elements of INT(p) as the *internal elements* of p. The special node $I_k = 12 \ldots k$ is called the *identity* node The elements of EXT(I_k) are called the *foreign elements*.

Hierarchical structure

Notice that there are $\frac{(n-1)!}{(n-k)!}$ nodes in $A_{n,k}$ which have element i in position j, for any fixed i and j ($1 \leq i \leq n$, $1 \leq j \leq k$). These nodes are interconnected in a manner identical to an $A_{n-1,k-1}$ graph. We denote by i_j this sub-graph of $A_{n,k}$ consisting of all nodes with element i in position j. For a fixed position j, the sub-graphs 1_j, 2_j, $\ldots$, n_j have disjoint sets of vertices and therefore form a partitioning of the set of vertices of $A_{n,k}$. *Figure 1* illustrates the partitioning of $A_{4,2}$ into four $A_{3,1}$'s : 1_2, 2_2, 3_2, and 4_2 (four triangles). This partitioning of $A_{n,k}$ into n copies of $A_{n-1,k-1}$ can be done in k different ways corresponding to the k possible values of j ($1 \leq j \leq k$) and can be carried out recursively so that each $A_{n-1,k-1}$ is partitioned into n-1 disjoint copies of $A_{n-2,k-2}$ and so on. It is shown in [10] that $A_{n,k}$ can be partitioned into $\frac{n!}{(n-p)!}$ node-disjoint copies of $A_{n-p,k-p}$ in $\frac{n!}{p!(n-p)!}$

different ways and that in total, $A_{n,k}$ contains $\binom{k}{p}\frac{n!}{(n-p)!}$ copies of $A_{n-p,k-p}$, for $1 \leq p \leq k-1$.

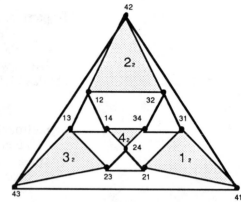

Figure 1. $A_{4,2}$ viewed as 4 interconnected $A_{3,1}$'s: 1_2, 2_2, 3_2, and 4_2 (the 4 triangles).

Graph embeddings

An embedding, f, of a graph G' = (V',E') into a graph G = (V,E) is a mapping of V' into V such that each vertex of G' is mapped to a distinct vertex of G (i.e., if $x \neq y$ then $f(x) \neq f(y)$). Note that for an embedding to exist we must have $|V| \geq |V'|$. Also if G is connected and $|V| \geq |V'|$ then an embedding of G' into G always exists. The ratio $\frac{|V|}{|V'|}$ is called the *expansion* of the embedding. The maximum distance in G between f(x) and f(y) for two adjacent nodes x,y of G' is called the *dilation* of the embedding. That is, the *dilation* of f is given by:

$$dilation = \text{Max } \{DISTANCE_G(f(x),f(y)) \mid (x,y) \in E'\},$$

where $DISTANCE_G(f(x),f(y))$ is the length of the shortest path in G between f(x) and f(y).

EMBEDDING OF GRIDS

In this section we describe unit expansion embedding in $A_{n,k}$, of a variety of d-dimensional grids ($1 \leq d \leq k$). We first present a useful lemma which determines an upper bound on the distance in $A_{n,k}$ between a node p and the node $SWAP_{i,j}(p)$ obtained by exchanging the elements i and j in p ($1 \leq i,j \leq n$).

Lemma 1. The distance from p to $SWAP_{i,j}(p)$, $1 \leq i,j \leq n$, is at most equal to 3.

Proof. Let $p = p_1 p_2 \ldots p_k$, there are four cases:

case 1: if i and j in EXT(p), then $SWAP_{i,j}(p)=p$. The distance from p to $SWAP_{i,j}(p)$ is then equal to 0.

case 2: if $i \in$ EXT(p) and $j = p_s$ for some s, $1 \leq s \leq k$, then SWAP$_{i,j}$(p)=$p_1 p_2 \ldots p_{s-1} i p_{s+1} \ldots p_k$, therefore the distance from p to SWAP$_{i,j}$(p) is equal to 1.

case 3: if $i = p_s$ for some s, $1 \leq s \leq k$, and $j \in$ EXT(p) (similar to case 2).

case 4: if $i = p_s$ and $j = p_t$, $1 \leq s,t \leq k$, then SWAP$_{i,j}$(p)=$p_1 p_2 \ldots p_{s-1} j p_{s+1} \ldots p_{t-1} i p_{t+1} \ldots p_k$. Therefore the distance from p to SWAP$_{i,j}$(p) is equal to 3 and the following is a minimum length path from p to SWAP$_{i,j}$(p): $p \rightarrow p_1 \ldots p_{s-1} y p_{s+1} \ldots p_k \rightarrow p_1 \ldots p_{s-1} y p_{s+1} \ldots p_{t-1} i p_{t+1} \ldots p_k \rightarrow$ SWAP$_{i,j}$(p), where y is any element of EXT(p). ◆

Theorem 1. An $(n-k+1) \times (n-k+2) \times \ldots \times (n-1) \times n$ grid can be embedded in $A_{n,k}$ with unit expansion and dilation 3.

Proof. We prove by induction on i, $n-k+1 \leq i \leq n$, that $A_{i,i-n+k}$ embeds an $(n-k+1) \times (n-k+2) \times \ldots \times (i-1) \times i$ grid with unit expansion and dilation 3. For $i = n-k+1$, $A_{i,i-n+k}$ is the fully connected graph with $n-k+1$ nodes, which trivially embeds a 1-dimensional $(n-k+1)$-grid with unit expansion and unit dilation. Assume $A_{i,i-n+k}$, for $n-k+1 \leq i < n$, embeds an $(n-k+1) \times (n-k+2) \times \ldots \times (i-1) \times i$ grid with unit expansion and dilation 3. Consider an $A_{i+1,i+1-n+k}$ graph partitioned into the i+1 copies: $1_{i-n+k}, 2_{i-n+k}, \ldots, i+1_{i-n+k}$ of $A_{i,i-n+k}$. By induction hypothesis, there exists a unit expansion and dilation 3 embedding of an $(n-k+1) \times (n-k+2) \times \ldots \times (i-1) \times i$ grid, G_1, in the sub-graph: 1_{i-n+k}. We obtain a grid, G_2, in 2_{i-n+k}, isomorphic to G_1 by exchanging the elements 1 and 2 in all the nodes of G_1. By lemma 1, each node p in G_1 is at most at a distance 3 from its image node SWAP$_{1,2}$(p) in G_2. Similarly we obtain a grid, G_3, in 3_{i-n+k}, isomorphic to G_2 by exchanging the elements 2 and 3 in all the nodes of G_2. We repeat this procedure i times to generate i+1 grids: $G_1, G_2, \ldots, G_{i+1}$ such that for each node p in G_j there exists a node p' in G_{j+1} which is at most at a distance 3 from p, for any j, $1 \leq j \leq i$. We combine the i+1 grids: $G_1, G_2, \ldots, G_{i+1}$ using the (p,p')-like connections to obtain an expansion 1, dilation 3 embedding of an $(n-k+1) \times (n-k+2) \times \ldots \times i \times (i+1)$ grid into $A_{i+1,i+1-n+k}$. ◆

Figure 2.a. Unit expansion, unit dilation embedding of a 1-dimensional grid in $A_{3,1}$.

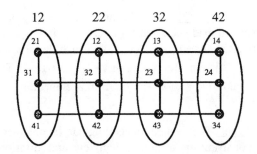

Figure 2.b. Unit expansion, dilation 3 embedding of a 3×4 grid in $A_{4,2}$.

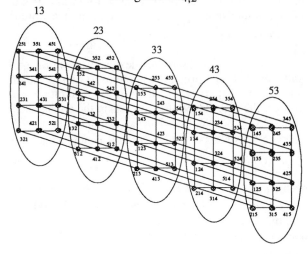

Figure 2.c. Unit expansion, dilation 3 embedding of a 3×4×5 grid in $A_{5,3}$

The construction of Theorem 1 is illustrated in Figure 2. In this example we first embed a 3 node 1-dimensional grid in $A_{3,1}$ with unit expansion and unit dilation (Figure 2.a). Then four isomorphic copies of this grid are combined to form a unit expansion and dilation 3 embedding of a 3×4 two-dimensional grid in $A_{4,2}$ (Figure 2.b). Finally five isomorphic copies of this 3×4 grid are combined to form a unit expansion and dilation 3 embedding of a 3×4×5 three-dimensional grid in $A_{5,3}$ (Figure 2.c). The following result has been proved in [16] and will allow us to extend the embeddable class of grids .

Lemma 2. Let G be an $n_1 \times n_2 \times \ldots \times n_d$ d-dimensional grid and let G' be an $(n_i \times n_j) \times n_1 \times \ldots \times n_{i-1} \times n_{i+1} \times \ldots \times n_{j-1} \times n_{j+1} \times \ldots \times n_d$ (d-1)-dimensional grid. There exists a unit expansion and unit dilation embedding of G' into G.

We now show how to realize unit expansion and dilation three embeddings in $A_{n,k}$ of a variety of d-dimensional grids ($2 \leq d \leq k$).

Theorem 2. Let G be an $n_1 \times n_2 \times \ldots \times n_d$ d-dimensional grid such that for every i, $1 \le i \le d$, n_i can be factorized as: $n_i = n_{i_1} n_{i_2} \ldots n_{im_i}$ and such that the set of all the n_{i_j}'s is the set: $\{n-k+1, n-k+2, \ldots, n-1, n\}$. There exists a unit expansion and dilation 3 embedding of G into $A_{n,k}$.

Proof. The desired embedding can be obtained by first constructing a unit expansion and dilation 3 embedding in $A_{n,k}$ of an $(n-k+1) \times (n-k+2) \times \ldots \times (n-1) \times n$ grid, using the construction of Theorem 1. Lemma 2 can then be applied repeatedly to combine for each i, $1 \le i \le d$, the m_i dimensions $n_{i_1}, n_{i_2}, \ldots, n_{im_i}$ into a unique dimension: n_i. ♦

Most parallel algorithms designed to run on grid configured multi-processors have assumed equal dimensions [22]. It would therefore be of interest to consider the embedding of such grids in the arrangement graphs. Let U denote the d-dimensional uniform grid of N nodes: $N^{1/d} \times N^{1/d} \times \ldots \times N^{1/d}$. Let R denote an $n_1 \times n_2 \times \ldots \times n_d$ d-dimensional grid such that $\prod_{i=1}^{d} n_i = N$. It has been shown in [3] that if $d = O(1)$ then R can simulate every step of U in $O((\max_i n_i)/N^{1/d})$ steps. A modified version of this result has been proposed in [22] which does not assume $d = O(1)$. This version states that R can simulate U in $O((\max_i n_i) 2^d/N^{1/d})$ steps. Using this result, we derive that the number of steps required to simulate the $N^{1/k} \times N^{1/k} \times \ldots \times N^{1/k}$ k-dimensional grid on the k-dimensional grid: $(n-k+1) \times (n-k+2) \times \ldots \times (n-1) \times n$, is in the worst case $O(n2^k/N^{1/k})$ steps, where $N = (n-k+1) \times (n-k+2) \times \ldots \times (n-1) \times n$.

EMBEDDING HYPERCUBES

In this section we study dilation 1, 2, 3, and 4 embeddings of hypercubes in arrangement graphs exploring the tradeoff between dilation and expansion in these embeddings. Let us denote by H_d the hypercube of dimension d. The following result is used in this section.

Lemma 3. If $m < n$, $t \le k$ and $m-t \le n-k$, then there exists a unit dilation embedding of $A_{m,t}$ into $A_{n,k}$.

Proof. $A_{n,k}$ can be partitioned either into n copies of $A_{n-1,k-1}$ or into k copies of $A_{n-1,k-1}$ and one copy of $A_{n-1,k}$ [10]. Therefore $A_{n,k}$ embeds with unit dilation both $A_{n-1,k-1}$ and $A_{n-1,k}$. Applying these embeddings recursively yields the claimed result. ♦

The simplest embedding of H_d is a unit expansion and unit dilation embedding into $A_{2d,1}$. However this embedding is not attractive since $A_{2d,1}$ is a non practical fully connected graph. A more interesting unit dilation embedding is described next.

Theorem 3. For every d, $d \ge 2$, there exists a unit dilation embedding of H_d into $A_{2d,d-1}$. The expansion of this embedding is $\dfrac{(2d)!}{2^d(d+1)!}$.

Proof. (Induction on d) $A_{4,1}$ is a fully connected graph with four nodes, therefore it embeds with unit expansion and unit dilation an H_2. Assume that $A_{2d,d-1}$ embeds H_d with unit dilation. Therefore there exists a mapping M of the set V_H^d of the 2^d nodes of H_d into the set of nodes of $A_{2d,d-1}$ such that:

(i) $\forall$ x,y $\in$ V_H^d, if $x \ne y$ then $M(x) \ne M(y)$

(ii) if (x,y) is an edge in H_d, then $(M(x), M(y))$ is an edge in $A_{2d,d-1}$.

Consider an H_{d+1} hypercube, and let $x = b_{d+1} b_d \ldots b_1$ be a node of H_{d+1}. We define a mapping M' of the nodes of H_{d+1} into the nodes of $A_{2d+2,d}$ as follows:

$$M'(x) = (2d+1+b_{d+1}) \parallel M(b_d b_{d-1} \ldots b_1) ,$$

where $\parallel$ denotes the concatenation(e.g. $5 \parallel 231 = 5231$). It can be easily verified that this mapping defines indeed a unit dilation embedding of H_{d+1} into $A_{2d+2,d}$. The expansion is trivially obtained by computing the ratio of the sizes (number of nodes) of the two graphs. ♦

The same approach used in the proof of Theorem 3 can be used to show the existence of a unit dilation embedding of H_d into $A_{2d,d}$, using as induction basis the case $d = 1$ (H_1 and $A_{2,1}$ are isomorphic). The expansion of this embedding is $\dfrac{(2d)!}{d!2^d}$ approximated by $\sqrt{2}(\dfrac{2d}{e})^d$ using Stirling's approximation. Using Lemma 3 we directly derive the following result.

Corollary. For every $d \ge 2$, there exists a unit dilation embedding of H_d into $A_{n,k}$ if $n \ge 2d$, $k \ge d-1$, and $n-k \ge d$.

We can obtain dilation 2 embeddings with smaller expansion. Expansion 1.5 and dilation 2 embeddings of H_2 in $A_{3,2}$ and of H_3 in $A_{4,2}$ are shown in Figure 3. This figure shows also an expansion 1.25 and dilation 2 embedding of H_4 in $A_{5,2}$. If we exclude embeddings in fully connected graphs, these embeddings are of minimum expansion since $A_{3,2}$, $A_{4,2}$, $A_{5,2}$ are the smallest non fully connected arrangement graphs with at least 4, 8, and

16 nodes respectively. We are unable to derive a general result for similar minimum expansion and dilation 2 embeddings for all hypercubes. It is however possible to obtain a dilation 3 embedding for any hypercube.

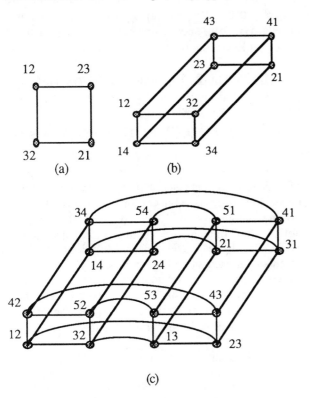

Figure 3. Dilation 2 embeddings of (a) H_2 in $A_{3,2}$, (b) H_3 in $A_{4,2}$ and (c) H_4 in $A_{5,2}$.

Theorem 4. For every $d > 3$ there exists a dilation 3 embedding of H_d into $A_{d+1,d-2}$. The expansion of this embedding is $\frac{(d+1)!}{6 \times 2^d}$.

Proof. (by induction on d) For d=4, we have obtained a dilation 2 embedding of H_4 in $A_{5,2}$ (Figure 3). Assume the result true for $d = m-1 > 3$. Consider a partitioning of $A_{m+1,m-2}$ into the m+1 copies of the $A_{m,m-3}$ sub-graphs: $1_{m-2}, 2_{m-2}, \ldots, (m+1)_{m-2}$. By induction hypothesis there exists a dilation 3 embedding of an (m-1)-hypercube in $A_{m,m-3}$. Let H_{m-1}^1 and H_{m-1}^2 denote two (m-1)-hypercubes embedded with dilation 3 in the $A_{m,m-3}$ sub-graphs: 1_{m-2} and 2_{m-2} respectively, such that H_{m-1}^1 and H_{m-1}^2 are obtained from each other by exchanging the elements 1 and 2 in the nodes. Using Lemma 1, each node of H_{m-1}^1 and its isomorphic node in

H_{m-1}^2 are at most at distance three from each other. Combining these two hypercubes by connecting all pairs of isomorphic nodes we obtain a dilation 3 embedding of H_m into $A_{m+1,m-2}$. ♦

H_5 node	$A_{6,3}$ node	H_5 node	$A_{6,3}$ node
00000	346	10000	345
00001	546	10001	645
00011	246	10011	245
00010	146	10010	145
00100	426	10100	425
00101	526	10101	625
00111	326	10111	325
00110	126	10110	125
01000	416	11000	415
01001	516	11001	615
01011	216	11011	215
01010	316	11010	315
01100	436	11100	435
01101	536	11101	635
01111	136	11111	135
01110	236	11110	235

Figure 4. Dilation 3 embedding of H_5 in $A_{6,3}$

Figure 4 shows a 3.75 expansion and dilation 3 embedding of H_5 in $A_{6,3}$. Smaller expansion factors can be obtained by using dilation 4 embeddings.

Theorem 5. If H_d can be embedded in $A_{n-1,k-1}$ with dilation 4, then H_{d+p}, where $p = \lfloor \log_2 n \rfloor$, can be embedded in $A_{n,k}$ with dilation 4.

Proof. H_{d+p} is composed of n hypercubes of dimension d. Let $H_d^1, H_d^2, \ldots, H_d^n$ be these hypercubes. Consider a partitioning of $A_{n,k}$ into the n copies: $1_k, 2_k, \ldots, n_k$ of $A_{n-1,k-1}$. By assumption, we can embed, with dilation 4, H_d^1 into 1_k. We obtain from this embedding a dilation 4 embedding for each H_d^i, $2 \le i \le n$, into the corresponding i_k by exchanging the elements 1 and i in all the nodes of H_d^1. Each node of H_d^i has therefore an isomorphic node in H_d^1 obtained from each other by exchanging the elements 1 and i. Let u and v be two adjacent nodes in H_{d+p}. If u and v are in the same H_d^i for some i, then they are at most 4 apart in $A_{n,k}$ (by assumption). If however $u \in H_d^i$, $v \in H_d^j$, and $i \neq j$ then

adjacency in H_{d+p} implies that u and v are isomorphic to some node $q = q_1 q_2 \ldots q_{k-1} 1$ in H_d^1. Notice that $u \in i_k$ and $v \in j_k$ and $q \in 1_k$. If $i \in EXT(q)$ or $j \in EXT(q)$ then either $u=q$ or $v=q$, and by Lemma 1 u and v are at most distance 3 apart. If however $i = q_a$ and $j = q_b$ for some a and b then u and v are given by:

$u = q_1 q_2 \ldots q_{a-1} 1 q_{a+1} \ldots q_{b-1} j q_{b+1} q_2 \ldots q_{k-1} i$, and
$v = q_1 q_2 \ldots q_{a-1} i q_{a+1} \ldots q_{b-1} 1 q_{b+1} q_2 \ldots q_{k-1} j$.

There exists a path: $u \to w1 \to w2 \to w3 \to v$ of length four between u and v such that:
$w1 = q_1 q_2 \ldots q_{a-1} 1 q_{a+1} \ldots q_{b-1} j q_{b+1} q_2 \ldots q_{k-1} y$
$w2 = q_1 q_2 \ldots q_{a-1} i q_{a+1} \ldots q_{b-1} j q_{b+1} q_2 \ldots q_{k-1} y$
$w3 = q_1 q_2 \ldots q_{a-1} i q_{a+1} \ldots q_{b-1} 1 q_{b+1} q_2 \ldots q_{k-1} y$,
where y is an arbitrary element of $EXT(u)$. Hence the above embedding of H_{d+p} into $A_{n,k}$ is of dilation 4. ♦

SPANNING TREES

Spanning trees are essential tools for many important communication problems such as broadcasting and personalized communication. A substantial amount of research has been directed towards embedding tree structures in various interconnection networks including hypercubes [4, 12, 13, 14, 26] and star graphs [11]. In this section we describe some spanning tree structures in arrangement graphs and investigate their properties. We construct spanning trees rooted at the identity node I_k of $A_{n,k}$. These constructions can be readily generalized to obtain spanning trees rooted at arbitrary nodes of $A_{n,k}$.

The Spanning Polynomial Tree $SPTn,k(I_k)$

We construct a spanning tree denoted $SPTn,k(I_k)$ rooted at the identity node I_k of $A_{n,k}$. We build $SPTn,k(I_k)$ recursively based on the hierarchical structure of $A_{n,k}$. We also show by induction that all the adjacent nodes of I_k in $A_{n,k}$ will be adjacent to I_k in $SPTn,k(I_k)$. For $k=1$ we obtain $SPTn,1(I_1)$ by connecting I_1 to all its neighbors ($A_{n,1}$ is a complete graph with n nodes). Clearly I_1 remains adjacent in $SPTn,1(I_1)$ to all its $A_{n,1}$ neighbors. For $k>1$, let $1_k, 2_k, \ldots, n_k$ be the decomposition of $A_{n,k}$ into n copies of $A_{n-1,k-1}$. For a given node p of $A_{n,k}$, let $ADJ_{x,y}(p)$ denote the adjacent node of p obtained by exchanging in p the element at position x with the element y from $EXT(p)$. We define the nodes: $t_1, t_2, \ldots, t_{k-1}$ and $r_1, r_2, \ldots, r_n$ as follows:

$t_i = ADJ_{i,k+1}(I_k)$, for $1 \leq i \leq k-1$,
$r_i = ADJ_{k,i}(t_i)$, for $1 \leq i \leq k-1$, (notice that $i \in EXT(t_i)$)

$r_k = I_k$, and $r_i = ADJ_{k,i}(I_k)$, for $k+1 \leq i \leq n$.

Notice that r_i, $1 \leq i \leq n$, $i \neq k$, is a node in i_k. Let $T^i = SPTn-1,k-1(r_i)$, $1 \leq i \leq n$, be the spanning polynomial tree in i_k rooted at r_i. By induction hypothesis T^k contains all the edges (I_k, t_i), $1 \leq i \leq k-1$, since the t_i's are adjacent to I_k in k_k. Let E^i, $1 \leq i \leq n$, be the set of edges of T^i.

Definition 2. The *Spanning Polynomial Tree* $SPTn,k(I_k)$ in $A_{n,k}$, rooted at the identity node I_k, is the spanning tree whose set of edges is:
$E = (\cup_{n}^{i=1} E^i) \cup \{(t_i, r_i), 1 \leq i \leq k-1\} \cup \{(I_k, r_i), k+1 \leq i \leq n\}$.

Figure 5 illustrates the recursive construction of $SPTn,k(I_k)$. For this construction to be valid, it remains to show that I_k is adjacent in $SPTn,k(I_k)$ to all its original neighbors in $A_{n,k}$. To prove this property, notice that during the recursive construction of $SPTn,k(I_k)$, I_k has been connected to n-k of its original neighbors at each step of the recursion. For instance, in the last induction step we connected I_k to its n-k original r_i neighbors for $k+1 \leq i \leq n$. Since there are k steps in the recursive construction therefore I_k is connected in $SPTn,k(I_k)$ to all its k(n-k) original neighbors in $A_{n,k}$.

$SPTn,k$ embeds n copies of $SPTn-1,k-1$. This hierarchical structure is desirable for solving many computation and communication problems. The sub-trees $SPTn-1,k-1$ of an $SPTn,k$ are isomorphic and can be obtained from each other by applying simple element mappings on the nodes.

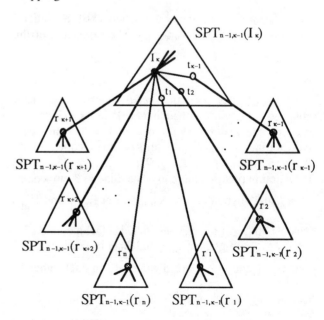

Figure 5. Recursive construction of $SPTn,k(I_k)$

It can be easily shown by induction that the height of $SPT_{n,k}$ is $2k-1$ which is of the same order as the diameter $\lfloor\frac{3}{2}k\rfloor$ of $A_{n,k}$.

<u>The Smallest Misplaced First spanning tree $SMF_{n,k}(I_k)$</u>

We now use another approach in building a spanning tree for $A_{n,k}$. We construct a greedy spanning tree (i.e., with minimum node-to-root path for all nodes) rooted at the identity node I_k of $A_{n,k}$. The construction can be easily generalized for trees rooted at any nodes.

<u>Definition 3.</u> The *Smallest Misplaced First* spanning tree of $A_{n,k}$, rooted at the identity node I_k and denoted *SMFn,k(Ik)*, is defined by the function Parent(p) given by:

$$Parent(p) = ADJx,y(p),$$

which for a node $p = p_1 p_2 .... p_k$ of $A_{n,k}$ uniquely defines its parent node in $SMF_{n,k}(I_k)$ and where $x = y =$ smallest non foreign external element of p, if it exists, or $y = k+1$ and x is such that p_x is the smallest misplaced[a] element among $p_1 p_2 .... p_k$ otherwise.

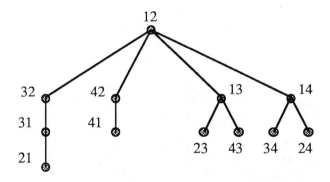

<u>Figure 6.</u> The SMF4,2(I2) spanning tree of $A_{4,2}$

In other words the parent node of p in $SMF_{n,k}(I_k)$ is obtained by correcting the smallest non foreign external element of p, if it exists, or by making external its smallest non foreign misplaced internal element by exchanging it with the foreign element $k+1$. Figure 6 shows the SMF4,2(I2) spanning tree of $A_{4,2}$. We now investigate some properties of $SMF_{n,k}(I_k)$.

<u>Theorem 6.</u> The $SMF_{n,k}(I_k)$ is a greedy spanning tree of $A_{n,k}$.

---

[a] p_x is called a misplaced element if $p_x \neq x$.

<u>Proof.</u> Routing from a node p to the root I_k of $SMF_{n,k}(I_k)$ using repeatedly the Parent function defined earlier satisfies the minimum distance routing conditions of arrangement graphs which are established in [10]. These conditions state that a path from p to I_k is of minimum length if every step of the routing from p to I_k along this path corresponds to correcting a non foreign external element or to making external a non foreign misplaced internal element. ♦

The following result is directly derived from the fact that $SMF_{n,k}(I_k)$ is greedy and that the diameter of $A_{n,k}$ is equal to $\lfloor\frac{3}{2}k\rfloor$.

<u>Corollary.</u> The $SMF_{n,k}(I_k)$ spanning tree has the optimal height $\lfloor\frac{3}{2}k\rfloor$.

The $SMF_{n,k}(I_k)$ has been defined using the Parent function, which for each node p of $A_{n,k}$ uniquely characterizes its parent node in $SMF_{n,k}(I_k)$. It is also useful to characterize the children of a node.

<u>Theorem 7.</u> The descendents in $SMF_{n,k}(I_k)$ of a node $p = p_1 p_2 .... p_k$ are given by the Children function:

Children(p) = {ADJx,y(p), such that $y \in$ EXT(p), $p_x=x$ and x is smaller than all non foreign external elements of p except possibly y} $\cup$ {ADJx,y(p), such that: (i) y is the smallest misplaced non foreign element, (ii) y is the only non foreign external element, (iii) $p_x=k+1$, and (iv) $x \neq y$}.

<u>Proof.</u> Directly obtained by reversing the Parent(p) function. ♦

CONCLUSION

We have further studied the topological properties of the newly proposed class of interconnection networks: the arrangement graphs. These graphs represent a good host for the embedding of other network topologies. This allows to simulate and make use of many available parallel and distributed algorithms which have been designed to solve different problems on various networks. We have obtained unit expansion embeddings of a variety of multi-dimensional grids and dilation 1, 2, 3 and 4 embeddings of hypercubes in arrangement graphs. We have also explored the tradeoff between dilation and expansion in these embeddings. Finally, We have described hierarchical and greedy spanning trees to support broadcasting and personalized communication in arrangement graph configured interconnection networks. These results confirm the viability of this class of graphs as an attractive interconnection scheme for parallel and distributed systems.

REFERENCES

[1] S. B. Akers, D. Harel and B. Krishnamurthy, "The Star Graph: An Attractive Alternative to the n-cube," *Intl. Conf. Parallel Processing*, (1987), pp. 393-400.

[2] S. B. Akers, and B. Krishnamurthy, "A Group Theoretic Model for Symmetric Interconnection Networks," *Intl. Conf. Parallel Processing*, (1986), pp. 216-223.

[3] M. J. Atallah, "On Multidimensional Arrays of Processors," *IEEE Trans. Comput.*, (October 1988), pp. 1306-1309.

[4] S. N. Bhatt, F. R. K. Chung, F. T. Leighton, and A. L. Rosenberg, "Optimal Simulations of tree machines," *Proc. 27th IEEE Symp. Foundations Comput. Sci.*, (1986), pp. 274-282.

[5] W. Bouknight et al., "The ILLIAC IV System," *Proc. IEEE 60*, (1972), pp. 369-388.

[6] M. Y. Chan, and F. Y. L. Chin, "On Embedding Rectangular Grids in Hypercubes," *IEEE Trans. Comput.*, (October 1988), pp. 1285-1288.

[7] T. F. Chan, and Y. Saad, "Multigrid algorithms on the hypercube multicomputers," *IEEE Trans. Comput.* (November 1986), pp. 969-977.

[8] K. Day, and A. Tripathi, "Embedding of Cycles in Arrangement Graphs," to appear in *IEEE Trans. on Comput.*

[9] K. Day, and A. Tripathi, *Characterization of Node Disjoint Paths in Arrangement Graphs*, Computer Science Dept., Univ. of Minnesota, TR 91-43, (October 1991) 21 pp.

[10] K. Day, and A. Tripathi, "Arrangement graphs: A Class of Generalized Star Graphs," *Information Processing Letters*, (July 1992) pp. 235-241.

[11] K. Day, and A. Tripathi, "A comparative study of Topological Properties of Hypercubes and Star Graphs," To appear in *IEEE Trans. on Parallel and Dist. Systems*.

[12] S. R. Deshpande, and R. M. Jenevin, "Scaleability of a binary tree on a hypercube," *Int. Conf. Parallel Processing*, (1986) pp. 661-668.

[13] S. L. Johnson, and C. T. Ho, "Optimum Broadcasting and Personalized Communication in Hypercubes," *IEEE Trans. Comput.*, (September 1989), pp. 1249-1268.

[14] S. L. Johnson, "Communication efficient basic linear algebra computations on hypercube architectures," *J. Parallel Dist. Comput.*, (April 1987), pp. 133-172.

[15] J. S. Jwo, S. Lakshmivarahan, and S. K. Dhall, "Characterization of Node-Disjoint (Parallel) Paths in Star Graphs," *5th intl. Parallel Processing Symp.*, (April 1991), pp. 404-409.

[16] J. S. Jwo, S. Lakshmivarahan, and S. K. Dhall, "Embedding of Cycles and Grids in Star Graphs," *Journal of Circuits, Systems, and Computers*, (No. 1, 1991), pp. 43-74.

[17] A. Menn, and A. K. Somani "An Efficient Sorting Algorithm for the Star Graph Interconnection Network," *Intl. Conf. Parallel Processing*, (1990) pp. 1-8.

[18] M. Nigam, S. Sahni, and B. Krishnamurthy, "Embedding Hamiltonians and Hypercubes in Star Interconnection Graphs," *Intl. Conf. Parallel Processing*, (1990) pp. 340-343.

[19] M. C. Pease, "The indirect Binary n-Cube Microprocessor Array," *IEEE Trans. Comput.*, (1977), pp. 458-473.

[20] F. P. Preparata, and J. Vuillemin, "The Cube-Connected Cycles: A Versatile Network for Parallel Computation," *CACM 24(5)*, (1981), pp. 300-309.

[21] K. Qiu, H. Meijer, and S. Akl, "Decomposing a Star Graph into Disjoint Cycles," *Information Processing Letters* (No. 39, 1991), pp. 125-129.

[22] S. Ranka, J. Wang, and N. Yeh, "Embedding Meshes on the Star Graph," *Intl. Conf. on Supercomputing*, (1990), pp. 476-485.

[23] Y. Saad, and M. H. Schultz, "Topological Properties of Hypercubes," *IEEE Trans. Comput.*, (July 1988), pp. 867-872.

[24] D. S. Scott, and J. Brandenburg, "Minimal Mesh Embeddings in Binary Hypercubes," *IEEE Trans. Comput.*, (October 1988), pp. 1284-1285.

[25] H. S. Stone, "Parallel Processing with the Perfect Shuffle," *IEEE Trans. Coumput.*, (C-20,1971), pp.153-161.

[26] A. Y. Wu, "Embedding of tree networks in hypercubes," *J. Parallel Dist. Comput.*, (1985), pp. 238-249.

ON EMBEDDINGS OF RECTANGLES INTO OPTIMAL SQUARES

Shou-Hsuan S. Huang, Hongfei Liu and Rakesh M. Verma*

Department of Computer Science, University of Houston, Houston TX 77204.

E-mail: rmverma@cs.uh.edu, s_huang@cs.uh.edu

Abstract – Let $G = h \times w$ be a rectangular grid and $H = s \times s$ be the optimal square grid for G, i.e., the least square grid which is no less than G in size. In this paper, an embedding scheme is presented for embedding G into H such that the dilation cost is at most 6. The significance of this result is that optimal expansion is always achieved, regardless of the aspect ratio of the rectangle, while keeping the dilation constant.

Key words: Embedding, grid, optimal, square, constant-dilation.

Introduction

Many important problems in computer science, such as VLSI circuit layout, simulating one parallel computer by another, scheduling in a distributed processing system and simulating one data structure by another, can be reduced to the problems of graph embeddings. Formally, an *embedding* of a graph $G = < V_G, E_G >$ into a graph $H = < V_H, E_H >$ is a one-to-one mapping $\phi : V_G \to V_H$. The quality of the embedding is measured by two parameters, the dilation and the expansion. The *dilation* of ϕ, denoted by $\delta(\phi)$, is defined as $\delta(\phi) = \max\{dist(\phi(u), \phi(v)) | (u, v) \in E_G\}$, where $dist(a, b)$ denotes the shortest path length between the nodes a and b in H. The *expansion* of ϕ, denoted by $\epsilon(\phi)$, is defined as $\epsilon(\phi) = |V_H| / |V_G|$. In this paper, graph G is a rectangular grid and graph H is the *optimal square grid* of G, i.e., the smallest square grid with at least as many grid points as in G.

Rectangular grid embedding has received much attention due to its fundamental importance. However, despite the attempts of many researchers, the important problem of embedding rectangles into *optimal* squares with constant dilation has remained open. We summarize below the previous results on this problem. Early attempts (for example, see Leiserson [4]) to embed rectangles into squares incurred an unbounded increase in the expansion, or an unbounded increase in the dilation. In 1982, Aleliunas and Rosenberg [1] introduced the ideas of line-compression and folding,

and produced embeddings with expansion cost 1.2 and dilation cost 15 (when the height of the rectangle is at least 25). For some ranges of the aspect ratio, better dilation is possible but with more expansion cost. In 1990, Melhem and Hwang [5] improved the results of embedding rectangles into squares by reducing dilation cost to 2 but keeping the expansion cost at 1.2 (when the height of the rectangle is at least 18). Independently, Ellis [2] gave a dilation-3 embedding into *near-optimal square* (with side length of the optimal side plus one).

In this paper we settle the long-standing open problem of constant-dilation embeddings of rectangles into optimal squares by giving a dilation 6 embedding method. Our approach combines two techniques: modified compression and modified folding. We use a modified form of compression to first compress the given rectangular grid G into one of its ideal rectangular grids G', where G' is chosen so that its width is a multiple of the side of the optimal square grid, H, for G and the compression ratio from G to G' is less than two. The modified compression stage is totally combinatorial and involves the construction of a (1,2) embedding matrix. We show that our embedding matrix gives a dilation 3 embedding for modified compression. Finally, we use a modified folding approach to fold G' into H. Folding introduces dilation 2 and hence the dilation for the entire process is $2 \times 3 = 6$.

The rest of the paper is organized as follows. In Section 2, we briefly review our matrix embedding scheme for a rectangular embedding method introduced by us earlier [3]. Section 3 gives the main result of embedding any rectangular grid into its optimal square grid. We conclude with remarks in Section 4.

Embedding matrices

Given a rectangular grid $G = h \times w$, without loss of generality, we will assume that $h < w$ throughout this paper. For a given $w' < w$ we say that $G' = h' \times w'$ is an *ideal rectangular grid* of G, if h' is the smallest integer such that $hw \leq h'w'$. The embedding of G into G' can be represented by an embedding ma-

*Partially supported by NSF CCR-9010366

trix $C_{h \times w'}$ of nonnegative integers. Each row of C denotes the distribution of the corresponding row of G inside G'. The connection pattern for each row can be arbitrary but we choose one to yield minimum dilation. On the other hand, given a matrix $C = (c_{ij})$ of size $h \times w'$, where $c_{ij} > 0$, each row sum equals w and each column sum is at most h', we have an embedding of $G = h \times w$ into $G' = h' \times w'$ by choosing the *snake-like* interconnection pattern, which will be used throughout this paper. Formally, let (i, j) denote the node in the ith row and j^{th} column of rectangular grid G, (k, l) denotes its image in G'. The embedding function is defined as follows: $\phi(i, j) = (k, l)$ where $l = \min\{\lambda : \sum_{s=1}^{\lambda} c_{is} \geq j\}$, and

$$k = \sum_{r=1}^{i-1} c_{rl} + \begin{cases} j - \sum_{s=1}^{l-1} c_{is} & \text{if } l \text{ is odd,} \\ 1 + \sum_{s=1}^{l} c_{is} - j & \text{if } l \text{ is even.} \end{cases}$$

Figure 1 shows the embedding matrix generated by the method in [3] for embedding a 5×7 grid into a 7×5 grid with dilation 2. This is used by us only when the compression ratio is no more than 3. For higher compression ratios we generate the embedding matrix as described below.

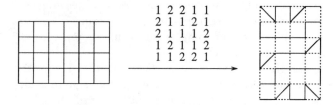

$$\begin{array}{ccccc} 1 & 2 & 2 & 1 & 1 \\ 2 & 1 & 1 & 2 & 1 \\ 2 & 1 & 1 & 1 & 2 \\ 1 & 2 & 1 & 1 & 2 \\ 1 & 1 & 2 & 2 & 1 \end{array}$$

Figure 1: Embedding of grid 5×7 into grid 7×5.

Suppose the embedding matrix C satisfies the following conditions:

(P1) $c_{ij} \in \{\lfloor \eta \rfloor, \lceil \eta \rceil\}$ for all $1 \leq i \leq h, 1 \leq j \leq w'$, where $\eta = w/w'$ is the *compression ratio*,

(P2) $c_{i+1,j+1} = c_{ij}$ for all $1 \leq i < h, 1 \leq j < w'$,

(P3) $\sum_{j=1}^{w'} c_{ij} = w$ for all $1 \leq i \leq h$,

(P4) $\sum_{i=1}^{h} c_{ij} \leq h'$ for all $1 \leq j \leq w'$,

(P5) Any two adjacent equal-length partial row sums differ by at most 1, i.e.,
$|\sum_{j=1}^{k} c_{ij} - \sum_{j=1}^{k} c_{i+1,j}| \leq 1$ for $1 \leq i < h$ and $1 \leq k \leq w'$,

(P6) Any two adjacent equal-length partial column sums differ by at most 1, i.e.,
$|\sum_{i=1}^{k} c_{ij} - \sum_{i=1}^{k} c_{i,j+1}| \leq 1$ for $1 \leq j < w'$ and $1 \leq k \leq h$.

Then, we can prove the following general theorem on rectangular grid embeddings.

Theorem 1. *Any embedding matrix $C = (c_{ij})$ satisfying (P1)-(P6) defines an embedding from $G = h \times w$ into $G' = h' \times w'$ with dilation at most $\lceil \eta \rceil + 1$.*

Proof. Consider two adjacent nodes (i, j) and (i', j') in G, they are either horizontally adjacent or vertically adjacent. We prove, by case analysis using definition of ϕ and the properties of C, that their images $\phi(i, j)$ and $\phi(i', j')$ are at most $\lceil \eta \rceil + 1$ edges away in G'. Let $\delta = |k - k'| + |l - l'|$.

Case A. Nodes are horizontally adjacent, $(i', j') = (i, j+1)$. Let $\phi(i, j) = (k, l)$ and $\phi(i, j+1) = (k', l')$. It is easy to see that $0 \leq l' - l \leq 1$.
(a) if $l = l'$, then $\delta = 1 < \lceil \eta \rceil + 1$. Because of the way ϕ is defined, if two horizontally adjacent nodes are mapped to the same column then they are vertically adjacent in the host grid G'.
(b) if $l' = l + 1$, then $\sum_{s=1}^{l} c_{is} = j$, (otherwise, if $\sum_{s=1}^{l} c_{is} > j$ then $\sum_{s=1}^{l} c_{is} \geq j+1$, hence $l' \leq l$) and

$$k = \sum_{r=1}^{i-1} c_{rl} + \begin{cases} c_{il} & \text{for } l \text{ odd,} \\ 1 & \text{for } l \text{ even,} \end{cases}$$

$$k' = \sum_{r=1}^{i-1} c_{r,l+1} + \begin{cases} 1 & \text{for } l \text{ even,} \\ c_{i,l+1} & \text{for } l \text{ odd,} \end{cases}$$

By (P6), we have $|k - k'| \leq 1$, hence $\delta \leq 2 \leq \lceil \eta \rceil + 1$.
Case B. Nodes are vertically adjacent, $(i', j') = (i+1, j)$. Let $\phi(i, j) = (k, l)$ and $\phi(i+1, j) = (k', l')$. It is easy to see that $-1 \leq l' - l \leq 1$.
(a) if $l' = l$, then

$$k' = \sum_{r=1}^{i} c_{rl} + \begin{cases} j - \sum_{s=1}^{l-1} c_{i+1,s} & \text{for } l \text{ odd,} \\ 1 + \sum_{s=1}^{l} c_{i+1,s} - j & \text{for } l \text{ even.} \end{cases}$$

By (P1) and (P5), we have $k' - k$

$$= c_{il} + \begin{cases} \sum_{s=1}^{l-1} c_{is} - \sum_{s=1}^{l-1} c_{i+1,s} & \text{for } l \text{ odd,} \\ \sum_{s=1}^{l} c_{i+1,s} - \sum_{s=1}^{l} c_{is} & \text{for } l \text{ even,} \end{cases}$$

$\leq \lceil \eta \rceil + 1$, hence $\delta \leq \lceil \eta \rceil + 1$.
(b) if $l' = l + 1$, by (P5) and the definition of ϕ, we must have $\sum_{s=1}^{l} c_{is} = j$ and $\sum_{s=1}^{l} c_{i+1,s} = j - 1$. Hence

$$k = \sum_{r=1}^{i-1} c_{rl} + \begin{cases} c_{il} & \text{for } l \text{ odd,} \\ 1 & \text{for } l \text{ even,} \end{cases}$$

$$k' = \sum_{r=1}^{i} c_{r,l+1} + \begin{cases} 1 & \text{for } l \text{ even,} \\ c_{i+1,l+1} & \text{for } l \text{ odd,} \end{cases}$$

Therefore, $k' - k =$

$$\sum_{r=1}^{i} c_{r,l+1} - \sum_{r=1}^{i-1} c_{rl} + \begin{cases} c_{i+1,l+1} - c_{il} & \text{for } l \text{ odd,} \\ 0 & \text{for } l \text{ even,} \end{cases}$$

$= c_{1,l+1}$ by (P2), which is at most $\lceil \eta \rceil$ by (P1). Hence $\delta \leq \lceil \eta \rceil + 1$.

(c) Similar to (b), omitted to save space.

Squaring up rectangles

Given an arbitrary rectangular grid, $G = h \times w$, its optimal square grid is $H = s \times s$, where $s = \lceil \sqrt{hw} \rceil$. The objective here is to embed G into H with dilation as small as possible. Simply using the folding technique in [1] we can obtain a dilation-2 embedding, but it only works for some input instances. By using the pure compression process in [3] we can obtain a dilation-η embedding, but recall that $\eta = \lceil w/s \rceil$, which can be very large.

Let $w = \rho h$, if $\rho \leq 9$, then the compression ratio from G to H $\eta = w/s = \rho h / \lceil \sqrt{\rho h^2} \rceil \leq \rho h / \sqrt{\rho h^2} = \sqrt{\rho} \leq 3$. Therefore, using the pure compression technique just mentioned we can embed G into H with dilation no more than 3. Hence we will focus on those grids whose aspect ratio (or, eccentricity), ρ, is greater than 9. The technique of folding followed by compression is used by Ellis [2] to get dilation cost 3, but the square cannot be guaranteed to be optimal. We can show that simple compression followed by folding also does not yield embeddings into optimal squares. Thus, it is clear that direct applications of earlier approaches are not likely to generate constant-dilation embeddings into optimal squares always and we need a different approach. Therefore, we develop a modified compression technique, and a variation of folding such that the combination of these two techniques yields constant-dilation embeddings into optimal squares. We first illustrate our approach through an example before formulating the general algorithm.

Consider embedding grid $G = 3 \times 61$ into its optimal square $H = 14 \times 14$. We first compress it into grid $G' = 4 \times 56$ (one of G's ideal rectangular grids) in such a way that all the fully-used (4 nodes) columns of G' are on one end (say left), and all the columns with 3 nodes used are on the other end and all the unused nodes are in the last row (Figure 2).

Then we fold grid G' into H so that the first 3 rows are completely folded while the last row is partially folded, in this example just enough to cover the first 15 used nodes (of the last row). There are 13 extra nodes in H. The compression process can be captured by the embedding matrix in Figure 2. It can be obtained by initializing a 3×56 matrix to $1's$ and then putting a 2 in the top left corner and moving it diagonally and

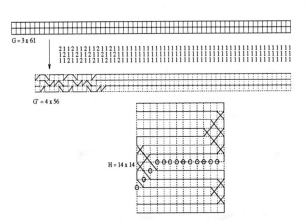

Figure 2: Embedding of grid 3×61 into grid 14×14.

circularly. We stop when we have 15 $2's$. The dilation from G to G' is 2. The dilation from G' to H is also 2. Hence we have a dilation-4 embedding of G into H for this example.

In general, let $k = \lfloor w/s \rfloor$ (if $w = ks$ or $(k+1)h \leq s$ then the simple folding scheme will give a dilation-2 result), and $w' = ks$. Then we have $h' = \lceil \frac{hw}{w'} \rceil$. Further, let $r = hw \bmod w'$, $a = h' - h$. Our objective is to generate an embedding matrix of size $h \times w'$ such that it embeds G into G' in such a way that the leftmost r columns of G' are fully used and the rightmost $w' - r$ columns are all left with the bottom (or top) nodes unused. We first show that if $\rho > 9$, then the compression ratio from G to G' $\eta = w/w' < 2$.

Lemma 1. *If $\rho > 9$, then the compression ratio η from $G = h \times \rho h$ to $G' = h' \times w'$ is less than 2.*

Proof. Clearly we can assume $h > 1$. Now for $\rho \geq 6.25$ we can derive

$$\begin{aligned}
\eta &= \frac{w}{w'} = \frac{\rho h}{ks} = \frac{\rho h}{\left\lfloor \frac{hw}{\lceil h\sqrt{\rho} \rceil} \right\rfloor \lceil h\sqrt{\rho} \rceil} \\
&\leq \frac{\rho h}{\left\lfloor \frac{h\rho}{\lceil h\sqrt{\rho} \rceil} \right\rfloor h\sqrt{\rho}} = \frac{\sqrt{\rho}}{\left\lfloor \frac{h\rho}{\lceil h\sqrt{\rho} \rceil} \right\rfloor} \\
&\leq \frac{\sqrt{\rho}}{\left\lfloor \frac{h\rho}{h\sqrt{\rho}+1} \right\rfloor} = \frac{\sqrt{\rho}}{\left\lfloor \frac{\rho}{\sqrt{\rho}+1/h} \right\rfloor} \\
&\leq \frac{\sqrt{\rho}}{\left\lfloor \frac{\rho - 1/4}{\sqrt{\rho}+1/2} \right\rfloor} \leq \frac{\sqrt{\rho}}{\left\lfloor \sqrt{\rho} - 1/2 \right\rfloor} \\
&< 2.
\end{aligned}$$

Therefore, the entries of the embedding matrix can be restricted to either 1 or 2. This embedding matrix should have $w - w'$ $2's$ and $2w' - w$ $1's$ in each row, we want it to have the left most r columns with column sum of h' and the right most $w' - r$ columns with column sum $h' - 1$. We use the following algorithm to generate the embedding matrix:

1. $\forall\ 1 \le i \le h$ and $1 \le j \le w'$, let $c_{ij} \leftarrow 1$;
2. $i \leftarrow 1$; $j \leftarrow 1$; counter$\leftarrow 0$;
    repeat
        if $c_{ij} = 2$ then
            repeat
                $i \leftarrow (i \bmod h) + 1$;
            until $c_{ij} <> 2$;
        else
            $c_{ij} \leftarrow 2$; counter $\leftarrow$ counter+1;
            $i \leftarrow (i \bmod h) + 1$; $j \leftarrow (j \bmod w') + 1$;
    until counter $= h(w - w')$.

It is not difficult to see that the above algorithm generates an embedding matrix C of size $h \times w'$ with exactly $w - w'$ $2's$ in each row, exactly a $2's$ in each of the left r columns and exactly $a - 1$ $2's$ in each of the remaining columns. For example, the embedding matrix for $G = 8 \times 176$ into $G' = 10 \times 152$ is generated by the above algorithm as shown in Figure 3.

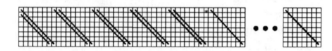

Figure 3: Embedding matrix for 8×176 into grid 10×152.

The matrix C generated by our algorithm satisfies properties (P1-P6) except that there exists at most one pair of (i_0, j_0) such that $c_{i_0+1,j_0+1} = c_{i_0,j_0} - 1$. This happens at the boundary between the rth and $(r+1)^{th}$ columns. We still have Theorem 1 by selecting either the top row or the bottom row if necessary for unused nodes (this means shifting the right $w' - r$ columns of images in G' down or leaving them as they are). Hence this embedding matrix defines an embedding from $G = h \times w$ into $G' = h' \times w'$ with dilation at most 3. Therefore, we have the following theorem:

Theorem 2. *Any matrix* $G = h \times w$, *where* $w = \rho h$ *($\rho > 9$), can be embedded into its optimal square grid* $H = s \times s$ *(where* $s = \lceil \sqrt{\rho} h \rceil$) *with dilation at most 6.*

Proof. Using the above notation, if $\rho \le 9$, then we embed G into H directly using the method in [3] to obtain a dilation cost of 3 (because the compression ratio from G to H is at most 3). If $\rho > 9$, then we first embed G into $G' = h' \times w'$ by the embedding matrix C generated by the algorithm. The dilation of this process is at most 3 by Theorem 1 (η from G to G' is less than 2). Then we fold G' into H so that we start folding all the rows of G' and drop the last row as soon as we cover the first ms (m is the smallest integer such that $ms \ge r$) nodes of the last row (ref. the example in Figure 1). Since $w' = ks$ is a multiple

of s, the folding will fit properly. Since dilation from G' to H is 2, the dilation from G to H is at most $2 \times 3 = 6$.

Conclusions and open problems

In this paper, we have presented an approach to embed rectangles into their optimal squares with dilation at most 6. No such result exists in the previous literature. Our approach can also be used for other embedding problems which are based on embedding matrices. All the known results on embedding rectangular grids into squares are summarized below in Table 1. It seems possible, although likely to be difficult, to reduce the dilation slightly. It also appears likely that dilation 2 can be achieved with expansion significantly smaller than 1.2, thus improving previous results.

Table 1: Progress on embedding rectangles into optimal squares.

Reference	Expansion	Dilation	Restriction
[1] '82	1.2	15	$h \ge 25$
[5] '90	1.2	2	$h \ge 18$
[2] '91	near-optimal	3	none
This paper	optimal	6	none

Acknowledgements: The authors thank the reviewers for their comments.

References

[1] Romas Aleliunas and Arnold L. Rosenberg, "On Embedding Rectangular Grids in Square Grids," *IEEE Trans. on Computers*, 31:9, pp. 907-913, 1982.

[2] John A. Ellis, "Embedding Rectangular Grids into Square Grids," *IEEE Trans. on Computers*, 40:1, pp. 46-52, 1991.

[3] Shou-Hsuan S. Huang, Hongfei Liu and Rakesh M. Verma, "A New Combinatorial Approach to Optimal Embeddings of Rectangles," Tech. Rep. UH-CS-92-24, University of Houston, Nov. 1992.

[4] C. E. Leiserson, "Area-efficient graph layouts (for VLSI)," in *Proc. 21st IEEE Symp. on Foundations of Computer Science*, 1980, pp. 270-281.

[5] R. G. Melhem and G. Y. Huang, "Embedding Rectangular Grids into Square Grids with Dilation Two," *IEEE Trans. on Computers*, 39:12, pp. 1446-1455, 1990.

Finding Articulation Points and Bridges of Permutation Graphs*

Oscar H. Ibarra and Qi Zheng

Department of Computer Science
University of California
Santa Barbara, CA 93106
ibarra@cs.ucsb.edu
zheng@cs.ucsb.edu

Abstract

We show that articulation points and bridges of permutation graphs can be found in $O(\log n)$ time using $O(n/\log n)$ processors on an EREW PRAM. The algorithms are optimal with respect to the time-processor product.

Keywords: parallel algorithm, EREW PRAM, articulation point, bridge, permutation graph

1. Introduction

Let $G = (V, E)$ be a connected graph with a set of vertices $V = \{1, 2, \cdots, n\}$ and a set of edges $E \subseteq V^2$. A vertex v in V is an *articulation point* if the deletion of v together with all the edges incident to it disconnects the graph into at least two nonempty components. An edge e is a *bridge* if its deletion results in G having two disconnected nonempty components.

Efficient parallel algorithms for finding articulation points and bridges in general graphs were reported in [9, 10]. Ramkumar and Rangan developed EREW PRAM parallel algorithms to find articulation points and bridges of interval graphs [7]. Using the traditional Depth-First-Spanning (DFS) tree method, the algorithms run in $O(\log n)$ time with $O(n^2/\log n)$ processors. Later, Sprague and Kulkarni proposed a non-DSF approach and showed that the problems for interval graphs can be solved in $O(\log n)$ time using $O(n/\log n)$ processors on an EREW PRAM [8]. In this paper, we show how to find articulation points and bridges in a different kind of graphs, called *permutation* graphs.

A graph $G = (V, E)$ is a *permutation graph* iff there exists a permutation function π on vertex set $V = \{1, 2, \cdots, n\}$ such that $(u - v)(\pi(v) - \pi(v)) < 0$ iff (u, v) is an edge in E. Given a graph G with n vertices, deciding whether G is a permutation graph and constructing a corresponding permutation function π can be done in $O(n^2)$ time [3, 6]. Permutation graphs are an important subclass of perfect graphs. They are used for modeling practical problems in many areas such as biology, genetics, VLSI design and network planning. Many algorithms have been developed for a variety of combinatorial optimization problems for permutation graphs [1, 2, 3].

Using some special properties of permutation graphs, we give an efficient way to characterize articulation points and bridges. Based on the new characterization, we develop parallel algorithms for finding articulation points and bridges of permutation graphs. The algorithms run in $O(\log n)$ time using $O(n/\log n)$ processors on an EREW PRAM. The $O(n)$ time-processor product is clearly optimal.

* Research supported in part by NSF Grant CCR89-18409 and a grant from the Institute for Scientific Computing Research - Lawrence Livermore National Laboratory.

In the next section, we give some preliminary observations, notations and the preprocessing steps. Section 3 presents the algorithm for finding articulation points, and Section 4 gives the algorithm for bridges.

2. Preliminary

Let $G = (V, E)$ be a permutation graph with $V = \{1, 2, \cdots, n\}$ and π be the permutation function of G. Permutation graphs have a nice geometric representation. Draw two parallel rows, each labeled with the integers $\{1, 2, \cdots, n\}$ in order. The lower row represents the vertex set V while the upper row represents the values of the permutation function π. For each $u \in V$ on the lower row draw a straight line segment connecting u with $\pi(u)$ on the upper row. In the geometric representation, the lines are vertices and (u, v) is an edge if the line segment for u crosses the line segment for v.

Geometric representations of permutation graphs have been extensively used to design efficient algorithms for various computational problems for permutation graphs [1, 2]. We will use the geometric representation and assume that the permutation function π is given as input.

Given a geometric representation P of a permutation graph G, deleting a vertex v and all the edges associated with v from G is the same as deleting the line segment v from P. A vertex v is called a *positive* vertex if $\pi(v) < v$. It is *neutral* if $\pi(v) = v$ and *negative* if $\pi(v) > v$.

Lemma 2.1. For any connected permutation graph G, no neutral vertex can be an articulation point.

Proof. Suppose a neutral vertex j is an articulation point. Since G is connected and j is neutral, there must exist a pair of vertices i and k such that $i < j < k$ and $\pi(k) < \pi(j) < \pi(i)$. Since j is an articulation point, removing j from G will result in a permutation graph G' with at least two disconnected nonempty components. Since line segment i crosses line segment k, $(i, k) \in E$. Because edge (i, k) is not removed, vertex i and k must be in the same component, say, C. In fact, every vertex w with $i < w < k$ and $w \neq j$ is in the component C. Suppose u is a vertex which is not in C and $u < i$. Since G is a connected graph, there exists a path from u to k in G. The path must go through the articulation point j. For any such path there is a vertex v in the path such that $(v, i) \in E$. Hence, there exists a path from u to k in G' and u is in C too. This contradicts the assumption that u is not in C. Similarly, we can prove that all vertices u with $u > k$ are in C. Therefore, G' is a connected graph and j is not an articulation point. $\square$

From Lemma 2.1, we need only check whether a positive or negative vertex is an articulation point. We shall call such vertices positive or negative articulation points. Because of symmetry in the geometric representation, the technique for finding the negative articulation points is similar to that of finding the positive articulation points. Hence, we only give the details of finding the positive articulation points.

Assume that v is a positive articulation point in a given permutation graph G with geometric representation P. By removing v, we have at least two disconnected nonempty components C_1 and C_2. Both components are permutation graphs. Since C_1 and C_2 are not connected, their geometric representation, say P_1 and P_2, do not intersect with each other. In the geometric representation P, the line segment $(v, \pi(v))$ has it v end in P_2 and $\pi(v)$ end in P_1. The two parts, P_1 and P_2, are connected through the single line v. If we can identify two nonempty components connected only through a single line segment then we can find a positive articulation point. Before we go further, we need following notation.

Definition 2.1. Given a permutation graph G and its permutation function π, functions $lv(u)$ and $lp(v)$ for $u \in V = \{1, 2, \cdots, n\}$ are defined as follows:

$$lv(u) = \max(\{u\} \cup \{v \mid (u, v) \in E\});$$

$lp(u) = w$ such that
$$\pi(w) = \max(\{\pi(u)\} \cup \{\pi(v) \mid (u, v) \in E\}).$$

In other words, $lv(u)$ is the vertex with the largest label among the vertices adjacent to vertex u, including u itself. Similarly, $lp(u)$ is the vertex with the largest permutation value among the vertices adjacent to vertex u, including u itself.

One can prove the following lemma.

Lemma 2.2. Let G be a permutation graph and π its permutation function. For all $u \in V$, $lv(u)$ and $lp(u)$ can be found in $O(\log n)$ time using $O(n/\log n)$ processors on an EREW PRAM.

Definition 2.2. Let G be a permutation graph and π its permutation function. For an integer $i \in \{1, 2, \cdots, n\}$, a vertex $v \leq i$ is *unmatched with respect to i* if $\pi(v) > i$. A π point $p \leq i$ is *unmatched with respect to i* if $\pi^{-1}(p) > i$.

Let $A(i)$ be the number of unmatched vertices with respect to i and $B(i)$ be the number of unmatched π points with respect to i, for $1 \leq i \leq n$. We show how to use the prefix sum method to compute $A(i)$ and $B(i)$ for $1 \leq i \leq n$.

For any vertex v in $V = \{1, 2, \cdots, n\}$, let

$$A_1(v) = \begin{cases} 1 & \text{if } \pi(v) > v \\ 0 & \text{otherwise} \end{cases}$$

$$B_1(v) = \begin{cases} -1 & \text{if } \pi(v) < v \\ 0 & \text{otherwise} \end{cases}$$

For any π point p in $\{1, 2, \cdots, n\}$, let

$$A_2(p) = \begin{cases} -1 & \text{if } \pi^{-1}(p) < p \\ 0 & \text{otherwise} \end{cases}$$

$$B_2(p) = \begin{cases} 1 & \text{if } \pi^{-1}(p) > p \\ 0 & \text{otherwise} \end{cases}$$

We claim that $A(i) = \sum_{j=1}^{i}(A_1(j) + A_2(j))$ and $B(i) = \sum_{j=1}^{i}(B_1(j) + B_2(j))$ for $1 \leq i \leq n$. The proof is straightforward.

Lemma 2.3. Let X be an array of elements and *prefix sum* be the problem of computing all the partial sums $S(i) = X(1) \oplus X(2) \oplus \cdots, \oplus X(i)$, where $\oplus$ is an associative operation. The prefix sum problem can be solved in $O(\log n)$ time using $O(n/\log n)$ processors on an EREW PRAM [4, 5].

Lemma 2.4 Given a permutation graph G with n vertices and its π function, all $A(i)$ and $B(i)$ can be computed in $O(\log n)$ time using $O(n/\log n)$ processors on an EREW PRAM.

Proof. Each element in arrays A_1, A_2, B_1 and B_2 only depends on local information, say v and $\pi(v)$ or p and $\pi^{-1}(p)$. They can be computed in $O(\log n)$ time with $O(n/\log n)$ processors. $A(i)$ and $B(i)$ are just prefix sum of $A_1 + A_2$ and $B_1 + B_2$. The rest of proof follows Lemma 2.3. □

3. Articulation points

In this section, we present a parallel algorithm **Art** for finding the articulation points. The input to the algorithm is a permutation function π of a connected permutation graph G with n vertices. The algorithm runs in $O(\log n)$ time using $O(n/\log n)$ processors on an EREW PRAM. First, we prove two lemmas.

Lemma 3.1. A positive vertex $v \in \{1, 2, \cdots, n\}$ is an articulation point of G iff there exists an integer i, $1 < i < n$, such that:

(1) $B(i) = 1$ and $A(i-1) = 1$,
(2) there is a vertex $u < i$ with $\pi(u) = i$ and
(3) $lv(u) = v$.

Proof. If there is a positive articulation point v, removing line segment v from the geometric representation P creates at least two nonempty subparts of P. Let P_1 be the part that contains vertex 1 and has $i-1$ line segments. That is the same as saying that the nonempty subgraph C_1 (corresponding to P_1) has $i-1$ vertices and contains vertex 1. The $i-1$ vertices must be the set of vertices $\{1, 2, \cdots, i-1\}$. Since each subpart of P is nonempty, we have $1 < i < n$. Further, P_1 resulted by removing line segment of a positive vertex v from the connected graph G. Therefore, $v > i$ and $\pi(v) < i$. The maximum π value for line segments in P_1 cannot be less than i; otherwise, the pigeon-hole principle is violated because there are $i-1$ lines to fit into $i-2$ π values (Note that line v has a π value less than i). The maximum π value for line segments in P_1 cannot be larger than i; otherwise, there exists at least one vertex k, where $k \neq v$ and $k > i-1$, such that $\pi(k) \leq i$. Hence, P_1 has more than $i-1$ lines. This is a contradiction. So, the maximum π value for line segments in P_1 is exactly i. Vertex v is the only unmatched π point with respect to i, i.e., $B(i) = 1$. Since there is a line segment in P_1 that has π value i, we have $A(i-1) \geq 1$. If $A(i-1) > 1$, then there are at least two line segments in P_1 that have π values larger than $i-1$ which cannot be true because the maximum π value for line segments in P_1 is i. Since the π point i belongs to a vertex $u \leq i-1$, condition (2) of the lemma is true. It is easy to see that $lv(u) = v$.

If the three conditions in Lemma 3.1 are true, the first $i-1$ vertices have π values in $\{1, 2, \cdots, i\}$. By removing $lv(u)$, the first $i-1$ vertices will be disconnected from rest of graph G. Therefore vertex $lv(u)$ is a positive articulation point of G. □

Lemma 3.2. A negative vertex $v \in \{1, 2, \cdots, n\}$ is an articulation point of G iff there exists an integer i, $1 < i < n$, such that:

(1) $A(i) = 1$ and $B(i-1) = 1$,
(2) vertex i has $\pi(i) < i$ and
(3) $lp(i) = v$.

Proof. Similar to the proof of Lemma 3.1. □

The outline of algorithm **Art** is given below:

Algorithm Art:

Input: Permutation function π of a permutation graph G with n vertices.

Output: Articulation points of G stored in arrays $Point_1$ and $Point_2$.

begin

/*Preprocessing steps */
(1) $Point_1(i) = Point_2(i) = 0$ for $1 \le i \le n$.
(2) Compute $lv(i)$ and $lp(i)$ for $1 \le i \le n$.
(3) Compute $A_1(i)$, $A_2(i)$, $B_1(i)$ and $B_2(i)$ for $1 \le i \le n$.
(4) Compute $A(i)$ and $B(i)$ for $1 \le i \le n$.
/* Finding positive articulation points */
(5) For each $i \in \{2, 3, \cdots, n-1\}$
    if $(B(i) = 1$ and $A(i) = 1$ and $\pi^{-1}(i) < i)$
    then $Point_1(i) = lv(\pi^{-1}(i))$
/* Finding negative articulation points */
(6) For each $i \in \{2, 3, \cdots, n-1\}$
    if $(A(i) = 1$ and $B(i-1) = 1$ and $\pi(i) < i)$
    then $Point_2(i) = lp(i)$
/* Post processing step */
(7) Compress arrays $Point_1$ and $Point_2$.

end.

Steps (1) - (4) are preprocessing steps described in Section 2, and they take $O(\log n)$ time using $O(n/\log n)$ processors. Steps (5) and (6) find the articulation points as described in the proof of Lemmas 3.1 and 3.2. With $O(n)$ processors, both steps can be done in constant time. With $O(n/\log n)$ processors, $O(\log n)$ time is sufficient. After step (6), arrays $Point_1$ and $Point_2$ may have many empty entries (i.e., 0) and repetitions of articulation points. Step (7) compresses arrays $Point_1$ and $Point_2$ to eliminate the empty entries and repetitions. Since articulation points are in a non-decreasing order in arrays $Point_1$ and $Point_2$, parallel prefix can be used to compress the two arrays [4, 8]. Step (7) also uses $O(\log n)$ time and $O(n/\log n)$ processors. After compressing, array $Point_1$ is a list of positive articulation points and array $Point_2$ is a list of negative articulation points of graph G. Hence, we have the following theorem.

Theorem 3.1. The problem of finding articulation points in permutation graphs can be solved in $O(\log n)$ time using $O(n/\log n)$ processors on an EREW PRAM.

4. Bridges

While removing an articulation point may create more than two nonempty disconnected components, removing a bridge creates exactly two nonempty disconnected components. A bridge is called a *simple bridge* if removing the bridge results in one of the components having only a single vertex. To find the simple bridges in a graph G, we only need to find the vertices in G with edge degree one and the edge attached to them. For permutation graphs, simple bridges can be found by using modified lv and lp functions which were given in Section 2. Modify $lv(i)$, for all i, such that $lv(i) = (u, v)$ where u is the largest vertex adjacent to i and v is the second largest vertex adjacent to i. Similarly, let the modified $lp(i) = (x, y)$ for all i, where x and y are the vertices adjacent to i such that x has the largest π value and y has the second largest π value. Therefore, the modified $lv(i)$ and $lp(i)$ points to four adjacent vertices of i if they exist. For a vertex with only one edge, three of the four pointers are empty and the nonempty one corresponds to a simple bridge. It is not difficult to see that the modified lv and lp can be computed in $O(\log n)$ time with $O(n/\log n)$ processors by using the parallel prefix sum operations. Hence, simple

bridges of permutation graphs can be found in $O(\log n)$ time using $O(n/\log n)$ processors.

An edge (u, v) is a non-simple bridge iff removing (u, v) results in two nonempty disconnected components, and each of the two components has at least two vertices. Note that after the removal of (u, v), vertex 1 and vertex n are always in different components. For any non-simple bridge (u, v), both u and v must be articulation points. Now we reduced the problem of finding non-simple bridges from all edges of G to only those edges between articulation points. Even so, there may exist $O(n)$ such edges for each articulation point. Thus, there may be $O(n^2)$ candidates for non-simple bridges. Fortunately, for permutation graphs, we can limit the number of candidates for non-simple bridges to $O(n)$. In other words, for each articulation point v, we only need check a constant number of edges attached to v to find the possible non-simple bridges.

Lemma 4.1. For any connected permutation graph G, no positive articulation point i is adjacent to any vertex j with $i < j$. Similarly, no negative articulation point i is adjacent to any vertex j with $j < i$.

Proof. Let i be a positive articulation point of G. If $(i, j) \in E$ and $i < j$, then $\pi(j) < \pi(i) < i$. Since G is connected, there must exist a vertex $u < i$ with $\pi(u) > \pi(i)$. Vertex u is adjacent to both i and j. Any vertex adjacent to i must also adjacent to u or j. Thus, all vertices adjacent to i are still connected through edge (u, j) even after removing i. Hence, i is not a positive articulation point which contradicts initial condition. Similarly, we can prove that (i, j) is not in E if i is a negative articulation point and $j < i$. □

Corollary 4.1. No two positive articulation points are adjacent and no two negative articulation points are adjacent.

Corollary 4.1 tells us that any non-simple bridge must be formed by a pair of positive and negative articulation points. Now, our task reduces to checking whether a positive articulation point is adjacent to a negative articulation point and whether the formed edge is a bridge. The next lemma provides a stronger condition for a pair of positive and negative articulation points to be a non-simple bridge.

Lemma 4.2. Let v be a positive articulation point. If an edge (u, v) is a non-simple bridge then u is a negative articulation point and either $\pi(u) = \pi(v) + 1$ or $u = v - 1$.

Proof. If (u, v) is a non-simple bridge, u must be a negative articulation point and $u < v$ by Lemma 4.1 and Corollary 4.1. Since (u, v) is an edge and $u < v$, we have $\pi(u) > \pi(v)$. Suppose that $\pi(u) \ne \pi(v) + 1$ and $u \ne v - 1$. There is a vertex i such that $u < i < v$. For any such i, $\pi(i)$ cannot be in the range of $(\pi(v), \pi(u))$; otherwise, u, i and v form a clique and (u, v) cannot be a bridge. Therefore, either $\pi(i) < \pi(v)$ or $\pi(i) > \pi(u)$. By the same argument, we can prove that $\pi^{-1}(p) < u$ or $\pi^{-1}(p) > v$ for every π point p in the range of $(\pi(v), \pi(u))$.

Assume that there is a vertex i where $u < i < v$ and $\pi(i) < \pi(v)$. For this case, every π point p in the range of $(\pi(v), \pi(u))$ must have $\pi^{-1}(p) > v$; otherwise, if there is a π point p has $\pi^{-1}(p) < u$, then the four vertices $\pi^{-1}(p)$, u, i and v form a cycle, and hence, (u, v) is not a bridge. For the same reason, every vertex i with $u < i < v$ must have $\pi(i) < \pi(v)$ (otherwise, u, i, v and $\pi^{-1}(p)$ form a cycle). Since (u, v) is a non-simple bridge, none of u and v can have edge degree one. There must exist a vertex x adjacent to v and $x \ne u$. Since every vertex i with $u < i < v$ has $\pi(i) < \pi(v)$ no such i is adjacent to v. We must have $x > v$ or $x < u$. If $x > v$, then $\pi(x) < \pi(v)$. This leads to a clique of u, v and x, and edge (u, v) cannot be a bridge. If $x < u$ then $\pi(x)$ must be greater than $\pi(u)$. Again, we have a clique of x, u and v and (u, v) cannot be a bridge. Hence, the assumption that $\pi(i) < \pi(v)$ cannot be true.

Now let us assume that $\pi(i) > \pi(u)$. Similar arguments as above can be used to prove that the assumption is false if (u, v) is a non-simple bridge.

Therefore, any non-simple bridge (u, v) must have $u = v - 1$ or $\pi(u) = \pi(v) + 1$ for a positive articulation point v and a negative articulation point u. □

Lemma 4.2 not only provides a stronger necessary condition for an edge to be a non-simple bridge but also reduces the number of possible non-simple bridges from $O(n)$ to exactly two for each positive articulation points. To find the bridges in a permutation graph, we only need check whether $(v-1, v)$ and $(\pi^{-1}(\pi(v)+1), v)$ are bridges for every positive articulation point v.

Lemma 4.3. Let v be a positive articulation point with $\pi(v) = p$. Let $u = \pi^{-1}(p+1)$. Then edge (u, v) is a non-simple bridge iff u is a negative articulation point and $A(p) = 1$.

Proof. If (u, v) is a non-simple bridge, then u must be a negative articulation point and $u < v$ by Corollary 4.1. By Lemma 4.1, $\pi(x) < \pi(u) = p + 1$ and $\pi(y) > \pi(v) = p$ for every $x < u$ and $y > v$. Since $\pi(u) = p + 1$ and $\pi(v) = p$, we have $\pi(x) < p$ and $\pi(y) > p + 1$ for any such x and y. Since u is a negative articulation point, $u \le p$. If $u = p$, then the given permutation graph is not connected because the set of vertices $\{1, 2, \cdots, u-1\}$ occupies the set of π values $\{1, 2, \cdots, p-1\}$. Hence, $u < p$ and there are $p - u$ number of π values in the set of $\{1, 2, \cdots, p-1\}$ is left for some vertices i with $u < i < v$. Similarly, we can prove that $v > p + 1$ and there $v - p - 1$ values in $\{p+2, p+3, \cdots, n\}$ is left for some vertices j with $u < j < v$. If there are two vertices i and j such that $\pi(i) > p + 1$ and $\pi(j) < p$ where $u < i < j < v$, then the four edges (i, j), (i, v), (u, j) and (u, v) are in E. The four edges form a cycle which means that edge (u, v) cannot be a bridge. Since (u, v) is a bridge, there exists a vertex i such that $\pi(x) < p$ and $\pi(y) > p + 1$ for any $u < x \le i$ and $i < y < v$. Further, we can see that $i = p$. Since every vertex in $\{1, 2, \cdots, p\}$ except vertex u has π value less than p, vertex u is the only unmatched vertex with respect to p. Therefore, $A(p) = 1$.

Now, let (u, v) be an edge, where u is a negative articulation point and v is a positive articulation point, such that $\pi(v) = \pi(u) - 1 = p$ and $A(p) = 1$. Since $A(p) = 1$, there is only one vertex i has $\pi(i) > p$ for any $i \le p$. Since u is a negative vertex, $u \le p$. Since $\pi(u) > p$, u is the only unmatched vertex with respect to p. Therefore, the first $p + 1$ π values are occupied by vertices v and i with $i \le p$. For any vertex j such that $j > p$, $\pi(j) > p + 1$. By removing edge (u, v), we have two disconnected components $\{1, 2, \cdots, p\}$ and $\{p+1, p+2, \cdots, n\}$. Hence, edge (u, v) is an non-simple bridge. □

Lemma 4.4. For any positive articulation point v, an edge $(v-1, v)$ is a non-simple bridge iff $v-1$ is a negative articulation point and $B(v) = 1$.

Proof. Similar to the proof of Lemma 4.3. □

We give algorithm **Bridge** for finding the bridges in a permutation graph below:

Algorithm **Bridge**:

Input: Permutation function π of a permutation graph G with n vertices.

Output: Bridges of G stored in arrays BR_1, BR_2 and BR_3.

begin

/* Finding simple bridges */
(1) Compute modified $lv(i)$ and $lp(i)$ for $1 \le i \le n$.
(2) $BR_1(i) = j$, for each vertex i with only one edge (i, j).

/* Finding non-simple bridges */
(3) Call **Art**
(4) **For** each positive articulation point i
    If $(\pi^{-1}(\pi(i+1))$ is a negative articulation point
        and $A(\pi(i)) = 1)$ **then**
            $BR_2(i) = \pi^{-1}(\pi(i+1))$.
    If $(i-1$ is a negative articulation point
        and $B(i) = 1)$ **then**
            $BR_3(i) = i-1$.

end.

Theorem 4.1. The bridges of permutation graphs can be found in $O(\log n)$ time using $O(n/\log n)$ processors on an EREW PRAM.

Proof. Steps (1) and (2) of algorithm **Bridge** find simple bridges. The details are given in the beginning of this section. Step (3) uses algorithm **Art** given in Section 3 to find articulation points. Step (4) implements the method in Lemmas 4.3 and 4.4 to find non-simple bridges. Each step can be carried out in $O(\log n)$ time using $O(n/\log n)$ processors. □

5. Conclusion

We have presented optimal parallel algorithms for finding articulation points and bridges of permutation graphs. The underlying technique involves efficient ways to characterize the problems. Our algorithms can be extended to solve the *biconnected components* problem. A biconnected component is a maximal subgraph without articulation points. It is interesting to know whether the problems of finding *k-connected components* or *k-edge-connected components* in permutation graphs can be solved in a similar way for arbitrary k.

References

[1] Arvind, K. and C. P. Regan, Connected Domination and Steiner Set on Weighted Permutation Graphs, *Information Processing Letters*, 41, 1992, pp. 215-220.

[2] Colbourn, C. J. and L. K. Stewart, Permutation Graphs: Connected Domination and Steiner Trees, *Discrete Mathematics*, 86, 1990, pp. 179-189.

[3] Golumbic, M. C., *Algorithmic Graph Theory and Perfect Graphs*, Academic Press, New York, 1980.

[4] Kruskal, C. P., L. Ruodolph and M. Snir, The Power of Parallel Prefix, *IEEE Transactions on Computers*, Vol. 34, 1985, pp. 965-838.

[5] Ladner, R. E. and M. J. Fischer, Parallel Prefix Computation, *Journal of ACM*, Vol. 27, 1980, pp. 831-838.

[6] Muller, J. H. and J. Spinrad, Incremental Modular Decomposition, *Journal of ACM*, 36, 1989, pp. 1-19.

[7] Ramkumar, G. D. S., and C. P. Rangan, Parallel Algorithms on Interval Graphs, *Proceedings of the 1990 International Conference on Parallel Processing*, 1990, pp. 72-74.

[8] Sprague, A. P. and K. H. Kulkarni, Optimal Parallel Algorithms for Finding Cut Vertices and Bridges of Interval Graphs, *Information Processing Letters*, 42, 1992, pp. 229-234.

[9] Tarjan R. E., and U. Vishkin, An Efficient Parallel Biconnectivity Algorithm, *SIAM Journal on Computing*, Vol. 14, 1985, pp. 862-874.

[10] Tsin, Y. H. and F. Y. Chin, Efficient Parallel Algorithms for a Class of Graph Theoretic Problems, *SIAM Journal on Computing*, Vol. 13, 1984, pp. 580-599.

SESSION 4C

IMAGE PROCESSING

PATTERN RECOGNITION USING FRACTALS

David W. N. Sharp* & R. Lyndon While**

*Dept. of Computing, Imperial College of Science, Technology & Medicine,
180, Queen's Gate, London SW7 2BZ. Email: dwns@doc.ic.ac.uk

**Dept. of Computer Science, University of Western Australia,
Nedlands, Perth, Western Australia 6009. Email: lyndon@cs.uwa.edu.au

. Abstract *We present a novel algorithm for the automatic recognition and classification of fractally-encoded images. The algorithm can recognise images from a fractal transformation library in the presence of scaling, displacement and rotation and is robust in the presence of noise. The algorithm forms the basis for a more general recognition mechanism if non-fractal images are encoded as fractal pictures using Barnsley's Collage Theorem. The algorithm is a member of a recently-discovered novel class of "Communication-Intensive Massively-Parallel" (CIMP) algorithms that have already been used for sorting, tessellation of the plane and pitch-period detection, and which show great promise for use in other application areas.*

1. INTRODUCTION

Despite much research effort, the goal of automatic computer recognition of arbitrary images has not yet been achieved. The problem has been attacked using various approaches, including sequential algorithms, parallel algorithms, mathematical transformations and neural networks [12, 18, 7]. One of the most notable landmarks on the way towards this goal is Aleksander's Wisard machine [1], which can distinguish between people's faces after appropriate training. The technology employed in the design of this machine does not lie firmly in any of the above paradigms: it is intermediate between a neural network and a massively parallel computer. The underlying technology used in Wisard is *connectionism*. Connectionist algorithms have the ability to associate one pattern with another by connecting together corresponding parts of a machine's software or hardware. This enables a particular input pattern, e.g. an array of pixel values, to be connected through a series of indirections to a particular output pattern, e.g. the name of the person whose face was input.

A massively parallel machine, such as Thinking Machines' Connection Machine CM2 [17], consists of thousands of processors that communicate through an interconnection network. A logically fully-interconnected network can deliver messages from any processor to any other. The hardware underlying the ALICE machine from Imperial College [6] contains a switching network [5] that enables Transputer-based MIMD machines to have these characteristics as well. Connectionist algorithms make use of this sort of network to set up connections between the data on different processors. However, use of the connectionist style is normally discouraged on such machines, because the machine's communication networks have always been much slower than their processors' calculation speed; thus a more conventional algorithm employing maximal computation and minimal communication is usually faster for solving conventional problems.

Ordinarily, problems are mapped onto parallel processors by considering what is to be calculated and distributing the work amongst the processors in a way that approximately balances the load on each. When this is done, it is usually the case that intermediate results calculated by one processor are required by another, and so communication is required to transfer results from where they are calculated to where they are needed. The destination of each communication is determined statically by the mapping of sub-computations to processors and is independent of the value of the intermediate result to be communicated.

As the source, destination and content of each communication is known in advance, the only information conveyed by the communication is the value transmitted. The communication is simply an overhead that arose when the tasks were distributed amongst the processors. This is why designers of algorithms for parallel machines try to minimize the amount of inter-processor communication. However, this approach has yet to yield an algorithm that can recognise an image drawn from a large library of possibilities.

An alternative approach to the utilisation of parallel machines is using "Communication-Intensive Massively-Parallel" (CIMP) algorithms [14]. In a CIMP algorithm, the destinations of messages are calculated at run-time, using the values of data local to the sending processor. When a processor receives a message, the very fact that the message was sent to that processor rather than one of the other N-1 processors gives the recipient log N bits of information about the local data on the sending processor - even if it was only sent a one bit message! The other processors can also derive information from the fact that they did *not* receive a message during that communications cycle. Processors therefore receive information *whether or not* they receive a message. Thus, in CIMP algorithms, complex calculations are transformed into simple, powerful communications. Techniques to derive these algorithms formally using functional language program transformation are described in [13, 15]. To date, CIMP algorithms have been successfully developed for parallel sorting, tessellation of the plane, fractal image generation and pitch extraction in speech [14, 16]. In this paper we develop a CIMP algorithm for fractal image recognition.

The successful application of fractal techniques to image processing has been made possible by Barnsley's Collage Theorem [4]. This theorem states that if a given image can be tiled using rotated, displaced, reduced copies of itself, then it is possible to reconstruct the image solely from the transformations used to produce the copies - it is not necessary to store a pixel map of the image. Compression ratios of up to 10000 to 1 have been claimed using this technique [2]. Barnsley's company "Iterated Systems Inc." is developing software to decompose *any* image into a tiling of fractal images [10]. Thus a fractal image recognition algorithm has great potential to be adapted to the recognition of arbitrary images.

Section 2 of the paper introduces the class of "iterated function system" fractals which we use and describes our CIMP algorithm for fractal image generation. Section 3 develops this algorithm into one which can discriminate between fractally-generated images, and can therefore be used as the basis of an image recognition system. Section 4 describes how we address the problems of rotation, displacement and scaling of images. Section 5 concludes the paper and discusses further work.

2. FRACTAL GENERATION

We have chosen to study the recognition of images that can be generated by an "iterated function system" (IFS) [2], as such images contain self-similarities that we hope to exploit in the recognition process.

The key feature of an IFS fractal image is that the image is made up solely of displaced, rotated and reduced copies of itself. For instance, the *curly* image in Fig. 1 is made from two transformed copies of itself: the blackened rightmost curl and the less black remainder. The blackened rightmost curl is a reduced, rotated, displaced copy of the whole image, produced by applying transformation T1. The remainder of the image is produced by applying transformation T2, which shrinks the image and rotates it anti-clockwise.

Each copy of the image is specified by an *affine* transformation, i.e. one that scales, rotates and displaces the image. The equation describing an affine transformation is given in Fig. 1, along with the values of the parameters a, b, c, d, e and f for each of the two transformations required to generate *curly*. The transformations to produce another fractal image, *fern*, from four copies of itself, are shown in Fig. 2. The bottom two leaves are each a transform of the whole image (T2 and T3), the bottom piece of stalk is a fern reduced into a bent line (T4), and the rest of the image (i.e. excluding the bottom piece of stalk and the bottom

two leaves) is the whole image reduced and displaced upwards (T1).

We observe four properties of IFS fractals.
- When any of the transformations is applied to any point in the image, the transformed point is also in the image.
- When a transformation is applied to two points in the image, the transformed points are closer together than the originals, because the transformations are *contractive*, i.e. they have a scaling factor of less than 1, and therefore map larger images to smaller ones.
- When a transformation is applied to a point outside the image and a point in the image, the two transformed points are closer together than the originals.
- When a transformation is applied to a point outside the image, the transformed point is nearer to (or inside) the image.

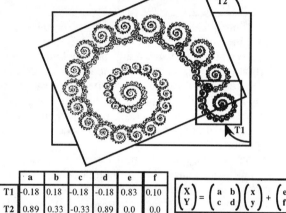

	a	b	c	d	e	f
T1	-0.18	0.18	-0.18	-0.18	0.83	0.10
T2	0.89	0.33	-0.33	0.89	0.0	0.0

$$\begin{pmatrix} X \\ Y \end{pmatrix} = \begin{pmatrix} a & b \\ c & d \end{pmatrix} \begin{pmatrix} x \\ y \end{pmatrix} + \begin{pmatrix} e \\ f \end{pmatrix}$$

Fig.1 The fractal image *curly* and its transformations.

One way of plotting a fractal [2] is to start from an arbitrary point and to apply a randomly-chosen transformation to the point, yielding a new point. The transformed point then has the same or another transformation applied to it, and so on, in an iterative fashion. After several transformation steps, the point will be inside the image. Subsequent transformations will then move the point to other points in the image. Plotting these points builds up the image in a non-deterministic fashion.

[11] describes a superior algorithm starting from the fixpoints of the transformations. The fixpoint of a transformation can be found by solving the transformation equation in Fig. 1 when X=x and Y=y. The other transformations are applied to the fixpoints to generate new points in the image. The new points can then be transformed to generate further points, some of which may already have been plotted. The transformations are applied iteratively until no more new points are reached. Using this method, the image is generated very quickly and uniformly. Another

deterministic sequential algorithm is given in [3]. An alternative deterministic parallel algorithm has been developed for the AMT Distributed Array Processor [9].

[14] describes a CIMP algorithm for fractal generation starting with a square of pixels inside which the image is known to lie (as shown in Fig. 2).

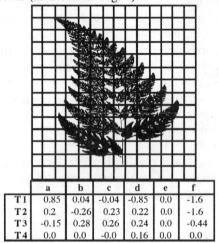

	a	b	c	d	e	f
T1	0.85	0.04	-0.04	-0.85	0.0	-1.6
T2	0.2	-0.26	0.23	0.22	0.0	-1.6
T3	-0.15	0.28	0.26	0.24	0.0	-0.44
T4	0.0	0.0	-0.0	0.16	0.0	0.0

Fig. 2 Transformations to produce *fern* (pixel squares containing the fractal image are not to scale).

Taking any pixel inside the image and applying all the transformations to it, we reach other pixels inside the image. If every transformation is applied to every pixel inside the image, the set of pixels reached is the same as the original set, i.e. the image is preserved: it is the fixpoint of the union of the images produced by the transformations.

Now consider applying the transformations to pixels outside the fractal image. All of the transformed pixels will be closer to (or inside) the image. However, some of the pixels near the edge of the square will not be reached, because there are no pixels outside the square to transform to them.

The algorithm allocates a processor to each pixel, using a trivial mapping between pixels and processors. Each processor calculates its *children*, i.e. those pixels P_1, P_2, ..., P_T (T is the number of transformations) which are reached by applying each of the transformations to its pixel. Each pixel is initially set to be black. Each processor sends a message to the processors responsible for P_1, P_2, ..., P_T, saying that these pixels should remain black.

Any processor that does not receive a message turns its pixel white. The communication process is then repeated, except that a processor whose pixel is white does not send any messages. A processor whose pixel is black sends exactly the same messages to exactly the same destinations.

The process terminates when there is no change in the set of black pixels. At this stage, the set of black pixels contains exactly those pixels which comprise the fractal image. The iteration process, and the number of iterations to each step, is shown in Fig. 3 for the *curly*, *fern* and *Sierpinski* fractals.

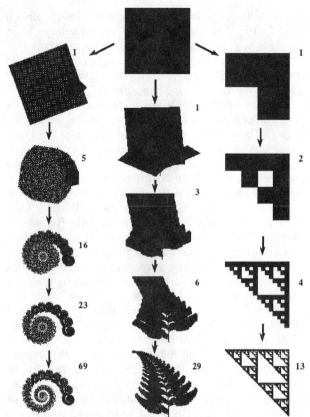

Fig. 3 Fractal images generated by iterative communication. Iteration numbers are shown.

We shall call the CIMP fractal generation algorithm Algorithm $\mathcal{B}$. Given sufficient processors, Algorithm $\mathcal{B}$ generates images in time which depends only on the number of iterations, which in turn depends on the resolution required and the contractiveness of the transformations.

Note that many processors may be sending a message to a single destination, but a destination is only interested in knowing if there were *any* messages destined for it. Thus, if a routing system like that in the Connection Machine is available, the messages can be ORed together en route, reducing the amount of traffic in the network. Also, since the pattern of communication is fixed once the children have been determined, it is possible to set the network into a fast fixed communication mode prior to the iteration process.

Fig. 4 Fractal mutation using Algorithm $\mathcal{B}$.

We have described the algorithm as starting from a black square enclosing the image. However, the final image depends only on the transformations which are applied: it is independent of the original image, provided the original contains at least one black pixel. For example, if the original image is a *curly* and we apply the *fern* transformations, the image converges to a *fern* (Fig. 4).

Note that starting from an image other than a black square requires making white pixels which receive messages black, as well as making black pixels which do not receive messages white. Even starting from a single pixel, applying the *fern* transformations generates the same *fern* (Fig. 5). Note that this generation will work from any pixel: this is guaranteed by the contractiveness of the transformations.

Fig. 5 Growing a single pixel into a fern using Algorithm $\mathcal{B}$.

3. FRACTAL RECOGNITION

The aim of fractal recognition is to take an unknown image and determine which, if any, of a library of fractals was used in its generation. We assume for the moment that the image has not been subject to any rotation, displacement or scaling: these issues are addressed in Section 4. We do, however, allow the presence of noise in the image.

An obvious first approach would be to apply the transformations of each fractal (separately) to copies of the unknown image and to count the number of iterations required for the transformations to generate their known, final image from the unknown image. Intuitively, we expect that if the unknown image is 'close to' the fractal being applied, it will take less iterations to generate the image of that fractal. In the degenerate case where the unknown image *is* the fractal image, it will take only one iteration. Comparing the number of iterations to the number required to generate the fractal from a black square enclosing the image should give us some measure of the 'closeness' of the unknown image to the fractal.

However, in the presence of noise, we find that the number of iterations required depends more on properties of the fractal transformations than on the closeness of the original image to the fractal. The number of iterations required to converge on a *curly* does not discriminate well between, say, a noisy *curly* and a *fern*. The problem is that Algorithm $\mathcal{B}$ is too powerful an algorithm for the purpose of discriminating images: given a set of transformations representing a fractal, it always converges on the image of that fractal, whatever the original image. For the purpose of discriminating images, we would like an algorithm which, given the *fern* transformations, would refine images 'close to' *ferns* towards a true *fern*, but would reduce images 'distant from' *ferns* to a white square. Such an algorithm would allow us to identify an unknown image by repeatedly applying the transformations of each fractal to copies of the image: all but one of the copies would disintegrate, leaving only the copy whose transformations corresponded to the fractal with which the image was generated. We quantify this 'disintegration' by measuring

the proportion of pixels in the image which survive the application of the fractal: this proportion will be far higher for the 'correct' fractal than for the others. Applying Algorithm $\mathcal{B}$ to this copy then generates the full, noise-free image of the correct fractal. Using sufficiently powerful hardware and/or a parallel implementation of the algorithm along the lines of that described in the previous section for Algorithm $\mathcal{B}$, we could then identify an unknown image in time independent of the size of the image.

An algorithm which satisfies these demands is Algorithm $\mathcal{L}$. Operationally, Algorithm $\mathcal{L}$ differs from Algorithm $\mathcal{B}$ in only one respect: it never makes a white pixel black, it only makes black pixels white. Starting from the unknown image, black pixels send messages to their children. A black pixel that did not receive a message turns white and plays no further part in the process. A white pixel which receives a message remains white. The algorithm iterates, and terminates when the image stabilises.

The superficial similarity between the two algorithms masks a fundamental difference, however. Whereas Algorithm $\mathcal{B}$ is a powerful algorithm which *generates* images, Algorithm $\mathcal{L}$ is a weaker algorithm which only attempts to *sustain* images. Applied to a black square, Algorithm $\mathcal{L}$ has the same effect as Algorithm $\mathcal{B}$. Applied to a single pixel, Algorithm $\mathcal{L}$ just eliminates the pixel (except in the special case when the pixel is the fixpoint of one of the transformations being applied, when the pixel survives).

The interesting case is when Algorithm $\mathcal{L}$ is applied to an unknown image. We identify four classes of pixels which are sustained by the algorithm.

Class 1 The fixpoints of the transformations. The fixpoints always sustain themselves (with one transformation) and all of their descendants (with the other transformations).

Class 2 Pixels whose ancestry includes a Class 1 pixel.

Class 3 Pixels whose ancestry includes themselves. These pixels form self-sustaining cycles (using a mixture of transformations) of arbitrary length in the final image.

Class 4 Pixels whose ancestry includes a Class 3 pixel.

Note that Class 1 is really a degenerate sub-class of Class 3, if we regard a fixpoint as a cycle of length zero.

The interesting question, then, is how the numbers of pixels in these classes vary with the nature of the unknown image. The survival of a pixel in Class 1 or Class 2 requires the presence of the relevant fixpoint and the chain of pixels leading from the fixpoint to the pixel.

Clearly an image is more likely to contain the fixpoints of the fractal with which it was generated than the fixpoints of other fractals, so we would expect

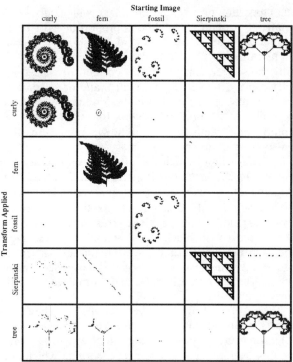

Fig. 6 The application of fractal transformations to various fractal images using Algorithm $\mathcal{L}$. The transform at the left of the row is repeatedly applied to the image at the top of the column.

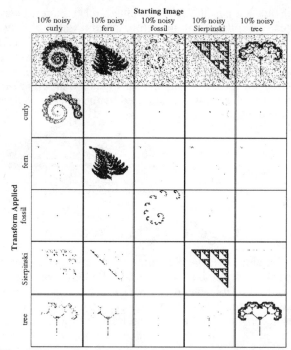

Fig. 8 Recognising original images that have been made noisy by randomly inverting 10% of the pixels. The transform at the left of the row is repeatedly applied to the image at the top of the column.

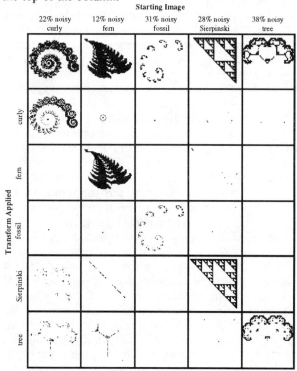

Fig. 7 The application of fractal transformations to systematically noisy images using Algorithm $\mathcal{L}$. The transform at the left of the row is repeatedly applied to the image at the top of the column.

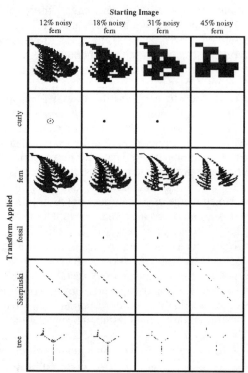

Fig. 9 The application of fractal transformations to various noisy fern images using Algorithm $\mathcal{L}$. The transform at the left of the row is repeatedly applied to the noisy fern at the top of the column.

classes 1 & 2 to be larger for images which are close to the image of the test fractal. Even more decisively, the survival of a pixel in Class 3 or Class 4 requires the presence of *every pixel in the relevant cycle*. This is very unlikely to be the case for images generated with the wrong transformations, and is sometimes not the case for noisy images generated with the correct transformations. Classes 3 & 4 will be very much larger for images close to the image of the test fractal: in addition, their size drops off gradually as the noise-level increases. This is shown in Fig. 9.

The net effect is that Algorithm $\mathcal{L}$ is an excellent discriminator between images. Images close to the fractal being tested will survive with far less 'pixel-loss' than more distant images. Perfect, noise-free images will survive intact. Fig. 6 illustrates the effect of applying a range of fractals to a range of fractal images using Algorithm $\mathcal{L}$. When the wrong transformations are applied to an image, the image collapses towards a white square. Figs. 7, 8 and 9 illustrate the effect of applying the fractal transformations to noisy images: they demonstrate that the algorithm is robust in the presence of quite high levels of random or systematic noise. We quantify the level of noise in an image as the number of pixels which are different from the perfect image of the fractal used in its generation.

We make two observations about Figs. 7, 8 and 9. At first sight, the tree/curly image (i.e. the tree transforms applied to the curly image) and the fossil/fossil image from Fig. 7 appear to have similar survival characteristics. However, it is important to remember that images in the figures should only be compared in columns, because we are applying several fractals in an attempt to recognise one unknown image. Therefore, the tree/curly image should be compared to the curly/curly image (which is obviously more dense) and the fossil/fossil image should be compared to the four sparse squares in its column. Secondly, it is noticeable that the noisier *fern* images in Fig. 9 actually look *more* like *ferns* after the transformations have been applied than before. The application of Algorithm $\mathcal{L}$ has to some degree 'cleaned up' the image.

4. IMAGE NORMALISATION

As it is described in Section 3, Algorithm $\mathcal{L}$ can only recognise images which are the correct size, in the correct place on the screen and with the correct orientation. By 'correct', here, we mean matching the image produced by applying the fractal transformations using Algorithm $\mathcal{B}$. To allow for the recognition of mutated copies of images, we will enhance the algorithm so that it 'normalises' the transformations to allow for the mutation before applying them to the image.

To derive the new transformations, we first work out a *correction matrix* which coerces the unknown image onto the known image. The correction matrix is composed of two sub-matrices, one compensating for displacement and

the other for rotation and scaling. Identifying these sub-matrices as M_d and M_{rs} respectively, the corrected pixel co-ordinates of the unknown image are given by

$$\begin{pmatrix} x' \\ y' \end{pmatrix} = M_{rs} \left[\begin{pmatrix} x \\ y \end{pmatrix} + M_d \right]$$

(M_{rs} is a 2×2 matrix and M_d is a 2-element vector.)

We then use the correction matrix to derive new values of the parameters for each transformation of the fractal, solving for the values a', b', c', d', e' and f' in the equation

$$\begin{pmatrix} a' & b' \\ c' & d' \end{pmatrix} \begin{pmatrix} x \\ y \end{pmatrix} + \begin{pmatrix} e' \\ f' \end{pmatrix} = \begin{pmatrix} a & b \\ c & d \end{pmatrix} M_{rs} \left[\begin{pmatrix} x \\ y \end{pmatrix} + M_d \right] + \begin{pmatrix} e \\ f \end{pmatrix}$$

The solutions of this equation are given in the Appendix. We use a fairly standard approach in generating the correction matrix.

M_d   We calculate the centre of mass, C, of an image using the standard equations with m=1 (i.e. black pixels have unit mass, white pixels have zero mass):

$$\bar{x} = \frac{\displaystyle\sum_{(x,y)\in B} m\, x}{\displaystyle\sum_{(x,y)\in B} m} \qquad \bar{y} = \frac{\displaystyle\sum_{(x,y)\in B} m\, y}{\displaystyle\sum_{(x,y)\in B} m}$$

B is the set of black pixels in the image. We make $M_d = C_k - C_u$, where $C_k = \begin{pmatrix} \bar{x}_k \\ \bar{y}_k \end{pmatrix}$ and $C_u = \begin{pmatrix} \bar{x}_u \\ \bar{y}_u \end{pmatrix}$ are the centres of mass of the known and unknown images, respectively.

M_{rs}   We scale the unknown image so that its second moments of area about its principal axes match those of the known image. This scaling is performed when its major principal axis is coincident with the x-axis and its minor principal axis is coincident with the y-axis. We therefore rotate the unknown image into this position, then scale the image, then rotate it to match the known orientation of the principal axes of the known image.

We identify the principal axes of an image, shown as Cx', Cy' in Fig. 10, using the standard formula in [8].

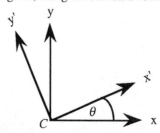

Fig. 10 Rotating the axes of an image by θ.

The orientation θ of the principal axes of an image relative to arbitrary axes Cx, Cy (for which the second moments of area I_{xx}, I_{yy} and I_{xy} are known) is given by

$$\tan 2\theta = \frac{-I_{xy}}{0.5(I_{xx} - I_{yy})} \qquad I_{xy} = \int xy\,dm = \sum_{(x,y) \in B} xy$$

$$I_{xx} = \int x^2 dm = \sum_{(x,y) \in B} x^2 \qquad I_{yy} = \int y^2 dm = \sum_{(x,y) \in B} y^2$$

To ensure that Cx' is the major principal axis, we take $0 < 2\theta < \pi$ if I_{xy} is negative, and $\pi < 2\theta < 2\pi$ if I_{xy} is positive. This fixes the angle of the major principal axis to the Cx axis, but does not fix its direction: the principal axis could be at θ or $\pi + \theta$ to the Cx axis. To resolve this ambiguity, we count the number of pixels on either side of the minor principal axis and normalise so that the major principal axis increases towards the 'heavy' side of the image.

Denoting the orientation of the principal axes of the unknown image to the arbitrary axes by α and that of the known fractal image by β, we make

$$M_{rs} = \begin{pmatrix} \cos\beta & \sin\beta \\ -\sin\beta & \cos\beta \end{pmatrix} M_s \begin{pmatrix} \cos\alpha & -\sin\alpha \\ \sin\alpha & \cos\alpha \end{pmatrix}$$

where M_s is the scaling matrix defined below. This corresponds to a rotation of $-\alpha$, then a scaling, then a rotation of β.

$\underline{M_s}$ We calculate the second moments of area I'_{xx}, I'_{yy} and I'_{xy} on the principal axes by

$$I'_{xx} = \frac{1}{2}(I_{xx} + I_{yy}) + \sqrt{\left(\frac{I_{xx} - I_{yy}}{2}\right)^2 + I_{xy}^2}$$

$$I'_{yy} = \frac{1}{2}(I_{xx} + I_{yy}) - \sqrt{\left(\frac{I_{xx} - I_{yy}}{2}\right)^2 + I_{xy}^2}$$

$$I'_{xy} = 0$$

Denoting the second moments of area of the principal axes of the unknown image by U'_{xx}, U'_{yy} and U'_{xy}, and those of the known image by K'_{xx}, K'_{yy} and K'_{xy}, we consider the effect of scaling the image by a factor s along both principal axis directions. Firstly, there is a factor of s^2 that arises from squaring distances in the calculation of the second moment of area along each principal axis. Secondly, the area of the image, and thus the number of pixels in the image, increases by a factor of s^2. Combining both factors we make

$$M_s = \begin{pmatrix} \sqrt[4]{\dfrac{K'_{xx}}{U'_{xx}}} & 0 \\ 0 & \sqrt[4]{\dfrac{K'_{yy}}{U'_{yy}}} \end{pmatrix}.$$ Of course, for a square scaling of the image $\dfrac{K'_{xx}}{U'_{xx}} = \dfrac{K'_{yy}}{U'_{yy}}$.

For each fractal being tested, we generate the correction matrix which fits the unknown image. The correction matrix is applied to each of the fractal's transformations: Algorithm $\mathcal{L}$ will then eliminate all of the candidate fractals except the one that was used to generate the unknown image.

All of these calculations can be sensitive to the distribution of noise in the unknown image. However, given that we have to use information gleaned from the unknown image to correct the transformations, this restriction applies to any approach to the problem. In effect, it constitutes a general limit on the amount of noise which an image can contain before it becomes effectively unrecognisable. Our approach bases the correction matrix on a summation of the properties of every pixel in the unknown image, therefore making it robust to non-systematic patterns of noise.

5. CONCLUSIONS

We have presented a novel communication-intensive massively-parallel algorithm for fractal image recognition. The algorithm works by applying the transformations of known fractals to copies of an unknown image. For a particular known fractal, only images generated using the transformations of that fractal will be sustained by the application of Algorithm $\mathcal{L}$: thus, all but one of the copies of the image will be eliminated. The application of our previously-presented fractal generation algorithm, Algorithm $\mathcal{B}$, to the surviving copy of the image will then return a perfect copy of the recognised image. The algorithm allows for the arbitrary displacement, scaling and rotation of images and is robust in the presence of noise. It is designed to exploit the connectionist structure of a massively parallel machine such as the Connection Machine CM2.

Barnsley's Collage Theorem and his later work with "Iterated Systems Inc." demonstrates a technology by which general images can be encoded as iterated function codes. We therefore expect that Algorithm $\mathcal{L}$ will be generally applicable in the field of image recognition.

The algorithm lies in the promising unexplored territory between classical artificial neural networks and massively parallel machines, the same area in which the technology behind the Wisard machine lies. However, whereas the Wisard machine employs random connectivity between image components, our CIMP approach dynamically generates a fractal connection structure based on the global self-similarity properties of the known images. This approach could potentially be extended so that connections are based on self-similarities in the *unknown* image. In this case, we would calculate the centre of mass, the second moments of area and the orientation of the principal axes of local conglomerations of pixels in an image. We would then use the equations for displacement, rotation and scaling in Section 4 to identify self-similarities between these conglomerations. The image library could then be expressed in terms of the local self-similarity properties of sub-components of the image rather than the more limited global properties expressed by fractal transforms. It would also have the effect of making the algorithm more tolerant of noise. This technique will lead to yet more powerful CIMP image recognition algorithms in the future.

ACKNOWLEDGEMENTS

We would like to thank David Lillie and Nick Merriam for their comments. This work was supported by the Science and Engineering Research Council of Great Britain by grant GR/H46299 "Communication-Intensive Massively-Parallel Algorithms".

REFERENCES

[1] I Aleksander, W.V. Thomas & P.A. Bowden, "Wisard: A Radical Step Forward in Image Recognition," *Sensor Review*, Vol. 4 No. 3, pp.120-124, July 1984.

[2] M.F. Barnsley & A.D. Sloan, "A Better Way to Compress Images," *Byte Magazine*, January 1988.

[3] M.F. Barnsley, *Fractals Everywhere*, Boston: Academic Press, 1988.

[4] M.F. Barnsley, *The Desktop Fractal Design System*, Academic Press, 1989.

[5] M.D. Cripps & A.J. Field, *The MARCH HARE Network Switching Device*, Research Report DoC 83/30, Dept. of Computing, Imperial College, March 83.

[6] M.D. Cripps, J. Darlington, A.J. Field, P.G. Harrison, M.J. Reeve, "The Design and Implementation of ALICE: A Parallel Graph Reduction Machine," *Selected reprints on dataflow and reduction architectures*, ed. S.S. Thakkar, IEEE Computer Society Press, 1987.

[7] M.C. Fairhurst, H.M.S. Abdel Wahab & P.S.J. Brittan, "Parallel Implementation of Image Classifier Architectures using Transputer Arrays," *3rd International Conference on Image Processing and its Applications*, pp. 136-40, July 1989.

[8] A.S. Hall, *An Introduction to the Mechanics of Solids*, John Wiley, p. 390, 1969.

[9] A.N. Horn, *Iterated Function Systems, the Parallel Progressive Synthesis of Fractal Tiling Structures, and their Applications to Computer Graphics*, Ph.D Thesis, Wolfson College, Oxford University, 1991.

[10] M. McGuire, *An Eye for Fractals*, Addison-Wesley, 1991.

[11] D.M. Monro, F. Dudbridge & A. Wilson, "Deterministic Rendering of Self-affine Fractals," *IEE Colloquium on the application of fractal techniques in image processing*, IEE Publications, December 1990.

[12] A. Rosenfield, "Survey - Image Analysis and Computer Vision," *CVGIP: Image Understanding*, Vol. 53, No. 3, pp. 322-365, May 1991.

[13] D.W.N. Sharp, *Functional Language Program Transformation for Parallel Computer Architectures*, Ph.D thesis, Imperial College, London, 1990.

[14] D.W.N. Sharp & M.D. Cripps, "Parallel Algorithms That Solve Problems by Communication," *Proc. 3rd IEEE Symposium on Parallel and Distributed Processing*, Dallas, Texas, pp. 87-94, December 1991.

[15] D.W.N. Sharp, H. Khoshnevisan & A.J. Field, "An Exercise in the Synthesis of Parallel Functional Programs for Message-Passing Architectures," *Parallel Computing: From Theory to Sound Practice*, eds. Wouter Joosen & Elie Milgrom, presented at the *European Workshops on Parallel Computing*, Stiges, Barcelona, March 1992.

[16] D.W.N. Sharp, R.L.While, "Determining The Pitch Period of Speech Using No Multiplications, " *Proc. Int. Conf. Acoustics, Speech & Signal Processing*, Minneapolis, April 1993.

[17] Thinking Machines, *Connection Machine Model CM-2 Technical Summary*, Thinking Machines Corporation Technical Report Series, HA87-4, 1987.

[18] W. Wang, S.S. Iyengar & J. Chen, "Massively Parallel Approach to Pattern Recognition," *9th Int. Conf. on Computing and Communications*, pp. 61-7, 1990.

APPENDIX

The solutions to the equation defining a', b', c', d', e' and f' in Section 4 are given below.

$$a' = \frac{\sqrt[4]{K'_{yy}}\, \sin\alpha\, (b\cos\beta + a\sin\beta)}{\sqrt[4]{U'_{yy}}} + \frac{\sqrt[4]{K'_{xx}}\, \cos\alpha\, (a\cos\beta - b\sin\beta)}{\sqrt[4]{U'_{xx}}}$$

$$e' = (\bar{x}_k - \bar{x}_u)\, a' + (\bar{y}_k - \bar{y}_u)\, b'$$

$$b' = \frac{\sqrt[4]{K'_{yy}}\, \cos\alpha\, (b\cos\beta + a\sin\beta)}{\sqrt[4]{U'_{yy}}} + \frac{\sqrt[4]{K'_{xx}}\, \sin\alpha\, (a\cos\beta - b\sin\beta)}{\sqrt[4]{U'_{xx}}}$$

$$f' = (\bar{x}_k - \bar{x}_u)\, c' + (\bar{y}_k - \bar{y}_u)\, d'$$

$$c' = \frac{\sqrt[4]{K'_{yy}}\, \sin\alpha\, (d\cos\beta + c\sin\beta)}{\sqrt[4]{U'_{yy}}} + \frac{\sqrt[4]{K'_{xx}}\, \cos\alpha\, (c\cos\beta - d\sin\beta)}{\sqrt[4]{U'_{xx}}}$$

$$d' = \frac{\sqrt[4]{K'_{yy}}\, \cos\alpha\, (d\cos\beta + c\sin\beta)}{\sqrt[4]{U'_{yy}}} + \frac{\sqrt[4]{K'_{xx}}\, \sin\alpha\, (c\cos\beta - d\sin\beta)}{\sqrt[4]{U'_{xx}}}$$

EFFICIENT IMAGE PROCESSING ALGORITHMS ON THE SCAN LINE ARRAY PROCESSOR*

(Extended Abstract)

David Helman and Joseph JáJá[†]

Department of Electrical Engineering

University of Maryland

College Park, MD 20742

helman@src.umd.edu and joseph@eng.umd.edu

[†]Also, Institute For Advanced Computer Studies and Institute For Systems Research.

ABSTRACT

We develop efficient algorithms for low and intermediate level image processing on the scan line array processor that handles images in a scan line fashion. For low level processing, we present algorithms for block DFT, block DCT, convolution, template matching, shrinking, and expanding. These algorithms run in real-time - that is, the output lines are generated at the rate of $O(m)$ time per line, where the required processing is based on neighborhoods of size $m \times m$. For intermediate level processing, we present efficient algorithms for scaling, translation, connected components, and convex hulls of multiple figures.

1   INTRODUCTION

The I/O bandwidth and the computational power required for real-time video and image processing are quite high. Consider, as an example, the I/O bandwidth needed to perform real-time HDTV simulation. Such a task involves the handling of $1K \times 1K$ frames at the rate of 60 frames per second and results in a bandwidth requirement of approximately 400 Mbytes/sec for a progressively scanned image.

A simple architecture that is primarily intended for low and medium level image processing is the scan line array processor, or SLAP [1] - [5]. The basic topology of this SIMD machine is a linear array of processors, in which the number of processors corresponds to the number of pixels in each row of the image. Incoming data is loaded line by line into the processor array, with a distinct processor receiving each column of the image. Each processor can communicate directly with its two immediate neighbors, whenever they exist. The major attractive features of this architecture include scalability, very high I/O bandwidth, the possibility of overlapping I/O and processing, and relatively low cost.

The design of algorithms for the linear array architecture has received only modest attention in the literature (see for example [6] - [8]). The image processing algorithms considered in [6] for the linear array are not suitable for the SLAP since they assume the shuffled row major distribution of the input image among the local memories of the processors. For an $n \times n$ image that is being processed in a scan line fashion, such an input distribution requires $\Omega(n^2)$ steps due to the bisection width of the linear array.

For our work, the problem of developing algorithms for low level image processing is treated seperately from that of intermediate level tasks. In the case of low level operations, it is assumed that we need to achieve the smallest possible latency consistent with an efficient running time. To this end, we require that the processing of a scan line begin immediately after it is received, and that the corresponding output line be generated in an amount of time that is independent of the image size. More formally, suppose that the required processing is based on neighborhoods of size $m \times m$. Then, we would like the output lines to be generated at a rate of $O(m)$ operations per line after an initial delay of $O(m)$ lines. If an algorithm attains this performance, then it is said to run in **real time**, which is clearly the best that can be achieved on this model.

*Supported in part by NSF, Grant No. CCR-9103135 and NSF Engineering Research Center Program CD 8803012

In the case of intermediate level operations, computational efficiency demands that all the relevant pixels be available for examination before processing can begin. Hence, we assume at the outset that the entire image is already stored in the local memories of the processors in the same way it was received, with each column stored at the processor of like index.

Our results can be summarized as follows.

- **Real-Time Low Level Operations**: We develop real-time algorithms for block discrete Fourier transform (DFT) and block discrete cosine transform (DCT), convolution, template matching, shrinking, and expanding. We also develop an algorithm for median filtering which requires $O(m \log m)$ time per scan line for an window of size $m \times m$.

- **Intermediate Level Operations**: We develop optimal algorithms for scaling and translation. We also develop algorithms for labelling the connected components and determining the convex hulls of multiple figures which run in $O(n \log n)$ and $O(n \log^2 n)$ time, respectively. Our algorithms are significantly simpler and easier to implement than those reported in the literature.

In this abstract, we only present sketches of our algorithms for median filtering, block DFT, convolution, and determining the convex hulls of multiple figures. For a complete treatment, see [9].

2   MEDIAN FILTERING

Let X be an input image of size $n \times n$ and let the window be of size $m \times m$, where m is assumed to be odd for convenience. The filtered image Y is defined by

$$
\begin{aligned}
Y[j,k] \;=\; \text{median} \;\; & \{X[j-r,k-s]: \\
& 0 \le j-r, k-s \le n-1 \\
& \text{and } -\lfloor \tfrac{m}{2} \rfloor \le r,s \le \lfloor \tfrac{m}{2} \rfloor \}, \\
& \hspace{6cm} (1)
\end{aligned}
$$

where $0 \le r,s \le n-1$. A straightforward approach solves this problem in $O(m^2)$ time per scan line, whereas our algorithm reduces the complexity to $O(m \log m)$ time per scan line.

Our method is based on the observation that the windows surround the image pixels $[j,k]$ and $[j+1,k]$ differ only by $2m$ values. This suggests that the m^2 values contained in the window around pixel $[j,k]$ should be stored in a suitable data structure that will allow us to (1) efficiently update the data structure to reflect the window about pixel $[j+1,k]$ and

(2) quickly find the median of the values stored in the data structure. One such data structure is the **order-statistic tree**, which allows one to dynamically delete an element, insert an element, or determine the pth smallest element in time proportional to the logarithm of the tree size. We will use such a tree in our algorithm.

Assume for our real-time implementation of this operation that the processors of the array have just received the jth row of the input image X. A right shift of this row $\lfloor \tfrac{m}{2} \rfloor$ positions followed by a left shift of this row $\lfloor \tfrac{m}{2} \rfloor$ positions allows each processor P_k to obtain the $m-1$ elements which immediately surround it. This communication step takes $O(m)$ time. Next, each processor P_k deletes the m oldest values in its order-statistic tree and then inserts both the value of $X[j,k]$ and the $m-1$ values received from its neighbors. As already noted, this can be accomplished in $O(m \log m)$ time. Finally, in $O(\log m)$ time, each processor P_k determines the median of its updated order-statistic tree and then outputs this result as the value of pixel $[(j - \lfloor \tfrac{m}{2} \rfloor), k]$ of the output image.

Theorem 1: Given a SLAP with n processors, the median filtering operation on an image of size $n \times n$ with a window of size $m \times m$ can be performed in real-time at a cost of $O(m \log m)$ operations per scan line and a latency of $\lfloor \tfrac{m}{2} \rfloor$ scan lines.

3   BLOCK DFT

Let the $n \times n$ input image X be decomposed into non-overlapping $m \times m$ blocks, where we assume for convenience that m divides n evenly. If we denote one such block as $X_{a,b}$, where $a, b \in \{0, 1, \ldots, \tfrac{n}{m}\}$, then the 2D-DFT of $X_{a,b}$ is defined as:

$$
Y_{a,b}[c,d] = \sum_{g=0}^{m-1} \sum_{h=0}^{m-1} X_{a,b}[c,d] e^{-\frac{i 2\pi c g}{m}} e^{-\frac{i 2\pi d h}{m}}, \quad (2)
$$

where $i = \sqrt{-1}$ and $0 \le c, d \le m-1$.

Assume for our real-time implementation of this operation that the input is the jth row of the image X. Initially, each processor broadcasts the value of its input pixel $X[j,k]$ to the $(m-1)$ other processors which share the same block, which requires no more than $O(m)$ time. Next, each processor P_k computes the $(k \bmod m)^{th}$ coefficient of the 1D-DFT of the $(j \bmod m)^{th}$ row of its block. Specifically, each processor P_k evaluates the expression:

$$Y'_k = \sum_{r=\lfloor k/m \rfloor m}^{\lfloor k/m \rfloor m + m - 1} X[j,r] e^{-\frac{i2\pi kr}{m}} \qquad (3)$$

Clearly, this involves no more than $O(m)$ operations. Next, each processor P_k uses this result to update the column of partially computed 2D-DFT coefficients held in the array $Partial-Sum_k$ at its local memory. More precisely, for each value of c, where $0 \leq c \leq (m-1)$, each processor P_k performs the following operation:

$$Partial-Sum_k[c] = Partial-Sum_k[c] + (Y'_k \times e^{-\frac{i2\pi cj}{m}}) \qquad (4)$$

Again, this clearly involves no more than $O(m)$ operations. When the last row of the block is processed, the values held in the array $Partial-Sum_k$ will be the $(k \bmod m)^{th}$ column of the of the 2D-DFT of the block just processed.

Theorem 2: Given a SLAP with n processors, the block DFT of an image of size $n \times n$ with each block of size $m \times m$ can be computed in real-time at a cost of $O(m)$ operations per scan line and a latency of m scan lines.

4   CONVOLUTION AND TEMPLATE MATCHING

Given an $n \times n$ image X and an $m \times m$ kernel W, the convolution of X and W is the $(n+m-1) \times (n+m-1)$ image Y defined by

$$Y[j,k] = \sum_{r=0}^{m-1} \sum_{s=0}^{m-1} X[j-r, k-s] W[r,s], \qquad (5)$$

where $0 \leq j, k \leq n+m-2$. Here we are assuming that $X[j-r, k-s]$ is equal to zero whenever $j-r$ or $k-s$ is not in the interval $[0, n-1]$. A straightforward computation of the convolution requires $\Theta(n^2 m^2)$ operations and therefore takes at least $\Omega(m^2)$ steps per scan line on the SLAP. By using the overlap-and-add strategy and then taking advantage of the properties of Fourier transforms, we can reduce the computational requirements to $O(m)$ steps per scan line. The same method can be extended to handle template matching within the same time bounds.

Suppose that we partition the input image X into $m \times m$ blocks $X_{a,b}$, where $a, b \in \{0, 1, \ldots, \frac{n}{m}\}$, and we let $Y_{a,b}$ be the convolution of $X_{a,b}$ and W, where $Y_{a,b}$

is of size $(2m-1) \times (2m-1)$. If we then let $a = \lfloor (\frac{j}{m}) \rfloor$, $b = \lfloor (\frac{k}{m}) \rfloor$, $c = j \bmod m$, and $d = k \bmod m$, then it is simple to verify that the value $Y[j,k]$ defined by (5) can now be obtained by simply summing the four entries $Y'_{a,b}[c,d]$, $Y'_{a-1,b-1}[c+m, d+m]$, $Y'_{a-1,b}[c+m, d]$, and $Y'_{a,b-1}[c, d+m]$. Hence, our matrix convolution problem can be reduced to that of block convolution.

Our block convolutions can be computed by appropriately padding zeros to $X_{a,b}$ and W to obtain the $2m \times 2m$ blocks $X'_{a,b}$ and W' and then performing their *circular* convolution. The circular convolution of $X'_{a,b}$ and W' can be efficiently computed by using the fact that

$$X'_{a,b} \bigodot W' = IDFT(\ DFT(X'_{a,b}) \bigotimes DFT(W') \ ) \qquad (6)$$

where $\bigodot$ denotes the circular convolution and $\bigotimes$ denotes the element-wise multiplication. The real-time implementation of the algorithm suggested by Eqn. (6) clearly requires the real-time computation of the DFT of each block $X'_{a,b}$. The algorithm described in the previous section can be easily modified to compute $DFT(X'_{a,b}) \bigotimes DFT(W')$ at the rate of $O(m)$ time per scan line. Further, since the computation of the IDFT is essentially analogous to the computation of the DFT we would expect that a modification of our algorithm for finding the block DFT would also yield the IDFT of $(DFT(X'_{(a,b)}) \bigotimes DFT(W'))$ in $O(m)$ operations per input line. Of course, we would need to complete our computation of $(DFT(X'_{(a,b)}) \bigotimes DFT(W'))$ before we could begin our computation of its IDFT. Hence, we would actually overlap our computation of $(DFT(X'_{(a,b)}) \bigotimes DFT(W'))$ not with the computation of the IDFT of $(DFT(X'_{(a,b)}) \bigotimes DFT(W'))$ but rather with the computation of the IDFT of $(DFT(X'_{((a-1),b)}) \bigotimes DFT(W'))$. Hence, we have the following theroem:

Theorem 3: Given a SLAP with n processors, the convolution of an $n \times n$ image and a $m \times m$ kernel can be computed in real-time at a cost of $O(m)$ operations per scan line and a latency of $2m$ scan lines. Similarly, the template matching problem can be solved at the same rate for an template of size $m \times m$.

5   CONVEX HULL

Consider an $n \times n$ input image in which each pixel can have any one of $O(n)$ possible labels. For each set of pixels which share a particular label, we wish to determine the convex hull, where the convex hull is informally defined as the smallest convex polygon which contains all the the points in the set. The algorithm presented here can also be modified to solve the case of $O(n^2)$ labels within the same complexity bounds, as long as we require that the pixels which belong to the same set form a connected component. The strategy of our algorithm is novel and amounts to making a number of left-to-right and right-to-left sweeps across the image for each label, each time eliminating a fraction of the non-extreme points. Extensive pipelining is used to insure that each sweep can be completed for all $O(n)$ labels in $O(n)$ time. In spite of the simplicity of our algorithm, its analysis requires a somewhat tricky geometric argument to establish that ($\lceil \log^2 n \rceil$) sweeps are sufficient, and this appears in the full paper [9].

To simplify our presentation, we concentrate on the problem of defining the upper hull, since the task of defining the lower hull is entirely analogous. We start with the following preprocessing steps:

Step1: For each column, we identify for each label the uppermost occurence of that label (if any) and designate that point as a candidate point for the upper hull of that set.

Step2: For each label, we make a left-to-right sweep across the image. As we pass each column k, we identify for the candidate point (if any) at that column the candidate point from columns 0 through k which has the minimum row index. Similarly, for each label, we also make a right-to-left sweep across the image to identify for each candidate point the candidate point to the right which has the minimum row index.

Step3: Each candidate point that lies below the line connecting its respective minima is eliminated from further consideration.

The candidate points left after these preprocessing steps form two "monotonic" sequences (one of which may be empty) with the property that the angle between any three consecutive points must be larger than 90°. We are now ready to describe the algorithm for determining the upper hull of each set;

I. Determine for each candidate point its successor with the same label.

II. repeat for $\lceil \log^2 n \rceil$ times:

Substep1: For each label, make a left-to-right pass across the image. For each candidate point Q, whose current predecessor is R and current successor is S, determine whether Q lies below the line connecting R and S. If it does, then remove Q from further consideration and make R the predecessor of S.

Substep2: For each label, make a right-to-left pass, performing the analogous operation as in Substep1.

Theorem4: Assume that we have a SLAP with n processors and an input image of size $n \times n$ distributed one column per processor. If we also assume that each pixel can belong to any one of $O(n)$ possible sets, then we can define the convex hulls of all these sets in $O(n \log^2 n)$ time.

References

[1] K. Chen and C. Svensson, "A 512-Processor Array Chip for Video/Image Processing," in *From Pixels to Features II*, H. Burkhardt et al. eds., Elsevier Science Publishers, (1991), pp. 187-199.

[2] D. Chin et al., "The Princeton Engine: A Real-Time Video System Simulator," *IEEE Trans. on Consumer Electronics*, (May 1988).

[3] A.I. Fisher and P.T. Highnam, "Real-Time Image Processing on Scan Line Array Processors," *IEEE Computer Society Workshop on Computer Architectures for Pattern Analysis and Image Database Management*, (1985), pp. 484-489.

[4] A.I. Fisher, "Scan Line Array Processors for Image Computations," *International Conference on Computer Architecture*, (1986), pp. 338-345.

[5] S. Knight et al., "The Sarnoff Engine: A Massively Parallel Computer for High Definition System Simulation," *Proceedings of Application Specific Array Processors*, (1992), pp. 342-357.

[6] H. M. Alnuweiri and V.K. Prasanna Kumar, "Optimal Geometric Algorithms for Fixed-Size Linear Arrays and Scan Line Arrays," *Proceedings IEEE Conference on Computer Vision and Pattern Recognition*, (1988), pp. 931-936.

[7] K. Doshi and P. Varman, "Optimal Graph Algorithms on a Fixed-Size Linear Array," *IEEE Transactions on Computers*, (1987), pp. 460-470.

[8] A.L. Fisher and P.T. Highnam, "Computing the Hough Transform on a Scan-Line Array Processor," *IEEE Trans. Pattern Anal. Machine Intell.*, (1989), pp. 262-265.

[9] D. Helman and J. JáJá, *Efficient Image Processing Algorithms on the Scan Line Array Processor*, Technical Report, Institute For Advanced Computer Studies, University of Maryland, College Park, (1993).

A Parallel Progressive Refinement Image Rendering Algorithm on a Scalable Multithreaded VLSI Processor Array

S.K. Nandy   Ranjani Narayan   V.Visvanathan
Supercomputer Education & Research Center
Indian Institute of Science
Bangalore - 560 012, INDIA
$< nandy, ranjani, vish > @cadl.iisc.ernet.in$

P. Sadayappan
Ohio State Univ.
Columbus, Ohio
U.S.A.
$saday@cis.ohio-state.edu$

Prashant S. Chauhan
C–DOT,
Bangalore - 560 052
INDIA

Abstract − *In this paper we develop a multithreaded VLSI processor linear array architecture to render complex environments based on the radiosity approach. The processing elements are identical and multithreaded. They work in Single Program Multiple Data (SPMD) mode. A new algorithm to do the radiosity computations based on the progressive refinement approach[2] is proposed. Simulation results indicate that the architecture is latency tolerant and scalable. It is shown that a linear array of 128 uni-threaded processing elements sustains a throughput close to 0.4 million patches/sec.*

INTRODUCTION

Radiosity is a method to model global illumination effects. It describes an equilibrium energy balance in the entire environment to be rendered. In this model all objects in the environment (also called patches) are sources of energy. Patches are assumed to be ideal diffuse with regard to emission and reflection processes. Therefore unlike ray tracing, the history or direction of a ray is lost once it is incident on a patch surface. Thus each patch surface can be treated as a secondary light source. Radiosity b_j of a patch j is defined as the total rate of energy leaving the surface and is equal to the sum of emitted energy e_j and reflected energy. The radiosity of patch j can be expressed as $b_j = e_j + \rho_j \sum_{i=1}^{N} b_i f_{ij}, \quad 1 \le j \le N$, where f_{ij} (form factor) denotes the fraction of energy that leaves patch i and lands on patch j, ρ_j is the reflectivity of patch j. For all patches j, the above expression can be written in matrix form as, $(I - \Lambda F)B = E$, where I is the $N \times N$ identity matrix, $B = \{b_j\}$, $E = \{e_j\}$, $\Lambda = diag(\rho_j)$ and $F = \{f_{ij}\}$. From the reciprocity relationship of form factors[2], we have, $f_{ij}A_i = f_{ji}A_j$ where A_k is the area of patch k. It may be noted that computation of both the form factor matrix and the solution B is compute intensive.

In [2] a progressive refinement approach to radiosity based rendering is proposed. The conventional Guass-Seidal iteration converges to the solution by solving systems of equations one row at a time. Cohen *et al.* use the reciprocity of form factors to reverse the process of solution by considering one column at a time. Thus rendering is carried out by doing a modified Guass-Seidal iteration to solve the radiosity equation $(I - \Lambda F)B = E$. By using such an approach, instead of computing one element b_i of the B vector, contribution to radiosities of $N - 1$ other

elements of vector B are incrementally updated. Hence the storage requirement is no more than $O(N)$ (primarily to store the vectors E and B). Form factors are computed every time and the new estimate of b_j is determined by shooting radiosity rays from a source patch i $(i \ne j)$ to all other patches in the environment. The process is repeated for every patch i in the environment and this constitutes one Gauss-Seidal iteration. It is reported by Cohen *et al.*, that by ordering the source patches in decreasing order of their radiosities and choosing a new source patch having the highest radiosity value from among the remaining patches to shoot radiosity rays, convergence of the B vector is achieved in steps less than that required for a single Gauss-Siedal iteration.

Cohen *et al.* report that for an environment comprising 30,000 patches, convergence was reached when only 2,000 source patches were considered. Despite the fast convergence, the actual time taken on a VAX 8700 was 5 hours. Clearly there is a need to accelerate this.

In this paper, we aim at solving the radiosity based rendering equation $(I - \Lambda F)B = E$ in a distributed manner through chaotic relaxation and hence converge on a B vector faster than the progressive refinement approach. We propose a multiprocessor solution for solving the radiosity equation and impose a restriction of $O(N/P)$ storage for each processor, where N is the number of patches in the environment and P is the number of processing elements, which is commensurate with the storage requirement in the progressive refinement approach.

Problems with rendering in any multiprocessor system are now those associated with overheads of data movement across processors, repeated movement of data within a processor's local memory and secondary store, and their duplication in the multiprocessor architecture. In our approach, we attempt at minimizing all the three.

RENDERING IN A MULTIPROCESSOR

A trivial solution to reduce repeated data movement across processors in a multiprocessor system, is to duplicate data in all processors. An alternate solution to this problem is shared memory systems with its associated problems. However, the key to minimizing data movement in a multiprocessor environment is determined by the manner in which data is partitioned, stored locally in the processing elements and used optimally and efficiently by the rendering algorithm. We develop a parallel algorithm to compute radiosities of patches in the environment and address the above issues.

In [2], form factors are recomputed every time a radiosity ray is shot, thereby eliminating the need to store F and reducing the storage requirement to $O(N)$ for storing only the B vector.

We consider a linear array of P homogeneous processors, each capable of independently carrying out radiosity computations. We restrict local storage on each processor to $O(N/P)$, sufficient to accomodate the following:

1. N/P patches and their radiosities,

2. All precomputed δ form factors[5, 4, 6] on the surface of a discretized hemisphere in a constant store,

We propose a partitioning scheme and a parallel algorithm for radiosity computations that are efficient both in terms of communication and computation. The overheads of communications is well overlapped with computations in the proposed latency tolerant multiprocessor architecture to be described in detail later.

Partitioning data

A naive partitioning scheme would be to geometrically divide the environment into P blocks of equal volume. We discard this scheme, since it is appropriate only for dense environments, with a uniform distribution of patches. Such environments are not common.

Instead we divide the environment into P blocks of equal number of patches, which tends to load balance the system. This scheme reduces interprocessor communication as well, by exploiting **spatial coherence**. This is because patches in a block are likely to have references to other patches within the same block and those that are not located within a block are likely to be in the occlusion zone.

Computing Radiosities

Each processor serves as a host for shooting radiosity rays between patches within and outside its block. Each patch serves as a source patch to the rest of the patches in the environment. Once an intersection between a radiosity ray (originating from a source patch i) and another patch j is found, radiosity computations have to be performed. This involves computations of form factor between patches i and j.

The *direct* form factor specifies the direct energy transfer between two diffuse patches and is proportional to the projection of the target patch onto a hemisphere placed on the source patch. The hemisphere covers the entire half space seen by the source patch. An efficient but approximate way to compute form factors is to discretize the hemisphere into R number of distinct regions[4, 5, 6], of equal surface area, depending on the accuracy required while rendering. The form factor between any two patches i and j is then the cumulative contribution of the r regions covered by the projection of patch j on the hemisphere centered over patch i. Contribution of each discrete region is called δ form factor. δ form factors of all discrete regions can be computed *a priori* and stored.

Since only those patches that are not occluded with respect to the source patch are considered in the radiosity computation, the number of patches that must be considered in the form factor computation is drastically reduced.

We exploit **ray coherence** to reduce the expensive intersection computation. We define two levels of discretization of the hemisphere – one that is coarse and the other that is fine. Radiosity rays are shot with origin at the center of the hemisphere (which is the center of the source patch). The coarse discretization of the sphere defines **sectorial** regions in 3-D along the direction of the radiosity rays. To start with, we attribute a single ray to represent one coarse sector. Assuming an underlying 3-D grid structure that spans the environment, a volume cell is the region enclosed between adjacent grid lines along the X, Y and Z axes[3]. A radiosity ray intersects a volume cell along the direction of its traversal.

In case the volume cell is empty, then the ray is allowed to proceed forward until it intersects with another volume cell which is not empty. Once an intersection with a non-empty volume cell is found, we switch to fine discretization of the hemisphere and compute form factors. (Intersection computations are carried out using the subdivision and bounding box algorithm[1, 4]). Form factors are computed using δ form factors corresponding to the fine discretization of the hemisphere. Those hemisphere rays corresponding to this fine discretization which do not intersect with any patch in the volume cell are advanced further and checked for intersection with a new volume cell. Note that a majority of the fine rays will find intersection with the patches in the volume cell because of ray coherence. However, this will lead to fragmentation of the half space as seen by the source patch into regions that are not occluded. We will henceforth refer to such regions as **sectors**. It is this fragmentation of the half space into sectors that we exploit (*a la ray coherence*) and consider only a subset of all patches for radiosity computations in the proposed multiprocessor architecture.

Shooting rays from a source patch to other patches that are local to the processor do not involve any additional data movement other than loading onto registers from local store. Whereas rays are shot to patches that are not local to the processor only along sectors (*i.e.* in the direction of the non-occluded regions). The communication involved here corresponds to the attributes of the source patch, its current radiosity value, the ray direction and its associated level of discretization.

In the proposed architecture, all communications are totally overlapped with computations and this will be elucidated further in a following section. Clearly, the overheads due to communication between processors can be kept within tolerable bounds by ensuring adequate communication bandwidth through proper simulation of the multiprocessor architecture. Further each processor is designed to handle multiple rays (corresponding to fine discretization) shot from a source patch simultaneously. Processing associated with a ray is henceforth referred to as a **thread of computation**. Hence each processor is capable of supporting multiple threads of computation.

Since the multiprocessor architecture has P processors, radiosity rays are shot from P different source patches at the same instant. The source patch chosen at each processor is the one with highest radiosity from among all patches local to it, which have not served as source patches. This avoids explicit synchronization among processors. The architecture is therefore latency tolerant. The mathematical analogue of this approach is a chaotic relaxation of the radiosity equation $(I - \Lambda F)B = E$.

Below we give a pseudo code of the parallel progressive refinement algorithm in the multiprocessor sytem.

Parallel Progressive Refinement();{

```
for all M processor do in parallel
repeat
for each patch i in a block assigned to processor
do
 {Define a coarse-discretized hemisphere on i;
 Shoot rays and find intersection with a
 volume cell;
 while (not ray having met boundary of a block)
 do
  {if (volume cell not empty)
  {Change hemisphere discretization to fine;
  for all non-occluded patch j in volume cell
  do in parallel
  /* Exploit ray coherence while executing
  K threads */
  {Calculate f_ij using delta form factors;
  ΔRad = ρ_j Δb_i F_ij A_i/A_j;
  Δb_j = Δb_j + ΔRad;
  /* Update change since last time patch j
  shot light */
  b_j = b_j + ΔRad;
  /* Update total radiosity of patch j */
  }
  Δb_i = 0;
  /* Reset unshot radiosity of patch i to 0 */
  }
  else
   if (ray within environment boundary)
    Forward ray to find new volume cell;
  }
 if (ray hits block boundary)
  Communicate ray and attributes to
  neighbour processor;
 if (requests pending from other processors)
  Update radiosity b_k of the target patch k;
 Broadcast radiosities of all patches in
 the block to other processors;
 }
until convergence;}
```

SYSTEM ARCHITECTURE

The multiprocessor system comprises P identical processing elements (PEs) connected as a linear array.

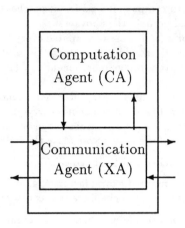

Figure 1: A Processing Element

Each PE comprises two basic units (see figure 1), *viz.*, a computation agent (CA) and a communication agent (XA). The two units communicate with each other asynchronously through messages. PEs communicate with each other using their respective XAs asynchronously. When a source PE p shoots a radiosity ray that must be forwarded to a neighbour processor, the ray and the source patch together with associated attributes are given to the XA by the CA with address of the destination PE q. XA of PE p sends this message over to the XA of a neighbour PE. On receipt of a message, a XA either forwards it to its CA, if the message is meant for it or forwards it to an appropriate neighbour. The CA of PE q, on receiving a request, computes the form factor and hence the contribution to radiosity of the target patch. Response to a request from a source PE terminates at the destination processor, after the corresponding radiosity value is computed. At the destination processor, the requests coming in from various source PEs need not be processed in the order of their arrival. This feature may be exploited for a solution of the radiosity equation through chaotic relaxation.

Figure 3 gives the architecture of the Computation Agent of a processing element. Form factor computation unit is replicated K times to support K threads of computation. The accuracy required while rendering determines K and is dependent on the ratio of fine to coarse discretization. In figure 3, the queues A, B, C, D and E facilitate asynchronous pipelined operation of the CA. Each queue stores tokens which are essentially information regarding the patch and ray.

SIMULATION RESULTS

A simulation platform for multiprocessor graphics system was built using C++. Simulations were carried out for a 2-dimensional scene upto 50000 randomly generated patches. The number of PEs was varied from 1 to 128.

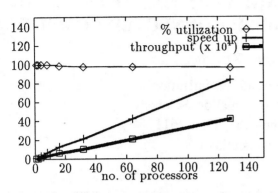

Figure 2: Performance Curves with 50,000 Patches

The following assumptions were made for simulation purposes. (See figure 2 for simulation results.)
(a) Each intersection computation consumes 200nsec.
(b) Delay on a communication link is $20\mu sec$.
(c) A ray crosses atmost one processor boundary before it is intersected by a patch.

CONCLUSIONS

In this paper we developed a parallel progressive refinement image rendering algorithm. We developed a linear array multiprocessor architecture, where each processor is capable of supporting multiple threads of computation.

Simulation studies reveal the following:
(a) The architecture is scalable, since the speed up shows a linear increase with increasing number of PEs.
(b) The system throughput is a linear function of the number of PEs. A 128-processor system sustains a system throughput of 0.4 million patches per second.
(c) A processor in a 128-processor system shows utilization of more than 90%.

ACKNOWLEDGMENTS

The authors thank Prof. Ed Deprettere and Prof. Patrick Dewilde for the technical discussions and suggestions. The authors acknowledge the systems support provided by Mr. S. Balakrishnan in developing the simulation platform for the graphic system. This work was partially supported by the TUD-IISc. collaboration funds and the KBCS project of the UNDP programme.

REFERENCES

[1] R. Pulleyblank and J. Kapenga, "The Feasibility of VLSI Chip for Ray Tracing Bicubic Patches", *IEEE Computer Graphics and Application*, (Vol. 7, No. 3, March 1987), pp. 33-44

[2] M. Cohen, S. Chen, R. Wallace and D. Greenberg, "A Progressive Refinement Approach to Fast Radiosity Image Generation", *Computer Graphics (SIGGRAPH 1988 Proceedings)*, (Vol. 22, No. 4, August 1988), pp. 75-84.

[3] C. E. Prakash and S. K. Nandy, "VOXEL based modeling and Rendering Irregular Solids", *Microprocessing and Microprogramming, (Proceedings of the EUROMICRO 90)*, (Vol. 30), pp. 341-346.

[4] A.C. Yilmaz, S. Hagestein, Ed. F. Deprettere and P. Dewilde, "A Hardware Algorithm for Fast Realistic Image Synthesis", *Fourth Eurographics Workshop on Graphics Hardware*, (January 1990).

[5] Jichun Bu and Ed. F. Deprettere, "A VLSI System Architecture for High Speed Radiative Transfer 3-D Image Systhesis", *The Visual Computer*, (No. 5, 1989), pp.121-133.

[6] S.K.Nandy, Ranjani Narayan, V. Visvanathan, "Computing Radiosities using an Adaptive Hemisphere Discretization Technique", *Technical Report*, (CAD Laboratory, IISc., March 93)

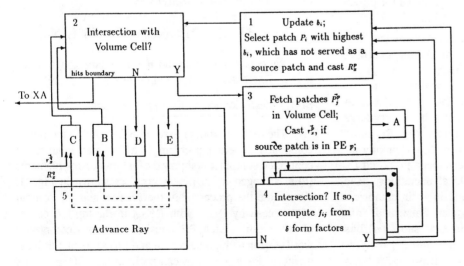

Figure 3: The Computation Agent of a PE p

$O(n)$–TIME AND $O(\log n)$–SPACE IMAGE COMPONENT LABELING WITH LOCAL OPERATORS ON SIMD MESH CONNECTED COMPUTERS

Hongchi Shi and Gerhard X. Ritter

Center for Computer Vision and Visualization
Department of Computer and Information Sciences
University of Florida, Gainesville, FL 32611
E-mail: hs@mosquito.cis.ufl.edu

Abstract — *A new parallel algorithm for image component labeling with local operators on SIMD mesh connected computers is presented, providing a positive answer to the open question posed by several researchers whether there exists an $O(n)$–time and $O(\log n)$–space local labeling algorithm on SIMD mesh connected computers. The algorithm uses a pipeline mechanism with stack–like data structures to achieve the lower bound of $O(n)$ in time complexity and $O(\log n)$ in space complexity. Furthermore, the algorithm has very small multiplicative constants in its complexities.*

INTRODUCTION

Labeling the connected components of a digitized image is a fundamental process in image analysis and machine vision [7]. The process of labeling assigns a unique label to the 1–pixels of each connected component in the image. Once an image has been labeled, the components which correspond to different objects can be studied, described, and possibly recognized by higher level image analysis algorithms. Labeling connected components has been intensively studied and many algorithms for different architectures have been proposed [2].

On an $n \times n$ SIMD mesh connected computer, Nassimi and Sahni [6] gave a well-known labeling algorithm which uses a divide-and-conquer technique with global operations and complex pointer operations to label an $n \times n$ binary image in $O(n)$ time and $O(\log n)$ bits of local memory. However, this algorithm has a very large multiplicative constant in its time complexity. Recently, some parallel algorithms based on the fast shrinking algorithm devised by Levialdi [5] have been proposed [3, 1]. These algorithms use only local operations, have very small multiplicative constants in their complexities, and are more practical. Cypher, Sanz, and Snyder's first algorithm [3] takes $O(n)$ time and $O(n)$ bits of local memory. Their second algorithm [3] requires only $O(\log n)$ bits of local memory. However, it takes $O(n \log n)$ time. Alnuweiri and Prasanna [1] provided an algorithm with a parameter k with $1 \le k \le \log(2n)$

which requires $O(kn^{1/k})$ bits of local memory and takes $O(kn)$ time. These well-known local labeling algorithms fail to achieve the lower bound of $O(n)$ in time comlexity and $O(\log n)$ in space complexity which can be achieved by the global labeling algorithm. Whether there exists a local labeling algorithm which can achieve that lower bound has remained an open question [1, 2]. Note that the label for each pixel has $O(\log n)$ bits and any algorithm based on Levialdi's shrinking algorithm has to propagate the labels $O(n)$ steps. The complexity in the open question should be assumed word complexity with a word having $O(\log n)$ bits.

Here, we propose an $O(n)$ time and $O(\log n)$ space local labeling algorithm which uses a pipeline mechanism with stack-like data structures and shrinks $O(\log n)$ images at each step. The algorithm achieves the complexity lower bound with small multiplicative constants, making it the first algorithm that is the most efficient in both practical and asymptotic complexity measures.

In the following presentation, we assume 8–connectivity among the pixels belonging to the same component. However, with simple modification, the algorithm can be extended to the 4–connectivity case.

LOCAL COMPONENT LABELING

We consider labeling an $n \times n$ binary image on an $n \times n$ SIMD mesh connected computer.

SIMD Mesh Connected Computer Model

The machine model used here for solving the image component labeling problem is a 2–dimensional mesh connected computer composed of n^2 processing elements (PEs) arranged in an $n \times n$ array as shown in Fig. 1. We denote the processor in the ith row and jth column of the mesh by $\text{PE}_{i,j}$ with $\text{PE}_{0,0}$ in the top left corner of the mesh. Each PE consists of a processor with word ($O(\log n)$ bits) data path and $O(\log n)$ bits of local memory. Each processor can perform any $O(\log n)$–bits logic and arithmetic operations in $O(1)$ time. The PEs operate in SIMD mode, with all control signals coming from a single control unit. Each PE that is not on the edge of the mesh is connected to each of its 4 neighbors via a word communication channel; and PEs on the edge are connected to fewer neighbors. When edge PEs are directed to read from nonexistent neighbors, they receive a 0; and data written to nonexistent neighbors are lost.

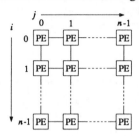

Fig. 1. Mesh connected computer model

To label an $n \times n$ binary image on an $n \times n$ mesh connected computer, we assign pixel (i,j) to $\mathrm{PE}_{i,j}$.

Parallel-Shrink and Label-Propagate Operators

Let a denote an $n \times n$ binary image on the point set $\mathbf{X} = \{(i,j) : 0 \leq i,j \leq n-1\}$ with $\mathbf{a}(i,j)$ being the value at pixel (i,j). A *neighborhood* $N(i,j)$ of pixel (i,j) is a set of image pixels around pixel (i,j). A *neighborhood operator* φ computes the new value of pixel (i,j) from the old values of the pixels in $N(i,j)$. If the distance between any pixel in $N(i,j)$ and pixel (i,j) is less than a small constant, the neighborhood operator φ is called a *local operator*.

We are interested in two basic local operators: parallel-shrink and label-propagate. The parallel-shrink operator uses the 2×2 neighborhood N_s and the label-propagate operator uses the 2×2 neighborhood N_p shown in Fig. 2.

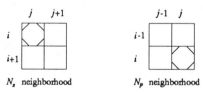

Fig. 2. Neighborhoods N_s and N_p

Levialdi [5] proposed a shrinking operator φ_s which is based on the Heaviside function and can be defined as

$$\varphi_s(i,j) = H(H(\mathbf{a}(i,j) + \mathbf{a}(i,j+1) + \mathbf{a}(i+1,j) - 1) + H(\mathbf{a}(i,j) + \mathbf{a}(i+1,j+1))),$$

where H is the Heaviside function defined by $H(t) = 0$ for $t \leq 0$ and $H(t) = 1$ if $t > 0$. This shrinking operator shrinks each component toward the top left corner of its bounding rectangle. Levialdi [5] has shown that a component with a bounding $r \times r$ rectangle will shrink

to a single black pixel after at most $2r - 2$ shrinking steps and then disappear in the next shrinking step. Thus, an $n \times n$ binary image can be shrunk to an all-zeros image in at most $2n - 1$ steps. The shrinking process preserves the connectivity property. Shi and Ritter [8] designed a new parallel shrinking operator which uses a 2×3 window and can shrink an $n \times n$ binary image to an all-zeros image in at most $\lceil 1.5n \rceil - 1$ shrinking steps. For simplicity, we will use Levialdi's shrinking operator here without affecting the asymptotic complexities of our local labeling algorithm. Since all the values involved in φ_s are binary, we can construct a truth table from the above formula for φ_s to obtain the following formula using the logic operations $\wedge$ (and) and $\vee$ (or).

$$\varphi_s(i,j) = (\mathbf{a}(i,j+1) \vee \mathbf{a}(i+1,j) \vee \mathbf{a}(i+1,j+1)) \\ \wedge \mathbf{a}(i,j) \vee \mathbf{a}(i,j+1) \wedge \mathbf{a}(i+1,j).$$

Let $\mathbf{a}_0$ denote the initial $n \times n$ binary image and $\mathbf{a}_l$ denote the image resulted from applying Levialdi's shrinking operator l times to $\mathbf{a}_0$. For an $n \times n$ image, we have a sequence of images $\mathbf{a}_{2n-1}, \mathbf{a}_{2n-2}, \ldots, \mathbf{a}_0$ with $\mathbf{a}_{2n-1}$ being an all-zeros image. We refer to l as the *index* of the image $\mathbf{a}_l$ in the sequence.

If $\mathbf{I}$ is a collection of k images with $\mathbf{I}[h]$ being some binary image $\mathbf{a}_l$, where $k = O(\log n)$ and $0 \leq h \leq k-1$, we can use

$$\varphi(i,j) = (\mathbf{I}(i,j+1) \vee_k \mathbf{I}(i+1,j) \vee_k \mathbf{I}(i+1,j+1)) \\ \wedge_k \mathbf{a}(i,j) \vee_k \mathbf{I}(i,j+1) \wedge_k \mathbf{I}(i+1,j)$$

to shrink k images simultaneously, where $\wedge_k$ and $\vee_k$ are bit-wise logic operations. This shrinking operator will be referred to as *parallel-shrink* operator. If $\mathbf{I}[h] = \mathbf{a}_{l_h}$, the new k images in $\mathbf{I}'$ resulted from applying parallel-shrink once to $\mathbf{I}$ is parallel-shrink($\mathbf{I}$) with $\mathbf{I}'[h] = \mathbf{a}_{l_h+1}$.

Note that the collection of $O(\log n)$ binary images is just an image with pixel values having $O(\log n)$ bits. It is not difficult to see that a parallel-shrink operation can be done in $O(1)$ time.

The *label-propagate* operator labels the 1–pixels of $\mathbf{a}_l$ from the already labeled image $\mathbf{a}_{l+1}$. Since the parallel-shrink operator may shrink different components into the same isolated pixel in different shrinking steps, we assign a unique new label $(i,j,l+1)$ to an isolated black pixel in $\mathbf{a}_l$ which represents a new component. Since $0 \leq i,j \leq n-1$ and $0 \leq l < 2n-1$, $(i,j,l+1)$ can be considered as a $(3\log n + 1)$–bits number with i in the first $\log n$ bits, j in the middle $\log n$ bits, and $l+1$ in the last $\log n + 1$ bits. Let $L_l(i,j)$ be the label of pixel (i,j) in $\mathbf{a}_l$ and $M_{l+1}(i,j) = \max\limits_{-1 \leq p,q \leq 0} \{L_{l+1}(i+p,j+q)\}$. Then,

$$L_l(i,j) = \begin{cases} M_{l+1}(i,j) & \text{if } \mathbf{a}_l(i,j) = 1, M_{l+1}(i,j) \neq 0 \\ (i,j,l+1) & \text{if } \mathbf{a}_l(i,j) = 1, M_{l+1}(i,j) = 0 \\ 0 & \text{otherwise.} \end{cases}$$

It is easy to see that a label-propagate operation can be done in $O(1)$ time.

Local Labeling Principle

The basic local labeling algorithm proceeds in two phases [4]. In the first phase, it applies Leviadi's shrinking operation φ_s to a binary $n \times n$ source image a_0 $2n-1$ times, resulting in $2n$ binary images $a_0, a_1, \ldots, a_{2n-1}$. Since a_{2n-1} is an all-zeros image, the label image for a_{2n-1} is also an all-zeros image. In the second phase, it applies a label-propagate operation to label image a_{2n-2} from a_{2n-1}, then to label a_{2n-3} from a_{2n-2}, and so on until a_0 is labeled. Note that the label-propagate operations are applied in the reverse order to the images generated by the shrinking operations. Intermediate images have to be stored or generated on the fly. If less than $2n$ bits of memory are allowed, then only a subset of the $2n$ intermediate images can be stored in the mesh at any time and the remaining images have to be generated in order during the labeling process. To label a_l from a_{l+1}, if a_l is in the memory, it can be labeled using a label-propagate operation; if it is not, we must generate it from the image a_r available in the memory with maximum index $r < l$.

A NEW FAST ALGORITHM

Let $2n = m^k$ for any integer k with $1 \le k \le \log(2n)$.

Data Structure

It is natural to have stack-like data structures to store the intermediate images. A *stack-like data structure* S of depth m shown in Fig. 3 can be considered as a data structure containing two regular stacks. When push pointer points to the bottom, S is empty; and when m elements have been pushed into an empty S, S is full. With pop(S), we obtain the top element and shift the elements in data 1 one position to the top. With push(a, S), we put the element a to the position pointed by push_pointer and make push_pointer point to the first unused element of S from the bottom.

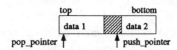

Fig. 3. Stack-like data structure

Each processing element $PE_{i,j}$ has $k+1$ stack-like data structures $S_{i,j}(k), S_{i,j}(k-1), \ldots, S_{i,j}(0)$, where $S_{i,j}(k)$ is 1-bit wide and 1-bit deep and each $S_{i,j}(h)$, $0 \le h \le k-1$, is 1-bit wide and m-bits deep. For any h, $0 \le h \le k$, $S_{i,j}(h)$'s with $0 \le i, j \le n-1$

form a 2-dimensional array of n^2 stack-like data structures, denoted by $S(h)$. Each stack-like data structure $S(h)$ stores images whose indices multiples of m^h in areas data 1 and data 2 with the highest index image on the top. Since all PEs in a SIMD computer perform the same instruction, any stack-like data structure operation is performed on the 2-dimensional array of all the corresponding stack-like data structures.

Pipeline Mechanism

In order to reduce the time complexity to $O(n)$ with only $\log n$ intermediate images stored, we introduce a pipeline mechanism to shrink k images at each time. The *pipeline* consists of k stages $Stage_0, Stage_1, \ldots, Stage_{k-1}$ with $Stage_h$ popping images from $S(h+1)$ and pushing images into $S(h)$. Given a full $S(h+1)$ by $Stage_{h+1}$, $Stage_h$ pops an image from it every m^{h+1} steps, shrinks the image step by step, and push the intermediate shrunk images every m^h steps into $S(h)$. Thus, each time when $Stage_h$ consumes a full $S(h+1)$, another full $S(h+1)$ has been produced by $Stage_{h+1}$. Each time when $Stage_h$ produces a full $S(h)$, the previous full $S(h)$ has been consumed by $Stage_{h-1}$. Therefore, we can organize the pipeline stages to work on k different images at each step. When $Stage_h$ produces its first full $S(h)$, $Stage_{h-1}$ starts to work. After $Stage_h$ produces its last full $S(h)$, it stops working. Whenever $Stage_0$ produces a full $S(0)$, the m images are popped from it and labeled by applying m label-propagate operations.

We use the step number to coordinate the pipeline stages. Let $Stage_{k-1}$ start to work at step $Start_{k-1} = 0$. Then, we can determine when each stage starts to work and when it stops working. $Stage_h$ starts to work at step $Start_h = \left(m^{k-(h+1)} - 1\right)m^{h+1}$ and stops working after step $Stop_h = \left(m^{k-(h+1)} - 1\right)m^{h+1} + \left(m^{k-h} - 1\right)m^h$, producing its first full $S(h)$ at step $Start_h + (m-1)m^h = \left(m^{k-h} - 1\right)m^h$ and its last full $S(h)$ at step $Start_h + \left(m^{k-h} - 1\right)m^h = \left(m^{k-(h+1)} - 1\right)m^{h+1} + \left(m^{k-h} - 1\right)m^h$. With this pipeline schedule, $Stage_h$ pops an image from $S(h+1)$ at a step which can be divided by m^{h+1} and push an image into $S(h)$ at a step which can be divided by m^h in its working period. It is easy to prove by induction that $Stage_h$ produces a full $S(h)$ exactly when $Stage_{h-1}$ needs one.

Labeling Algorithm

The new fast algorithm is presented below. The input is an $n \times n$ binary image a0 and the output is the label image la. Variable step for coordination of all the pipeline stages. Variable I holds all the current binary images of the k pipeline stages. Two more variables `high_s` and `low_s` are used to indicate that only $stage_{high\_s}$ through $Stage_{low\_s}$ are in their working periods. Initially, all stack-like data structures are empty and `la` contains all zeros.

```
Algorithm Local_component_labeling
  step = 0; low_s = high_s = k-1;
  push(a0, S(k));
  while (high_s >= 0) {
    if (step % m**low_s == 0) {
      last_s = max{s: step % m**s ==0
                   low_s<=st<=high_s};
      for (h=last_s; h>=low_s; h--) {
        if (step % m**(h+1) == 0)
          I[h] = pop(S(h));
        if (step % m**h == 0)
          push(I[h], S(h));
      }
      if ( (low_s == 0) &&
           (step+1 % m == 0) )
        for (l=0; l<m; l++)
          la = label_propagate(la,
                   pop(S(0)));
      if (step >= stop(high_s))
        high_s = high_s-1;
      if ( (low_s > 0) &&
           (step >= start(low_s-1)) )
        low_s = low_s-1;
    }
    I = parallel_shrink(I);
    step = step+1;
  }
end Algorithm.
```

Algorithm Correctness and Complexities

We define the *output* of a pipeline stage $Stage_h$ to be the sequence of images popped from $S(h)$ in the order they are popped. The following theorem can be proved by induction.

Theorem 1: *In the above algorithm, the output of* $Stage_h$ *is a sequence of* m^{k-h} *images* $a_{(m^{k-h}-1)m^h}$, $a_{(m^{k-h}-2)m^h}, \ldots, a_{lm^h}, \ldots, a_0$ *for* $0 \le h \le k-1$.

From Theorem 1, we know that the output of $Stage_0$ is $a_{m^k-1}, a_{m^k-2}, \ldots, a_0$. This is exactly the sequence of images the label-propagate procedure should use. Thus, the algorithm is correct.

Theorem 2: *The space complexity of the above algorithm is* $O(kn^{1/k})$ *and the time complexity is* $O(n)$.

It is not difficult to see that the space complexity of the above algorithm is $O(kn^{1/k})$. The time com-

plexity is dominated by the number of parallel-shrink operations which is $2m^k - m - 1$, the number of label-propagate operations which is $2n$, and the number of stack-like data structure push/pop operations which is

$$\sum_{h=0}^{k-1} \left(m^{k-(h+1)} + m^{k-h} \right) + 2n = \frac{m+1}{m-1}\left(m^k - 1\right) + 2n.$$

Thus, the time complexity is $O(n)$.

When $k = O(\log n)$, we have $O(kn^{1/k}) = O(\log n)$. Thus, the local labeling algorithm is an $O(n)$–time and $O(\log n)$–space algorithm.

The detail analysis of the algorithm correctness and complexities is given in [9]. Also, an illustrative example of the algorithm is given there.

CONCLUSION

The algorithm presented in this paper is the most efficient algorithm for image component labeling on mesh connected computers in both theoretical and practical complexity measures.

REFERENCES

1. H. Alnuweiri and V. Prasanna, *Fast image labeling using local operators on mesh-connected computers*, IEEE Transactions on Pattern Analysis & Machine Intelligence, PAMI-13 (1991), pp. 202–207.

2. ——, *Parallel architectures and algorithms for image component labeling*, IEEE Transactions on Pattern Analysis & Machine Intelligence, PAMI-14 (1992), pp. 1014–1034.

3. R. Cypher, J. Sanz, and L. Snyder, *Algorithms for image component labeling on SIMD mesh connected computers*, IEEE Transactions on Computers, 39 (1990), pp. 276–281.

4. F. Leighton, *Introduction to Parallel Algorithms and Architectures: Arrays, Trees, Hypercubes*, Morgan Kaufmann Publishers, Inc., San Mateo, CA, 1992.

5. S. Levialdi, *On shrinking binary picture patterns*, Communications of the ACM, 15 (1972).

6. D. Nassimi and S. Sahni, *Finding connected components and connected ones on a mesh-connected parallel computer*, SIAM Journal on Computing, 9 (1980), pp. 744–757.

7. A. Rosenfeld and A. Kak, *Digital Image Processing*, Academic Press, New York, 2nd ed., 1982.

8. H. Shi and G. Ritter, *A fast parallel binary image shrinking algorithm*. Submitted, 1993.

9. ——, *O(n)-time and O(log n)-space image component labeling with local operations on SIMD mesh connected computers*. Preprint, 1993.

Solving the Region Growing Problem on the Connection Machine

Nawal Copty, Sanjay Ranka, Geoffrey Fox, and Ravi Shankar

School of Computer and Information Science
Syracuse University, Syracuse, NY 13244
Email: nkcopty, ranka, rshankar@top.cis.syr.edu, gcf@npac.syr.edu

Abstract − *This paper presents a parallel algorithm for solving the region growing problem based on the split and merge approach. The algorithm was implemented on the CM-2 and the CM-5 in the data parallel and message passing models. The performance of these implementations is examined and compared.*

Keywords: Region growing, Split and merge, Data parallel, Message passing, Connection machine.

THE REGION GROWING PROBLEM

Region growing is a general technique for image segmentation. Image characteristics are used to group adjacent pixels together to form regions. Regions are then merged with other regions to *grow* larger regions. A region might correspond to a world object or a meaningful part of one [2].

The merging of pixels or regions to form larger regions is usually governed by a *homogeneity criterion* that must be satisfied. A variety of homogeneity criteria have been investigated for region growing. The *pixel range* homogeneity criterion requires that the difference between the minimum and maximum intensities within a region not exceed a threshold value T.

There are many approaches for solving the region growing problem [1, 2, 10]. This paper presents a parallel algorithm for solving the problem based on the split and merge approach [5]. The algorithm aims to reduce the number of merge steps required to identify the regions in the image by using a preprocessing split stage.

While previous parallel implementations of the split and merge approach have used dynamic or tree structures to represent the regions in the image [8, 9], our implementations use only one and two-dimensional arrays to solve the problem. Moreover, we introduce an element of randomness to the merging of regions. For a detailed presentation, refer to [3].

THE SPLIT AND MERGE APPROACH

The split and merge approach solves the region growing problem in two stages: the *split stage* and the *merge stage*.

The Split Stage

In the split stage, an $N \times N$ image is partitioned into square regions which satisfy the homogeneity criterion. At first, each pixel is considered a homogeneous square region of size 1×1. Then every group of four adjacent pixels are tested for homogeneity. If the homogeneity criterion is satisfied, the pixels are combined into one larger square region of size 2×2, and so on. The split stage terminates when the whole image is one square region of size $N \times N$, or when no more square regions can be merged. Figure 1 shows the square regions produced by the split stage for a 4×4 image, where the threshold value $T = 3$.

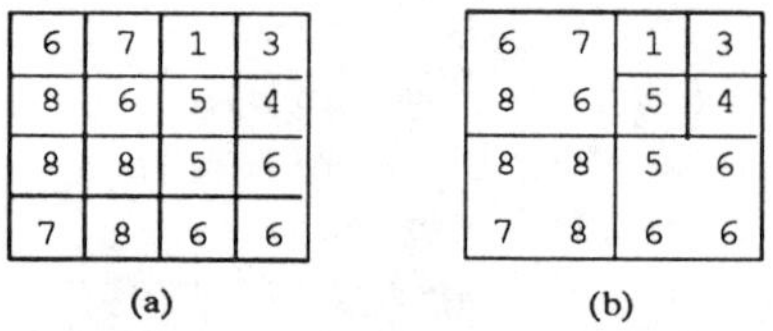

Square regions: (a) at start of the split stage; (b) after first and final split iteration

Figure 1: The Split Stage

The Merge Stage

In the merge stage, the square regions are iteratively merged into larger and larger regions which satisfy the homogeneity criterion. The merge continues until no more merges are possible.

The merge is achieved by reformulating the region growing problem as a weighted, un-directed graph problem, where the vertices of the graph represent the regions in the image, and the edges represent the neighboring relationships among these regions. That is, an edge e exists between two vertices v and w of the graph, if and only if the regions represented by v and w share a common boundary. The weight of the edge e is the difference between the maximum and minimum pixel values in the union of the two regions represented by v and w.

Obviously, only vertices connected by edges satisfying the homogeneity criterion can be merged. In one merge iteration, each region selects for merging its neighbor that *best satisfies* the homogeneity criterion. A tie may be broken by selecting the neighbor with the smallest (largest) ID, or by selecting a neighbor at random. Two regions actually merge if they select each other for merging. Once two regions merge, the vertices and edges of the graph are updated to reflect the new regions in the image.

Figure 2 shows the different regions obtained and their corresponding graphs in each iteration of the merge stage, for the 4×4 image of Figure 1. Ties are broken by selecting the neighbor with the smallest ID. The small numbers appearing in the upper left-hand corners of the regions denote the region IDs.

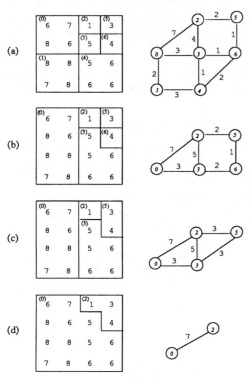

Regions: (a) at start of the merge stage; (b) after first merge iteration; (c) after second merge iteration; (d) after third and final merge iteration

Figure 2: The Merge Stage

PARALLEL IMPLEMENTATIONS

The region growing problem is a representative of a type of loosely synchronous problems, known as adaptive irregular problems, whose data objects evolve during the computation in a time synchronized manner [4]. The problem exhibits a dynamic behavior that starts with a high degree of parallelism that very rapidly diminishes to a much lower degree of parallelism.

The split and merge algorithm for solving the region growing problem was implemented in both the data parallel and message passing models. In the data parallel model, the CM Fortran programming language was used, and the same program was executed on both the CM-2 and the CM-5. In the message passing model, on the other hand, sequential Fortran 77 supplemented with message passing library routines (CMMD) was used, and the program was executed on the CM-5. Only one and two-dimensional arrays were used to represent the various data items required. Two-dimensional arrays were used to store the intensities and other information pertaining to the pixels, while one-dimensional arrays were used to store information about the vertices and edges of the graph modeling the problem.

Data Parallel Implementation

The data parallel implementation of the split and merge algorithm consists of the following steps:

1. The two-dimensional pixel image is repeatedly split into homogeneous square regions. The split stage stops when the whole image is one homogeneous square region, or when no more merges are possible.

2. For each square region in the pixel image, a corresponding graph vertex is created, and for each pair of neighboring square regions, an edge is created. Edges that do not satisfy the homogeneity criterion are de-activated.

3. A region determines its neighboring region that best satisfies the homogeneity criterion. In the case of a tie, one of the neighboring regions is chosen at random. Two regions merge if their merge choices are mutual. In one merge iteration, several region pairs can merge at the same time without conflicting with each other.

4. When two regions merge, the region with the smaller ID becomes the representative of the two. The vertices and edges of the graph are updated to reflect the new regions in the image. Edges that do not satisfy the homogeneity criterion are de-activated.

5. If there still exist active edges, then steps 3 and 4 are repeated. Otherwise, the program terminates.

Message Passing Implementation

The message passing implementation of the split and merge algorithm is a hand-coded translation of the data parallel one and consists of the following steps:

0. The image is mapped to the node processor grid such that each processor receives an $\frac{N}{P1} \times \frac{N}{P2}$ sub-image of the original image. This partitioning maintains adjacency between neighboring blocks of the image.

1. Each node processor splits independently its $\frac{N}{P1} \times \frac{N}{P2}$ sub-image and determines the homogeneous square regions within it.

2. Each node processor sets up the vertices and edges of the graph associated with its sub-image. Boundary information is exchanged so that edges connected to vertices in other processors are created.

3. The node processors cooperate to merge the homogeneous square regions.

4. The node processors cooperate to update the vertices and edges of their graphs.

5. If there still exist any active edges in any of the node processors, then steps 3 and 4 are repeated. Otherwise, the node programs terminate.

At several points in the message passing implementation, irregular communication is required, where each of the node processors sends zero or more messages to other processors in an irregular fashion.

Two different communication schemes were investigated. The first, called **Linear Permutation (LP)** [7], uses synchronous (blocked) message passing. In this scheme, each node obtains a copy of the communication matrix, using a global concatenation operation. Then in step i, $0 < i < Q$, processor p_k sends a message to processor $p_{(k+i)}$ MOD Q, and receives a message from processor $p_{(k-i)}$ MOD Q, where Q is the total number of node processors. The second communication scheme uses **asynchronous** message passing.

Resolving Ties at Random

In order to achieve a higher degree of parallelism, we introduced an element of randomness in our parallel implementations. In case of a tie during the merge stage, the tie is broken by selecting a neighbor *at random* instead of selecting the neighbor with the smallest (largest) ID, since the latter approach imposes a serialization on the order of the merges. The random approach in breaking ties was shown to be significantly faster than the approach of selecting the neighbor with the smallest (largest) ID, since it generally results in a larger number of merges per merge iteration.

Complexity

Given an $N \times N$ pixel image, the complexity of the parallel split and merge algorithm depends on the number of processors (P) used and the number of iterations required to find the regions in the image.

The Split Stage: In the best case, when every pixel is a region by itself, only one split iteration is required. In the worst case, when the whole image is one homogeneous square region, $log(N)$ split iterations are required.

The time complexity of the split stage in the data parallel implementation on the CM-2 is $O(\frac{N^2}{P} + logP)$, while that of the data parallel and message passing implementations on the CM-5 is $O(\frac{N^2}{P} + (\tau \times logP))$, where τ is the communication set up time for one split iteration.

The Merge Stage: In the best case, a region consisting of R sub-regions will require $logR$ iterations to merge. In the worst case, when only one pair of regions is merged in each iteration, it will require $R-1$ merge iterations.

Let R_i denote the number of homogeneous square regions found in the image at the end of the split stage, and let R_t denote the number of regions found at the end of the merge stage. If we assume that the number of regions is reduced by a factor of k at every step in the merge stage, then the time complexity of the merge stage in the data parallel implementation on the CM-2 is $O(\frac{R_i \times logR_i}{P} + log_k \frac{R_i}{R_t} \times logP)$. For details of the analysis and assumptions made, refer to [3].

The time complexity of the the data parallel and message passing implementations on the CM-5, on the other hand, is difficult to analyze, as the number of messages sent by the processors in each step of the merge stage depends on the image.

PERFORMANCE

The data parallel implementation (using CM Fortran) of the split and merge algorithm was executed on both a 16K CM-2 and a 32-node CM-5, while the message passing implementation (using F77 + CMMD) was executed on a 32-node CM-5. A variety of images were used. The performance of the implementations for images of sizes 128×128 and 256×256 is presented below. LP refers to the Linear Permutation communication scheme and Async refers to the asynchronous one.

Image 1: 128×128 image composed of two nested rectangular regions
No. of square regions found at end of split stage = 436
No. of regions found at end of merge stage = 2

	Split (secs)	Split Iters	Merge (secs)	Merge Iters
CM Fortran on :				
CM-2 (8K procs)	0.200	4	9.511	19
CM-2 (16K procs)	0.112	4	7.027	20
CM-5 (32 nodes)	0.361	4	33.013	19
F77 + CMMD on :				
CM-5 (32 nodes, LP)	0.022	4	6.914	24
CM-5 (32 nodes, Async)	0.021	4	4.025	20

Image 2: 128×128 image composed of a collection of rectangles
No. of square regions found at end of split stage = 193
No. of regions found at end of merge stage = 7

	Split (secs)	Split Iters	Merge (secs)	Merge Iters
CM Fortran on :				
CM-2 (8K procs)	0.200	4	8.184	18
CM-2 (16K procs)	0.112	4	5.345	17
CM-5 (32 nodes)	0.360	4	31.615	20
F77 + CMMD on :				
CM-5 (32 nodes, LP)	0.022	4	9.236	35
CM-5 (32 nodes, Async)	0.021	4	6.441	35

Image 3: 128×128 image composed of a collection of circles
No. of square regions found at end of split stage = 1732
No. of regions found at end of merge stage = 11

	Split (secs)	Split Iters	Merge (secs)	Merge Iters
CM Fortran on :				
CM-2 (8K procs)	0.200	4	13.711	24
CM-2 (16K procs)	0.112	4	9.538	25
CM-5 (32 nodes)	0.361	4	42.570	27
F77 + CMMD on :				
CM-5 (32 nodes, LP)	0.022	4	9.454	33
CM-5 (32 nodes, Async)	0.021	4	5.516	28

Image 4: 256×256 image composed of two nested rectangular regions
No. of square regions found at end of split stage = 823
No. of regions found at end of merge stage = 2

	Split (secs)	Split Iters	Merge (secs)	Merge Iters
CM Fortran on :				
CM-2 (8K procs)	1.008	5	13.882	26
CM-2 (16K procs)	0.529	5	10.381	28
CM-5 (32 nodes)	2.052	5	37.588	25
F77 + CMMD on :				
CM-5 (32 nodes, LP)	0.097	5	16.512	37
CM-5 (32 nodes, Async)	0.097	5	10.942	29

Image 5: 256×256 image composed of a collection of rectangles
No. of square regions found at end of split stage = 298
No. of regions found at end of merge stage = 7

	Split (secs)	Split Iters	Merge (secs)	Merge Iters
CM Fortran on :				
CM-2 (8K procs)	1.008	5	9.287	19
CM-2 (16K procs)	0.529	5	6.633	20
CM-5 (32 nodes)	2.046	5	24.471	16
F77 + CMMD on :				
CM-5 (32 nodes, LP)	0.099	5	14.388	35
CM-5 (32 nodes, Async)	0.098	5	6.640	35

Image 6: 256×256 image of a "tool"

No. of square regions found at end of split stage = 2248

No. of regions found at end of merge stage = 4

	Split (secs)	Split Iters	Merge (secs)	Merge Iters
CM Fortran on :				
CM-2 (8K procs)	1.008	5	19.530	34
CM-2 (16K procs)	0.529	5	13.426	33
CM-5 (32 nodes)	2.066	5	75.582	45
F77 + CMMD on :				
CM-5 (32 nodes, LP)	0.098	5	12.192	36
CM-5 (32 nodes, Async)	0.098	5	7.236	38

The bar chart of Figure 3 gives a visual comparison of the times taken by the merge stage in the various implementations.

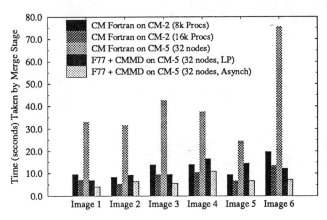

Figure 3: Comparison of Times Taken by the Merge Stage (Images 1-6)

OBSERVATIONS

In examining the performance of the different parallel implementations of the split and merge algorithm, we make the following observations:

- The number of merge iterations required to find the regions in an image are not identical in all cases. The random numbers generated, as well as the order in which messages are received affect the actual merges that take place and hence the number of merge iterations required to solve the problem.

- Asynchronous communication on the CM-5 is faster than Linear Permutation, since in Linear Permutation the nodes must loop a larger number of times to complete the required communications.

- The CM Fortran version on the CM-2 runs faster than that on the CM-5. The SIMD hardware of the CM-2 directly supports the data parallel model, while compilers, assemblers, and other system software of the CM-5 have to deal with the many "housekeeping" details such as load balance and synchronization.

- The message passing implementation runs significantly faster than the data parallel one on the CM-5. The data parallel implementation relies on the CM Fortran compiler as well as the run-time system to lay out the data and to provide communication among the nodes, while, in the message passing implementation, the programmer exercises control over synchronization, data partitioning, and load balancing.

- CM Fortran allows only limited ways of distributing data on different processors. With the availability of new data distribution directives in High Performance Fortran, the performance of the data parallel implementation is expected to be closer to the message passing one.

Acknowledgements: We would like to thank Paul Coddington, Pablo Tamayo, and Jhy-Chun Wang for interesting and helpful discussions, and Gregor von Laszewski for help in preparing the manuscript.

References

[1] H. Alnuweiri and V. Prasanna, "Parallel Architectures and Algorithms for Image Component Labeling", *IEEE Trans. Patt. Anal. Machine Intell.*, Vol. 14, 1992, pp. 1014-1034.

[2] D. Ballard and C. Brown, *Computer Vision*, Prentice Hall, Englewood Cliffs, New Jersey, 1982.

[3] N. Copty, S. Ranka, G. Fox, and R. Shankar, "Solving the Region Growing Problem on the Connection Machine", Technical Report, January 1993, Northeast Parallel Architectures Center, Syracuse University.

[4] G. Fox et al, "Software support for irregular and loosely synchronous problems", Technical Report, May 1992, Northeast Parallel Architectures Center, Syracuse University.

[5] S. L. Horowitz and T. Pavlidis, "Picture Segmentation By a Directed Split-and-Merge Procedure", *Proc. 2nd International Joint Conference on Pattern Recognition*, pp. 424-433, August 1974.

[6] S. Ranka and S. Sahni, *Hypercube Algorithms.* Springer-Verlag, New York, 1990.

[7] S. Ranka, J. Wang, and G. Fox, "Static and runtime algorithms for all-to-many personalized communication on permutation networks", *Proc. International Conference on Parallel and Distributed Systems*, December 1992.

[8] J. C. Tilton, "Image segmentation by iterative parallel region growing with applications to data compression and image analysis", *Proc. 2nd Symposium on the Frontiers of Massively Parallel Computation*, 1988.

[9] M. Willebeek-LeMair and A. Reeves, "Solving nonuniform problems on SIMD computers: Case study on region growing", *Journal of Parallel and Distributed Computing*, Vol. 8, 1990, pp. 135-149.

[10] Zucker, "Region growing: Childhood and adolescence", *Computer Graphics and Image Processing*, Vol. 5, 1976, pp. 382-399.

SESSION 5C

NUMERICAL ANALYSIS

MINIMUM COMPLETION TIME CRITERION
FOR PARALLEL SPARSE CHOLESKY FACTORIZATION

Wen-Yang Lin and Chuen-Liang Chen
Department of Computer Science and Information Engineering
National Taiwan University
Taipei, Taiwan, ROC
E-mail:clchen@csie.ntu.edu.tw

Abstract -- *It is well known that a judicious choice of ordering has great impact on the sparse matrix factorization. Many proposed reordering algorithms attempt to minimize the corresponding elimination tree height, which is, however, not an accurate indication of the actual parallel factorization time. We will illustrate the appalling discrepancy with a contrived example. To alleviate this deficiency, we define a parallel timing function. Using it as criterion, we will present a reordering algorithm such that the obtained filled graph is preserved and the completion time for parallel factorization is minimum among all equivalent reorderings.*

1. INTRODUCTION

Many scientific applications give rise to systems of linear equations for which the coefficient matrices are large, sparse, positive definite, and symmetric. Solving such linear systems has had great impact on the design of new computer architectures, compilers, etc. In [5] the solution process of such systems of equations is separated into four stages. Of which, the first stage is to find an ordering of the coefficient matrix A to reduce the amount of fill-ins that occur during the factorization process. However, simply reducing fill-ins may not provide an ordering appropriate for parallel factorization. An approach to generate low fill-in orderings suitable for parallel sparse factorization is to decouple the reduction of fill-in and enhancement of parallelism into separate phases. First a good fill-reducing ordering, such as minimum degree [8,11], is applied, then based on this initial ordering an equivalent reordering is produced such that the reordered matrix can be factored effectively in parallel.

In 1982, Jess and Kees [7] proposed an algorithm that produces a reordering whose corresponding elimination tree height is minimum over all perfect orderings of the filled graph. Many methods adapted from them then were introduced [12,13,14]. These methods all use the height of the elimination tree as the criterion. That implies the execution time associated with each node is equivalent. This criterion, however, is not an accurate indication of the actual parallel factorization time [13]. To alleviate this deficiency, we define a parallel completion time function adapted from Liu [9]. Using it as the criterion, we will present a reordering algorithm such that the obtained filled graph is preserved and show that its completion time for parallel factorization is minimum among all equivalent reorderings.

An outline of the paper is as follows. First, we will describe some necessary background material for parallel sparse factorization. In Section 3, we will define the parallel completion time criterion and give an example to show that minimizing the height of an elimination tree does not necessarily minimize the parallel completion time. In the extreme, the completion time of a minimum-height elimination tree will be infinite-fold of the minimum value. Our algorithm of finding an ordering with minimum completion time is described and proved in Section 4. Finally, Section 5 gives our conclusions.

2. BACKGROUND

Let A be an $n \times n$ sparse positive definite symmetric matrix and, after applying a minimum degree ordering P (or permutation), the resulting matrix be PAP^T. There is an unique $n \times n$ lower triangular matrix L with positive diagonal such that $PAP^T = LL^T$. This is the Cholesky factorization of PAP^T. In this paper, we will assume the column Cholesky factorization, a method that has been widely implemented in many sparse matrix software packages [16], is used.

The *filled graph* of PAP^T, denoted as G, is used to indicate the nonzeros and possible fill-ins in PAP^T during factorization, i.e., a node v_i denotes the ith column and an edge (v_i, v_j) denotes a nonzero or fill-in,

This work is supported in part by the National Science Council of R.O.C., under contract number NSC82-0408-E-002-400.

for $1 \le i, j \le n$.

An *equivalent reordering* of PAP^T refers to a permutation on PAP^T such that the filled graph is preserved. This means the permuted matrix will suffer the same fill-in and require the same number of operations for factorization. From the work of Rose [1,2] and Liu [10], it is known that finding an equivalent reordering of PAP^T is equal to finding a perfect ordering on G. Moreover, the process to find an ordering on G is usually modeled by a sequence of elimination steps. Initially, let $G^{(0)} = G$. In the ith step, for $1 \le i \le n$, a node v is eliminated from $G^{(i-1)}$ and its adjacent set is mutually connected. The reduced graph is then called $G^{(i)}$.

A clique is a set of nodes with the property that all its members are pairwise adjacent and, moreover, if no other node can be added while preserving the pairwise adjacent property then the clique is called *maximal*. A node v in graph G is called *simplicial* if its adjacency set $Adj_G(v)$ is a clique and hence the elimination of a simplicial node creates no filled edge [4].

To study the potential parallelism existing in sparse matrix factorization, a commonly used structure is *elimination tree*, which was introduced in [6] and was first used to study parallel elimination in [7]. Following [15], the elimination tree T associated with G contains the same nodes as G and for each v_k with $k < n$, its parent node is v_p, where $p = \min \{j \mid j > k$ and $l_{jk} \ne 0\}$ and l_{jk} is the entry in the jth row and kth column of the matrix L. Figure 1 shows a matrix PAP^T, filled graph G and elimination tree T, where a symbol '×' denotes a nonzero or fill-in due to factorization.

It is worth to mention that, for column Cholesky factorization, a node v_j in the elimination tree represents the task associated with column j in PAP^T that is defined as

$$Tcol(j) = \left\{ cmod(j,k) \mid k < j \wedge l_{jk} \ne 0 \right\} \cup \left\{ cdiv(j) \right\},$$

where $cmod(j, k)$ denotes modification of column j by column k and $cdiv(j)$ denotes division of column j by a scalar [9]. Jess and Kees [7] ignore the fact that a variant number of operations exist in each column for a sparse matrix and assume all $Tcol(j)$ spend the same execution time, for $1 \le j \le n$. The primary deficiency of this assumption is that it can not reflect the real completion time for parallel factorization. Thought their algorithm guarantees a minimum height elimination tree, it does not imply the completion time is minimized. To avoid this problem, Liu [9,13] suggested using the parallel completion time as the criterion.

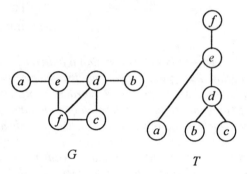

$$PAP^T = \begin{pmatrix} a & & & & \times & \\ & b & & \times & & \\ & & c & \times & & \times \\ \times & \times & d & \times & \times \\ \times & & & \times & e & \times \\ & \times & \times & \times & & f \end{pmatrix}$$

FIG 1. *An 6 × 6 matrix, filled graph and elimination tree.*

3. PARALLEL COMPLETION TIME CRITERION

Definitions

Adapted from Liu's definition, we modify the graph model used by Jess and Kees such that each node v_j in the elimination tree remains representing the task associated with a column j, i.e., $Tcol(j)$, but the operations in $Tcol(j)$ are taken into account. With this graph model, we define the execution time of a node v_j as

$$\begin{aligned} time_T(v_j) &= \text{\# of multiplicative operations in } Tcol(j) \\ &= \sum_{k < j \wedge l_{jk} \ne 0} \left| \{cmod(j,k)\} \right| + |cdiv(j)| \\ &= \sum_{k \le j \wedge l_{jk} \ne 0} \left| \{ l_{ik} \mid l_{ik} \ne 0, \text{ for } i \ge j \} \right| \end{aligned}$$

and the completion time of v_j in T as

$$Ctime_T(v_j) = \begin{cases} time_T(v_j), & v_j \text{ is terminal} \\ time_T(v_j) + mchtime_T(v_j), & \text{else} \end{cases}$$

where $mchtime_T(v_j)$ denotes the maximum completion time of children of v_j and equals to

$$\max \left\{ Ctime_T(x) \mid x \in child_T(v_j) \right\}$$

Note that the completion time represents the time required to complete factoring jth column at the disposal of unlimited number of processors. Under this definition, the parallel completion time of the whole factorization, denoted as $Ctime_T$, is equal to $Ctime_T(v_n)$ and corresponds to a critical weighted path in the elimination tree T. Figure 2 shows the $time_T(v)$ / $Ctime_T(v)$ for each node v of the elimination tree T in Figure 1.

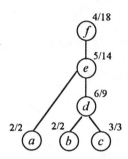

FIG 2. *The $time_T(v)$ / $Ctime_T(v)$ for each node v of the elimination tree T in FIG 1.*

Evaluation

In studying sparse matrix's factorization, the *clique* concept is useful. It has been shown that G can be represented as a set of maximal cliques [4] and we assume G comprises $K_1, K_2, ..., K_q$ maximal cliques. Similar to the meaning of $G^{(i)}$, we denote the *residual clique* [14] of K_j after i elimination steps as $K_j^{(i)}$. For a node v in G, let the *monotone (prior)* adjacency set $Madj(v)$ ($Padj(v)$) denote the set of all nodes adjacent to numbered higher (lower) than v, and $Mdeg(v)$ ($Pdeg(v)$) the size of the set. The following Lemmas give a basis to evaluate the execution time of a node.

Lemma 1 [14]. *A node is simplicial if and only if it belongs to only one maximal clique.* □

Lemma 2 [14]. *For a maximal clique K, if v is the lowest-numbered node, then $K = \{v\} \cup Madj(v)$.* □

Lemma 3. *A node becoming simplicial will remain simplicial until it is eliminated.* □

Corollary 4. *For a perfect ordering of G, a node can be eliminated in the i-th elimination step if and only if it is simplicial in $G^{(k)}$ for some $k < i$.* □

Consider a perfect ordering on G and the ith node v_i. From Lemma 1 and Corollary 4, v_i is simplicial in $G^{(i-1)}$, and contained in single maximal residual clique, which is denote as $K_{\Omega(v_i)}^{(i-1)}$. Then by definition and Lemma 2 we can deduce the following equation:

$$|cdiv(i)| = |\{v_i\} \cup Madj(v_i)|$$
$$= Mdeg(v_i) + 1 \qquad (1)$$
$$= |K_{\Omega(v_i)}^{(i-1)}|$$

Now, let us consider the execution time of v_i. According to the definition, it is obviously that $time_T(v_i)$ equals to the number of nonzeros in the submatrix, say M, with row denoting $\{v_i\} \cup Madj(v_i)$ and column $Padj(v_i) \cup \{v_i\}$, meaning that the modification corresponds to the nonzeros in columns $Padj(v_i)$ and division as stated in Eq. (1). From Eq. (1), the number of nonzeros in column j of M, however, equals to the size of the residual clique that contains v_j in the $(j-1)$th elimination step, i.e., $|K_{\Omega(v_j)}^{(j-1)}|$, minus the number of nodes in $K_{\Omega(v_j)}^{(j-1)}$ that eliminated during steps j to $i-1$. In other words, it equals to $|K_{\Omega(v_j)}^{(i-1)}|$ and hence we obtain the following theorem.

Theorem 5. *For a perfect ordering on G, the execution time of the i-th node v_i is*

$$time_T(v_i) = \sum_{v_j \in Padj(v_i)} \left| K_{\Omega(v_j)}^{(i-1)} \right| + \left| K_{\Omega(v_i)}^{(i-1)} \right|. \qquad □$$

To make a clear understanding of Theorem 5, we illustrate the evaluation of the execution time of each node in Table 1.

TABLE 1. *The evaluation of the time for each node in FIG 1.*

step i	v	$Padj(v)$	$K_{\Omega(v)}$	$time_T(v)$	$K_1^{(i)}$	$K_2^{(i)}$	$K_3^{(i)}$	$K_4^{(i)}$
0	-	-	-	-	$\{a\ e\}$	$\{b\ d\}$	$\{c\ d\ f\}$	$\{d\ e\ f\}$
1	a	ϕ	K_1	2	$\{a\ e\}$	$\{b\ d\}$	$\{c\ d\ f\}$	$\{d\ e\ f\}$
2	b	ϕ	K_2	2	$\{e\}$	$\{b\ d\}$	$\{c\ d\ f\}$	$\{d\ e\ f\}$
3	c	ϕ	K_3	3	$\{e\}$	$\{d\}$	$\{c\ d\ f\}$	$\{d\ e\ f\}$
4	d	$\{b\ c\}$	K_4	6	$\{e\}$	$\{d\}$	$\{d\ f\}$	$\{d\ e\ f\}$
5	e	$\{a\ d\}$	K_4	5	$\{e\}$	ϕ	$\{f\}$	$\{e\ f\}$
6	f	$\{c\ d\ e\}$	K_4	4	ϕ	ϕ	$\{f\}$	$\{f\}$

Comparison with Minimum Height Criterion

In this subsection, we will construct a class of filled graphs $\flat_n$, $n \geq 2$, such that for $\flat_n$ the completion time of its corresponding minimum-height elimination tree will be n times the completion time of another elimination tree.

Consider $\flat_n = (V_n, E_n)$, where V_n and E_n represent the set of nodes and edges in $\flat_n$, respectively. We define V_n and E_n as follows.

$$V_n = \{a_i | 1 \leq i \leq n\} \cup \{b_i | 1 \leq i \leq n\} \cup$$
$$\left(\bigcup_{i=1}^{n} \{c_{ij} | 1 \leq j \leq 2i - 1\} \right) \cup$$
$$\left(\bigcup_{i=1}^{n} \{d_{ijk} | 1 \leq j \leq m, 1 \leq k \leq 2i\} \right),$$

$$E_n = \left\{(a_i, a_{i+1}) \middle| 1 \le i \le n-1\right\} \cup$$
$$\left(\left\{(b_i, a_i) \middle| 1 \le i \le n\right\} \cup \left\{(b_i, a_{i+1}) \middle| 1 \le i \le n-1\right\}\right) \cup$$
$$\left\{(c_{i1}, a_i) \middle| 1 \le i \le n\right\} \cup$$
$$\left\{(c_{ij}, c_{i,j+1}) \middle| 1 \le i \le n, 1 \le j \le 2i-2\right\} \cup$$
$$\left\{(d_{ij1}, b_i) \middle| 1 \le i \le n, 1 \le j \le m\right\} \cup$$
$$\left\{(d_{ijk}, d_{ij,k+1}) \middle| 1 \le i \le n, 1 \le j \le m, 1 \le k \le 2i-1\right\},$$

where m denotes some integer.

We first consider the case for b_2 as shown in Figure 3. Its minimum-height elimination tree and reduced completion time elimination tree, denoted as T_2 and T'_2 respectively, are shown in Figure 4. The value in each node denotes its execution time that is calculated according to Theorem 5. It is easy to verify that $Ctime_{T_2} / Ctime_{T'_2} = (2m + 17)/(m + 18) \approx 2$ for large m.

For b_n, the minimum height elimination tree T_n and the elimination tree with reduced completion time T'_n can be constructed as shown in Figure 5, where for simplicity the subtrees containing nodes cs and ds are omitted. The following equations related to the execution times for all nodes in T_n and T'_n are straightforward.

$$time_{T_n}(a_i) = \begin{cases} 4, & \text{if } i = 1, \\ 5, & \text{if } i = n, \\ 6, & \text{otherwise}. \end{cases}$$

$$time_{T_n}(b_i) = \begin{cases} m + 2, & \text{if } i = n, \\ m + 4, & \text{otherwise}. \end{cases}$$

$$time_{T_n}(c_{ij}) = \begin{cases} 2, & \text{if } j = 2i - 1, \\ 3, & \text{otherwise}. \end{cases}$$

$$time_{T_n}(d_{ijk}) = \begin{cases} 2, & \text{if } k = 2i, \\ 3, & \text{otherwise}. \end{cases}$$

$$time_{T'_n}(a_i) = \begin{cases} 4, & \text{if } i = 1, \\ 5, & \text{if } i = n, \\ 7, & \text{otherwise}. \end{cases}$$

$$time_{T'_n}(b_i) = \begin{cases} m + 4, & \text{if } i = 1, \\ m + 2, & \text{if } i = n, \\ m + 3, & \text{otherwise}. \end{cases}$$

$$time_{T'_n}(c_{ij}) = \begin{cases} 2, & \text{if } j = 2i - 1, \\ 3, & \text{otherwise}. \end{cases}$$

$$time_{T'_n}(d_{ijk}) = \begin{cases} 2, & \text{if } k = 2i, \\ 3, & \text{otherwise}. \end{cases}$$

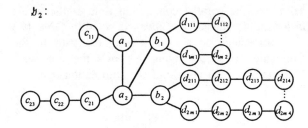

FIG 3. *An illustration of b_2*

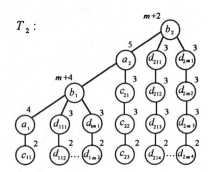

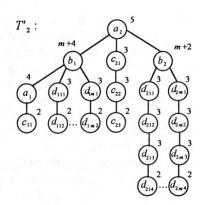

FIG 4. *The minimum height and reduced completion time elimination trees of b_2.*

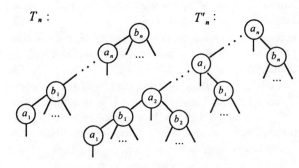

FIG 5. *The constructions of T_n and T'_n.*

For T_n the critical weighted path is $c_{11}, a_1, b_1, ..., a_n, b_n$ and for T'_n it is through the subtree rooted with b_2 to a_n. According to the above equations, it then follows that $Ctime_{T_n} / Ctime_{T'_n} = (mn + 10n - 3) / (m + 6n + 8)$ $\approx n$ for $m \gg n$ and $\lim_{n \to \infty} Ctime_{T_n} / Ctime_{T'_n} = \infty$.

Reordering Under Completion Time Criterion

Under the parallel timing function defined previously, we will show an important property associated with the perfect orderings on G, which is useful for later construction and proving of our reordering algorithm.

Theorem 6. *Let $J = (u_1, u_2, ..., u_n)$ be a perfect ordering of G. Assume there exist two nodes u_{k-1}, u_k such that they are simplicial in $G^{(k-2)}$ and*
(1) they belong to different maximal cliques, or
(2) they belong to the same maximal clique and

$$\max\{Ctime_{T_J}(x) | x \in child_{T_J}(u_{k-1})\} \geq$$
$$\max\{Ctime_{T_J}(y) | y \in child_{T_J}(u_k) - \{u_{k-1}\}\}.$$

Then, the ordering $L = (u_1, ..., u_{k-2}, u_k, u_{k-1}, u_{k+1}, ..., u_n)$ is also a perfect ordering of G and $Ctime_Q \leq Ctime_P$, where P, Q represent the corresponding elimination trees associated with orderings J and L, respectively.

Proof. Since u_{k-1}, u_k are simplicial in $G^{(k-2)}$, then from Corollary 4 L is also a perfect ordering. To prove $Ctime_Q \leq Ctime_P$, let us consider two cases: u_{k-1} and u_k belong to the same maximal clique in $G^{(k-2)}$ or not.

In the former case, no edge exists between u_{k-1} and u_k, so they must be on different paths to the root of the tree. The interchanging of u_{k-1} and u_k thus will not change the original tree structure and the completion time as well.

In the latter case, there exists a direct edge between them in $G^{(k-2)}$. According to the definition of the elimination tree, u_k must be the father of u_{k-1}. Suppose that the father of u_k is u_j for some j, $k + 1 \leq j \leq n$, and the maximal clique containing u_{k-1}, u_k in $G^{(k-2)}$ is K. Then we can construct the subtree associated with u_j as Figure 6(a), where S_j is the set of the subtree rooted with the children of u_j excluding u_k, and S_{k-1}, S_k are defined similarly. In addition, let t_j, t_k and t_{k-1} denote the completion time associated with S_j, S_k and S_{k-1}.

First, if all nodes in S_{k-1} are not adjacent to u_k in G, then, after interchange of u_{k-1}, u_k, this subtree will become as Figure 6(b). Note as J is transformed into L, only the orders of u_{k-1} and u_k within the cliques containing them both are interchanged, which means, recalled Theorem 5, the sizes of the corresponding residual cliques in computing the execution times of u_{k-1} and u_k are exchanged. Let the essentially exchanged parts for u_k and u_{k-1} be α and β, respectively and assume $time_P(u_j) = c$, $time_P(u_{k-1}) = a + \alpha$ and

$time_P(u_k) = b + \beta$, it then follows that $time_Q(u_j) = c$, $time_Q(u_{k-1}) = a + \beta$, and $time_Q(u_k) = b + \alpha$. From the known condition $t_k \leq t_{k-1}$, we have

$$Ctime_P(u_j) = c + \max\{Ctime_P(u_k), t_j\},$$
$$= c + \max\{b + \beta +$$
$$\max\{Ctime_P(u_{k-1}), t_k\}, t_j\},$$
$$= c + \max\{b + \beta + a + \alpha + t_{k-1}, t_j\},$$

and for ordering L

$$Ctime_Q(u_j) = c + \max\{Ctime_Q(u_{k-1}), t_j\},$$
$$= c + \max\{a + \beta +$$
$$\max\{Ctime_Q(u_k), t_{k-1}\}, t_j\},$$
$$= c + \max\{a + \beta +$$
$$\max\{b + \alpha + t_k, t_{k-1}\}, t_j\}.$$

Now, depending on whether t_{k-1} is greater than or less than $b + \alpha + t_k$, $Ctime_Q(u_j)$ may be $c + \max\{a + \beta + t_{k-1}, t_j\}$ or $c + \max\{a + \beta + b + \alpha + t_k, t_j\}$. Both of them are less than or equal to $Ctime_P(u_j)$.

On the other hand, if there exist some nodes in S_{k-1} adjacent to u_k in G, then, by the definition of elimination tree, only those also being children of u_{k-1} will become children of u_k in Q. Let the set of subtrees rooted with these nodes be S_a and let t_a denote its completion time while $t_{a'}$ the completion time of $S_{k-1} - S_a$. The subtree rooted with u_j in Q is as Figure 6(c). Now, if S_a contains the node with maximum completion time among those in S_{k-1}, then $\max\{t_a, t_k\} = t_a = t_{k-1} \geq t_{a'}$ and, following the previous analysis, we can deduce that $Ctime_Q(u_j) = Ctime_P(u_j)$. Otherwise, $t_{a'} = t_{k-1} \geq \max\{t_a, t_k\}$ and again it follows that $Ctime_Q(u_j) \leq Ctime_P(u_j)$.

From the forgoing, since only the subtree rooted with u_j are modified and $Ctime_Q(u_j) \leq Ctime_P(u_j)$, we conclude that $Ctime_Q \leq Ctime_P$. $\square$

4. REORDERING ALGORITHM

Description

The basic concept of our algorithm is a kind of greedy method [3]. Given an initial ordering of A and its correspondent filled graph G, we will find an equivalent reordering to minimize the parallel completion time of factorization. We assume in the ith elimination step, the elimination tree composed of the previous $i - 1$ nodes has been constructed. Then the node eliminated next must satisfy
(1) it is a simplicial node in the current graph and

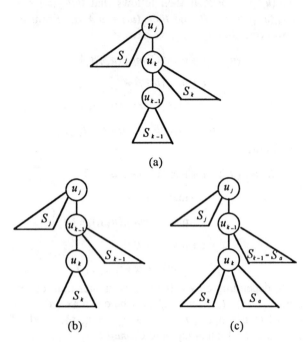

FIG. 6. *Subtrees rooted with u_j for orderings J and L.*

(2) its completion time is the smallest among all the simplicial nodes.

Under this basic concept, our algorithm follows.

Algorithm. *Parallel Reordering.*
begin
    given an initial ordering of A and its filled graph G;
    $i \leftarrow 0$; $G^{(i)} \leftarrow G$;
    $\ell \leftarrow$ a sorted list of all simplicial nodes in $G^{(i)}$ by
        their completion times in increasing order;
    while $\ell \neq \phi$ **do**
        delete the first node v from ℓ and number it;
        $i \leftarrow i + 1$; $G^{(i)} \leftarrow G^{(i-1)} - \{v\}$;
        for $u \in Adj_{G^{(i-1)}}(v)$ **do**
            if u is new simplicial in $G^{(i)}$ **then**
                compute $Ctime(u)$ and insert it to ℓ;
                for $w \in Adj_{G^{(i)}}(u)$ **do**
                    if $Ctime(u) > mchtime(w)$ **then**
                        $mchtime(w) \leftarrow Ctime(u)$;
        endfor
    endwhile
end

In the while loop, the evaluation of the completion time of a new simplicial node deserves more deliberation. Recall from Theorem 5, only if we know the relative orders of a node within all the cliques containing it can we calculate its execution time. That is, the execution time is determined as being eliminated

but in our algorithm we have to determine it as a node first becoming simplicial. Fortunately, the fact that a node in ℓ is simplicial means it is only in a maximal clique and all nodes in the other cliques containing it have been eliminated. Therefore, its relative orders within the maximal cliques containing it except the current one have been determined. All we have to do is to determine the order of a simplicial node within the current maximal clique containing it. The following Lemmas give this determination.

Lemma 7. *The nodes belong to the same maximal clique are on the same path to the root of the elimination tree.* □

Lemma 8. *In the previous algorithm, as a node is deleted from ℓ, its completion time is no less than those eliminated before it and is no greater than those eliminated after it.*

Proof. It is straight forward from the fact that a node v deleted from ℓ has the minimum completion time among all nodes in ℓ and the new simplicial nodes to be inserted into S are ancestors of v. □

Lemma 9. *Assume in the i-th elimination step, after eliminating node v there are some nodes in $G^{(i)}$, say u_1, u_2, ..., u_s, becoming simplicial. Let the maximal clique containing them is K, then,*
(1) the relative orders between u_1, u_2, ..., u_s in K are irrelevant and
(2) they should be ordered before those nonsimplicial nodes in K.

Proof. Clearly, v and u_1, u_2, ..., u_s are in the same maximal clique of $G^{(i-1)}$ and by Lemma 7, they should form a linear path on the elimination tree. To prove the first case, let us assume u_1, u_2, ..., u_s are ordered before those nonsimplicial nodes in K.

According to Lemma 8, the completion times of all children of u_i, $1 \leq i \leq s$, that eliminated before v are no greater than the completion time of v. But simultaneously, there maybe exist some nodes in K that becoming simplicial earlier and remaining in ℓ. Let the one with maximum completion time be x, then clearly $Ctime_T(x) \geq Ctime_T(v)$. Without loss of generality, let u_1, u_2, ..., u_s are numbered as the ordering they are. Depending on the existence of x, the completion time of u_s will equal to $Ctime_T(x) + \sum_{i=1}^{s} time_T(u_i)$ or $Ctime_T(v) + \sum_{i=1}^{s} time_T(u_i)$. From the derivation of Theorem 5, however, it can be seen that $\sum_{i=1}^{s} time_T(u_i)$ is the same for all permutations of u_1, u_2, ..., u_s. Hence, the orders between u_1, u_2, ..., u_s are irrelevant.

Now assume there exists a nonsimplicial node w in K not eliminated later than u_1, u_2, ..., u_s and ordered immediately before u_j, $1 \leq j \leq s$. The fact that w is nonsimplicial implies it has some children ordered later than children of u_j excluding w. Then according to Lemma 8 and Theorem 6, w should be ordered after u_j. This will repeat till w is ordered after $u_1, u_2, ..., u_s$ and is true for all nonsimplicial nodes in K. Hence, we prove the second case. $\square$

On the basis of Lemma 9, once a node become simplicial then we can calculate its execution time and so its completion time. Furthermore, it can be proved that the complexity of our algorithm is $O(n \log n + e)$, where n is the number of nodes and e the number of edges in G.

Minimum Completion Time Property

In this subsection, we will show that given a filled graph G, the completion time associated with the ordering obtained from our algorithm is minimum among all perfect orderings on G.

Lemma 10. *The ordering obtained from our algorithm is a perfect ordering on G.* $\square$

Theorem 11. *The elimination tree associated with the ordering obtained from our algorithm has the minimum completion time among all elimination trees associated with perfect orderings on G.*

Proof. Let $I = (u_1, u_2, ..., u_n)$ be the ordering obtained by our algorithm and $J = (v_1, v_2, ..., v_n)$ be an optimal ordering. Then I and J are both perfect orderings. We will show that J can be transformed into I without any increase in the completion time. Hence, I must also be optimal.

Assume $I \neq J$. Then let k be the least index such that $u_k \neq v_k$. Since I and J are both perfect orderings, then by Corollary 4 u_k and v_k are both simplicial in $G^{(k-1)}$. Let us assume $v_j = u_k$ for some j, $k < j \leq n$, and L is a reordering of J such that v_j is ordered immediately before v_k while other nodes are unchanged, i.e., $L = (v_1, v_2, ..., v_{k-1}, v_j, v_k, ..., v_{j-1}, v_{j+1}, ..., v_n)$. We will show that L can be obtained from a sequence of reorderings on J, for which v_j is moved left one position each time. The sequence is defined as follows:

$$R_j^{(0)}(J) = J,$$

$$R_j^{(i)}(J) = R_j(R_j^{(i-1)}(J))$$
$$= (v_1, v_2, ..., v_{j-i-1}, v_j, v_{j-i}, ..., v_{j-1}, v_{j+1}, ..., v_n), \quad \text{for } 1 \leq i \leq j - k.$$

Note that v_j is simplicial in $G^{(k-1)}$, then by Lemma 3 it remains simplicial in $G^{(i)}$, for $i \geq k$. Hence the reorderings of J defined above are also perfect.

Now, let us consider $R_j^{(i)}(J)$. For simplicity we let P be the corresponding elimination tree of $R_j^{(i-1)}(J)$ and Q the elimination tree of $R_j^{(i)}(J)$. To show that $Ctime_Q \leq Ctime_P$, we only have to deliberate, according

to Theorem 6, the case that v_j and v_{j-i} belong to the same maximal clique in $G^{(j-i-1)}$. It should be true that v_j become simplicial earlier than v_{j-i}. Otherwise, since v_j is simplicial in $G^{(k-1)}$, so v_j, v_{j-i} should be in the same maximal clique in $G^{(k-1)}$. From Lemma 9, v_{j-i} will be ordered before v_j by our algorithm, i.e., $v_{j-i} \in \{v_1, v_2, ..., v_{k-1}\}$, which contradicts to our assumption.

Let the subtree rooted with v_j in T_R be as follows.

We assume v_j and v_{j-i} are becoming simplicial in the pth and qth step, respectively, for $p \leq q \leq j - i - 1$ and $p \leq k - 1$. Since v_j and v_{j-i} are in the same maximal clique of $G^{(k-1)}$, by the definition of elimination tree and Lemma 7, it follows that $S_j \subseteq \{v_1, v_2, ..., v_p\}$ and $v_q \in S_{j-i}$. If $q \leq k - 1$, then clearly $t_j \leq t_{j-i}$. Otherwise, by Lemmas 7 and 8, we can deduce $Ctime_p(v_q) \geq t_j$ and so $t_j \leq t_{j-i}$. Therefore, $Ctime_Q \leq Ctime_P$ by Theorem 6. This shows that J can be transformed to L without increasing the completion time and repeating this transformation, L will ultimately become I. $\square$

Example

Take the graph b_2 constructed in Figure 3 for example and assume $m = 2$, we show some critical derivations in Table 2, where the value below each node in the list ℓ represents its completion time. It can verify that the corresponding elimination tree is as T'_2 in Figure 3. Similarly, for general b_n, it can be deduced that the completion time of T'_n is minimum.

5. CONCLUSIONS

We have defined the parallel completion time criterion and shown it is more realistic than the elimination tree height criterion. Using the parallel timing function, we have presented an algorithm for finding an equivalent reordering for sparse matrices. We have also shown the equivalent reordering obtained from our method has the minimum completion time. Note that we have assumed the column Cholesky factorization is used through all discussions in this article. Our next step is to incorporate the other forms of Cholesky factorization into the completion time criterion, find the optimal reorderings and sketch the relation between the reorderings under different Cholesky factorizations.

TABLE 2. *The list ℓ for b_2 and $m = 2$.*

step i	max_child of a_1	b_1	a_2	b_2	list ℓ
0	c_{11}	-	-	-	$\begin{array}{cccccc} c_{11} & d_{112} & d_{122} & c_{23} & d_{214} & d_{224} \\ 2 & 2 & 2 & 2 & 2 & 2 \end{array}$
1	c_{11}	a_1	a_1	-	$\begin{array}{cccccc} d_{112} & d_{122} & c_{23} & d_{214} & d_{224} & a_1 \\ 2 & 2 & 2 & 2 & 2 & 6 \end{array}$
$\vdots$	$\vdots$				$\vdots$
10	c_{11}	a_1	c_{21}	-	$\begin{array}{ccccc} d_{111} & a_1 & c_{21} & d_{212} & d_{222} \\ 5 & 6 & 8 & 8 & 8 \end{array}$
11	c_{11}	a_1	b_1	-	$\begin{array}{ccccc} a_1 & c_{21} & d_{212} & d_{222} & b_1 \\ 6 & 8 & 8 & 8 & 12 \end{array}$
$\vdots$	$\vdots$				$\vdots$
16	c_{11}	a_1	b_1	d_{211}	$\begin{array}{cc} d_{211} & b_1 \\ 11 & 12 \end{array}$
17	c_{11}	a_1	b_2	d_{211}	$\begin{array}{cc} b_1 & b_2 \\ 12 & 15 \end{array}$
18	c_{11}	a_1	b_2	d_{211}	$\begin{array}{cc} b_2 & a_2 \\ 15 & 20 \end{array}$
$\vdots$	$\vdots$				$\vdots$

REFERENCES

[1] D.J. Rose, "Triangulated graphs and the elimination process," *Journal of Mathematical Analysis and Application,* vol. 32 (December, 1970), pp. 597-609.

[2] D.J. Rose, "A graph-theoretic study of the numerical solution of sparse positive definite systems of linear equations," *Graph Theory and Computing*, Academic Press, New York, (1972), pp. 183-217.

[3] E. Horowitz, and S. Sahni, *Fundamentals of Computer Algorithms*, Computer Science Press, Potomac, Maryland, (1978).

[4] M. Golumbic, *Algorithmic Graph Theory and Perfect Graph*, Academic Press, New York, (1980).

[5] A. George, and J.W.H. Liu, *Computer Solution of Large Sparse Positive Definite Systems*, Prentice Hall, Englewood Cliffs, New Jersey, (1981).

[6] I.S. Duff, "Full matrix techniques in sparse Gaussian elimination," *Lecture Notes in Mathematics (912)*, G. Watson, ed., Springer-Verlag, Berlin, (1982), pp. 71-84.

[7] J.A.G. Jess, and H.G.M. Kees, "A data structure for parallel L/U decomposition," *IEEE Transactions on Computers*, vol. 31 (March, 1982), pp. 231-239.

[8] J.W.H. Liu, "Modification of the minimum-degree algorithm by multiple elimination," *ACM Transactions on Mathematical Software*, vol. 11 (June, 1985), pp. 141-153.

[9] J.W.H. Liu, "Computational models and task scheduling for parallel sparse Cholesky factorization," *Parallel Computing*, vol. 3 (October, 1986), pp. 327-342.

[10] J.W.H. Liu, "Equivalent sparse matrix reordering by elimination tree rotations," *SIAM Journal on Scientific and Statistical Computing*, vol. 9 (May, 1988), pp. 424-444.

[11] A. George, and J.W.H. Liu, "The evolution of the minimum degree ordering algorithm," *SIAM Review*, vol. 31 (March, 1989), pp. 1-19.

[12] J.W.H. Liu, and A. Mirzaian, "A linear reordering algorithm for parallel pivoting of Chordal graphs," *SIAM Journal on Discrete Mathematics*, vol. 2 (February, 1989), pp. 100-107.

[13] J.W.H. Liu, "Reordering sparse matrices for parallel elimination," *Parallel Computing*, vol. 11 (July, 1989), pp. 73-91.

[14] J.G. Lewis, B.W. Peyton, and A. Pothen, "A fast algorithm for reordering sparse matrices for parallel factorization," *SIAM Journal on Scientific and Statistical Computing*, vol. 10 (November, 1989), pp. 1146-1173.

[15] J.W.H. Liu, "The role of elimination trees in sparse factorization," *SIAM Journal on Matrix Analysis and Application*, vol. 11 (January, 1990), pp. 134-172.

[16] M.T. Heath, E. Ng, and B.W. Peyton, "Parallel algorithms for sparse linear systems," *SIAM Review*, vol. 33 (September, 1991), pp. 420-460.

Scalability of Parallel Algorithms for Matrix Multiplication*

Anshul Gupta   and   Vipin Kumar

Department of Computer Science,
University of Minnesota
Minneapolis, MN - 55455

agupta@cs.umn.edu and *kumar@cs.umn.edu*

Abstract

A number of parallel formulations of dense matrix multiplication algorithm have been developed. For arbitrarily large number of processors, any of these algorithms or their variants can provide near linear speedup for sufficiently large matrix sizes and none of the algorithms can be clearly claimed to be superior than the others. In this paper we analyze the performance and scalability of a number of parallel formulations of the matrix multiplication algorithm and predict the conditions under which each formulation is better than the others. We present a parallel formulation for hypercube and related architectures that performs better than any of the schemes described in the literature so far for a wide range of matrix sizes and number of processors. The superior performance and the analytical scalability expressions for this algorithm are verified through experiments on the Thinking Machines Corporation's CM-5^{TM}† parallel computer for up to 512 processors.

1   Introduction

Matrix multiplication is widely used in a variety of applications and is often one of the core components of many scientific computations. Since dense matrix multiplication algorithm is highly computation intensive, there has been a great deal of interest in developing parallel formulations of this algorithm and testing its performance on various parallel architectures [1, 2, 4, 5, 6, 8, 9, 12, 13, 18, 7].

Some of the early parallel formulations of matrix multiplication were developed by Cannon [4], Dekel, Nassimi and Sahni [8], and Fox *et. al.* [9]. Variants and improvements of these algorithms have been presented in [2, 13]. In particular, Berntsen [2] presents an algorithm which has a strictly smaller communication overhead than Cannon's algorithm, but has a smaller degree of concurrency [11].

For arbitrarily large number of processors, any of these algorithms or their variants can provide near linear speedup for sufficiently large matrix sizes, and none of the algorithms can be clearly claimed to be superior than the others. Scalability analysis is a an effective tool for predicting the performance of various algorithm-architecture combinations. Hence a great deal of research has been done to develop methods for scalability analysis [15]. In this paper, we use the isoefficiency metric [10, 15] to analyze the scalability of a number of parallel formulations of the matrix multiplication algorithm for hypercube and related architectures. We analyze the performance of various parallel formulations of the matrix multiplication algorithm for different matrix sizes and number of processors, and predict the conditions under which each formulation is better than the others. We present a parallel algorithm for the hypercube and related architectures that performs better than any of the previously described schemes for a wide range of matrix sizes and number of processors. The superior performance and the analytical scalability expressions for this algorithm are verified through experiments on the CM-5 parallel computer for up to 512 processors.

In this paper we assume that on a message passing parallel computer, the time required for the complete transfer of a message containing m words between two adjacent processors is given by $t_s + t_w m$, where t_s is the message startup time, and t_w (per-word communication time) is equal to $\frac{y}{B}$ where B is the bandwidth of the communication channel between the processors in bytes/second and y is the number of bytes per word. For the sake of simplicity, we assume that each basic arithmetic operation (*i.e.*, one floating point multiplication and one floating point addition in case of matrix multiplication) takes unit time. Therefore, t_s and t_w are relative data communication costs normalized w.r.t. the unit computation time.

2   The Isoefficiency Metric of Scalability

It is well known that given a parallel architecture and a problem instance of a fixed size, the speedup of a parallel algorithm does not continue to increase

*This work was supported by IST/SDIO through the Army Research Office grant # 28408-MA-SDI to the University of Minnesota and by the University of Minnesota Army High Performance Computing Research Center under contract # DAAL03-89-C-0038.

†CM-5 is a trademark of the Thinking Machines Corporation.

with increasing number of processors but tends to saturate or peak at a certain value. For a fixed problem size, the speedup saturates either because the overheads grow with increasing number of processors or because the number of processors eventually exceeds the degree of concurrency inherent in the algorithm. For a variety of parallel systems, given any number of processors p, speedup arbitrarily close to p can be obtained by simply executing the parallel algorithm on big enough problem instances [15]. The ease with which a parallel algorithm can achieve speedups proportional to p on a parallel architecture can serve as a measure of the scalability of the parallel system.

Let us define the size W of a problem as the time taken by an optimal (or the best known) sequential algorithm to solve the given problem on a single processor. Let $T_o(W, p)$ be the sum total of all the overheads incurred by all the p processors during the parallel execution of the algorithm. Now the efficiency of a parallel algorithm-architecture combination (henceforth referred to as a parallel system) is given by $E = \frac{W}{W + T_o(W, p)} = \frac{1}{1 + \frac{T_o(W, p)}{W}}$. For a class of parallel systems called **scalable** parallel systems, the efficiency can be maintained at a desired value (between 0 and 1) for increasing p, provided W is also increased. In order to maintain a fixed efficiency, W should be proportional to $T_o(W, p)$ or the following relation must be satisfied :

$$W = KT_o(W, p), \qquad (1)$$

where $K = \frac{E}{1-E}$ is a constant depending on the efficiency to be maintained. The isoefficiency function [10, 15] of a parallel system is determined by abstracting W as a function of p through algebraic manipulations on Equation (1). If the problem size needs to grow as fast as $f_E(p)$ to maintain an efficiency E, then $f_E(p)$ is defined as the isoefficiency function of the parallel system for efficiency E. The smaller the isoefficiency function, the more scalable the parallel system is considered.

The isoefficiency function of a combination of a parallel algorithm and a parallel architecture relates the problem size to the number of processors necessary to maintain a fixed efficiency or to deliver speedups increasing proportionally with increasing number of processors. For a given parallel algorithm, for different parallel architectures, W may have to increase at different rates w.r.t. p in order to maintain a fixed efficiency. A small rate or isoefficiency function indicates a high scalability. Isoefficiency analysis has been found to be very useful in characterizing the scalability of a variety of parallel systems [15]. An important feature of isoefficiency analysis is that in a single expression, it succinctly captures the effects of characteristics of the parallel algorithm as well as the parallel architecture on which it is implemented.

3 Parallel Matrix Multiplication Algorithms

In this section we briefly describe some well known parallel matrix multiplication algorithms give their parallel execution times.

3.1 A Simple Algorithm

Consider a logical two dimensional mesh of p processors (with $\sqrt{p}$ rows and $\sqrt{p}$ columns) on which two $n \times n$ matrices A and B are to be multiplied to yield the product matrix C. Let $n \geq \sqrt{p}$. The matrices are divided into sub-blocks of size $\frac{n}{\sqrt{p}} \times \frac{n}{\sqrt{p}}$ which are mapped naturally on the processor array. The algorithm can be implemented on a hypercube by embedding this processor mesh into it. In the first step of the algorithm, each processor acquires all those elements of both the matrices that are required to generate the $\frac{n^2}{p}$ elements of the product matrix which are to reside in that processor. This involves an all-to-all broadcast of $\frac{n^2}{p}$ elements of matrix A among the $\sqrt{p}$ processors of each row of processors and that of the same sized blocks of matrix B among $\sqrt{p}$ processors of each column which can be accomplished in $2t_s \log p + 2t_w \frac{n^2}{\sqrt{p}}$ time. After each processor gets all the data it needs, it multiplies the $\sqrt{p}$ pairs of sub-blocks of the two matrices to compute its share of $\frac{n^2}{p}$ elements of the product matrix. Assuming that an addition and multiplication takes a unit time, the multiplication phase can be completed in $\frac{n^3}{p}$ units of time. Hence, the parallel execution time is:

$$T_p = \frac{n^3}{p} + 2t_s \log p + 2t_w \frac{n^2}{\sqrt{p}} \qquad (2)$$

This algorithm is memory-inefficient. The memory requirement for each processor is $O(\frac{n^2}{\sqrt{p}})$ and thus the total memory requirement is $O(n^2 \sqrt{p})$ words as against $O(n^2)$ for the sequential algorithm.

3.2 Cannon's Algorithm

A parallel algorithm that is memory efficient and is frequently used is due to Cannon [4, 1]. Again the two $n \times n$ matrices A and B are divided into square submatrices of size $\frac{n}{\sqrt{p}} \times \frac{n}{\sqrt{p}}$ among the p processors of a wrap-around mesh (which can be embedded in a hypercube). The sub-blocks of A and B residing with the processor (i, j) are denoted by A^{ij} and B^{ij} respectively, where $0 \leq i < \sqrt{p}$ and $0 \leq j < \sqrt{p}$. In the first phase of the execution of the algorithm, the data in the two input matrices is aligned in such a manner that the corresponding square submatrices at each processor can be multiplied together locally. This is done by sending the block A^{ij} to processor

$(i, (j + i)mod\sqrt{p})$, and the block B^{ij} to processor $((i + j)mod\sqrt{p}, j)$. The copied sub-blocks are then multiplied together. Now the A sub-blocks are rolled one step to the left and the B sub-blocks are rolled one step upward and the newly copied sub-blocks are multiplied and the results added to the partial results in the C sub-blocks. The multiplication of A and B is complete after $\sqrt{p}$ such steps. On a hypercube with cut-through routing, the time spent in the initial alignment step can be ignored w.r.t. to the $\sqrt{p}$ shift operations during the multiplication phase, as the former is a simple one-to-one communication along non-conflicting paths. Since each sub-block movement in the second phase takes $t_s + t_w \frac{n^2}{p}$ time, the total parallel execution time for all the movements of the sub-blocks of both the matrices is given by the following equation:

$$T_p = \frac{n^3}{p} + 2t_s\sqrt{p} + 2t_w \frac{n^2}{\sqrt{p}} \qquad (3)$$

3.3 Berntsen's Algorithm

Due to nearest neighbor communications on the $\sqrt{p} \times \sqrt{p}$ wrap-around array of processors, Cannon's algorithm's performance is the same on both mesh and hypercube architectures. In [2], Berntsen describes an algorithm which exploits greater connectivity provided by a hypercube. The algorithm uses $p = 2^{3q}$ processors with the restriction that $p \leq n^{3/2}$ for multiplying two $n \times n$ matrices A and B. Matrix A is split by columns and B by rows into 2^q parts. The hypercube is split into 2^q subcubes, each performing a submatrix multiplication between submatrices of A of size $\frac{n}{2^q} \times \frac{n}{2^{2q}}$ and submatrices of B of size $\frac{n}{2^{2q}} \times \frac{n}{2^q}$ using Cannon's algorithm. It is shown in [2] that the time spent in data communication by this algorithm on a hypercube is $2t_s p^{1/3} + \frac{1}{3}t_s \log p + 3t_w \frac{n^2}{p^{2/3}}$, and hence the total parallel execution time is given by the following equation:

$$T_p = \frac{n^3}{p} + 2t_s p^{1/3} + \frac{1}{3}t_s \log p + 3t_w \frac{n^2}{p^{2/3}} \qquad (4)$$

The terms associated with both t_s and t_w are smaller in this algorithm than the algorithms discussed in Sections 3.1 to 3.2. It should also be noted that this algorithm, like the one in Section 3.1 is not memory efficient as it requires storage of $2\frac{n^2}{p} + \frac{n^2}{p^{2/3}}$ matrix elements per processor.

3.4 The DNS Algorithm

3.4.1 One Element Per Processor Version

An algorithm that uses a hypercube with $p = n^3 = 2^{3q}$ processors to multiply two $n \times n$ matrices was proposed by Dekel, Nassimi and Sahni in [8, 17].

The p processors can be visualized as being arranged in an $2^q \times 2^q \times 2^q$ array. In this array, processor p_r occupies position (i, j, k) where $r = i2^{2q} + j2^q + k$ and $0 \leq i, j, k < 2^q$. Thus if the binary representation of r is $r_{3q-1}r_{3q-2}...r_0$, then the binary representations of i, j and k are $r_{3q-1}r_{3q-2}...r_{2q}$, $r_{2q-1}r_{2q-2}...r_q$ and $r_{q-1}r_{q-2}...r_0$ respectively. Each processor p_r has three data registers a_r, b_r and c_r, respectively. Initially, processor p_s in position $(0,j,k)$ contains the element $a(j,k)$ and $b(j,k)$ in a_s and b_s respectively. The computation is accomplished in three stages. In the first stage, the elements of the matrices A and B are distributed over the p processors. As a result, a_r gets $a(j,i)$ and b_r gets $b(i,k)$. In the second stage, product elements $c(j,k)$ are computed and stored in each c_r. In the final stage, the sums $\Sigma_{i=0}^{n-1} c_{i,j,k}$ are computed and stored in $c_{0,j,k}$.

The above algorithm accomplishes the $O(n^3)$ task of matrix multiplication in $O(\log n)$ time using n^3 processors. Since the processor-time product of this parallel algorithm exceeds the sequential time complexity of the algorithm, it is not cost-optimal. In the following sub-sections we present two cost-optimal variations of this algorithm which use fewer than n^3 processors.

3.4.2 Variant With More Than One Element Per Processor

This variant proposed in [8, 17] can work with $n^2 r$ processors, where $1 < r < n$, thus using one processor for more than one element of each of the two $n \times n$ matrices. The algorithm is similar to the one above except that a logical processor array of r^3 (instead of n^3) superprocessors is used, each superprocessor comprising of $(n/r)^2$ hypercube processors. In the second step, multiplication of blocks of $(n/r) \times (n/r)$ elements instead of individual elements is performed. This multiplication of $(n/r) \times (n/r)$ blocks is performed according to the algorithm in Section 3.2 on $\frac{n}{r} \times \frac{n}{r}$ subarrays (each such subarray is actually a subcube) of processors using Cannon's algorithm for one element per processor. This step will require a communication time of $2(t_s + t_w)\frac{n}{r}$.

In the first stage of the algorithm, each data element is broadcast over r processors. In order to place the elements of matrix A in their respective positions, first the buffer $a_{(0,j,k)}$ is sent to $a_{(k,j,k)}$ in $\log r$ steps and then $a_{(k,j,k)}$ is broadcast to $a_{(k,j,l)}, 0 \leq l < r$, again in $\log r$ steps. By following a similar procedure, the elements of matrix B can be transmitted to their respective processors. In the second stage, groups of $(n/r)^2$ processors multiply blocks of $(n/r) \times (n/r)$ elements each processor performing n/r computations and $2n/r$ communications. In the final step, the elements of matrix C are restored to their designated processors in $\log r$ steps. The communication time can thus be shown to be equal to $(t_s + t_w)(5\log r + 2\frac{n}{r})$ resulting in the parallel run time given by the following equation:

$$T_p = \frac{n^3}{p} + (t_s + t_w)(5\log(\frac{p}{n^2}) + 2\frac{n^3}{p}) \quad (5)$$

If $p = \frac{n^3}{\log n}$ processors are used, then the parallel execution time of the DNS algorithm is $O(\log n)$. The processor-time product is now $O(n^3)$, which is same as the sequential time complexity of the algorithm.

3.5 Our Variant of the DNS Algorithm

Here we present another scheme to adapt the single element per processor version of the DNS algorithm to be able to use fewer than n^3 processors on a hypercube. In the rest of the paper we shall refer to this algorithm as the GK variant of the DNS algorithm. As shown later in Section 5, this algorithm performs better than the DNS algorithm for a wide range of n and p. Also, unlike the DNS algorithm which works only for $n^2 \leq p \leq n^3$, this algorithm can use any number of processors from 1 to n^3. In this variant, we use $p = 2^{3q}$ processors where $q < \frac{1}{3}\log n$. The matrices are divided into sub-blocks of $\frac{n}{2^q} \times \frac{n}{2^q}$ elements and the sub-blocks are numbered just the way the single elements were numbered in the algorithm of Section 3.4.1. Now, all the single element operations of the algorithm of Section 3.4.1 are replaced by sub-block operations; *i.e.*, matrix sub-blocks are multiplied, communicated and added.

Let t_{mult} and t_{add} is the time to perform a single floating point multiplication and addition respectively, and $t_{mult} + t_{add} = 1$. In the first stage of this algorithm, $\frac{n^2}{p^{2/3}}$ data elements are broadcast over $p^{1/3}$ processors for each matrix. In order to place the elements of matrix A in their respective positions, first the buffer $a_{(0,j,k)}$ is sent to $a_{(k,j,k)}$ in $\log p^{1/3}$ steps and then $a_{(k,j,k)}$ is broadcast to $a_{(k,j,l)}, 0 \leq l < p^{1/3}$, again in $\log p^{1/3}$ steps. By following a similar procedure, the elements of matrix B can be sent to the processors where they are to be utilized in $2\log p^{1/3}$ steps. In the second stage of the algorithm, each processor performs $(\frac{n}{p^{1/3}})^3 = \frac{n^3}{p}$ multiplications. In the third step, the corresponding elements of $p^{1/3}$ groups of $\frac{n^2}{p^{2/3}}$ elements each are added in a tree fashion. The first stage takes $4t_s \log p^{1/3} + 4t_w \frac{n^2}{p^{2/3}} \log p^{1/3}$ time. The second stage contributes $t_{mult}\frac{n^3}{p}$ to the parallel execution time and the third stage involves $t_s \log p^{1/3} + t_w \frac{n^2}{p^{2/3}} \log p^{1/3}$ communication time and $t_{add}\frac{n^3}{p}$ computation time for calculating the sums. The total parallel execution time is therefore given

by the following equation:

$$T_p = \frac{n^3}{p} + \frac{5}{3}t_s \log p + \frac{5}{3}t_w \frac{n^2}{p^{2/3}} \log p \quad (6)$$

This execution time can be further reduced by using a more sophisticated scheme for one-to-all broadcast on a hypercube [14]. This is discussed in detail in [11].

4 Scalability Analysis

Recall from Section 2 that the isoefficiency function for a certain efficiency E can be obtained by equating W with $\frac{E}{1-E}T_o$ (Equation (1)) and then solving this equation to determine W as a function of p. In most of the parallel algorithms described in Section 3, the communication overhead has two different terms due to t_s and t_w. When there are multiple terms in T_o of different order, it is often not possible to obtain the isoefficiency function as a closed form function of p. As p and W increase in a parallel system, efficiency is guaranteed not to drop if none of the terms of T_o grows faster than W. Therefore, if T_o has multiple terms, we balance W against each individual term of T_o to compute the respective isoefficiency function. The component of T_o that requires the problem size to grow at the fastest rate w.r.t. p determines the overall isoefficiency function of the entire computation. Sometimes, the isoefficiency function for a parallel algorithm is due to the limit on the concurrency of the algorithm. For instance, if for a problem size W, an algorithm can not use more than $h(W)$ processors, then as the number of processors is increased, eventually W has to be increased as $h^{-1}(p)$ in order to keep all the processors busy and to avoid the efficiency from falling due to idle processors. If $h^{-1}(p)$ is greater than any of the isoefficiency terms due to communication overheads, then $h^{-1}(p)$ is the overall isoefficiency function and determines the scalability of the parallel algorithm. Thus it is possible for an algorithm to have little communication overhead, but still a bad scalability due to limited concurrency.

We now determine the isoefficiency functions for all the algorithms discussed in Section 3. The problem size W is taken as n^3 for all the algorithms.

4.1 Isoefficiency Analysis of Cannon's Algorithm

From Equation (3), it follows that the total overhead over all the processors for this algorithm is $2t_s p\sqrt{p} + 2t_w n^2\sqrt{p}$. In order to determine the isoefficiency term due to t_s, W has to be proportional to $2Kt_s p\sqrt{p}$ (see Equation (1)), where $K = \frac{1}{1-E}$ and E is the desired efficiency that has to be maintained. Hence the following isoefficiency relation results:

$$n^3 = W \propto 2Kt_s p\sqrt{p} \quad (7)$$

Similarly, to determine the isoefficiency term due to t_w, n^3 has to proportional to $2Kt_w n^2 \sqrt{p}$. Therefore,

$$n^3 \propto 2Kt_w n^2 \sqrt{p}$$

$$\Rightarrow \quad n \propto 2Kt_w \sqrt{p}$$

$$\Rightarrow \quad n^3 = W \propto 8K^3 t_w^3 p^{1.5} \qquad (8)$$

According to both Equations (7) and (8), the asymptotic isoefficiency function of Cannon's algorithm is $O(p^{1.5})$. Also, since the maximum number of processors that can be used by this algorithm is n^2, the isoefficiency due to concurrency[1] is also $O(p^{1.5})$. Thus Cannon's algorithm is as scalable on a hypercube as any matrix multiplication algorithm using $O(n^2)$ processors can be on any architecture.

The above analysis also applies to the simple algorithm because both the degree of concurrency and the communication overheads (due to the t_w term which determines the overall isoefficiency function) are the same for these two algorithms.

4.2 Isoefficiency Analysis of Berntsen's Algorithm

The overall overhead function for this algorithm can be determined from the expression of the parallel execution time in Equation (4) to be $2t_s p^{4/3} + \frac{1}{3}t_s p \log p + 3t_w n^2 p^{1/3}$. By an analysis similar to that in Section 4.1, it can be shown that the isoefficiency terms due to t_s and t_w for this algorithm are given by the following equations:

$$n^3 = W \propto 2Kt_s p^{4/3} \qquad (9)$$

$$n^3 = W \propto 27K^3 t_w^3 p \qquad (10)$$

Recall from Section 3.3 that for this algorithm, $p \leq n^{3/2}$. This means that $n^3 = W \propto p^2$ as the number of processors is increased. Thus the isoefficiency function due to concurrency is $O(p^2)$, which is worse than any of the isoefficiency terms due to the communication overhead. Thus this algorithm has a poor scalability despite little communication cost due to its limited concurrency.

4.3 Isoefficiency Analysis of the DNS Algorithm

It can be shown that the overhead function T_o for this algorithm is $(t_s + t_w)(\frac{5}{3}p \log p + 2n^3)$. Since W is $O(n^3)$, the terms $2(t_s + t_w)n^3$ will always be balanced w.r.t. W. This term is independent of p and does not contribute to the isoefficiency function. It does however impose an upper limit on the efficiency that this algorithm can achieve. Since, for this algorithm, $E = \frac{1}{1 + \frac{5/3p \log p}{n^3} + 2(t_s + t_w)}$, an efficiency higher

---

[1] $n^2 \propto p \Rightarrow n^3 = W \propto p^{1.5}$.

than $\frac{1}{1 + 2(t_s + t_w)}$ can not be attained, no matter how big the problem size is. Since t_s is usually a large constant for most practical MIMD computers, the achievable efficiency of this algorithm is quite limited on such machines. The other term in T_o yields the following isoefficiency function for the algorithm:

$$n^3 = W \propto \frac{5}{3}Kt_s p \log p \qquad (11)$$

The above equation shows that the asymptotic isoefficiency function of the DNS algorithm on a hypercube is $O(p \log p)$. It can easily be shown that an $O(p \log p)$ scalability is the best any parallel formulation of the conventional $O(n^3)$ algorithm can achieve on any parallel architecture [3] and the DNS algorithm achieves this lower bound on a hypercube.

4.4 Isoefficiency Analysis of the GK Algorithm

The total overhead T_o for this algorithm is equal to $\frac{5}{3}t_s p \log p + \frac{5}{3}t_w n^2 p^{1/3} \log p$ and the following equations give the isoefficiency terms due to t_s and t_w respectively for this algorithm:

$$n^3 = W \propto \frac{5}{3}Kt_s p \log p \qquad (12)$$

$$n^3 = W \propto \frac{125}{27}K^3 t_w^3 p(\log p)^3 \qquad (13)$$

5 Relative Performance of the Four Algorithms on a Hypercube

Subsections 4.1 through 4.4 give the overall isoefficiency functions of the four algorithms on a hypercube architecture. The asymptotic scalabilities and the range of applicability of these algorithms is summarized in Table 1.

Note that Table 1 gives only the asymptotic scalabilities of the four algorithms. In practice, none of the algorithms is strictly better than the others for all possible problem sizes and number of processors. Further analysis is required to determine the best algorithm for a given problem size and a certain parallel machine depending on the number of processors being used and the hardware parameters of the machine. A detailed comparison of these algorithms based on their respective total overhead functions is presented in the next section.

We compare a pair of algorithms by comparing their total overhead functions (T_o) as given in Table 1. For instance, while comparing the GK algorithm with Cannon's algorithm, it is clear that the t_s term for the GK algorithm will always be less than that for Cannon's algorithm. Even if $t_s = 0$, the t_w term of the GK algorithm becomes smaller than that of Cannon's algorithm for $p > 130$ million. Thus, $p = 130$ million is the cut-off point beyond which the GK algorithm will perform better than Cannon's algorithm irrespective of the values of n. For $p < 130$ million, the performance of the GK algorithm

Algorithm	Total Overhead Function, T_o	Asymptotic Isoeff. Function	Range of Applicability
Berntsen's	$2t_s p^{4/3} + \frac{1}{3}t_s p \log p + 3t_w n^2 p^{1/3}$	$O(p^2)$	$1 \le p \le n^{3/2}$
Cannon's	$2t_s p^{3/2} + 2t_w n^2 \sqrt{p}$	$O(p^{1.5})$	$1 \le p \le n^2$
GK	$\frac{5}{3}t_s p \log p + \frac{5}{3}t_w n^2 p^{1/3} \log p$	$O(p(\log p)^3)$	$1 \le p \le n^3$
DNS	$(t_s + t_w)(\frac{5}{3}p \log p + 2n^3)$	$O(p \log p)$	$n^2 \le p \le n^3$

Table 1: *Communication overhead, scalability and range of application of the four algorithms on a hypercube.*

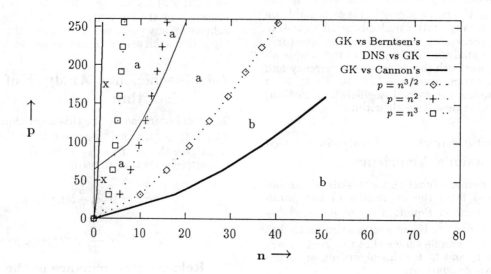

Figure 1: *A comparison of the four algorithms for $t_w = 3$ and $t_s = 150$.*

will be better than that of Cannon's algorithm for values of n less than a certain threshold value which is a function of p and the ration of t_s and t_w. A hundred and thirty million processors is clearly too large, but we show that for reasonable values of t_s, the GK algorithm performs better than Cannon's algorithm for very practical values of p and n.

In order to determine ranges of p and n where the GK algorithm performs better than Cannon's algorithm, we equate their respective overhead functions and compute n as a function of p. We call this $n_{Equal-T_o}(p)$ because this value of n is the threshold at which the overheads of the two algorithms will be identical for a given p. If $n > n_{Equal-T_o}(p)$, Cannon's algorithm will perform better and if $n < n_{Equal-T_o}(p)$, the GK algorithm will perform better. If we equate T_o for the two algorithms, then $T_o^{(Cannon)} = 2t_s p^{3/2} + 2t_w n^2 \sqrt{p} = T_o^{(GK)} = \frac{5}{3}t_s p \log p + \frac{5}{3}t_w n^2 p^{1/3} \log p$. Therefore,

$$n_{Equal-T_o}(p) = \sqrt{\frac{(5/3 p \log p - 2p^{3/2})t_s}{(2\sqrt{p} - 5/3 p^{1/3} \log p)t_w}} \quad (14)$$

Similarly, equal overhead conditions can be determined for other pairs of algorithms too and the val- ues of t_w and t_s can be plugged in depending upon the machine in question to determine the best algorithm for a give problem size and number of processors. We have performed this analysis for three practical sets of values of t_w and t_s. In the rest of the section we demonstrate the practical importance of this analysis by showing how any of the four algorithms can be useful depending on the problem size and the parallel machine available.

Figures 1, 2 and 3 show the regions of applicability and superiority of different algorithms.

The plain lines represent equal overhead conditions for pairs of algorithms. For a curve marked "X vs Y" in a figure, algorithm X has a smaller value of communication overhead to the left of the curve, algorithm Y has smaller communication overhead to the right side of the curve, while the two algorithms have the same value of T_o along the curve. The lines with symbols $\diamond$, $+$ and $\square$ plot the functions $p = n^{3/2}$, $p = n^2$ and $p = n^3$, respectively. These lines demarcate the regions of applicabilities of the four algorithms (see Table 1) and are important because an algorithm might not be applicable in the region where its overhead function T_o is mathematically superior than others. In all the figures in this section, the region marked with an **x** is the one

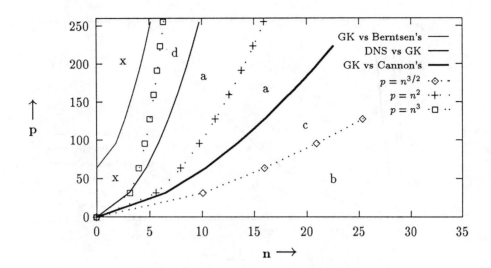

Figure 2: *A comparison of the four algorithms for $t_w = 3$ and $t_s = 10$.*

where $p > n^3$ and none of the algorithms is applicable, the region marked with an **a** is the one where the GK algorithm is the best choice, the symbol **b** represents the region where Berntsen's algorithm is superior to the others, the region marked with a **c** is the one where Cannon's algorithm should be used and the region marked with a **d** is the one where the DNS algorithm is the best.

Figure 1 compares the four algorithms for $t_w = 3$ and $t_s = 150$. These parameters are very close to that of a currently available parallel computer like the nCUBE2^{TM*}. In this figure, since the $n_{Equal-T_o}$ curve for the DNS algorithm and the GK algorithm lies in the **x** region, and the DNS algorithm is better than the GK algorithm only for values of n smaller than $n_{Equal-T_o}(p)$. Hence the DNS algorithm will always[2] perform worse than the GK algorithm for this set of values of t_s and t_w and the latter is the best overall choice for $p > n^2$ as Berntsen's algorithm and Cannon's algorithm are not applicable in this range of p. Since the $n_{Equal-T_o}$ curve for GK and Cannon's algorithm lies below the $p = n^{3/2}$ curve, the GK algorithm is the best choice even for $n^{3/2} \leq p \leq n^2$. For $p < n^{3/2}$, Berntsen's algorithm is always better than Cannon's algorithm, and for this set of t_s and t_w, also than the GK algorithm. Hence it is the best choice in that region in Figure 1.

In Figure 2, we compare the four algorithms for a hypercube with $t_w = 3$ and $t_s = 10$. Such a machine could easily be developed in the near future by using faster CPU's (t_w and t_s represent relative communi-

cation costs w.r.t. the unit computation time) and reducing the message startup time. By observing the $n_{Equal-T_o}$ curves and the regions of applicability of these algorithms, the regions of superiority of each of the algorithms can be determined just as in case of Figure 1. It is noteworthy that in Figure 2 each of the four algorithms performs better than the rest in some region and all the four regions **a**, **b**, **c** and **d** contain practical values of p and n.

In Figure 3, we present a comparison of the four algorithms for $t_w = 3$ and $t_s = 0.5$. These parameters are close to what one can expect to observe on a typical SIMD machine like the CM-2. For the range of processors shown in the figure, the GK algorithm is inferior to the others[3]. Hence it is best to use the DNS algorithm for $n^2 \leq p \leq n^3$, Cannon's algorithm for $n^{3/2} \leq p \leq n^2$ and Berntsen's algorithm for $p < n^{3/2}$.

6 Experimental Results

We verified a part of the analysis of this paper through experiments of the CM-5 parallel computer. On this machine, the fat-tree [16] like communication network on the CM-5 provides simultaneous paths for communication between all pairs of processors. Hence the CM-5 can be viewed as a fully connected architecture which can simulate a hypercube connected network. We implemented Cannon's algorithm described in Section 3.2 and the algorithm described in Section 3.5.

Since the CM-5 can be considered as a fully connected network of processors, the expression for the parallel execution time for the algorithm of Section 3.5 will have to be modified slightly. The first part

---

*nCUBE2 is a trademark of the Ncube corporation.

[2]Actually, the $n_{Equal-T_o}$ curve for DNS vs GK algorithms will cross the $p = n^3$ curve for $p = 2.6 \times 10^{18}$, but clearly this region has no practical importance.

[3]The GK algorithm does begin to perform better than the other algorithms for $p > 1.3 \times 10^8$, but again we consider this range of p to be impractical.

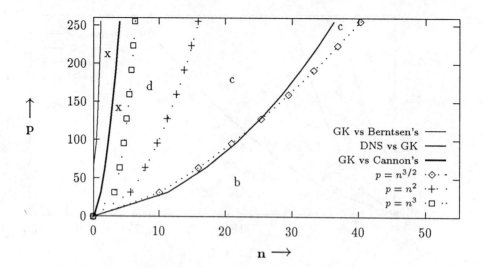

Figure 3: *A comparison of the four algorithms for $t_w = 3$ and $t_s = 0.5$.*

of the procedure to place the elements of matrix A in their respective positions, requires sending the buffer $a_{(0,j,k)}$ to $a_{(k,j,k)}$. This can be done in one step on the CM-5 instead of $\log(p^{1/3})$ steps on a conventional hypercube. The same is true for matrix B as well. It can be shown that the following modified expression gives the parallel execution time for this algorithm on the CM-5:

$$T_p = \frac{n^3}{p} + t_s(\log p + 2) + t_w\frac{n^2}{p^{2/3}}(\log p + 2) \quad (15)$$

Computing the condition for equal T_o for this and Cannon's algorithm by deriving the respective values of T_o from Equations (15) and (3), it can be shown that for 512 processors, Cannon's algorithm should perform better that our algorithm for $n > 295$. Since the number of processors has to be a perfect square for Cannon's algorithm on square matrices, in Figure 4, we draw the efficiency vs n curve for $p = 484$ for Cannon's algorithm and for $p = 512$ for the GK algorithm[4]. The cross-over point closely matches the predicted value. These experiments suggest that the algorithm of Section 3.5 can outperform the classical algorithms like Cannon's for a wide range of problem sizes and number of processors. Moreover, as the number of processors is increased, the cross-over point of the efficiency curves of the GK algorithm and Cannon's algorithm corresponds to a very high efficiency. As seen in Figure 4, the cross-over happens at $E \approx 0.93$ and Cannon's algorithm can not outperform the GK algorithm by a wide margin at such high efficiencies. On the other hand, the GK algorithm achieves an efficiency of 0.5

for a matrix size of 112×112, whereas Cannon's algorithm operates at an efficiency of only 0.28 on 484 processors on 110×110 matrices. In other words, in the region where the GK algorithm is better than Cannon's algorithm, the difference in the efficiencies is quite sig..ificant.

7 Concluding Remarks

In this paper we have presented the scalability analysis of a number of matrix multiplication algorithms described in the literature [4, 8, 2, 13]. Besides analyzing these classical algorithms, we show that the GK algorithm that we present in this paper outperforms all the well known algorithms for a significant range of number of processors and matrix sizes. The scalability analysis of all these algorithms provides several important insights regarding their relative superiority under different conditions. None of the algorithms discussed in this paper is clearly superior to the others because there are a number of factors that determine the algorithm that performs the best. In this paper we predict the precise conditions under which each formulation is better than the others. It may be unreasonable to expect a programmer to code different algorithms for different machines, different number of processors and different matrix sizes. But all the algorithms can stored in a library and the best algorithm can be pulled out by a smart preprocessor/compiler depending on the various parameters.

We show that an algorithm with a seemingly small expression for the communication overhead is not necessarily the best one because it may not scale well as the number of processors is increased. For instance, the best algorithm in terms of communication overheads (Berntsen's algorithm described in Section 3.3) turns out to be the least scalable one with an isoefficiency function of $O(p^2)$ due its lim-

---

[4]This is not an unfair comparison because the efficiency can only be better for smaller number of processors.

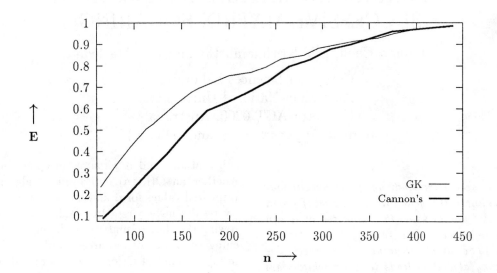

Figure 4: *Efficiency vs matrix size for Cannon's algorithm (p = 484) and the GK algorithm (p = 512).*

ited degree of concurrency.

References

[1] S. G. Akl. *The Design and Analysis of Parallel Algorithms.* Prentice-Hall, 1989.

[2] Jarle Berntsen. Communication efficient matrix multiplication on hypercubes. *Parallel Computing,* 12:335 – 342, 1989.

[3] Dimitri P. Bertsekas and John N. Tsitsiklis. *Parallel and Distributed Computation.* Prentice Hall, 1989.

[4] L. E. Cannon. A cellular computer to implement the Kalman Filter Algorithm. Technical report, Ph.D. Thesis, Montana State University, 1969.

[5] V. Cherkassky and R. Smith. Efficient mapping and implementations of matrix algorithms on a hypercube. *The Journal of Supercomputing,* Vol. 2:7 – 27, 1988.

[6] N. P. Chrisopchoides, M. Aboelaze, E. N. Houstis, and C. E. Houstis. The parallelization of some lavel 2 and 3 BLAS operations on distributed memory machines. In *Proceedings of the 1st International Conference of the Austrian Center of Parallel Computation.* Springer-Verlag Series Lacture Notes in Computer Science, 1991.

[7] Eric F. Van de Velde. Multicomputer matrix computations: Theory and practice. In *Proceedings of the 1989 Conferece on Hypercubes, Concurrent Computers, and Applications,* pages 1303 – 1308, 1989.

[8] Eliezer Dekel, David Nassimi, and Sartaj Sahni. Parallel matrix and graph algorithms. *SIAM Journal of Computing,* 10:657 – 673, 1981.

[9] G.C. Fox, S.W. Otto, and A.J.G. Hey. Matrix algorithms on a hypercube I : Matrix multiplication. *Parallel Computing,* 4:17 – 31, 1987.

[10] Ananth Grama, Anshul Gupta, and Vipin Kumar. Isoefficiency function: A scalability metric for parallel algorithms and architectures. Technical report, Computer Science Department, University of Minnesota, April 1993.

[11] Anshul Gupta and Vipin Kumar. The scalability of Matrix Multiplication Algorithms on parallel computers. Technical Report TR 91-54, Computer Science Department, University of Minnesota, Minneapolis, MN 55455, 1991.

[12] Paul G. Hipes. Matrix multiplication on the JPL/Caltech Mark IIIfp hypercube. Technical Report C3P 746, Concurrent Computation Program, California Institute of Technology, Pasadena, CA - 91125, 1989.

[13] Ching-Tien Ho, S. Lennart Johnsson, and Alan Edelman. Matrix multiplication on hypercubes using full bandwidth and constant storage. In *Proceedings of the 1991 International Conference on Parallel Processing,* pages 447 – 451, 1991.

[14] S. Lennart Johnsson and Ching-Tien Ho. Optimum broadcasting and personalized communication in hypercubes. *IEEE Transactions on Computers,* 38(9):1249 – 1268, September 1989.

[15] Vipin Kumar and Anshul Gupta. Analyzing scalability of parallel algorithms and architectures. Technical report, TR-91-18, Computer Science Department, University of Minnesota, June 1991. A short version of the paper appears in the Proceedings of the 1991 International Conference on Supercomputing, Germany, and as an invited paper in the Proc. of 29th Annual Allerton Conference on Communication, Control and Computing, Urbana,IL, October 1991.

[16] Charles E. Leiserson. Fat-trees : Universal networks for hardware efficient supercomputing. In *Proceedings of the 1985 International Conference on Parallel Processing,* pages 393 – 402, August 20 - 23, 1985.

[17] S. Ranka and S. Sahni. *Hypercube Algorithms for Image Processing and Pattern Recognition.* Springer-Verlag, New York, 1990.

[18] Walter F. Tichy. Parallel matrix multiplication on the connection machine. Technical Report RIACS TR 88.41, Research Institute for Advanced Computer Science, NASA Ames Research Center, Ames, Iowa, 1988.

GENERALISED MATRIX INVERSION BY SUCCESSIVE MATRIX SQUARING

Lujuan Chen, E V Krishnamurthy and Iain Macleod

Computer Sciences Laboratory
Australian National University
Canberra ACT 0200, Australia
(e-mail: lujuan@cslab.anu.edu.au)

Abstract

This paper uses successive squaring of a composite matrix to approximate the generalised inverse of an m by n matrix A. For $m \approx n$, the g-inverse of A can be computed in parallel time ranging from $O(\log n)$ to $O(\log^2 n)$. The simple structure of the successive matrix squaring algorithm leads to a straightforward parallel implementation. Test results are provided.

1 Introduction

In 1955 Penrose [1] showed that for an arbitrary matrix $A \in C^{m \times n}$, there is a unique matrix X satisfying the set of equations:

$$AXA = A, \quad XAX = X, \quad (AX)^* = AX, \quad (XA)^* = XA \quad (1)$$

where * represents the complex conjugate. X is commonly known as the Moore-Penrose or generalised inverse and is often denoted by A^+, which reduces to the conventional matrix inverse A^{-1} when A is square and non-singular.

The g-inverse has many applications in statistics, prediction theory, control system analysis, curve fitting, etc. Sequential algorithms for computing the g-inverse have been known since 1962 [2], but implementation of a parallel algorithm has only recently been reported when Krishnamurthy and Ziavras [3] described their adaptation of the Ben-Israel and Greville (B-IG) algorithm [2] for the CM-2 Connection Machine. In contrast, parallel computation of non-singular matrix inverses has been studied for nearly twenty years. It was originally thought that parallel computation of a matrix inverse required a time linear with the order of the input matrix. However, in 1976 Csanky [4] proved that given a non-singular matrix A of order n, A^{-1} could be computed in parallel time $O(\log^2 n)$ if sufficiently many processors were available. Attempts to prove that $O(\log^2 n)$ was a lower time bound for parallel matrix inversion continued until the discovery [5] of algorithms with a time bound of $O(\log^\alpha n)$, where $1 \leq \alpha < 2$. Csanky's result for general non-singular matrices has not been improved

substantially and it remains an open question as to whether an $O(\log n)$ deterministic algorithm exists. Improved values for α and many interesting complexity results have arisen from this quest, through narrowing the class of matrices to be inverted and broadening the range of resources available. For example, in [6] a bound of $O(\log n)$ was obtained in this manner. In a recent paper, Codenotti [7] presents an iterative algorithm for solution of linear systems and inversion of matrices based on a repeated matrix squaring scheme. For very well conditioned matrices this algorithm achieves a time bound $O(\log n)$. To date, all such repeated squaring methods have required that $\rho(P) < 1$ for convergence, where $P = I - \beta A^T A$ is the iterative correction matrix in schemes such as $X_{K+1} = PX_K + Q$ [4,7]. Previously published algorithms also have not specified ranges for β within which their convergence is assured.

In the following sections we present a deterministic iterative algorithm for computing the g-inverse A^+ of matrix $A \in C^{m \times n}$, the set of m by n complex matrices. We call this algorithm successive matrix squaring (SMS). Each iterative step in the computation comprises (parallel) squaring of a composite matrix T. For very well conditioned matrices this algorithm enables calculation of the g-inverse of A in time $O(\log(m+n))$ with relative error bounded by any $\epsilon \in (0,1)$, independent of matrix size. Compared with the algorithm in [7], our iterative scheme $X_{K+1} = PX_K + Q$ (derived from equation (1)) permits $\rho(P) \leq 1$ rather than $\rho(P) < 1$. This means that A can have reduced rank.

2 Successive Matrix Squaring

We start from (1) to construct a suitable matrix equation. Since $AXA = A$ and $(AX)^* = AX$, we have $A^* = (AXA)^* = A^*(AX)^* = A^*AX$. Therefore $X = X - \beta(A^*AX - A^*) = (I - \beta A^*A)X + \beta A^*$, where β is a relaxation parameter. Letting

$$P = (I - \beta A^*A) \quad \text{and} \quad Q = \beta A^*,$$

the solution of matrix equation $X = PX + Q$ can be approximated by the following iterative scheme:

$$\begin{cases} X_{K+1} = PX_K + Q \\ X_1 = Q \end{cases} \qquad (2)$$

We can compute (2) in parallel given the following observations. Consider the $m + n$ by $m + n$ matrix T, defined as

$$T = \begin{bmatrix} P & Q \\ 0 & I \end{bmatrix}$$

and notice that the matrix $T^K, K = 1, 2, \cdots$, is given by

$$T^K = \begin{bmatrix} P^K & \sum_{i=0}^{K-1} P^i Q \\ 0 & I \end{bmatrix}.$$

By direct inspection, it is not difficult to see that $\sum_{i=0}^{K-1} P^i Q$ is the Kth approximant to A^+, as given in (2), that is, $X_K = \sum_{i=0}^{K-1} P^i Q$. It immediately follows that the calculation of X_K can be reduced to the computation of T^K. In turn, T^K can be computed by "repeated squaring," that is

$$\begin{cases} T_0 = T \\ T_{i+1} = T_i^2 \quad i = 0, 1, \cdots, J; \ J \text{ such that } 2^J \geq K \end{cases} \qquad (3)$$

Note that K steps of (3) are equivalent to 2^K steps of (2). The following theorems prove that the iterative solution of (2) converges to A^+ using scheme (3).

Theorem 1: If T_0 and T_K are as given in scheme (3), we have

$$T_K = T_0^{2^K} = \begin{bmatrix} P^{2^K} & \sum_{i=0}^{2^K-1} P^i Q \\ 0 & I \end{bmatrix}.$$

Proof: If $K = 0$, $T_0 = \begin{bmatrix} P & Q \\ 0 & I \end{bmatrix}$ true. Assume this conclusion is true for $K - 1$. Observing that $T_K = T_{K-1} \cdot T_{K-1}$, the proof follows by mathematical induction.

The next theorem gives the necessary and sufficient conditions for the SMS algorithm to converge to A^+ in terms of the L_2 norm, denoted here as $\| \cdot \|$.

Theorem 2: The sequence of approximations

$$X_{2^K} = \sum_{i=0}^{2^K-1} (I - \beta A^* A)^i \beta A^*$$

determined by the SMS algorithm (3) converges to the g-inverse A^+ in the matrix norm of A if, and only if, β is a fixed real number such that

$$0 < \beta < 2/\|A\|^2 = 2/\lambda_1(A^* A).$$

In the case of convergence we have the error estimate

$$\frac{\| A^+ - X_{2^K} \|}{\| A^+ \|} \leq (1 - \beta \lambda_r)^{2^K}, \qquad (4)$$

as derived in [8]. Furthermore, the optimum parameter is $\beta_{\text{op}} = 2/(\lambda_1 + \lambda_r)$ [9,10]. In this case the error estimate is given by

$$\frac{\| A^+ - X_{2^K} \|}{\| A^+ \|} \leq \left(\frac{\lambda_1 - \lambda_r}{\lambda_1 + \lambda_r} \right)^{2^K}, \quad K = 0, 1, 2, \cdots, (5)$$

where λ_1 and λ_r are the nonzero largest and smallest eigenvalues of $A^* A$ respectively. Note that this is a tighter bound by a factor of $\sqrt{\lambda_1/\lambda_r}$ than that previously established by Petryshyn [9]. If $\lambda_1 = \lambda_r$ we can check directly that $A^+ = A^*/\lambda_1$. We can use a method similar to that of Petryshyn to prove the necessary and sufficient convergence condition of the SMS method.

Estimation of β based on eigenvalues becomes difficult for larger matrices. In this case we can estimate β using $\beta = 2/\text{tr}(A^* A)$; this value will usually be suboptimal, but guarantees the convergence of the algorithm since

$$0 < 2/\text{tr}(A^* A) \leq 2/(\lambda_1 + \lambda_r).$$

The original B-IG algorithm is given by

$$X_{K+1} = X_K(2I - AX_K),$$

starting with $X_0 = \beta A^*$, where $0 < \beta < 2/\text{tr}(AA^T)$. To compare this scheme with SMS, we first rearrange the iteration such that $X_{K+1} = (2I - X_K A)X_K$, which gives

$$\begin{cases} X_{K+1} = (I + P_K)X_K \\ P_{K+1} = P_K^2 \\ X_0 = \beta A^*, \quad P_0 = I - \beta A^* A. \end{cases} \qquad (6)$$

Using mathematical induction we can prove that

$$X_K = \sum_{i=0}^{2^K-1} (I - \beta A^* A)^i \cdot \beta A^*$$

which shows the equivalence of SMS to the B-IG method. Thus, they both have the same time bound.

3 Complexity Analysis

This section analyses implementation of the SMS method in terms of the number of parallel steps and the number of parallel processors needed to compute the g-inverse to a pre-specified precision.

Theorem 3: Let $A \in C^{m \times n}$, then $T_K = T_0^{2^K}$ of the iterations (3) can be computed in time $O(K \log(m + n))$ on $M(m + n)$ processors, where $M(m + n)$ is the number of processors necessary to support matrix multiplication in time $O(\log(m + n))$.

Proof: Since the exponent of T doubles at each squaring step of (3), $\log_2 2^K = K$ matrix multiplications are sufficient to compute T^{2^K}. With regard to the number of processors, [11] shows that $M(n) \leq n^\omega$ where the exponent ω is such that sequential matrix multiplication can be performed with $O(n^\omega)$ arithmetic operations. The lowest bound established to date is $\omega \approx 2.38$ [12].

In the remainder of this section we determine the number of iterations K as a function of the condition number $\lambda_1(A^* A)/\lambda_r(A^* A)$ and the required accuracy ϵ, letting $\mu = \lambda_r(A^* A)/\lambda_1(A^* A)$. Since A is not a zero matrix, it is easy to see that $0 < \mu \leq 1$.

Theorem 4: Let $Q = \beta A^*$, $P = I - \beta A^* A$ and $0 < \beta < \dfrac{2}{\| A \|^2}$. Then by use of the SMS algorithm (3), an approximate g-inverse $X_{2\kappa}$, such that the relative error is less than ϵ, can be computed

(a) in $K = \left\lceil \log_2 \dfrac{\ln(1/\epsilon)}{\ln 2} \right\rceil$ iterations, if $\dfrac{1}{3} \le \mu < 1$

(b) in $K = \left\lceil \log_2 \left(-\dfrac{\ln \epsilon}{2\mu} \right) \right\rceil$ iterations, if $0 < \mu < \dfrac{1}{3}$.

The proof of this theorem is given in [8].

When calculating K, in common with other researchers [7] we assume an upper bound to the condition number or equivalently a lower bound σ to μ. The following theorem (also proved in [8]) summarises:

Theorem 5: An approximate g-inverse $X_{2\kappa}$ of $A \in C^{m \times n}$, with $n > m$ and $\mu \ge \sigma$, can be obtained in time $T = O(\log(n)(\log \log(1/\epsilon) + \log(1/\sigma)))$ on $O(M(m+n))$ processors. In particular $T = O(\log(n))$ if $\sigma = O(1)$ and $T = O(\log^2(n))$ if $\sigma = O(n^{-\alpha})$, where both ϵ and α (≥ 1) are independent of m and n. Similar results apply in the case $n \le m$.

4 Numerical Examples

The following examples compare the observed rate of convergence with that predicted from (5), contrasting the relative error obtained after the number of iterations estimated from Theorem 4 with the specified error ϵ. Calculations were performed using Mathematica 2.1 on a Sun 4 ($MachinePrecision = 16$ digits).

4.1 Square Non-Singular Matrix

The 8 by 8 real test matrix A used here was formed by constructing a symmetric matrix with off-diagonal elements chosen from a uniform random distribution with range $(0, 1)$. The diagonal elements were calculated as ten times the absolute sum of elements in the corresponding row. The resulting matrix was thus of full rank and reasonably well conditioned, as demonstrated by the calculated eigenvalues of $A^T A$ (2726.3, 2309.2, 1953.7, 1835.7, 1767.1, 1640.9, 1613.8, 627.12) giving $\mu = \lambda_8/\lambda_1 = 0.2300$. The number of iterations necessary to achieve a specified relative error $\epsilon = 10^{-6}$ is calculated as $K = \lceil 4.91 \rceil = 5$. Using $\beta_{op} = 2/(\lambda_1 + \lambda_8) = 5.96 \times 10^{-4}$, after four and five iterations the relative error $\dfrac{\| A^+ - X_{2\kappa} \|}{\| A^+ \|}$ was 5.69×10^{-4} and 3.16×10^{-7} respectively, verifying the estimate of K.

In the absence of calculated eigenvalues, we can use $\beta_{tr} = 2/\text{tr}(A^T A) = 1.38 \times 10^{-4}$. In this case, eight iterations were required to achieve a specified accuracy

of $\epsilon = 10^{-6}$, with a relative error in computing A^+ of 8.78×10^{-11}. (After seven iterations the relative error was 9.60×10^{-6}.)

4.2 Rectangular Complex Matrix with Reduced Rank

Consider the following matrix (with rank = 2):
$$A = \begin{bmatrix} 0 & 2i & i & 0 & 4+2i & 1 \\ 0 & 0 & 0 & -3 & -6 & -3-3i \\ 0 & 2 & 1 & 1 & 4-4i & 1 \end{bmatrix}.$$

$A^* A$ has largest and smallest eigenvalues $\lambda_1 = 112.46$ and $\lambda_r = 15.544$, giving $\mu = 0.1382$. For a specified error of $\epsilon = 10^{-6}$, we can calculate K as $\lceil 5.64 \rceil = 6$. In this case there are only two non-zero eigenvalues, so $\beta_{op} = \beta_{tr} = 0.0156$. After five and six iterations, the relative errors in A^+ were 1.36×10^{-4} and 1.85×10^{-8} respectively, indicating that rapid convergence is achieved for moderately well conditioned matrices.

4.3 Ill-Conditioned Square Matrix

Elements of the 8 by 8 random test matrix used here were chosen from a uniform random distribution with range $(0, 1)$. Several matrices were constructed and their eigenvalues calculated. From these we selected one which had full rank but was rather ill-conditioned, and for which $A^T A$ had eigenvalues of 17.197, 0.9980, 0.8457, 0.4939, 0.3606, 0.06760, 0.01226 and 0.0004218, giving $\mu = 2.45 \times 10^{-5}$ and $\beta_{op} = 0.116$. For a specified relative error of $\epsilon = 10^{-4}$ we can estimate K as $\lceil 17.52 \rceil = 18$. Using β_{op}, after 17 and 18 iterations the relative errors obtained in A^+ were 1.62×10^{-3} and 2.62×10^{-6}. Using $\beta_{tr} = 0.100 \approx \beta_{op}$ the errors after 17 and 18 iterations were 3.96×10^{-3} and 1.56×10^{-5}.

4.4 Matrix With Equal Eigenvalues

Consider the following matrix:
$$A = \begin{bmatrix} 0 & -2+2i & -2-2i & 0 \\ -2-2i & 0 & 0 & 2-2i \\ -2+2i & 0 & 0 & 2+2i \\ 0 & 2+2i & 2-2i & 0 \end{bmatrix}.$$

$A^* A$ has four equal eigenvalues of 16.0 giving $\mu = 1.0$ and $\beta_{op} = 1/16$. We can quickly show that $A^* A = 16I$ (a scalar matrix), that $P = I - \beta_{op} A^* A$ is a null matrix and that $X_0 = Q = \beta_{op} A^*$ is already equal to A^+.

If we use $\beta_{tr} = 1/32 = \beta_{op}/2$, then we need to iterate to obtain the generalised inverse of A within a specified accuracy. In this case, we find that the relative error in computing A^+ after four and five iterations is 1.53×10^{-5} and 2.33×10^{-10} respectively.

5 Discussion

The above numerical experiments suggest that our tighter upper bound on the error in the iterative estimate of A^+ provides a realistic estimate of the number of iterations required in practice. The example of Section 4.2 also demonstrates that the SMS method is applicable to matrices with reduced rank.

The scheme which underlies the SMS algorithm (squaring of matrix T) requires two multiplications and one addition of the block matrices in the upper half of T per iterative step. These matrices are of similar size to those in the iterative scheme of the B-IG method, $X_{K+1} = X_K(2I - AX_K)$, so the number of elementary operations per step is also similar. If we compute T^2 directly, there are additional (trivial) operations but the simpler structure could well be attractive in hardware implementations. Assuming that we have a parallel matrix squaring architecture which leaves the elements of the squared matrix in their original locations, we can then simply repeat the squaring operation K times. The obvious advantage here is that there is no need to save the results or to load the individual processors with fresh data items at the end of each cycle of successive squaring, global data movement of this type being a relatively time consuming operation on typical parallel architectures.

The disadvantage of the directly expressed SMS algorithm is that the matrix T is of size $m+n$ by $m+n$ whereas the matrices in the B-IG method are of size n by m. Thus, in the case where $m \approx n$, there are $\approx (2n)^3$ operations (one multiplication and addition each) per iteration for a directly expressed SMS algorithm versus $\approx 2n^3$ operations in the B-IG scheme. To the extent that the parallel architecture employed has a large number of locally-connected processors and the accompanying matrix squaring algorithm avoids global data communication, then the smaller data communication overheads in the SMS algorithm will offset the approximate factor of four increase in arithmetical operations.

As noted above, the number of non-trivial arithmetical operations in each iterative step of the SMS algorithm is similar to that of the B-IG scheme; the difference in actual operations is of course accounted for by the null and identity block matrices in the lower half of T. Working within the context of VLSI systolic architectures, we are currently exploring strategies for taking advantage of the simple structure of the SMS algorithm while at the same time avoiding overheads arising from these unnecessary operations [13].

6 References

[1] R Penrose, "A generalized inverse for matrices," Proc Cambridge Philos Soc, **51** (1955), pp406-413.

[2] A Ben-Israel and T N E Greville, Generalized Inverses: Theory and Applications, Wiley, New York, 1974.

[3] E V Krishnamurthy and S G Ziavras, "Matrix g-inversion on the Connection Machine," Report CS-TR-2123, University of Maryland, October 1988.

[4] L Csanky, "Fast parallel matrix inversion algorithm," SIAM J Comput, **5** (1976), pp618-623.

[5] S A Cook, "A taxonomy of problems with fast parallel algorithms," Inform and Control, **64** (1985), pp2-22.

[6] B Codenotti and M Leoncini, Parallel Complexity of Linear System Solutions, World Scientific Publ Co, Singapore, 1991.

[7] B Codenotti, M Leoncini and G Resta, "Repeated matrix squaring for the parallel solution of linear systems," pp725-732 in PARLE '92 (Parallel Architectures and Languages in Europe), Lecture Notes in Computer Science, **605**, Springer-Verlag, New York, 1992.

[8] L Chen, E V Krishnamurthy and I Macleod, "Generalised matrix inversion and rank computation by successive matrix powering," submitted to Parallel Computing.

[9] W V Petryshyn, "On generalised inverses and on the uniform convergence of $(I - \beta K)^n$ with application to iterative methods," J Math Anal & App, **18** (1967), pp417-439.

[10] E V Krishnamurthy and S K Sen, Numerical Algorithms: Computations in Science and Engineering, East-West Press, New Delhi, 1986.

[11] V Pan, "Complexity of parallel matrix computations," Theor Comput Sci, **54** (1987), pp65-85.

[12] D Coppersmith and S Winograd, "Matrix multiplication via arithmetic progression," pp1-6 in Proc 19th Annual ACM Symposium on Theory of Computing, Berkeley, California, Springer-Verlag, 1987.

[13] L Chen, B B Zhou and I Macleod, "Fast parallel solution of linear systems via successive matrix squaring," submitted to 1993 International Conference on Application Specific Array Processors.

PARALLEL COMPUTATION OF THE SINGULAR VALUE DECOMPOSITION ON TREE ARCHITECTURES

Zhou B. B. and Brent R. P.

Computer Sciences Laboratory

The Australian National University

e-mail:{bing,rpb}@cslab.anu.edu.au

Abstract *We describe a new Jacobi ordering for parallel computation of SVD problems. The ordering uses the high bandwidth of a perfect binary fat-tree to minimise global interprocessor communication costs. It can thus be implemented efficiently on fat-tree architectures.*

1 Introduction

Let A be a real $m \times n$ matrix. Without loss of generality we assume that $m \geq n$. The singular value decomposition (SVD) of A is its factorization into a product of three matrices

$$A = U \Sigma V^T,$$

where U is an $m \times n$ matrix with orthonormal columns, V is an $n \times n$ orthogonal matrix, and Σ is an $n \times n$ non-negative diagonal matrix, say $\Sigma = diag(\sigma_1, \cdots, \sigma_n)$.

There are various ways to compute the SVD [2]. To achieve efficient parallel SVD computation the best approach may be to adopt the Hestenes one-sided transformation method [3] as advocated in [1].

The Hestenes method generates an orthogonal matrix V such that

$$AV = H,$$

where the columns of H are orthogonal. The nonzero columns $\tilde{H}$ of H are then normalised so that

$$\tilde{H} = U_r \Sigma_r$$

with $U_r^T U_r = I_r$, $\Sigma_r = diag(\sigma_1, \cdots, \sigma_r)$ and $r \leq n$ is the rank of A.

The matrix V can be generated as a product of plane rotations. As in the traditional Jacobi algorithm, the rotations are performed in a fixed sequence called a *sweep*, each sweep consisting of $n(n-1)/2$ rotations, and every column in the matrix is orthogonalised with every other column exactly once per sweep. The iterative procedure terminates if one complete sweep occurs in which all columns are orthogonal and no columns are interchanged. If the rotations in a sweep are chosen in a reasonable, systematic order, the convergence rate is ultimately quadratic [2].

Since one Jacobi plane rotation operation only involves two columns, there are disjoint operations which can be executed simultaneously. In a parallel implementation, we want to perform as many non-interacting operations as possible at each parallel time step.

In this paper we present a new parallel Jacobi ordering. This ordering may be called a *fat-tree ordering* because it uses the high bandwidth of a fat-tree to minimise global interprocessor communication costs. Thus it can be implemented efficiently on the fat-tree architectures.

The paper is organised as follows: Section 2 briefly describes fat-tree architectures. Our fat-tree ordering is described in Section 3 and compared with the (different) fat-tree ordering of [4]. Our conclusions are given in Section 4.

2 Fat-Tree Architectures

A fat-tree, based on a complete binary tree, is a routing network for parallel communication [5]. In a fat-tree a set of processors is located at the leaves of the tree and there are two channels corresponding to each edge, that is, one from parent to child and the other from child to parent. The number of wires in a channel is called the *capacity* of the channel. If the levels from bottom (the leaves) up are numbered $1, 2, \ldots$ and the capacity of the channels at level 1 is γ, the capacity of the channels at level k is given by $2^{k-1}\gamma$ for a (perfect) binary fat-tree. In other words, the capacity of the channels in the tree is increased by a factor of two for each increase in level. Thus, the overall communication bandwidth at each level is constant. If a factor of less than two is used (as in the CM5), we say that the tree is a *skinny* fat-tree.

A problem which is compute-bound on a serial computer may be communication-bound on a parallel computer. Thus a key issue in designing a parallel algorithm for a given problem is how to minimise the communication cost so that the computational capability of a parallel machine can be exploited to the full. Experimental results on the CM5 [6] suggest that, in order to achieve high performance on a skinny fat-tree architecture, communication should be kept local (especially for large messages) and contention should be avoided as far as possible.

3 Fat-Tree Ordering

In the following discussion we assume for convenience that n is a power of 2. We say that a communication is a *level-r* communication if the number of levels that a message from one leaf to another has to move up through the fat-tree (before coming down to its destination) is r. Thus, nearest neighbour communication between siblings in a tree architecture is level-one communication.

A fat-tree ordering was recently introduced in [4]. In the ordering of [4], most communications are local, and global communications are minimised. However, the disadvantages of the scheme recommended in [4] are –

1. Convergence may be slower than usual, because the number of rotations between any fixed pair (i, j) is variable rather than constant.

2. The logic to generate forward and backward sweeps is more involved than the logic to generate just a forward sweep.

3. On average an extra half-sweep has to be performed as the number of sweeps to termination has to be an *even* integer.

In this section we introduce a new fat-tree ordering. The communication cost is about the same as for the ordering of [4]. Only one procedure is required for every sweep, and the original order of the indices is maintained after the completion of each sweep. Therefore, our ordering avoids all three problems noted above for the ordering of [4].

Our fat-tree ordering is made up from two basic orderings, the *two-block* ordering and the *four-block* ordering, which are defined in Sections 3.1-3.2.

3.1 The two-block ordering

Suppose that there are two blocks, each containing 2^k indices. The objective of the two-block ordering

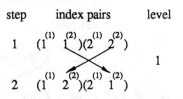

Figure 1: Basic module for two-block ordering.

is to let each index in one block meet each index in the other block once, so 2^{2k} different index pairs are generated. In the discussion below an ordering is called an ordering of size 2^k (or size 2^k ordering) if each block holds 2^k indices.

The basic module for our two-block ordering is depicted in Fig. 1. In the figure each block contains only two indices. The superscript (i) on each index in the figure indicates to which block that particular index belongs. Since there are only two indices in each block, the procedure (or a sweep of the ordering) takes only two steps to complete. At the first step, the indices from the two blocks are interleaved, forming two index pairs. The two indices in block 2 (or block 1) are then interchanged so that another two index pairs are generated at the second step.

We have assumed that each leaf on the tree holds only two indices. Communication is required if indices from different leaves are to be interchanged. It can easily be seen that our basic module requires only level-one communication if the block size is two, which results in minimal communication cost on a tree architecture. Therefore, in the derivation of our fat-tree ordering we always divide a large problem into a number of problems of size 2 in order to minimise the total communication cost. Also, the two indices in block 2 are exchanged after a sweep. If the same procedure is repeated once, the order of indices will be restored.

We now consider the case where one block holds more than two indices. We apply the divide and conquer technique, that is, a large problem is first divided into smaller sub-problems, the sub-problems are solved, and the sub-results are combined to obtain a result for the original problem. In the following a block is called a *rotating* block if the two indices (or two sub-blocks of indices) in the block exchange their positions during a two-block ordering. For example, see block 2 in Fig. 1.

We only consider the ordering of size $4 = 2^2$. The idea can easily be extended to the general case. In this ordering each block is first divided into two sub-blocks, each containing two indices. If each sub-block is considered as a super-index, the basic module may be applied. Since one super-index contains two indices, each super-index pair forms a sub-problem of size 2. Therefore, we have actually divided the orig-

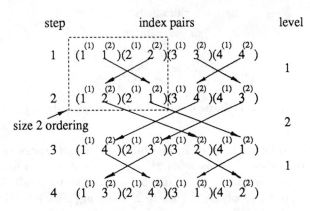

Figure 2: The two-block ordering of size 4.

inal problem into four half-size sub-problems. These sub-problems are solved in two super-steps, two at a time. The ordering is illustrated in Fig. 2.

Since there are interchanges of indices between sub-blocks (or super-indices), a level-two communication is required between the two super-steps. It can be seen from Fig. 2 that the two sub-blocks (1, 2) and (3, 4) in the second block have exchanged their positions after one sweep. However, the original order of the indices within each sub-block is maintained. This is because we always let the sub-blocks from the original second block be the rotating blocks when the ordering of size 2 is applied, and these sub-blocks are rotated twice during the computation. If the same procedure is executed once again the level-two communication is performed twice. Thus the order of the indices in block 2 will be restored. The indices in block 1 do not change their positions during the computation.

3.2 The four-block ordering

Suppose that we have four blocks, each containing 2^k indices. Our aim is to let each of the 2^{k+2} indices meet each other exactly once in a sweep of the ordering, to generate a total number of $2^{k+1}(2^{k+2} - 1)$ different index pairs.

We now consider the simplest case, where there are only four indices involved in the ordering. To generate six different index pairs one sweep of the ordering requires three steps. There are many ways to do this; two of them are depicted in Fig. 3.

If we enumerate the indices from the left, starting with 1, the original order of the indices will be (1, 2, 3, 4). This order is maintained after a sweep with the first ordering depicted in Fig. 3(a). However, with the second ordering depicted in Fig. 3(b) the positions of indices 3 and 4 are reversed after the first sweep, and the order is only restored after two consecutive sweeps of the ordering.

step	index pairs	level	index pairs
1	(1 2)(3 4)		(1 2)(3 4)
		1	
2	(1 3)(2 4)		(1 4)(3 2)
		1	
3	(1 4)(2 3)		(1 3)(4 2)
		1	
1	(1 2)(3 4)		(1 2)(4 3)
	(a)		(b)

Figure 3: Basic modules for four-block ordering.

The first algorithm has another advantage. It can be seen from Fig. 3(a) that the left index in any index pair is always smaller than the right index. If we store the column with larger norm on the left after each step of the SVD computation, then the singular values are obtained in nonincreasing order.

Note that in Fig. 3(a) there is a left-right arrow in an index pair in step 3. This indicates that the two indices in that pair have to be swapped before the communication between index pairs takes place for the next step. This implies that the two associated columns have to be exchanged in the SVD computation, which may degrade performance. However, this problem can easily be avoided. (See [8] for details.)

3.3 The merge procedure

Our fat-tree ordering algorithm is derived by using the following merge procedure. Suppose that there is a total number of 2^n indices. To begin the procedure these indices are first organised into 2^{n-2} groups, each holding only four indices. The four-block ordering is then applied so that the indices in each group will meet each other once. Next each pair of two consecutive groups is combined to form a super-group. Each group in a super-group is also divided into two blocks, so there are four blocks in each super-group. If each block is considered as a super-index, the four-block ordering may be applied. Each two consecutive super-groups may further form a super-supergroup and the four-block ordering is once again applied. The operation terminates if the 2^n indices are just in a big group and the four-block ordering applied to this big group is completed.

It should be noted that our objective is to let the 2^n indices meet each other exactly once in a sweep. Thus the two indices are not allowed to meet if they have met at a previous stage of the same sweep. In the following we give an example to illustrate the merge procedure. The method is easily extensible to problems of larger sizes.

step   index pairs   level

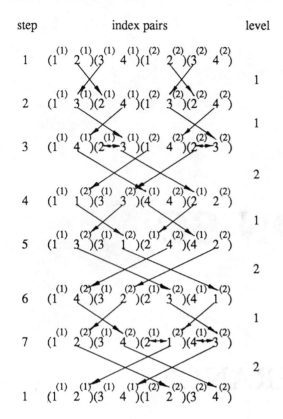

Figure 4: The four-block ordering for eight indices.

4 Conclusions

A new Jacobi ordering algorithm for parallel computation of SVD problems on fat-tree architectures has been introduced. It is currently being implemented on a 32-node CM5 at the Australian National University. Since the CM5 has a skinny fat-tree architecture, it is expected that the hybrid ordering described in [8], which is a combination of our fat-tree ordering and a ring ordering, will be the most efficient one, since that ordering does not cause any contention and reduces the number of global communications required by the ring ordering. If the CM5 used a perfect fat-tree, then our fat-tree ordering would be more attractive.

References

[1] R. P. Brent and F. T. Luk, "The solution of singular-value and symmetric eigenvalue problems on multiprocessor arrays", *SIAM J. Sci. and Statist. Comput.*, 6, 1985, pp. 69-84.

[2] G. H. Golub and C. F. Van Loan, *Matrix Computations*, The Johns Hopkins University Press, Baltimore, MD, second ed., 1989.

[3] M. R. Hestenes, "Inversion of matrices by biorthogonalization and related results", *J. Soc. Indust. Appl. Math.*, 6, 1958, pp. 51-90.

[4] T. J. Lee, F. T. Luk and D. L. Boley, *Computing the SVD on a fat-tree architecture*, Report 92-33, Department of Computer Science, Rensselaer Polytechnic Institute, Troy, New York, November 1992.

[5] C. E. Leiserson, "Fat-trees: Universal networks for hardware-efficient supercomputing", *IEEE Trans. Computers*, C-34, 1985, pp. 892-901.

[6] R. Ponnusamy, A. Choudhary and G. Fox, "Communication overhead on CM5: an experimental performance evaluation", in *Frontiers '92*, Proc. Fourth Symp. on the Frontiers of Massively Parallel Computation, IEEE, 1992, pp. 108-115

[7] J. H. Wilkinson, *The Algebraic Eigenvalue Problem*, Clarendon Press, Oxford, 1965, pp. 277-278.

[8] B. B. Zhou and R. P. Brent, "Parallel computation of the singular value decomposition on tree architectures", Technical Report, Computer Sciences Laboratory, Australian National University, Canberra, Australia, 1993.

Consider the case $n = 3$, or $2^n = 8$. We first divide the indices into two groups. Each group holds four consecutive indices. After a four-block ordering procedure applied to each group, the indices in the same group meet each other once. The two groups are then merged to form a super-group. The four blocks of indices in the super-group are organised in such a way that the indices in blocks 1 and 2 from the left group are interleaved and the indices in block 3 and 4 from the right group are organised in the same manner. To be specific, blocks 1, 2, 3 and 4 contain indices $(1^{(1)}, 3^{(1)})$, $(4^{(1)}, 2^{(1)})$, $(1^{(2)}, 3^{(2)})$ and $(4^{(2)}, 2^{(2)})$, respectively (see Fig. 4). Note that the indices in each original group have already been combined with each other in the previous stage, which is exactly the computation required in super-step 1 of the four-block ordering of Section 3.2. Thus, only super-steps 2 and 3 remain to be performed. Since the blocks are interleaved in each super-index pair, the two-block ordering procedure may be applied to let the indices from different blocks in each super-index pair meet each other once, which completes the merge procedure. The details are illustrated in Fig. 4. It is clear that the order of the indices is unchanged by the merge procedure.

SESSION 6C

FAULT-TOLERANCE

A Fault-Tolerant Parallel Algorithm for Iterative Solution of the Laplace Equation *

Amber Roy-Chowdhury and Prithviraj Banerjee

Coordinated Science Lab

1308 W Main Street

University of Illinois

Urbana IL 61801

Abstract

Algorithm based fault tolerance is an inexpensive method of achieving fault tolerance without requiring any hardware modifications. Algorithm-based schemes have been proposed for a wide variety of numerical applications. However, for a particular class of numerical applications, namely those involving the iterative solution of linear systems, there exist almost no fault-tolerant algorithms in the literature. In this paper, we describe a fault-tolerant version of a parallel algorithm for iteratively solving the Laplace equation over a grid. The fault-tolerant algorithm is based on the popular successive overrelaxation scheme with red-black ordering. We use the Laplace equation merely as an illustration; fault-tolerant versions of other iterative schemes for solution of linear systems arising from discretizations of other partial differential equations may be similarly derived.

We also present a new way of dealing with the roundoff errors which complicate the check phase of algorithm-based schemes. Our approach is based on error analysis incorporating some simplifications and gives high fault coverage and no false alarms for a large variety of data sets, as shown by our results.

The timing overheads of our fault-tolerant algorithm over the basic SOR algorithm involving no fault tolerance decrease with increasing problem dimension and become negligible for large data sizes.

1 Introduction

Algorithm-based fault tolerance (ABFT) is a well established technique which is used to develop reliable versions of numerical algorithms [1][2]. Basically, an algorithm is modified in order to preserve an invariant before and after all computations involving data elements. A lack of preservation of the invariant at the end of the algorithm indicates the presence of errors involving data computations. The algorithm-based fault tolerance technique for a particular application may be so designed to preserve the invariant not only at the end of all computation on data elements but also at suitable intermediate points during the course of the algorithm which may then serve as intermediate checkpoints. Reliable versions of such numerical algorithms as matrix multiplication [2][3], gaussian elimination [3], fast fourier transform [4][5], QR factorization [6], singular value decomposition [7] and several others have been developed in the past. Recently, ABFT techniques have been applied to a parallel bitonic sort algorithm [8], thus demonstrating that nonnumerical applications can also be made reliable by the use of algorithm-based checks.

In this paper, we develop a reliable, low overhead version of a successive overrelaxation algorithm to iteratively solve the Laplace equation for discrete points over a grid. An ABFT scheme for this application was reported in [9] for systolic arrays. The algorithm described in this paper is suitable for implementation on general purpose distributed memory multicomputers and has several advantages over the algorithm discussed in [9], which we point out in the paper. We use the Laplace equation as an illustrative example of how one could go about developing reliable versions of other algorithms for iteratively solving systems of linear equations which arise from the discretization of other partial differential equations. We also use our reliable version to demonstrate a new technique for dealing with roundoff errors which can complicate the invariant check step in the algorithm. Our method for dealing with roundoff error uses simplified error analysis techniques and gives much better error coverage results than earlier techniques [7][3][10].

The organization of this paper is as follows. We first discuss the simple serial algorithm in order to motivate the reader for the reliable parallel algorithm which is our goal. We demonstrate our low overhead invariant calculation and error analysis for the reliable version of the serial algorithm. We then present the reliable version of the parallel algorithm involving invariant and error bound computation. Finally, we present experimental results on an Intel iPSC/2 hypercube indicating error coverage and timing overhead for the parallel algorithm.

*Acknowledgement: This research was supported in part by the Office of Naval Research under contract N00014-91-J-1096 and in part by the Joint Services Electronics Program under contract N00014-84-C-1049.

```
for(k=0;k<iter;k++)
{
    for(i=1;i<=n;i++)
    for(j=1;j<=n;j++)
    t[i][j] = 0.25*(u[i][j-1]+u[i][j+1]+u[i-1][j]+u[i+1][j]);

    for(i=1;i<=n;i++)
    for(j=1;j<=n;j++)
    u[i][j] = t[i][j];
}
```

Figure 1: Code for the solution of the Laplace equation by the iterative Gauss-Jacobi technique

2 Serial Algorithm for Solving the Laplace Equation Using Successive Overrelaxation

2.1 Serial SOR algorithm

The Laplace equation is a second-order elliptic partial differential equation described by the following equation

$$\frac{\partial^2 u}{\partial x^2} + \frac{\partial^2 u}{\partial y^2} = 0 \qquad (1)$$

We may "solve" the Laplace equation numerically over a region by discretizing it in the x and y directions to obtain a grid of points and then computing the approximate solution values at these points. Boundary conditions are usually specified by providing the values of $u(x, y)$ at the boundary points of the grid. The algorithm is shown in Fig. 1 and is often referred to as the iterative Gauss-Jacobi technique [11]. Here, u is assumed to be a 2-dimensional array, each element of which stores the value computed at a grid point. It is assumed that the grid is discretized so that it has $n + 2$ points along each dimension, though the more general case of having a different number of grid points along each dimension does not add any complication to subsequent analyses. It is also assumed that rows 0 and $n + 1$ and columns 0 and $n + 1$ are initialized to store the boundary conditions. Also note that the algorithm is assumed to execute the outermost loop a predefined *iter* times at the end of which the interior grid points are assumed to be close to their final values. Often this termination condition is determined at runtime by specifying that the outer loop continue until the maximum difference over all grid points of a point value at the current iteration from its value at the previous iteration drops below a threshold. Again, the latter condition for termination makes no difference whatsoever to subsequent analyses.

We may use the updated values of the $u(i, j)$'s as soon as they are available in the update statement of the Gauss-Jacobi loop to obtain the code in Fig. 2, which is referred to as the iterative Gauss-Seidel technique. This modification results in asymptotically faster convergence to the steady state values. We may further accelerate the rate of convergence to the steady state value by introducing a

```
for(k=0;k<iter;k++)
{
    for(i=1;i<=n;i++)
    for(j=1;j<=n;j++)
    u[i][j] = 0.25*(u[i][j-1]+u[i][j+1]+u[i-1][j]+u[i+1][j]);
}
```

Figure 2: Code for the solution of the Laplace equation by the iterative Gauss-Seidel technique

```
for(k=0;k<iter;k++)
{
    for(i=1;i<=n;i++)
    for(j=1;j<=n;j++)
    u[i][j] = (1-w)*u[i][j]+0.25*w*(u[i][j-1]+u[i][j+1]+u[i-1][j]+u[i+1][j]);
}
```

Figure 3: Successive overrelaxation code for the solution of the Laplace equation

parameter ω to overrelax the computation of Fig. 2, so that each updated point is now computed as an average of its old value and the values of its four neighboring points, as shown in Fig. 3. The code of Fig. 3 is often referred to as a successive overrelaxation iterative method (SOR) and ω is referred to as the relaxation parameter. A careful choice of ω results in much faster convergence rates than the previous two schemes. The code of Fig. 3 is hard to parallelize since each update of $u(i, j)$ involves new values of $u(i, j - 1)$ and $u(i - 1, j - 1)$ and old values of $u(i, j + 1)$ and $u(i + 1, j)$. However, a modified SOR algorithm may be devised with the same asymptotic rate of convergence as the algorithm of Fig. 3 which is well suited for parallel implementation. In this algorithm, the grid points are divided into two disjoint sets. We refer to those $u(i, j)$ for which $i + j$ is even as red points and those for which $i + j$ is odd as black points. Then, each update of Fig. 3 may be split up into two updates, one involving only the red points and the other involving only the black points, as shown in Fig. 4. The code of Fig. 4 is referred to as a Red-Black SOR method. The update of each red point utilizes the values of its four neighbors, which are black points, and the update of each black point utilizes the values of its four neighbors, which are red points. Thus, parallelizing the code of Fig. 4 poses no problem since we may first update all red points in parallel and then update all black points in parallel. The code of Fig. 4 forms the basis of our fault-tolerant serial and parallel algorithms.

2.2 Serial Red-Black SOR algorithm with checks

The idea for a fault-tolerant algorithm for the Red-Black SOR algorithm described in the previous section stems from the idea of maintaining checksums based on the linearity property [7]. We assume that n is even, though the case when n is odd may be treated similarly. For each red-black update, we make the following observations:

(i) Every red interior point is used in the update of

```
for(k=0;k<iter;k++)
{
    /* update red points */
    for(i=1;i<=n;i++)
    {
    for(j=2-(i%2);j<=n;j+=2)
    u[i][j] = (1-w)*u[i][j] + 0.25*w*(u[i][j-1]+u[i][j+1]+u[i-1][j]+u[i+1][j]);
    }
    /* update black points */
    for(i=1;i<=n;i++)
    {
    for(j=1+(i%2);j<=n;j+=2)
    u[i][j] = (1-w)*u[i][j] + 0.25*w*(u[i][j-1]+u[i][j+1]+u[i-1][j]+u[i+1][j]);
    }
}
```

Figure 4: Red-Black SOR code

exactly one red point and every black interior point is used in the update of exactly one black point.

(ii) Every black boundary point is used in the update of exactly one red point and every red boundary point is used in the update of exactly one black point.

(iii) The black points $u(1, n)$ and $u(n, 1)$ are used in the update of exactly two red points. The red points $u(1, 1)$ and $u(n, n)$ are used in the update of exactly two black points.

(iv) The black points for which $i = 1$ or $i = n$ or $j = 1$ or $j = n$ except for $u(1, n)$ and $u(n, 1)$ are used in the update of exactly three red points. The red points for which $i = 1$ or $i = n$ or $j = 1$ or $j = n$ except for $u(1, 1)$ and $u(n, n)$ are used in the update of exactly three black points.

(v) All other black points are used in the update of exactly four red points. All other red points are used in the update of exactly four black points.

Observation (i) follows directly from the statement for updating the red points in Fig. 4. The remaining observations are illustrated in Fig. 5 for the black points. Observations (i) through (v) lead us to maintain a sum on all the red points and update this sum at the end of each update of the red points. Let us denote the sum of all red points by S_R, the sum of all red boundary points by S_{RB}, the sum of $u(1, 1)$ and $u(n, n)$ by S_{RC} the sum of the red points of (iv) by S_{RO} and the sum of the red points of (v) by S_{RI}. The corresponding sums of the black points are denoted by $S_B, S_{BB}, S_{BC}, S_{BO}$ and S_{BI}. Each red interior point contributes $(1 - \omega)$ times its value to the updated value of a red point while each black point contributes 0.25ω times its value to each red point update it is involved in. Similarly, each black interior point contributes $(1 - \omega)$ times its value to the updated value of a black point while each red point contributes 0.25ω times its value to each black point update it is involved in. These last two observations, along with observations (i) through (v), allow us to update S_R and S_B following the updates of the red and black points respectively in terms of the other sums as follows

$$S_R = (1 - \omega)S_R + \omega(S_{BI} + 0.75S_{BO} + 0.5S_{BC} + 0.25S_{BB})$$
(2)

$$S_B = (1 - \omega)S_B + \omega(S_{RI} + 0.75S_{RO} + 0.5S_{RC} + 0.25S_{RB})$$
(3)

Following the update of the red points, besides updating S_R as shown in Eq. (2), we also need to recompute S_{RO}, S_{RC} and S_{RI} since these have changed also. Similarly, following the update of the black points we need to recompute S_{BO}, S_{BC} and S_{BI}. The boundary elements are constant and so we do not need to recompute S_{RB} or S_{BB}. (Note that in the parallel implementation, the boundary elements for each processor's portion of the grid may need to be recomputed, since these may be interior points in the global grid). S_{RO} and S_{BO} require $O(n)$ operations to compute, while computing S_{RC} and S_{BC} require just one addition operation. Since S_R is the sum of all red interior points, i.e., $S_R = S_{RO} + S_{RC} + S_{RI}$, and similarly S_B is the sum of all black interior points, i.e., $S_B = S_{BO} + S_{BC} + S_{BI}$, we may compute S_{RI} and S_{BI} in constant time once the red and black updates, respectively, and the updates of all other red and black sums, respectively, have been completed, using the following equations

$$S_{RI} = S_R - S_{RO} - S_{RC}$$
(4)

$$S_{BI} = S_B - S_{BO} - S_{BC}$$
(5)

Finally, after the required number of iterations have been executed or convergence has been attained, the sum of the red interior points is compared with S_R, and the sum of the black interior points is compared with S_B. In the absence of computational faults, these values should be equal to within a tolerance. (The tolerance is necessary due to roundoff error accumulation).

2.3 Error bound derivation

Due to differences in roundoff error accumulation in the red and black points and S_R and S_B, one has to allow for a threshold when comparing between the sum of the red points and S_R and the sum of the black points and S_B. Previous researchers have either suggested an experimental evaluation of the threshold [7][3] which is at best useful for data sets of fixed size and limited data range or a mantissa checksum test [10] which suffers from the disadvantage of not being able to check floating point additions without recourse to duplication or experimental determination of the threshold. We deal with the thresholding problem in a different manner by computing the error expressions for the variables involved in the computation. In order to simplify our error analysis, we compute the global error at the end of one iteration of the outer loop in terms of the local error for the loop and the global error accumulation upto the end of the previous iteration. A more detailed discussion of this approach and its application to several other algorithms can be found in [12] and [13]. Error analysis of the SOR code with checks forms the subject of the following paragraphs.

The statements which we need to analyze for error are of four types

(i) $a = b + c$

(ii) $a = b - c - d$

(iii) $a = Ab + BC(c + d + e + f)$

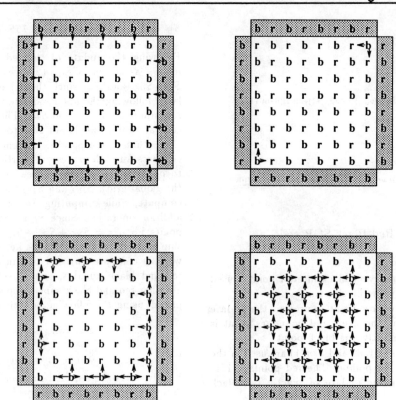

Figure 5: Interactions of black points with red points

(iv) $a = Ab + B(c + Cd + De + Ef)$
where lowercase letters denote variables and capitals denote constants. In subsequent discussions, we use the notation $|err(v)|$ to denote an upper bound on the absolute error associated with a variable v. We further use $\hat{v}$ to denote the value of a variable v including the floating point errors accumulated in computing v. Thus, we can relate $\hat{v}$, v and $err(v)$ by

$$|\hat{v}| \leq |v| + |err(v)| \qquad (6)$$

We now proceed to derive error bounds for the expressions in (i),(ii),(iii) and (iv) above using the following fundamental lemma concerning error accumulation in floating point computations

Lemma 1 *A floating point computation of z, denoted by $z = fl(x \oplus y)$, results in an error of $err(z) = (x \oplus y)\delta$, where $|\delta| \leq \epsilon = 2^{-t}$, where t denotes the mantissa size.*

Proof: Refer to [14], page 113 □

In subsequent derivations we will assume that all elements taking part in the floating point computations are positive. Thus, the absolute values of each element are the same as their actual values. Also, in this case, the δ of Lemma 1 is always positive. We will indicate in subsection 2.4 how to initially transform the values on the boundary and interior of the grid so that at all subsequent steps, the value of each grid point is positive.

The error expression for (i) is given by the following theorem

Theorem 1 *The error accumulation in the computation of the floating point expression*

$$\hat{a} = fl(\hat{b} + \hat{c}) \qquad (7)$$

is given by

$$err(a) \leq a\epsilon + err(b) + err(c) \qquad (8)$$

Proof: By Lemma 1 we then have $\hat{a} = (\hat{d} + \hat{b})(1 + \delta_1)$. Using Lemma 1 to get the bound $\delta_i \leq \epsilon$ and discarding the $O(\epsilon^2)$ terms which arise under the assumption that they are negligible compared to the $O(\epsilon)$ terms, we obtain the following relation

$$\hat{a} \leq \hat{b} + \hat{c} + (\hat{b} + \hat{c})\epsilon \qquad (9)$$

Using Eq. (6) to substitute for the hatted variables in Eq. (9) and once again using the simplification of discarding the $O(\epsilon^2)$ terms which arise, we get the following bound on the value of $\hat{a}$

$$\hat{a} \leq a + (b + c)\epsilon + err(b) + err(c) \qquad (10)$$

from where we obtain the following relation for $err(a)$

$$err(a) \leq (b + c)\epsilon + err(b) + err(c) \qquad (11)$$

Substituting $a = b + c$ in the above expression completes the proof □

The error expression for (ii) is given by the following theorem

Theorem 2 *The error accumulation in the computation of the floating point expression*

$$\hat{a} = fl(\hat{b} - \hat{c} - \hat{d}) \tag{12}$$

is given by

$$err(a) \leq (2b + 2c + d)\epsilon + err(b) + err(c) + err(d) \tag{13}$$

Proof: The proof is similar to Theorem 1 and is omitted due to lack of space. Detailed proof appears in [12]. □
The error expression for (iii) is given by the following theorem

Theorem 3 *The error accumulation in the computation of the floating point expression*

$$\hat{a} = fl(A\hat{b} + BC(\hat{c} + \hat{d} + \hat{e} + \hat{f})) \tag{14}$$

is given by

$$\begin{aligned} err(a) \leq & \ 2a\epsilon + BC(4c + 4d + \\ & 3e + 2f)\epsilon + Aerr(b) + BC(err(c) + \\ & + err(d) + err(e) + err(f)) \end{aligned} \tag{15}$$

Proof: The proof is similar to Theorem 1 and is omitted due to lack of space. Detailed proof appears in [12]. □

The error expression for the floating point computation of (iv) is as follows

Theorem 4 *The error accumulation in the computation of the floating point expression*

$$\hat{a} = A\hat{b} + B(\hat{c} + C\hat{d} + D\hat{e} + E\hat{f}) \tag{16}$$

is given by

$$\begin{aligned} err(a) \leq & \ (2a + B(3c + 4Cd + 3De \\ & + 2Ef))\epsilon + Aerr(b) + B(err(c) + \\ & + Cerr(d) + Derr(e) + Eerr(f)) \end{aligned} \tag{17}$$

Proof: The proof is similar to Theorem 1 and is omitted due to lack of space. Detailed proof appears in [12]. □

2.4 Modified algorithm with checks and error bounding

The derivation of the error expressions now enables us to write a fault-tolerant version of the algorithm in Fig. 6 which also computes the error bounds on the fly. We make two initial transformations to the original problem. Initially, before starting the iterative solution process, we determine the value of the negative boundary element of largest magnitude, say M and add $|M|$ to all points on the grid (boundary and interior). This results in all points on the grid taking on positive initial values. It may easily be verified that from this point on, the $u(i, j)$'s computed in each iteration exceed the corresponding $u(i, j)$'s of the unmodified problem by exactly $|M|$. Thus, the solution to the original problem may be recovered by subtracting $|M|$ from each of the $u(i, j)$'s computed in the modified problem. Also note that the $u(i, j)$ values in the modified problem are always positive, so that the error expressions of subsection 2.3 may be directly applied. Another initial modification which we introduce is scaling the elements by dividing each element by the largest boundary element after translating them by the value of the negative boundary element of largest magnitude. Again, it may be easily verified that following the latter modification, the magnitude of every element in subsequent iterations of the algorithm lies between 0 and 1, leading to an easy correctness check at intermediate points in the algorithm. The solution to the original problem at any point may be recovered by simply scaling the elements with the reciprocal of the boundary element of largest magnitude.

Using Theorems 1 through 4 to compute error expressions, we get the code shown in Fig. 6. There are three points to note here.

First, we store the error bound for a variable called v in a variable called $errv$. For each of the sum variables, the error variables may be updated by using one of the three error expressions derived in subsection 2.3. However, we need to introduce two extra error variables $errSR_-$ and $errSB_-$ which keep track of the error accumulation of the red and black points respectively. The expressions for updating $errSR_-$ and $errSB_-$ in each iteration shown in Fig. 6 may be derived by summing over all red and black points, respectively, the error expressions for each individual red or black point which may be obtained by an application of Theorem 3. The sums of the errors over the red points are bounded by the old value of $errSR_-$ while the sums of the errors over the black points are bounded by the old value of $errSB_-$ since the values of the black boundary points never change and thus do not incorporate any roundoff errors.

Second, wherever error bounds for individual elements of $u(i, j)$ arise in our error expressions, we drop them since maintaining error bounds for individual grid elements would require too much computation overhead. This approximation still leads to the error expressions acting as upper bounds on the error in practice.

Third, we postpone the multiplication of the error variables by ϵ to the very end, when the error variables are actually used to compute the tolerance for the check on the sums of the red and black points. This saves some extra computations.

3 Parallel Algorithm for Solving the Laplace Equation

3.1 Parallel SOR algorithm

The serial Red-Black SOR code with checks lends itself easily to efficient implementation on a parallel machine. We assume that the underlying architecture is a 2-dimensional mesh with N_1 and N_2 processors in each dimension. For simplicity, we assume that the grid dimensions (assumed to be n in each direction) are divisible by $2N_1$ and $2N_2$. Then, the initial data distribution is blockwise with the processor in the mth row and lth column in the mesh receiving the block containing the interior points $u(i,j)$, $\frac{mn}{N_1} + 1 \leq i \leq \frac{(m+1)n}{N_1}$, $\frac{ln}{N_2} + 1 \leq j \leq \frac{(l+1)n}{N_2}$. The boundary points are distributed to processors holding the adjacent interior points. At the end of each iteration, each processor in the mesh receives from its north and south neighbors (if they exist) messages containing the elements of u adjoining the elements in its top and bottom rows respectively. Similarly, each processor also receives from its east and west neighbors messages containing the elements of u adjoining the elements in its rightmost and leftmost columns respectively. Each processor then updates its portion of the grid in a manner identical to the serial algorithm, and then proceeds with the next iteration of the algorithm.

3.2 Parallel algorithm with checks and error bounding

From the discussion in subsection 3.1 we note that from the point of view of a single processor, the computations performed by it are exactly the same as the serial algorithm being used to solve a problem of smaller dimensions, except possibly for nonconstant boundary elements received from neighboring processors. Thus we may incorporate fault tolerance into the parallel algorithm in the same manner as for the serial algorithm by introducing the variables for keeping track of the sums of the red and black points of each processor's local block of u and using the same expressions as for the serial algorithm. Our error bound expressions for the sum variables which were developed for the serial algorithm via Theorems 1 through 4 are valid here as well since now each processor performs the same kind of operations as in the serial case, except that the boundary elements for the processor may not be fixed in this case. Thus, we need to recompute the sum of the red and black boundary elements following the red and black updates in each iteration, unlike the serial algorithm where these sums only needed to be computed once before the beginning of the iterative updates. Besides dropping error variables involving individual elements of u, as in the serial algorithm, we make the further simplification of dropping error variables involving the boundary elements wherever they arise in updating the error expressions for a particular processor since otherwise we would have to maintain error variables for each element of u, an unacceptably large overhead. (Recall that in the serial algorithm, the boundary

```
for(k=0;k<iter;k++)
{
    /* update red points */
    for(i=1;i<=n;i++)
    {
    for(j=2-(i%2);j<=n;j+=2)
    u[i][j] = (1-w)*u[i][j] + 0.25*w*(u[i][j-1]+u[i][j+1]+u[i-1][j]+u[i+1][j]);
    }
    /* update red sums and error variables */
    SR = (1-w)*SR + w*(SBI+0.75*SBO+0.5*SBC+0.25*SBB);
    errSR_ = 2*SR+3.25*w*SB+(1-w)*errSR_+w*errSB_;
    errSR = (1-w)*errSR+w*(errSBI+0.75*errSBO+0.5*errSBC+0.25*errSBB)
            +2*w*SR+w*(2*SBI+3*SBO+1.5*SBC+0.5*SBB);
    SRO = errSRO = 0;
    for(j=3;j<=n;j+=2)
    {
    SRO += u[1][j];
    errSRO += SRO;
    }
    for(j=2;j<=n;j+=2)
    {
    SRO += u[n][j];
    errSRO += SRO;
    }
    for(i=3;i<=n;i++)
    {
    SRO += u[i][1];
    errSRO += SRO;
    }
    for(i=2;i<n;i+=2)
    {
    SRO += u[i][n];
    errSRO += SRO;
    }
    SRC = u[1][1]+u[n][n];
    errSRC = SRC;
    SRI = SR-SRO-SRC;
    errSRI = errSR+errSRO+errSRC+2*(SR+SRO)+SRC;

    /* similarly update black points and black sums */
        .
        .
        .

}


/* check sum of red points */
errSUM =  SUM = 0;
for(i=1;i<=n;i++)
for(j=2-(i%2);j<=n;j+=2)
{
SUM += u[i][j];
errSUM += SUM;
}

if(abs(SUM-SR)>(errSR+errSR_+errSUM)*pow(2.0,-mantissa_size))
error();

/* similarly check sum of black points */
    .
    .
```

Figure 6: Reliable SOR code incorporating error expression computation

elements were not computed on and thus did not accumulate any roundoff errors). We have found that even with this simplification, our error expressions act as an upper bound on the error in practice. Also, there is a possibility that boundary elements received from a neighbor might be in error. We therefore translate and scale the initial data as for the serial algorithm. Recall that this ensures that the values of all elements involved in the computations lie between 0 and 1 at any point in the algorithm. This property may be used to check the correctness of the boundary data received from neighbors. Gross errors which cause any element to take on values less than 0 or greater than 1 can be immediately detected. More subtle errors are likely to be detected when we compare the values of SR and SB and the sum over all red and black points of u, respectively, on every processor at the conclusion of the algorithm.

3.3 Relation to prior work

At this point we would like to mention an earlier work on developing a fault-tolerant SOR algorithm [9]. The algorithm is based on suitably choosing the initial values on the grid so that at subsequent time steps, solutions at each point are monotonically increasing. The algorithm uses a fine-grained data distribution in which each processor computes on only two adjacent data points and is thus unsuitable for implementation on a general purpose distributed memory parallel computer such as the testbed we ran our experiments on. Also, the monotonic behavior is preserved only for a certain range of the relaxation parameter ω, unlike our algorithm which does not restrict the range of ω in any way in order to achieve fault-tolerance. The SOR algorithm in [9] requires individual checking of each element of the grid following each SOR iteration to determine whether the monotonicity property holds, which represents a large and constant overhead. In contrast, our algorithm only requires checking the boundary elements of the grid sub-block on each processor, which requires only $O(n)$ steps as compared to $O(n^2)$ SOR update operations for each iteration, and a final check step after convergence has been achieved (or the specified number of SOR iterations have been performed), which requires $O(n^2)$ steps. Thus the checking overhead for our algorithm diminishes both with increasing grid size and increasing the number of iterations.

4 Experimental Results

The effectiveness of an algorithm-based fault tolerance method can be judged by evaluating it according to three criteria. First, the percentage of *false alarms*, which are defined as the percentage of times roundoff errors caused an algorithm-based check to fail, must be low. Second, the error coverage, which may be defined as the percentage of time hardware errors were detected due to a failed check, must be high. Third, the overhead of the fault-tolerant algorithm over the basic algorithm due to extra computations involving the checks and their tolerance expressions, must be as small as possible.

We implemented the fault-tolerant parallel Red-Black SOR algorithm on a 16-processor Intel iPSC/2 hypercube.

Table 1: Error coverage results for fault-tolerant SOR algorithm

	Error Coverage	Significant Error Coverage
Transient Bit-level	65	98
Transient Word-level	100	100
Permanent Bit-level	69	98
Permanent Word-level	100	100

Meshes of various dimensions were simulated on the hypercube. Subsection 4.1 gives the false alarm and error coverage results. Subsection 4.1 gives the timing overhead of the fault-tolerant algorithm over the basic algorithm for various grid and mesh dimensions.

4.1 Error coverages

In order to determine the false alarm percentage, we ran our fault-tolerant parallel version of the SOR algorithm on 1350 data sets without injecting hardware errors. Any failed checks on these data sets could then be assumed to be false alarms caused by the magnitude of roundoff errors exceeding the upper bound estimate on roundoff errors provided by the tolerance expression. We varied grid sizes from 32x32 to 192x192, the number of iterations from 200 to 1000 and mesh dimensions from 1x1 to 4x4. (Recall that our scaling and translating of initial data ensured that the data range being computed on was between 0 and 1 in all cases). However, not even a single case of false alarm was observed over this set of runs involving data sets with widely different characteristics.

Error coverages were determined by simulation of transient and permanent bit and word level errors over 1800 runs over different data sets. The grid size and mesh dimensions were varied over the same range as for the false alarm calculation, while the number of iterations were varied from 100 to 500. Error coverage of word level errors was always 100%, whether the errors were permanent or transient, indicating the effectiveness of our method for detecting errors of a gross nature. Error coverages of transient and permanent bit level errors were around 65%. However, most of these undetected errors were in the less significant bits. In order to illustrate the effectiveness of our method in detecting serious errors, we computed new error coverages for only those errors which caused a change of 0.1% or more in the element values they were injected in. Subject to this restriction, we found error coverages of bit level errors to be around 98%, showing the effectiveness of our method in detecting errors that resulted in significant data corruption. The results are summarized in Table 1. The column heading "Significant Error Coverage" refers to errors which caused a change of more than 0.1% in the elements they were injected in and were detected by our algorithm.

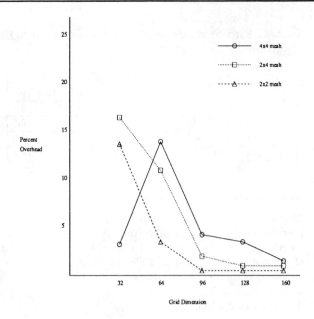

Figure 7: Timing overhead of the fault-tolerant algorithm over the basic algorithm

4.2 Timing overheads

Fig. 7 shows the timing overheads of the fault-tolerant Red-Black SOR algorithm with checks and error bounding over the basic Red-Black SOR algorithm incorporating no fault tolerance for various mesh dimensions and grid sizes. The overheads become negligible (less than 1%) as we increase the grid size, which is only to be expected since we perform only $O(n)$ check operations in each iteration of Red-Black update, compared to $O(n^2)$ operations required for the updates of each red and black point in the grid.

5 Conclusions

In this paper, we have presented a fault-tolerant method for a popular iterative scheme for solving a partial differential equation. Minor extensions of the method can be used to solve discretizations of other problems, such as the Poisson equation, and other methods of discretization, such as using a 9-point stencil and different coloring schemes instead of a 5-point stencil with Red-Black ordering as was the case here. We have presented a new method of dealing with the roundoff accumulation problem which complicates the invariant check step in algorithm-based fault-tolerant encodings based on error analysis incorporating some simplifications in the interest of keeping overheads low. We have shown by our results on a wide variety of data sets that our method yields a truly low overhead fault-tolerant algorithm with high error coverage.

References

[1] K.-H. Huang and J. A. Abraham, "Algorithm-based fault tolerance for matrix operations," *IEEE Trans. Comput.*, vol. C-33, pp. 518–528, June 1984.

[2] J.-Y. Jou and J. A. Abraham, "Fault-tolerant matrix operations on multiple processor systems using weighted checksums," *SPIE Proceedings*, vol. 495, August 1984.

[3] P. Banerjee, J. T. Rahmeh, C. Stunkel, V. S. Nair, K. Roy, V. Balasubramanian, and J. A. Abraham, "Algorithm-based fault tolerance on a hypercube multiprocessor," *IEEE Trans. Comput.*, vol. 39, pp. 1132–1145, September 1990.

[4] J.-Y. Jou and J. A. Abraham, "Fault-tolerant fft networks," *IEEE Trans. Comput.*, vol. 37, pp. 548–561, May 1988.

[5] Y.-H. Choi and M. Malek, "A fault-tolerant fft processor," *IEEE Trans. Comput.*, vol. 37, pp. 617–621, May 1988.

[6] V. Balasubramanian and P. Banerjee, "Tradeoffs in the design of efficient algorithm-based error detection schemes for hypercube multiprocessors," *IEEE Trans. Softw. Eng.*, vol. 16, pp. 183–194, February 1990.

[7] V. Balasubramanian, "*The Analysis and Synthesis of Efficient Algorithm-Based Error Detection Schemes for Hypercube Multiprocessors.*" Ph.D. dissertation, Univ. of Illinois, Urbana-Champaign, February 1991. Tech. Report no. CRHC–91–6, UILU–ENG–91–2210.

[8] B. M. McMillin and L. M. Ni, "Reliable distributed sorting through the application-oriented fault tolerance paradigm," *IEEE Trans. Parallel Distrib. Systems*, vol. 3, pp. 411–420, July 1992.

[9] K.-H. Huang, "*Fault-Tolerant Algorithms for Multiple Processor Systems.*" Ph.D. dissertation, Univ. of Illinois, Urbana-Champaign, November 1983. Tech. Report no. CSG-20.

[10] F. T. Assaad and S. Dutt, "More robust tests in algorithm-based fault-tolerant matrix multiplication," *Proc. FTCS-22*, pp. 430–439, June 1992.

[11] J. M. Ortega, *Introduction to Parallel and Vector Solution of Linear Systems.* New York: Plenum Publishing Corp., 1988.

[12] A. Roy-Chowdhury, "*Evaluation of Algorithm Based Fault-Tolerance Techniques on Multiple Fault Classes in the Presence of Finite Precision Arithmetic.*" M.S. Thesis, Univ. of Illinois, Urbana-Champaign, August 1992. Tech. Report no. CRHC–92–15, UILU–ENG–92–2228.

[13] A. Roy-Chowdhury and P. Banerjee, "Tolerance determination for algorithm-based checks using simplified error analysis techniques," *Proc. FTCS-23*, June 1993. To Appear.

[14] J. H. Wilkinson, *The Algebraic Eigenvalue Problem.* Oxford: Clarendon Press, 1965.

Emulating Reconfigurable Arrays for Image Processing Using the MasPar Architecture*

Jose Salinas and Fabrizio Lombardi
Texas A&M University
Department of Computer Science
College Station, TX, 77843-3112
joses@cs.tamu.edu, lombardi@cs.tamu.edu

Abstract -- This paper examines a fault tolerant scheme for two-dimensional arrays of processors which functionally reconfigures the array without the use of spares. Reconfiguration approaches for different interconnection networks are analyzed. Also, three approaches are proposed for mapping image data to and from the array, depending on the type of array and computational power available in each processing element. The proposed reconfiguration approaches have been emulated on a 32×64 processor MasPar array computer.

INTRODUCTION

The need for reliable architectures which also meet the computing power required for image processing applications, has recently shifted attention to 2-dimensional processor arrays [2]. These types of architectures are commonly referred to as single instruction multiple data (SIMD) computers and are based on the principle of having a single instruction to affect several active processors in parallel. SIMD array architectures are composed of a set of processing elements (PEs) arranged in a 2-dimensional grid with an interconnection network provided between processing elements. The structure of a PE ranges from a very simple element (such as ALU) to a highly complex microprocessor [5]. Many applications have been proposed for image processing, such as image restoration and computer vision [1][2]. Since many of these applications can be parallelized and are based mainly on nearest-neighbor operations, they are highly suited for implementation on a 2-D SIMD array. Each PE in the array is programmed to operate on a set of localized image pixels and nearest-neighbor operations are performed by communicating with neighbor PEs [5].

With parallel architectures, as the number of processing elements increases, so does the probability that the entire system will fail due to a fault in at least one of those elements. Therefore, to increase the reliability of the systems as well as improve the manufacturing yield, defect-tolerance and fault-tolerance must be implemented efficiently. Several approaches have been proposed for implementing fault tolerant schemes on 2-D systems [3][7]. One of the most popular approaches for fault-tolerance is to include a set of spare PEs to replace the faulty PEs [7].

A fault-tolerant array structure which does not use spare elements is proposed based on [9]. The proposed structure has three main objectives: to reconfigure the array in a short time, to minimize the loss of image data mapped into the array, and to reduce the amount of additional hardware for array reconfiguration. The proposed reconfiguration strategies have been emulated on a MasPar architecture.

EMULATION ON A MASPAR ARCHITECTURE

The proposed algorithms for array reconfiguration have been emulated on a MasPar architecture instead of the typical implementation using simulation techniques. Emulation is defined as a software implementation of a particular hardware on an underlying architecture. The emulation on a MasPar architecture has several advantages over simulation approaches for array reconfiguration. The most important difference between emulation and simulation is the amount of program translation necessary to achieve functional correctness for a particular algorithm. With an emulation approach, the objective of the software is to create a fully functional hardware architecture using software. The mesh connected architecture of the MasPar and the corresponding interconnection network has many similarities to most array architectures proposed in the literature. Additionally, emulation techniques result in a more accurate performance evaluation and better real life data results.

The 32x64 MasPar computer used for the emulation of the proposed algorithms is an SIMD architecture with 2048 processors [6]. There are two types of PE communication protocols: each PE can either communicate with any other PE or with the ACU. An ACU-PE bus is used for communications between the PEs and the ACU. Data or instructions are broadcast

---

*This research is supported in part by a grant from NSF.

to the PEs in parallel when the communication is from the ACU to the PEs.

If there is communication between neighbor PEs, then an *xnet* primitive (or global router) can be used. The *xnet* procedures are used when communication is between any two PEs which lie in any of the eight neighbor directions, otherwise the router is used. The architecture of the MasPar computer makes it highly suitable for emulation of the proposed approaches since the algorithms can be directly mapped into the MasPar array of PEs. Faulty PEs are emulated as inactive PEs in the MasPar and fault-free PEs are kept as active processors. If a PE is faulty (e.g. the PE is inactive) the *xnet* procedures are able to skip that PE and communicate directly to the next PE which is active.

When emulating a *4-way* connected array, the north, east, west, and south PE connections are used to communicate with other PEs in the array. The diagonal PE connections northwest, northeast, southwest and southeast are also used to emulate an *8-way* connected array. Any other type of communication is done using the router. "Virtual" processors around the array are used to avoid the toroidal wraparound and achieve the correct number of neighbors when reconfiguring the PEs on the boundaries. The virtual processors are implemented in software with fixed positions.

HARDWARE MODEL

A fault tolerant architecture which does not use spare PEs and hence overcomes many of their problems, has been proposed in [9]. The architecture consists of a set of PEs arranged in a two dimensional mesh, where each PE is connected to four neighbors.

This structure can be expanded to form a new architecture where each PE is connected to n neighbors. It is defined as an *n-neighbor connected array of processing elements* composed of an SIMD type architecture arranged into a two dimensional grid with each processing element in the array connected to n neighbor elements using bi-directional links.

The interconnection network between PEs is restricted to a very simple structure. If a PE has been previously found to be faulty, a set of local switches in the network bypass the faulty PE and effectively remove it from the array. The network consists of a set of *2n* switches per PE for an *n*-way connected array. Each switch is made of three sets of bus lines and one control input. Figure 1(a) shows how the switches are implemented for the horizontal PE connections. The bus lines consist of one set of data lines connected to the neighbor PE and two sets of output data lines. When the

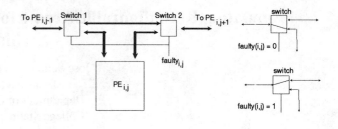

Figure 1. (a) Switch placement for horizontal PE connections, (b) Switch states for faulty and fault-free PEs.

faulty line is set to 0, the PE has been found to be not faulty, and the input data lines are connected to the lower output data lines which logically include the PE in the network. When the *faulty* line is 1, the input data lines are connected to the upper output lines to effectively bypass the PE. Figure 1(b) shows the switch states when the PE is faulty and fault-free respectively.

A FAULT-TOLERANT SCHEME FOR ARRAYS OF PROCESSORS

The proposed fault tolerant scheme is implemented as a three-step process. The first step is the *testing phase* and includes all procedures for diagnosing and locating all faulty PEs. The second step is the *reconfiguration phase*. Once the PEs have been diagnosed in the previous step, the array is reconfigured to reduce the impact of the faulty PEs. Finally, the last step is the *operational phase*. Once the array has been reconfigured, the actual image processing applications are mapped into the array and executed.

RECONFIGURATION STRATEGIES

Reconfiguration Phase

Once faulty processors have been diagnosed, the reconfiguration phase attempts to minimize their impact on the functionality of the array. The first step is to set the *faulty* lines for each of the faulty PEs to 1. This will bypass the faulty PEs. Initially, each PE stores its physical position in its local memory.

[9] proposes a simple iterative reconfiguration procedure that "fills" in the gaps left by the faulty PEs with fault-free PEs. Each PE is given a "conceptual" position which may be altered to move the PE conceptually in the desired direction. The net effect of this process is to distribute the loss of processing elements to a loss of resolution over the whole array. The reconfiguration algorithm is based on the following equations [9]:

$$X_{i,j} = \frac{X_{i,j+1} + X_{i,j-1} + X_{i+1,j} + X_{i-1,j}}{4}$$
$$Y_{i,j} = \frac{Y_{i,j+1} + Y_{i,j-1} + Y_{i+1,j} + Y_{i-1,j}}{4} \qquad (1)$$

where $X_{i,j}(Y_{i,j})$ are the conceptual row and column positions of $PE_{i,j}$ after one iteration of the reconfiguration procedure and $X_{i,j+k}(Y_{i,j+k})$ or $X_{i+k,j}(Y_{i+k,j})$ are the conceptual positions of the four fault-free neighbors of $PE_{i,j}$ (assuming a 4-way connected array). Equations (1) can be generalized as follows for an n-way connected array:

$$X_{i,j} = \frac{\sum\limits_{p=1}^{n} X_p}{n} \qquad Y_{i,j} = \frac{\sum\limits_{p=1}^{n} Y_p}{n} \qquad (2)$$

where $X_p(Y_p)$ is the $X(Y)$ coordinate of the pth working (fault-free) neighbor of $PE_{i,j}$. Since the *faulty* lines have been set, the pth neighbor is just the next PE along a particular direction.

Reconfiguration Results

The proposed reconfiguration algorithms are implemented iteratively on each fault-free PE and are executed until the conceptual positions converge. New conceptual positions are computed based on (2). A new parameter (referred to as the *stopping parameter* ε) is defined as the maximum difference between the old PE conceptual positions and the new PE conceptual positions for all PEs. When the difference between old and new positions for each PE is less than ε, then the array is assumed to have converged and the procedure stops. This is the first difference between the proposed approach and [9] and is required to guarantee the termination of the reconfiguration process. The algorithms are then measured in terms of the number of reconfiguration steps taken by the array before all PEs have converged to within the specified value for ε.

A reconfiguration step is defined formally as follows. Given an NxM array of processors, a *reconfiguration step* corresponds to the time taken by all PEs in the array to compute a set of new conceptual values and check for convergence. Therefore, a reconfiguration step is given as:

$$R^t = \max(R_{i,j}^t) \ \forall \ 0 \le i \le N\text{-}1 \ , 0 \le j \le M\text{-}1 \qquad (3)$$

where R^t is the reconfiguration time at step t and $R_{i,j}^t$ is the time taken by $PE_{i,j}$ to compute the value of the reconfiguration equation and verify convergence. Using (3), the overall reconfiguration time corresponds to the time taken by all the reconfiguration steps until all PEs in the array have converged within the specified value of ε. Hence, the overall reconfiguration step is given as:

$$R = \sum_{t=0}^{T} R^t \ | \ C_{i,j}^T - C_{i,j}^{T-1} \le \varepsilon \qquad (4)$$

where R is the total reconfiguration time the array, and $C_{i,j}^T$ is the value of the computation of the conceptual position at time t.

Uniform Random Fault Results

Under a uniform random fault distribution, the probability that a $PE_{i,j}$ is faulty, depends on the uniform distribution over the entire array. Fault injection through emulation was performed for 4-way and 8-way connected arrays. A uniform random number generator was used to generate a set of rows and columns corresponding to faulty PEs ranging from a single faulty PE to 1000 faulty PEs (corresponding to 49% of the array being faulty for a *32x64* MasPar structure). Additionally, emulations where performed using different values of ε. Figure 2 shows the results of the reconfiguration algorithm on 4-way and 8-way connected arrays.

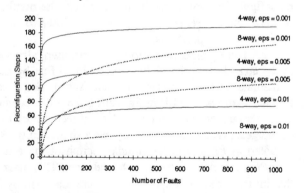

Figure 2. Reconfigurable 4-way and 8-way connected arrays with uniform random faults.

Reconfiguration with Cluster Faults

Cluster faults are defined as faults which are concentrated at a particular area (or site) of the array. Typically, cluster faults are caused by defects during the manufacturing process. Figure 3 shows the results of the emulations with cluster faults on a 4-way and 8-way array for different values of ε.

Reconfiguration With Initial Guess

The purpose of the proposed reconfiguration procedure is to manipulate the fault-free processing

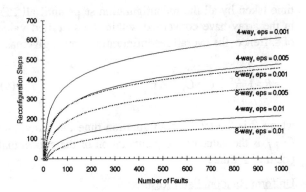

Figure 3. Reconfigurable 8-way and 4-way connected arrays with cluster faults.

elements in the array to fill in the gaps left by the faulty PEs. The major problem with this approach (in the absence of the stopping parameter it is the same as [9]) is that a large number of iterations may be needed when the array has large gaps (i.e. when cluster faults are present). To solve this problem, instead of initializing the PEs to their physical positions, the PEs are initialized to a conceptual position which is an approximation of the final positions of the converged PE values after reconfiguration. Hence, when the reconfiguration procedure is implemented, the number of iterations are reduced since PEs are closer to their final converging positions.

During reconfiguration, if (2) are used, all gaps in the array left by the faulty PEs are filled first by the immediate neighbors (which are not faulty). Neighbor PEs are then used to fill in the gaps left by PEs which were just moved, and so on until array reconfiguration converges. Assume that the array has a single faulty PE (denoted as $PE_{m,k}$) in the middle section of the array (i.e. $PE_{m,k}$ is not on the array boundaries). This means that initially, the neighbors of $PE_{m,k}$ are separated by 2 conceptual positions and any other PEs are separated by 1 unit. In the first iteration of the reconfiguration procedure, the n immediate neighbors of $PE_{m,k}$ are moved closer to $PE_{m,k}$. In the next iteration, the neighbor elements are moved again, and so are the PEs which are physically two steps away from $PE_{m,k}$. At the termination of the reconfiguration procedure, all PEs are separated by $1+\Delta_{i,j}$ units, where $\Delta_{i,j}$ is the distance added or subtracted to $PE_{i,j}$ to make up the gap created by $PE_{m,k}$. By guessing an appropriate value for a $\Delta_{i,j}$ of each PE in the array, then the number of reconfiguration steps to achieve convergence can be lowered. The value of $\Delta_{i,j}$ is defined as the *difference in conceptual positions for $PE_{i,j}$*. Depending on the relative position of each PE with respect to $PE_{m,k}$, then the value of $\Delta_{i,j}$

is either positive or negative. When there is more than one faulty PE, the value of $\Delta_{i,j}$ is given by the sum of the individual values of $\Delta_{i,j}$ for each faulty PE. Hence the value of $\Delta_{i,j}$ depends on two values: the distance of $PE_{i,j}$ from the faulty PEs and the number of faulty PEs. In order to determine the initial conceptual positions for the fault-free PEs, a value of $\Delta_{i,j}$ for each $PE_{i,j}$ must be computed. However, this may involve computing n different values of $\Delta_{i,j}$ for each $PE_{i,j}$ (assuming there are n faulty PEs). To solve this problem, the average of the conceptual positions of the faulty PEs is first computed. This value is then used to compute $\Delta_{i,j}$ for each fault-free $PE_{i,j}$.

The procedure to guess the values of $\Delta_{i,j}$ is based on an inverse exponential distribution function as corresponding to the lower bounds presented in [8] for wire length and channel width. This choice is explained as follows. Each faulty PE leaves an empty position in the conceptual positions of the array indices by 2 units. This means that the neighbor positions have to be brought in closer to distribute the loss of one conceptual position. The amount of conceptual movement depends on the number of communication links between the fault-free PEs and the faulty PEs, where the closer fault-free PEs are moved more than PEs which are farther away. An exponential function with a base of 2 therefore, maps easily to the needed conceptual movement as follows:

$$\Delta_{i,j}=2^{-l} \tag{5}$$

where l is the number of communication links to a faulty PE. This equation only works if the loss of a faulty PE is distributed on the PEs in one direction (left, right, etc.). To distribute the loss over two directions, (5) is modified as follows:

$$\Delta_{i,j}=2^{-(l+1)} \tag{6}$$

Since each PE is addressed according to two directions (x and y directions), (6) is directly used to determine the initial values of $\Delta_{i,j}$. The procedure to derive an initial value for each PE is straightforward.

THEORETICAL BACKGROUND

Consider initially the problem of array convergence. Let the array A have a size of N rows by M columns. For simplicity, assume that A is connected as a *4-way* mesh. The approach for an *8-way* (or *n-way*) connected array can be derived in a similar manner. If A is fault-free, then the conceptual values for each $PE_{i,j}$ after t reconfiguration steps are given by the following equation which is derived directly from (2).

$$a_{i,j}^{t} = \frac{a_{i+1,j}^{t-1}}{4} + \frac{a_{i-1,j}^{t-1}}{4} + \frac{a_{i,j+1}^{t-1}}{4} + \frac{a_{i,j-1}^{t-1}}{4} \qquad (7)$$

where a is either the x or y conceptual positions of the 4 neighbors of $PE_{i,j}$ for reconfiguration step t.

Equation (7) is a recurrence relation and can be expanded further by substituting the values for each a^{t-1} with the values for a^{t-2} as follows:

$$a_{i,j}^{t} = \frac{\frac{a_{i+2,j}^{t-2}}{4} + \frac{a_{i,j}^{t-2}}{4} + \frac{a_{i+1,j+1}^{t-2}}{4} + \frac{a_{i+1,j-1}^{t-2}}{4}}{4} + \frac{\frac{a_{i,j}^{t-2}}{4} + \frac{a_{i-2,j}^{t-2}}{4} + \frac{a_{i-1,j+1}^{t-2}}{4} + \frac{a_{i-1,j-1}^{t-2}}{4}}{4} + \cdots \qquad (8)$$

which simplifies to:

$$a_{i,j}^{t} = \frac{a_{i-1,j-1}^{t-2}}{4} + \frac{a_{i+1,j+1}^{t-2}}{8} + \frac{a_{i-1,j+1}^{t-2}}{8} + \frac{a_{i-1,j-1}^{t-2}}{8} + \frac{a_{i+1,j}^{t-2}}{8} + \frac{a_{i+2,j}^{t-2}}{16} + \frac{a_{i-2,j}^{t-2}}{16} + \frac{a_{i,j+2}^{t-2}}{16} + \frac{a_{i,j-2}^{t-2}}{16} \qquad (9)$$

Substitutions can be performed on each new equation with the values for each a^{t-i} until $t - i = 0$. However, special consideration has to be taken when the equation reaches the conceptual values of the boundary PEs. Conceptual values of boundary PEs are determined by the 2 or 3 neighbor PEs and 2 or 1 constant values depending if the PE is on a boundary or on a corner of the array.

The reconfiguration computation for each $PE_{i,j}$ given in (7) can be also mapped into a tree: each node in the tree represents the conceptual reconfiguration value of a $PE_{l,m}$ at some time t depending on the level of the node. The root of the tree represents the conceptual reconfiguration value of $PE_{i,j}$. Each node in the tree has n children, where n is the number of neighbor connections per each PE (i.e. 4 for a 4-way connected array). Figure 4 shows an example of a computation tree for $PE_{i,j}$ in a 4-way connected array.

The tree in Figure 4 has an exponential number of nodes; however, some of the nodes are repeated at each level and are computed only once. Also, the reconfiguration computation at each node is performed in parallel with the remaining nodes at the same level. Hence, only one reconfiguration step is performed per level of the tree. The depth of the tree gives the maximum number of conceptual reconfiguration calculations which are computed in other PEs before the conceptual value for $PE_{i,j}$ is derived. Hence, the total

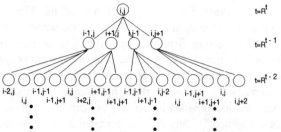

Figure 4. Tree structure for computation of $PE_{i,j}$

number of reconfiguration steps for the array is given by:

$$R = \max\{depth(PE_{i,j})\} \times f(\varepsilon) \qquad (10)$$

where $depth(PE_{i,j})$ is the depth of the reconfiguration tree for $PE_{i,j}$ and $f(\varepsilon)$ is a function dependent on the value of ε which determines the convergence of the conceptual position of $PE_{i,j}$.

One of the bounds on the number of reconfiguration steps is the depth of the reconfiguration tree with the maximum size. If a maximum depth can be found for the set of trees which make up the PEs in the array, then it can be used to find an upper bound on the number of reconfiguration steps for the array. This upper bound is given next based on the results of (10). Given an n-way connected array with N rows and M columns, the following theorem establishes the upper bound for the number of reconfiguration steps.

Theorem *The number of reconfiguration steps needed for conceptually reconfiguring the array using (2) and converging within ε, has an upper bound of:*

$$O(\max\{M,N\} \times \log(1/\varepsilon)) \qquad (11)$$

Only a sketch of the proof is provided due to space limitations.

Assume that the number of reconfiguration steps for an NxM array with no link faults has a bound which is greater than (11). Therefore, there is at least one $PE_{i,j}$ in the array which needs more than $O(\max\{M,N\} \times \log(1/\varepsilon))$ steps to converge within ε. The number of reconfiguration steps for a PE is directly related to the depth of the reconfiguration tree for that PE. Hence, the bound on the depth of the reconfiguration tree for $PE_{i,j}$ has to be greater than (11).

One of the limiting factors for the reconfiguration tree is the number of PEs in the array. The number of nodes at any level of the tree is at most equal to the number of PEs in the array. Hence, in the worst case, the number of reconfiguration steps for a PE is equal to the number of conceptual position computations needed at another level (before reaching the top node). For example, if one of the PEs were faulty at the boundary of the array, in the worst case, all the PEs starting from the opposite side of the faulty boundary may have to compute a new conceptual position before a new position for the faulty PE is determined. Hence, the maximum number of computations needed before any conceptual position can be computed for any PE , is bounded by the maximum dimension of the array given by $max\{N,M\}$.

However, the new conceptual position computed by a PE is not guaranteed to have converged within the specified ϵ. The set of computations may have to be repeated several times before convergence is achieved. If the array is n-way connected, the result of the computation at each reconfiguration step tends by a factor proportional to n to be more exact than the previous computation in the worst case, because n values are used to determine the value of the new conceptual position and (7) is a monotonically increasing function. Hence, the number of computations needed before a converging value is obtained, is bounded by the function $\log_n(1/\epsilon)$.

In the worst case, all sets of computations have to converge before the computation of the top node in the tree has converged within ϵ; otherwise, if any of the previous computations has not converged below ϵ, then when the values are propagated up and the root node will not have converged within ϵ. Therefore, if the two previous bounds are combined, the result on the number of reconfiguration steps is given by:

$$\text{Bound} = max\{M,N\} \times \log_n(1/\epsilon) \qquad (12)$$

Given in O notation and deleting the n base of the log function, (12) is the same as (11). This contradicts the original assumption which states that there was at least one PE which had an upper bound greater than (11). Therefore, all PEs in the array reconfigure with an upper bound of (11). ♦

DATA MAPPING APPROACHES

Provided the array has completed the reconfiguration phase, the next step is to map the image processing applications into the array. Running an application on an array consists of three steps: image input mapping, computation steps, and image output mapping.

Input Mapping

Assuming that the array is fault-free then, one simple type of input mapping consists of mapping one pixel in the image to a corresponding PE in the array if the number of PEs is equal to the number of pixels in the image. This type of mapping is very simple and does not require a computation overhead. However, this mapping is only valid provided the entire array is fault-free. If the array is faulty and has been reconfigured, the conceptual positions of the PEs are evenly distributed throughout the array. In this situation, to map the image into the array, all PEs round their final conceptual positions to the nearest integer. Each PE then inputs the image pixel at that position. However, some of the pixels may not be mapped into the array because there could be more pixels than available PEs.

Output Mapping

Once the image processing application has completed its execution, the image is output from the array and reconstructed. The output mapping defines how the image is extracted to minimize the impact of the faulty PEs. For a fault-free array with a one-pixel-per-processor mapping, the output mapping is the opposite of the input mapping. The image is reconstructed by extracting the pixel values of each of the PEs, thus forming the output image.

If the array has faulty PEs and a one-pixel-per-processor input mapping is used, then the reconstructed image will not have the pixels which were lost during the input mapping. This means that the image will have empty pixels at the positions of the unmapped pixels. This corresponds to the output mapping proposed by [9]. Two new methods are proposed to minimize this problem. The easiest approach for implementation is to use the background of color of the image as the color of the empty pixels.

A better approach is to use a majority neighbor fill method to determine what color to use on unmapped pixels. The procedure is as follows: for each pixel that is not mapped out of the array, examine all the non-empty neighbor pixels. From these neighbors, choose the color which appears more often, except for the empty neighbors.

Results for Image Mapping

Figures 5,6,7,8 show the results of the different types of emulated mapping approaches. The images are of size 300x300 gray-scale pixels.

Figure 5 shows examples of the original image to be mapped into the array. Figure 6 shows the images with 40% of faulty PEs and no output mapping. Figure 7 shows the images when 40% of the PEs are faulty and the output mapping is done with background fill. Figure 8 shows the images when 40% of the PEs are faulty and the output mapping is done with a majority neighbor fill. Figure 9 shows the comparison in relative distortion between the no-fill, background and the majority neighbor fill mappings. The relative distortion is computed by counting the number of pixels whose values are different from the original image divided by the total number of pixels in the image. The majority neighbor fill has a lower distortion than the background and the no-fill mappings. However, distortion values are image dependent and differ from image to image.

The average difference in pixel values was also computed for the background and nearest neighbor fills (they were not computed for the no-fill mapping since unmapped pixels are assumed to be undefined). For each pixel which is different in the output image from the original image, the difference in pixel values has been measured. Figure 10 shows the results for the majority neighbor technique. The average maximum difference corresponds to pixels which have a value which is less than the pixel in the original image. The average minimum difference corresponds to pixels which have a values which is more than the pixel in the original image. The average difference is defined as the value of the difference between maximum and minimum. For the example of Figure 5, the pixels range from 0 to 100 gray values for the black and white image, although, the same measurements apply for color images. Figure 11 shows the results when the background fill is used. In this case, the pixels which have a different value from the original image have been filled with the background color black. Hence, only the average minimum difference can be measured.

By comparing Figure 10 and 11, the majority neighbor fill has a lower average pixel difference for any set of faulty PEs. Hence, differences in PE colors are less noticeable and produce less distortion of the output image.

CONCLUSIONS

This paper has described several new reconfiguration strategies for fault-tolerant SIMD arrays of processors. These strategies are based on an architecture which does not use spare rows or columns, instead it distributes the loss of computation due to faulty array elements over the entire array. A hardware model is described; this model is able to disconnect a faulty array element from the rest of the array without affecting the communication between fault-free sets of processors. The reconfiguration approaches were emulated for a 4-way and an 8-way connected arrays of processors using a *32x64* MasPar architecture.

After the image processing applications have been implemented, the image is then mapped out of the array and reconstructed. When mapping out the image from an array with faulty PEs, some pixels are left empty. The proposed approach solves this problem by using the image background or the set of neighbors to fill in the empty pixels.

REFERENCES

[1] R. H. T. Bates and M. J. McDonnell, "Image restoration and reconstruction", Clarendon Press, 1986.

[2] V. Cantoni et al., "Parallel image processing primitives", *Image analysis and processing*, V. Cantoni, S. Levialdi, and G. Musso, eds., Plenum Press, 1986.

[3] R. A. Evans, J. V. McCanny and K. W. Wood, "Wafer scale integration based on self organization", *Wafer Scale Integration*, C. R. Jesshipe and W. R. Moore Eds., Adam Hilger 1986.

[4] E. Hall, *Computer image processing and recognition*, Academic Press, 1979.

[5] P. F. Leonard and R. N. Mudge, "System design for local neighborhood processing", *Proceedings of SPIE*, Vol. 534, 1985, pp. 44 - 50.

[6] MasPar Computer Corporation, *MasPar MP-1 Standard Programming Manual*, MasPar Computer Corporation, 1990.

[7] M. Sami and R. Stefanelli, "Reconfigurable architectures for VLSI processor arrays", *Proc. IEEE*, Vol. 74, No. 5, May 1986.

[8] T. Leighton, "A Survey of algorithms for integrating wafer-scale systolic arrays", *Wafer Scale Integration*, G. Saucier and J. Trilhe, eds., Elsevier Science Publishers, North-Holland, IFIP, 1986, pp. 177 - 194.

[9] J. A. Trotter and W. R. Moore, "Imperfectly connected 2D arrays for Image Processing", *FTCS*, January 1989, pp. 88 - 92.

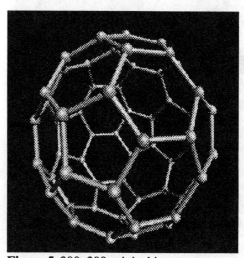

Figure 5. 300×300 original image.

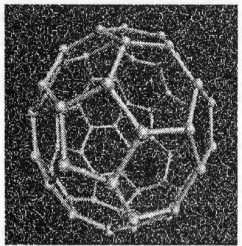

Figure 6. Image with no fill on a MasPar with 40% faulty PEs.

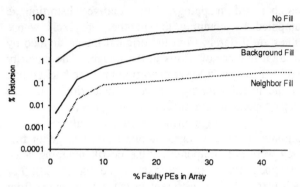

Figure 9. Distortion comparison between the background fill procedure and the majority fill procedure.

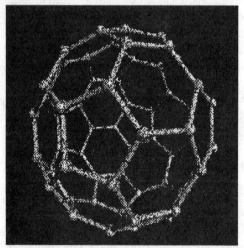

Figure 7. Image with background fill on a MasPar with 40% faulty PEs.

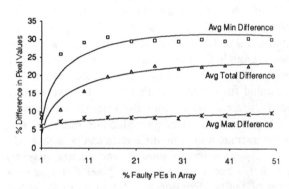

Figure 10. Average pixel difference for the majority neighbor fill.

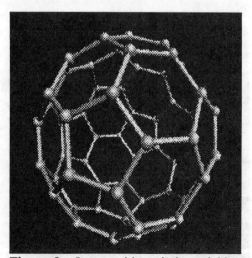

Figure 8. Image with majority neighbor fill on a MasPar with 40% faulty PEs.

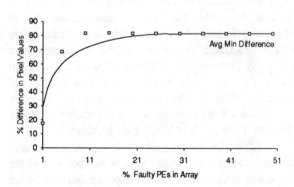

Figure 11. Average minimum pixel difference for the background fill.

Ring Embedding in an Injured Hypercube*

Yu-Chee Tseng and Ten-Hwang Lai

Department of Computer and Information Science
Ohio State University, Columbus, Ohio 43210
E-mail: {tseng, lai}@cis.ohio-state.edu

Abstract — *We consider the problem of embedding a ring in a hypercube that contains possible faulty nodes. Existing algorithms allow the number of faulty nodes to be at most $2n - \Theta(\sqrt{n \log n})$, where n is the dimension of the hypercube. We propose an embedding scheme that can tolerate up to $\Theta(2^{n/2})$ faulty nodes, largely increasing the number of tolerable faulty nodes in a ring embedding.*

1 Introduction

The graph embedding problem has received a lot of attentions, especially when the host graph is a fault-free ([4, 7, 8]) or faulty ([1, 2, 3, 5, 10, 11]) hypercube. In an injured hypercube, one fundamental problem is to construct a longest possible ring that does not contain any faulty node or faulty link. In the literature, both *link-fault model* and *node-fault model* have been studied. Searching for a 2^n-ring[‡] under the link-fault model when the number of faulty links is $\leq n - 2$ is feasible [5].

In the node-fault model, it is normally assumed that once a processor fails all links connected to it immediately stop working. Adopting this model, Chan and Lee [2] proposed an efficient algorithm that can construct a ring of size at least $2^n - 2f$, provided that the hypercube has at most $f \leq \lfloor (n+1)/2 \rfloor$ faulty nodes. The *penalty factor* (i.e., the ratio of the number of nodes not in the ring to the number of faulty nodes) is ≤ 2. Recently, Wang and Cypher [10] showed how to find a ring of size $2^n - 2 \max\{f_{odd}, f_{even}\}$, where f_{odd} and f_{even} (both $\leq n - \Theta(\sqrt{n \log n})$) are respectively the number of faulty nodes with odd and even number of 1's (or called *odd* and *even* nodes, respectively) in their string representations. As a ring must alternate between odd and even nodes, their ring is longest possible.

Unfortunately, none of these algorithms can handle the case where $f > 2n - \Theta(\sqrt{n \log n})$, tremendously limiting their applicability.

The purpose of this paper is to present a ring embedding algorithm that allows up to $\Theta(\sqrt{2^n})$ faulty nodes in an n-cube. Specifically, when the number of faulty nodes f is less than $\gamma(k)$, where $\gamma(k) = (n - 2k)2^k$, the penalty factor is $p = 2$ for $k = 0$ and $p = 2^{k+2}/3$ for $k \geq 1$. Thus, a ring of size at least $2^n - pf$ can be constructed. The variable k is any integer satisfying $0 \leq k \leq (n-4)/2$. The penalty factor incurred by the algorithm grows in a slower speed (in a factor of $\leq 4/(3(n - 2k))$ with respect to k) than the number of tolerable faults increases. The maximum number of faults that can be handled by our

scheme is $\gamma(\lfloor (n-4)/2 \rfloor) = \Theta(2^{n/2})$. As $\Theta(2^{n/2}) \gg 2n - \Theta(\sqrt{n \log n})$, our algorithm largely increases the number of faults tolerable in a ring embedding.

2 Definitions

A binary n-cube is an undirected graph with 2^n nodes each labeled with a distinct binary string $b_n \ldots b_1$. Node $b_n \ldots b_{i+1} b_i b_{i-1} \ldots b_1$ and node $b_n \ldots b_{i+1} \overline{b_i} b_{i-1} \ldots b_1$ are joined by an edge and this edge is said to be *along* dimension i, $1 \leq i \leq n$. A subcube of dimension $d \leq n$ is represented by a ternary string $x_n \ldots x_1$, where $x_i \in \{0, 1, *\}$, with exactly d occurrences of $*$ (the "don't-care" bit).

Let $[n]$ denote the set $\{1, 2, \ldots, n\}$. Let Q be any n-cube and k any integer between 0 and n. If D is a subset of $[n]$ with cardinality $|D| = k$, then D specifies k dimensions of the cube. One may use D as a template and partition the n-cube into k-cubes as follows. Define an equivalence relation $\approx_D$ between any two nodes $b_n \ldots b_1$ and $c_n \ldots c_1$ of Q by

$$b_n \ldots b_1 \approx_D c_n \ldots c_1 \text{ iff } \forall i \in [n] - D : b_i = c_i.$$

The relation $\approx_D$ partitions Q into 2^{n-k} disjoint sets, each set constituting a subcube of dimension k. These k-cubes each have exactly k *'s in their string representations on the bit positions as specified by D. Two k-cubes are said to be *adjacent* if their string representations differ in exactly one bit position. Note that there are 2^k edges between any pair of adjacent k-cubes. If we regard each k-cube as a *supernode* and the edges between two adjacent k-cubes altogether as a single *superedge*, then the n-cube Q becomes an "$(n-k)$-cube of supernodes and superedges." Such a cube is sometimes called a *supercube* so as to emphasize the fact that it contains supernodes and superedges. The supercube obtained from Q via template D is denoted as $Q(D)$. Note that $Q(D) = Q$ when $D = \emptyset$. Thus, a binary cube is a special case of a supercube.

When we say that a graph G is a subgraph of a supercube $Q(D)$, it is understood that G itself contains supernodes and superedges. In this paper, we will deal with rings and extended rings on a supercube $Q(D)$ as defined below.

A *ring* is a connected graph in which every node has degree 2. We define a special topology called *extended ring* (or *e-ring*) as a ring of any length extended with zero or more branches; each branch is an edge joining a new node to some node in the ring, and there is at most one extended branch at any node of the original ring. Figure 1 shows an e-ring of length 6 with 2 branches. In this paper, we will denote a ring and its corresponding e-ring (if the correspondence is clear) by R and $\hat{R}$, respectively. If node $x \in \hat{R} - R$,

---

*This research was supported by NSF Grant CCR-9010589.
[‡]An i-ring is a ring of i nodes. i is called the size or length of the ring.

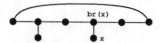

Figure 1: An e-ring of length 6 with 2 branches.

we denote by $br(x)$ the node in R at which x is attached to R.

For clarity, supernodes, superedges, rings, and e-ring that are constructed from k-cubes may be referred to as k-nodes, k-edge, k-rings, and k-e-rings, respectively. We state the following simple fact as a lemma for ease of later reference.

Lemma 1 *Any supercube $Q(D)$ is a bipartite graph and any ring on $Q(D)$ is of even length.*

3 Basic Constructs

This section describes algorithms that construct a k-ring and/or k-e-ring out of a given $(k + 2)$-ring and/or $(k + 2)$-e-ring. These algorithms will be used in the next section to construct a ring in an injured hypercube.

Let $D \subset D' \subseteq [n]$ be two templates such that $D = k$ and $|D'| = k + 2$. Consider the supercubes $Q(D)$ and $Q(D')$. Each supernode in $Q(D)$ is a k-node, and each in $Q(D')$ a $(k + 2)$-node. Throughout this section, k-nodes and $(k + 2)$-nodes thus refer to supernodes in $Q(D)$ and in $Q(D')$, respectively. One may readily observe that each $(k + 2)$-node consists of four k-nodes that are connected by k-edges as a 2×2 mesh

3.1 Constructing k-Rings from $(k + 2)$-Rings

Suppose that each k-node in $Q(D)$ as well as each $(k+2)$-node in $Q(D')$ is defined to be either *damaged* or *healthy* in a way such that a $(k + 2)$-node is healthy iff all of its four k-nodes except possibly one are healthy (or, equivalently, a $(k + 2)$-node is damaged iff it contains more than one damaged k-nodes). Call a $(k + 2)$-node *extremely healthy* if it contains no damaged k-node.

Under these conditions, we are able to obtain a k-ring with some desired properties from a given $(k + 2)$-ring. Specifically, suppose we are given a $(k + 2)$-ring $X = (X_0, X_1, \ldots, X_{m-1})$ of length m with all X_i's being healthy and at least one of them extremely healthy. Each X_i contains four k-nodes and so X contains a total of $4m$ k-nodes. Assuming that f out of the $4m$ nodes are damaged, we show how to construct a k-ring containing at least $4m - 2f$ healthy nodes.

Lemma 2 *Suppose that $X = (X_0, X_1, \ldots, X_{m-1})$ is a $(k + 2)$-ring of length m with these properties: 1) each X_i is healthy and 2) at least one of the X_i's is extremely healthy. It is possible to construct from X a k-ring that contains only healthy k-nodes and that has length at least $4m - 2f$, where f is the number of damaged k-nodes in X.*

Proof. We prove the lemma by construction. We first present a procedure $Path(X_i, a, b)$ that, given a healthy k-node a in X_i, will construct a k-path within X_i starting at a and then extend the path to some healthy k-node b in

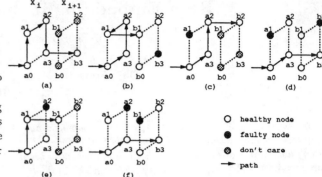

Figure 2: The path establishing rules in procedure $Path()$. X_i is the current $(k + 2)$-node and a_0 is the staring k-node.

X_{i+1}[§]. Node b is part of the output of the procedure. The constructed path will contain at least two healthy k-nodes in X_i (or four if X_i is extremely healthy). Then we develop an algorithm $Ring(X)$ that, starting with some X_i, repeatedly applies this procedure to X_i, X_{i+1}, $\ldots$, X_{m-1}, X_0, $\ldots$, X_{i-1} to obtain a k-ring. (There is some subtlety that needs to be taken care of when the construction reaches X_{i-1}.)

Below, procedure $Path(X_i, a, b)$ intends to construct a path $a \to \cdots \to b$. We will denote by a_j and b_j, $j = 0..3$, the nodes on X_i and X_{i+1}, respectively, with a_j adjacent to b_j (see Fig. 2). Node a must be healthy and, without loss of generality, we assume $a = a_0$. The output b will be in $\{b_0, \ldots, b_3\}$.

Procedure $Path(X_i, a, b)$
1) If all a_1, a_2 and a_3 are healthy, then
   a) if b_3 is healthy, construct the path $a_0 \to a_1 \to a_2 \to a_3 \to b_3$ (Fig. 2(a));
   b) otherwise, the path $a_0 \to a_3 \to a_2 \to a_1 \to b_1$ (Fig. 2(b)).
2) If a_1 is damaged, then
   a) if b_2 is healthy, establish $a_0 \to a_3 \to a_2 \to b_2$ (Fig. 2(c));
   b) otherwise, $a_0 \to a_3 \to b_3$ (Fig. 2(d)).
3) If a_2 is damaged, then
   a) if b_1 is healthy, establish $a_0 \to a_1 \to b_1$ (Fig. 2(e));
   b) otherwise, $a_0 \to a_3 \to b_3$ (Fig. 2(f)).
4) If a_3 is damaged, the scenario is symmetric to case 2. Construct a path accordingly.
5) Let b be the last node in the path constructed above.

Algorithm $Ring(X)$
1) If $f = 0$, construct a ring from X (of length $4m$) in an obvious way and return.
2) Otherwise, let X_{r-1} be any extremely healthy $(k+2)$-node in X such that X_r is not extremely healthy. Let a be any of the two healthy k-nodes in X_r that are adjacent to X_r's only damaged k-node.
3) for $i := r, r + 1, \ldots, m - 1, 0, \ldots, r - 2$ do
   a) call $Path(X_i, a, b)$;

[§]Apply modulo-m to $i + 1$ as appropriate.

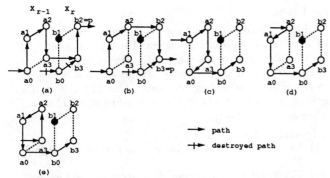

Figure 3: The path establishing rules in step 4 of $Ring()$ for X_{r-1}. It is assumed that b_0 is the starting k-node on X_r and b_1 is damaged.

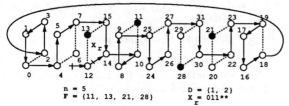

Figure 4: An execution of $Ring()$ in a faulty 5-cube.

    b) $a := b$;

4) Let a_j and b_j, $j = 0..3$, be the k-nodes of X_{r-1} and X_r, respectively, with a_j and b_j being adjacent (see also Fig. 3). Without loss of generality, assume that b_0 is the k-node where the path in X_r starts and b_1 the damaged k-node in X_r. Apply one of the following rules according to the location of a as resulting from step 3:

    a) If $a = a_0$, consider the ending point p ($\in \{b_2, b_3\}$) of the path on X_r.

        i) If $p = b_2$, establish the path $a_0 \rightarrow a_1 \rightarrow a_2 \rightarrow a_3 \rightarrow b_3$ and destroy the path $b_0 \rightarrow b_3$ (Fig. 3(a));

        ii) otherwise, establish $a_0 \rightarrow a_1 \rightarrow a_2 \rightarrow b_2 \rightarrow b_3$ and destroy $b_0 \rightarrow b_3$ (Fig. 3(b)).

    b) If $a = a_1$, establish $a_1 \rightarrow a_2 \rightarrow a_3 \rightarrow a_0 \rightarrow b_0$ (Fig. 3(c)).

    c) If $a = a_2$, establish $a_2 \rightarrow a_1 \rightarrow a_0 \rightarrow b_0$ (Fig. 3(d)).

    d) If $a = a_3$, establish $a_3 \rightarrow a_2 \rightarrow a_1 \rightarrow a_0 \rightarrow b_0$ (Fig. 3(e)).

The above algorithm starts at some X_r and proceeds to $X_{r+1}, X_{r+2}, \ldots, X_{m-1}, X_0, \ldots, X_{r-2}$. Each call to $Path()$ determines the starting k-node for the next call. Step 4 uses special rules for X_{r-1} and it may modify the path (at most one edge) already constructed for X_r. Figure 4 shows a sample execution of $Ring()$ in a faulty 5-cube where $D = \emptyset$ and $D' = \{1, 2\}$. The routing starts at $X_r = 011**$ and proceeds to the right until reaching $X_{r-2} = 000**$. In the final step for $X_{r-1} = 001**$, the case of 4.a.ii is applied. The resulting ring is of size $32 - 8$.

We now derive the minimum length of the k-ring so constructed. Consider all possibilities in Fig. 2; in (d), a_1 "injures" a_2; in (e), a_2 "injures" a_3; and in (f), a_2 "injures"

a_1. Hence, each X_i except $i = r - 1$ contributes 4 nodes to the ring if it is extremely healthy and at least 2 nodes otherwise. Observe in Fig. 3 that one more healthy node in X_{r-1} may be sacrificed without being injured by any damaged node. We conclude that the ring size is at least $4m - 2f - 1$. As this number must be even (by Lemma 1), the lemma is proved. $\qquad\square$

We make a useful observation below concerning the k-ring constructed by $Ring()$. It can be easily verified by considering each case in Fig. 2 and Fig. 3.

Observation 1 *Each X_i contributes at least two* adjacent *k-nodes to the ring, and thus if a healthy k-node in X_i is not included in the ring then it must be adjacent to some other k-node in X_i that is in the ring.*

3.2 Constructing k-E-rings from $(k+2)$-E-rings

Given a $(k+2)$-e-ring $\hat{X}$, we show how to construct from it a k-e-ring $\hat{Y}$ of healthy k-nodes. Below, the algorithm E-$ring()$ will make use of $Ring()$.

Algorithm E-$ring(\hat{X})$

1) Let X be the ring corresponding to $\hat{X}$. Call $Ring(X)$ and let the output be T, a k-ring.

2) Extend T to a k-e-ring $\hat{T}$ by connecting each healthy k-node $x \in X$ but $x \notin T$ to a k-node in T with a k-edge. (To prevent two x's from connecting to a same $br(x)$, we choose to let each x and its corresponding $br(x)$ belong to the same $(k+2)$-node $\in X$. This is possible by Observation 1.)

3) For each extremely healthy $(k+2)$-node $x \in \hat{X} - X$, break an edge in $\hat{T}$ contributed by $br(x)$ and "join" all four k-nodes of x into $\hat{T}$ from the broken edge. Let the joining result be $\hat{Y}$. $\qquad\square$

Step 1 generates a k-ring and step 2 extends it to a k-e-ring. For example, using the scenario in Fig. 4, T will be the ring indicated in the figure. Connecting 6 to 7, 12 to 14, 29 to 31, and 20 to 22 forms $\hat{T}$. In step 3, we combine those extremely healthy $(k+2)$-nodes in $\hat{X}$ into $\hat{T}$. Note that the existence of such a "breakable" k-edge is due to Observation 1. Figure 5 shows the scenarios before and after the joining.

Let's use $|G|$ to denote the number of (super)nodes in a graph G. The following lemma, which describes the "size" of $\hat{Y}$, can be easily obtained from Lemma 2.

Lemma 3 *Let f be the number of damaged k-nodes in X and f' be the number of $(k+2)$-nodes in $\hat{X} - X$ which are not extremely healthy. The constructed $\hat{Y}$ has size $|\hat{Y}| \geq 4|\hat{X}| - f - 4f'$.*

4 Ring Embedding on Hypercubes

Given an n-cube and a set F of faulty nodes in the cube, we want to find a ring that contains no faulty node. It is desirable that the ring be as large as possible. For brevity, a faulty node is simply referred to as a fault and let $f = |F|$.

We have presented in the preceding section an algorithm that can be used to construct a $(k-2)$-e-ring out of a k-

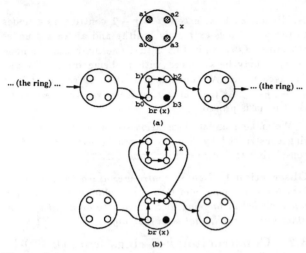

Figure 5: Example of the "joining process" in step 3 of *E-ring()*. (a) is before the joining and (b) is after the joining.

e-ring. Thus, starting with some k-e-ring, if we can repeatedly apply this algorithm to obtain a $(k-2)$-e-ring, a $(k-4)$-e-ring, and so on, then we will have at the end a regular e-ring (0-e-ring) in the n-cube. Along this line of reasoning, we first define damaged and healthy supernodes.

A k-node is said to be *damaged* if it contains $2^{k/2}$ or more faults, and *healthy* otherwise. When $k = 0$, this definition is consistent with the state of a single node. Also, recall that in Section 3.1 the state of a $(k+2)$-node is defined based on the states of its four k-nodes. The following lemma shows that under our definition the desired property is satisfied.

Lemma 4 *Let x be a $(k+2)$-node and let $\{x_0, x_1, x_2, x_3\}$ be any partition of x into k-nodes. If x is healthy then all but at most one of the four k-nodes in x are healthy.*

Suppose that an n-cube Q has f faults. If f is not extremely large, then Q can be partitioned into sufficiently large k-nodes such that none of them is damaged. Furthermore, all these k-nodes can be put into a sequence to form a k-ring. This is proved in the following.

Lemma 5 *Let Q be an n-cube with f faults, where $n \geq 6$. If $f < (n-2k)2^k$ for some integer k such that $0 \leq k \leq (n-4)/2$, then Q can be partitioned into $2(k+1)$-nodes in such a way that 1) all of them are healthy and 2) they together form a $2(k+1)$-ring (of length $2^{n-2(k+1)}$).*

Proof. It has been proved in [1] that in any k-cube with $\leq k-1$ faults, there exists a template D such that $|D| = 2$ and such that any 2-node in $Q(D)$ contains at most 1 fault. In our case, any template D with $|D| = 2k$ will partition the n-cube into $2k$-nodes, of which at most $f/2^k \leq n - 2k - 1$ nodes are damaged. As $Q(D)$ is an $(n-2k)$-supercube, it follows immediately that there exists some $D' \in [n] - D$, with $|D'| = 2$, such that in $Q(D \cup D')$ each $2(k+1)$-node contains at most one damaged $2k$-node (and thus is healthy).

Using Gray codes, one may readily construct from $Q(D \cup D')$ a full-size $2(k+1)$-ring. The reader is referred to [6]

for details. □

Having established the above two lemmas, we are ready to describe our embedding scheme. The input is a set F of faulty nodes contained in the n-cube. Due to the constraint in Lemma 5, it is assumed $n \geq 6$ (when $n \leq 5$ it is easy to solve the ring embedding problem). Also, $|F| < 2^{n/2}$ if n is even and $|F| < (5/4)2^{\lfloor n/2 \rfloor}$ if n is odd (this is obtained when $k = \lfloor (n-4)/2 \rfloor$).

Algorithm *Embed(F)*

1) Let k be the smallest integer such that $|F| < (n - 2k)2^k$.

2) Construct a $2(k+1)$-ring R_{k+1} according to Lemma 5 and let $\hat{R}_{k+1} = R_{k+1}$ (i.e., regard R_{k+1} as an extended ring).

2) **for** $i := k$ **downto** 0 **do**

   Call *E-ring*($\hat{R}_{i+1}$) and let the output be $\hat{R}_i$ (a $2i$-e-ring).

3) Return R_0, the ring contained in $\hat{R}_0$. □

The correctness of *Embed()* and the length of the obtained ring are guaranteed by the following results. Detailed proofs are omitted and can be found in [9].

Lemma 6 *For each i, $0 \leq i \leq k$, R_{i+1} contains at least one extremely healthy $2(i+1)$-node.*

Theorem 1 *If $f < (n-2k)2^k$ and $0 \leq k \leq (n-4)/2$, algorithm Embed() constructs a ring of size at least $2^n - 2f$ for $k = 0$, and at least $2^n - (2^{k+2}/3)f$ for $k \geq 1$.*

References

[1] J. Bruck, R. Cypher and D. Soroker, "Tolerating Faults in Hypercubes using Subcube Partitioning," *IEEE TC*, Vol. 41, No. 5, May 1992, pp. 599-605.

[2] M. Y. Chan and S.-J. Lee, "Distributed Fault-Tolerant Embeddings of Rings in Hypercubes," *J. of Parallel and Distrib. Comput.* **11**, pp. 63-71 (1991).

[3] J. Hastad, T. Leighton and M. Newman, "Reconfiguring a Hypercube in the Presence of Faults," *ACM Symp. on Theory of Computing*, 1987, pp. 274-284.

[4] T.-H. Lai and W. White, "Mapping Pyramid Algorithms into Hypercubes,," *J. of Parallel and Distrib. Comput.* **9**, pp. 42-54 (1990).

[5] S. Latifi, S. Zheng and N. Bagherzadeh, "Optimal Ring Embedding in Hypercubes with Faulty Links," *Fault-tolerant Computing Symp.*, 1992, pp. 178-184.

[6] E. M. Reingold, J. Nievergelt, and N. Deo, *Combinatorial Algorithm*. Prentice-Hall, 1977.

[7] Y. Saad and M. H. Schultz, "Topological Properties of Hypercubes," *IEEE TC*, July 1988, pp. 867-872.

[8] Y.-C. Tseng, T.-H. Lai and L.-F. Wu, "Matrix Representation of Graph Embedding in a Hypercube," accepted to *J. of Parallel and Distrib. Comput.*

[9] Y.-C. Tseng and T.-H. Lai, "Ring Embedding in an Injured Hypercube," Tech. Rep., *OSU-CISRC-3/93-TR12*, Dept. of Comp. & Info. Sci., Ohio State Univ.

[10] A. Wang and R. Cypher, "Fault-tolerant Embeddings of Rings, Meshes, and Tori in Hypercubes," *Symp. on Paral. and Distrib. Process.*, 1992, pp. 20-29.

[11] P.-J. Yang, S.-B. Tien and C. S. Raghavendra, "Embedding of Multidimensional Meshes on to Faulty Hypercubes," *ICPP*, 1991, Vol. I, pp. 571-574.

An Adaptive System-Level Diagnosis Approach for Mesh Connected Multiprocessors

C. Feng, L. N. Bhuyan and F. Lombardi

Computer Science Department
Texas A&M University
College Station, TX 77843-3112
Email: cfeng@cs.tamu.edu

Abstract: *Traditional adaptive centralized system diagnosis assumes a fully connected network topology, hence it can not be used in a number of classes of multiprocessor systems, such as meshes. This paper proposes an adaptive system-level diagnosis algorithm for meshes with wraparound, such as Intel Paragon machine. It is proved that the diagnosis cost required by the proposed approach is lower than the known diagnosis algorithms which can be applied to mesh architectures. Also over-d fault problem can be efficiently solved by our method, where d is the diagnosability.*

1   Introduction

To ensure high reliability along with high performance [2], periodic diagnosis is usually built in a multiprocessor system to identify faulty units, such that reconfiguration and recovery can be undertaken. To avoid large degradation in performance of the system, system-level diagnosis must be efficient and have a low overhead [4]. There are many theoretical results on system-level diagnosis, but few results [8] [3] [11] have addressed real world multiprocessor systems.

A mesh-connected processor array is a major class of multiprocessor system, which is well studied and implemented [6] [1] [7]. The purpose of this paper is to develop a system-level diagnosis algorithm for a mesh with wraparound that is more efficient than the existing algorithms applied to meshes.

We use the *PMC* model [9] as the fault model. This model establishes a directed weighted testing graph $G^*(V, E^*)$. An ordered pair $(v_i, v_j) \in E^*$ means that unit v_i (as a tester) tests unit v_j (as a tested unit). The binary weight on the edge $w(v_i, v_j)$ denotes the test outcome: 0 if v_i evaluates v_j as fault-free or 1 if v_i evaluates v_j as faulty. The test outcome is reliable when the tester is fault-free; it is unreliable when the tester is faulty, no matter whether the tested unit is faulty or fault-free. Communication links are assumed to be fault-free as they are generally passive components. The host node in the target system is used as the syndrome analyzer.

We consider all the three measures of cost for on-line system-level diagnosis, namely, diagnosis time, number of tests, and number of test links. In this paper, diagnosis time is measured by the number of rounds. Within a round, tests are assumed to be performed in parallel using one unit of time. Also a unit cannot be tested by and test its neighbors at the same time.

2   The Proposed Algorithm

A diagnosis is defined as *correct* if a faulty PE will not be diagnosed to be fault-free and a fault-free PE will not be diagnosed to be faulty; while a diagnosis is defined as *complete* if the status of all PE's can be identified either fault-free or faulty [11]. To achieve correct and complete diagnosis, the number of faulty PE's must be bounded by assumption [9]. The diagnosability d of a correct and complete diagnosis algorithm is generally determined by the connectivity of the testing graph. Usually the underlying testing graph is t-fault diagnosable [9], and $t \le d$. If there are more than d faulty PE's in the system, correct and complete diagnosis can not be guaranteed but correct diagnosis should be pursued. Before we present the new approach, let us critically review a few existing diagnosis algorithms that can be applied to mesh connected multiprocessors.

The connectivity of a mesh with wraparound is 4. Thus, the diagnosability d for a correct and complete diagnosis is 4. Independent of the number of faulty PE's, by [9], the traditional diagnosis algorithm requires that every PE must test and be tested by 4 other PE's. For the adaptive safe diagnosis algorithm proposed in [12], however, the diagnosis cost depends on the number of faulty PE's f, but the growth of test assignment is quite static in the sense that the algorithm increases the connectivity of the testing graph rather than the necessary tests in every round. Both the traditional diagnosis algorithm [9] and the adaptive safe diagnosis algorithm [12] cannot deal with the scenario when more than d faulty PE's exist. This is referred to as the *over-d fault scenario* in this paper.

The diagnosis algorithm LDA1 [11] has been proposed to diagnose a regularly interconnected VLSI/WSI processor array in the presence of a large number of faulty PE's. When applied to an $r \times c$

mesh with wraparound, correct and complete diagnosis is guaranteed if $f \leq 4$ while correct diagnosis can be achieved with high probability if f is over 4 but grows slower than $O(r \times c)$ when $r \times c \to \infty$. The diagnosis cost for LDA1 is the same as for the traditional diagnosis algorithm [9], which is independent of the fault size.

To overcome the drawbacks while preserving the diagnosis capability of the above existing algorithms, we propose below a new diagnosis algorithm for mesh architectures.

Consider an $r \times c$ mesh with wraparound, where r and c are the number of rows and the number of columns, respectively. A *faulty row* is a row in which at least 1 PE is faulty, while a *fault-free row* is a row in which all the PE's are fault-free. Over-d fault scenario is intended to be dealt with in the proposed algorithm by generally assuming a fault bound $t < min(r, c)$. Due to the wraparound connections, an $r \times c$ mesh consists of r independent loops along the rows each with c PE's. The basic idea of the proposed approach is as follows. Initially, the single loop test [9] is performed in each row, and the counter-clockwise direction in the single loop test is assumed without loss of generality. After the single loop test, the row without any test outcome of 1 is a fault-free row, otherwise it is a faulty row. The reason is given in the correctness proof of the proposed algorithm in the next section. Next, a faulty row searches for a fault-free neighbor row to obtain parallel tests. If a fault-free neighbor row exists and has not been involved in testing within the same round, then it can be used to test this faulty row in parallel within one round and the status of all the PE's in this faulty row can be identified. If no fault-free neighbor row exists, this faulty row will wait for the fault-free PE's in its neighbor rows to be identified, then let them test its PE's in the following round. However, while waiting, this faulty row is not left idle. Instead, it will try to perform tests, which were not performed before, among its PE's as much as possible. The resulted test outcomes, together with those obtained in the single loop test, may be used later when the testers are identified to be fault-free such that diagnosis time may be saved. The iteration continues until the status of all PE's have been identified or no more PE's can be identified for their status. The proposed algorithm is illustrated below.

Step 1: Each PE tests its next neighbor in the same row along the counter-clockwise direction.

Step 2: if the test outcomes of a row have no 1, **then** this row is a fault-free row; **else then** this row is a faulty row. Add the faulty row into the set A which is initially empty.

Step 3: for each faulty row $row_i \in A$ **do:**
    Begin
    Let *upper* be the number of PE's in row_i with unknown status which can be identified by the PE's of its upper row with known status as fault-free, which have not yet been involved in testing during the current round and are stored in the set

U_i; and let *lower* be the number of PE's in row_i with unknown status which can be identified by the PE's of its lower row with known status as fault-free, which have not yet been involved in testing during the current round and are stored in the set L_i.
    if *upper* > 0 **or** *lower* > 0, **then**
    Begin
      if *upper* $\geq$ *lower*, **then** let these PE's in U_i test their neighbors in row_i in parallel. Those tested PE's then have their status identified;
      else then let these PE's in L_i test their neighbors in row_i in parallel. Those tested PE's then have their status identified;
    End
    else, for each $PE_a \in row_i$ **do:**
    Begin
      if PE_a and one of its neighbors PE_b in row_i have not yet been involved in testing during the current round **and** the status of PE_b is not known **and** PE_a has never tested PE_b before, **then** let PE_a test PE_b;
    End
    End

Step 4: for each faulty row $row_i \in A$ **do:**
    Begin
    if row_i has some of the identified PE's as fault-free, **then** use their test outcomes obtained before to identify their neighbors in the same row;
    if the status of all the PE's in row_i have been identified, **then** remove row_i from A and continue;
    End
    if $|A| \neq 0$ **and** $|A|$ has been reduced during the current round, **then** go to Step 3;
    else terminate.

Example: Consider a 5×5 mesh with wraparound as shown in Figure 1(a). For clarity, only partial wires of wraparound are depicted. Assume PE_1, PE_3, PE_6 and PE_{12} are faulty. Figures 1(b), 1(c) and 1(d) illustrate the diagnosis process using the proposed algorithm, where a black box, a shaded box and a white box represent an unidentified PE, an identified faulty PE and an identified fault-free PE, respectively. In Figure 1(b), each PE in every row tests one of its neighbors along the counter clockwise direction and the test outcome is labeled beside the arc. It takes 3 rounds to finish this step as we assume that a PE can not be tested by or test another PE within the same round. As a result, row_0, row_1 and row_2 all have certain test outcomes of 1's which indicate that they are faulty rows, while row_3 and row_4 have no test outcome of 1 which implies that they are fault-free rows. Therefore, PE_{15}, PE_{16}, PE_{17}, PE_{18}, PE_{19}, PE_{20}, PE_{21}, PE_{22}, PE_{23} and PE_{24} are diagnosed to be fault-free now. In the next round as shown in Figure 1(c), all the PE's in row_4 and row_3, which are known as fault-free, test in parallel their neighbors in row_0 and row_2. The faulty PE's of PE_1, PE_3 and PE_{12} are thus located and the fault-free PE's of PE_0,

PE_2, PE_4, PE_{10}, PE_{11}, PE_{13} and PE_{14} are thus identified. Within the same round, the faulty row row_1 can not be tested by either row_0 or row_2. So the PE's in row_1 try to perform tests, which were not executed before, as much as possible. Here, PE_5 and PE_8 test PE_9 and PE_7, respectively. Although the test outcomes can not be used to identify the tested PE's now, they may be used later when the testers are known as fault-free.

In the following round as shown in Figure 1(d), row_2 is chosen to test row_1 because it can identify more PE's in row_1 than row_0 can. PE_5, PE_8 and PE_9 are thus known as fault-free and PE_6 is identified to be faulty. Because PE_{12} is faulty, it can not be used to test PE_7. However, recall that in the last round, PE_8 tested PE_7 with test outcome of 0 and PE_8 is now known as fault-free, therefore we can conclude that PE_7 is fault-free. The status of all the PE's have been identified and the diagnosis process is thus terminated. Totally, 41 tests, 39 test links and 5 testing rounds (diagnosis time) have been used. If a traditional diagnosis algorithm [9] or LAD1 [11] or the adaptive safe diagnosis algorithm [12] is used, 100 tests, 50 test links and 12 testing rounds must be required.

3   Theoretical Analysis

This section gives the analysis for the three measures of on-line diagnosis cost for the proposed algorithm compared with the existing approaches. These results are only stated due to lack of space. The formal proofs are in [5].

Theorem 1: The proposed diagnosis algorithm on an $r \times c$ mesh with wraparound, where the fault bound $t < min(5, r, c)$, is correct and complete.

Proof: (*Sketch*). After the single loop test at each row, the row without any test outcome of 1 must be fault-free. Also, there must exist at least 1 fault-free row as we assume that $r > t$. There must be a fault-free path from a faulty PE to a PE in a fault-free row. □

From Theorem 1, we know that the diagnosability d of a correct and complete diagnosis for the proposed algorithm is $min(5, r, c)$ - 1. The following lemma gives the upper bound of the diagnosis cost for the proposed diagnosis algorithm when the fault bound $t \leq d$.

Lemma 1: Consider an $r \times c$ mesh with wraparound, where the fault bound $t < min(5, r, c)$. The number of test links, the number of tests and the diagnosis time (rounds) required by the proposed algorithm are TL, TN and TR, respectively. Then,

$$TL \leq (r + 5) \cdot c - 5$$
$$TN \leq (r + 8) \cdot c - 2$$
$$TR \leq \begin{cases} 7 & c \text{ is even} \\ 9 & c \text{ is odd} \end{cases}$$

Theorem 2: Consider an $r \times c$ mesh with wraparound, where the fault bound $t < min(5, r, c)$. The proposed diagnosis algorithm requires less diagnosis cost than the traditional diagnosis approach [9]

and the algorithm LDA1 [11] applied to a mesh with wraparound.

Theorem 3: Consider an $r \times c$ mesh with wraparound, where the fault bound $t < min(5, r, c)$. Assume that the number of tests, the number of test links and the diagnosis time rounds required by the adaptive safe diagnosis algorithm [12] applied to the mesh are s_TN, s_TL and s_TR, respectively. Then,

$$TN \begin{cases} = s\_TN & f = 0 \\ < s\_TN & f = 1, 2, 3, 4 \end{cases}$$

$$TL \begin{cases} = s\_TL & f = 0 \\ > s\_TL & f = 1 \\ < s\_TL & f = 2, 3, 4 \end{cases}$$

$$TR \begin{cases} \geq s\_TR & f = 0 \\ \leq s\_TR & f = 1 \\ < s\_TR & f = 2, 3, 4 \end{cases}$$

Theorem 3 indicates that the proposed algorithm, in most of the cases, requires less diagnosis cost than the adaptive safe diagnosis approach [12] applied to the mesh. When the fault bound $t \leq d$, Theorems 2 and 3 analytically illustrate the superiority of the proposed algorithm over the three existing approaches [9] [11] [12] in terms of the diagnosis cost. However, when $t > d$, the traditional diagnosis algorithm [9] and the adaptive safe diagnosis algorithm [12] are not applicable.

4   Over-d Fault Diagnosis

Any deterministic diagnosis algorithm has its diagnosability d. The scenario when over d faulty PE's exist has not been addressed in depth in the existing literature on adaptive system-level diagnosis. The reason is that most of the existing adaptive diagnosis algorithms assume that all the PE's are completely connected. A completely connected network maximizes the possible diagnosability. However, a 2D mesh with wraparound only has connectivity of 4. Therefore, the diagnosability of the proposed adaptive diagnosis algorithm on a mesh cannot be as high as those for completely connected architectures. Somani, Agarwal and Avis have addressed the issue of over-d fault diagnosis in [10], but they have only considered the non-adaptive diagnosis.

In our proposed diagnosis algorithm, when the fault bound is over d, it can be still possible to detect and locate all the faulty PE's given by the following conditions.

Theorem 4: The proposed diagnosis algorithm for an $r \times c$ mesh with wraparound, is a correct and complete diagnosis *iff*
(1) the number of faulty PE's $f < min(r, c)$; and
(2) there is a fault-free path between each of the PE's in $V - Z$ and one of the PE's in Z, where V represents all the PE's in the mesh and Z represents the PE's in the identified fault-free rows after the single loop test at each row.

In Theorem 4, Condition (1) is set to guarantee correct diagnosis. Combined with Condition (2), correct and complete diagnosis can be guaranteed. For our approach, from Condition (1), correct diagnosis is guaranteed when $d < f < min(r, c)$. For the algorithm LDA1 [11], however, correct diagnosis is not guaranteed whenever $f > d$.

We define the *probability of correct and complete diagnosis P* as

$$P = \frac{\text{Diagnosable fault sets}}{\text{Total fault sets}}$$

Theorem 5: Assume that the probabilities of correct and complete diagnosis for the proposed algorithm and the algorithm LDA1 [11] are P_{new} and P_{LDA1}, respectively. Let $f < min(r, c)$. Then, $P_{new} \geq P_{LDA1}$.

Since there is at least a fault-free row in our scheme provided $f < min(r, c)$, the clockwise direction test along this fault-free row can be saved. Thus, the number of tests required by our algorithm is less than in the algorithm LDA1 [11]. We conjecture that this is also true for the number of test links and the diagnosis time. The simulation study on the average performance presented in the next section will support this conjecture.

5   Average Performance

In this section, we show that the average diagnosis cost of our algorithm is much better than those of the existing ones. Monte Carlo simulation is performed for the proposed algorithm to evaluate the average number of tests, the average number of test links and the average diagnosis time (rounds). We randomly generate 10,000 fault patterns to derive every average case.

Consider different numbers of faulty PE's (up to 4) in a 10×10 mesh with wraparound. Figures 2(a), 2(b) and 2(c) show the number of tests, the number of test links and the diagnosis time (rounds), respectively required by our approach against the traditional diagnosis algorithm [9], the algorithm LDA1 [11] and the adaptive safe diagnosis algorithm [12] applied to this mesh, which are denoted by TDA, LDA1 and ASDA, respectively for brevity. In Figures 2(a), 2(b) and 2(c), the shaded areas stand for the over-d fault scenario. For this scenario, TDA and ASDA are not applicable. By simulation for the over-d fault scenario, we found that the probabilities of correct and complete diagnosis for the proposed algorithm and the algorithm LDA1 are the same and marginally close to 1 when $f < 10$. For example, when $f = 9$, $P = 0.999800$. From Figure 2, it is clear that the diagnosis cost of our scheme is significantly lower than the existing approaches.

6   Conclusions

System-level diagnosis is a very important technique to preserve high reliability in multiprocessor systems. However, it must be efficient and must have a very low overhead. An adaptive system-level diagnosis approach for an $r \times c$ mesh with wraparound, has been proposed in this paper. Three measures for diagnosis cost (diagnosis time, number of tests and number of test links) are analyzed for the proposed algorithm. It is proved that the diagnosis cost required by this algorithm is significantly lower than in the traditional diagnosis algorithm and is most of the cases

lower than in the adaptive safe diagnosis algorithm applied to meshes. For the over-d fault scenario, our algorithm outperforms the algorithm LDA1 [11] when the number of faulty PE's is moderately large, i.e., $d < f < min(r, c)$.

References

[1] D. P. Bertsekas and J. N. Tsitsiklis, *Parallel and Distributed Computation, Numerical Methods*, Prentice-Hall International, Inc., 1989.

[2] L. N. Bhuyan and C. R. Das, "Dependability Evaluation of Multicomputer Networks," *Proc. Int. Conf. on Parallel Processing*, pp. 576-583, Aug. 1986.

[3] R. Bianchini, Jr. and R. Buskens, "An Adaptive Distributed System-Level Diagnosis Algorithm and Its Implementation", *Proc. 21st FTCS*, pp. 222-229, 1991.

[4] A. T. Dahbura and G. M. Masson, "An $O(n^{2.5})$ Fault Identification Algorithm for Diagnosable Systems," *IEEE Trans. Comput.*, Vol. C-33, pp. 486-492, June, 1984.

[5] C. Feng, "Fault Detection and Location for Parallel and Distributed Computer Systems", *Ph.D Dissertation* (in progress), Texas A&M University.

[6] K. Hwang and F. A. Briggs, *Computer Architecture and Parallel Processing*, New York: McGraw-Hill, 1984.

[7] *Paragon*™ *XP/S Product Overview*, Intel Corp., 1991

[8] A. Kavianpour and K. H. Kim, "Diagnostic Power of Four Basic System-Level Diagnosis Strategies for Hypercubes", *1990 Int. Conf. on Parallel Processing*, Vol. 1, pp 267-271, 1990.

[9] F. P. Preparata, G. Metze, and R. T. Chien, "On the Connection Assignment Problem of Diagnosable Systems, " *IEEE Trans. Electron. Comput.*, Vol. EC-16, No. 12, pp. 848-854, Dec. 1967.

[10] A. K. Somani, V. K. Agarwal and D. Avis, "A Generalized Theory for System-Level Diagnosis," *IEEE Trans. Comp.*, Vol C-36, No. 5, pp. 538-546, May 1987.

[11] A. K. Somani, and V. K. Agarwal, "Distributed Diagnosis Algorithms for Regular Interconnected Structures", *IEEE Trans. on Computers*, Vol. 41, No. 7, pp. 899-906, July, 1992.

[12] N. H. Vaidya and D. K. Pradhan, "System-Level Diagnosis for Recoverability and Safety", Technical Report (Submitted for publication), Department of Electrical and Computer Engineering, University of Massachusetts, May, 1991.

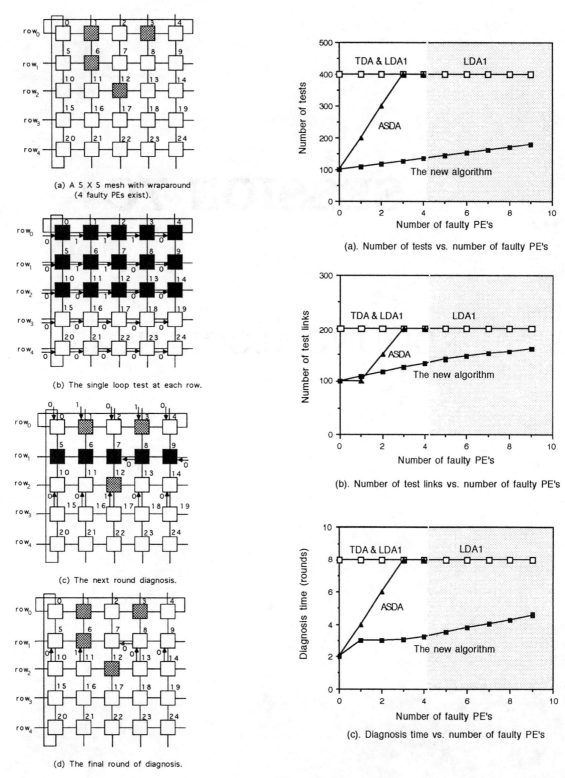

(a) A 5 X 5 mesh with wraparound
(4 faulty PEs exist).

(b) The single loop test at each row.

(c) The next round diagnosis.

(d) The final round of diagnosis.

Figure 1: An example of the proposed algorithm.

(a). Number of tests vs. number of faulty PE's

(b). Number of test links vs. number of faulty PE's

(c). Diagnosis time vs. number of faulty PE's

Figure 2: Diagnosis cost vs. number of faulty PE's in a 10×10 mesh with wraparound.

SESSION 7C

ROUTING ALGORITHMS

FAST PARALLEL ALGORITHMS FOR ROUTING ONE-TO-ONE ASSIGNMENTS IN BENES NETWORKS*

Ching-Yi Lee and A. Yavuz Oruç

Electrical Engineering Department

and

Institute for Advanced Computer Studies

University of Maryland

College Park, MD 20742-3025

Abstract– *This paper presents new results on routing unicast (one-to-one) assignments over Benes networks. Parallel routing algorithms with polylogarithmic routing times have been reported earlier [10,8], but these algorithms can only route permutation assignments unless unused inputs are assigned to dummy outputs. This restriction is removed in this paper by using techniques that permit bypassing idle or unused inputs without any increase in the order of routing cost or routing time. We realize our routing algorithm on two different topologies. The algorithm routes a unicast assignment involving $O(k)$ pairs of inputs and outputs in $O(\log^2 k + \lg n)$ time[1] if every pair of processors is interconnected by a direct link, and in $O(\lg^4 k + \lg^2 k \lg n)$ time if the processors are interconnected by an extended shuffle-exchange network. The same algorithm can be pipelined to route α unicast assignments, each involving $O(k)$ pairs of inputs and outputs, in $O(\lg^2 k + \lg n + (\alpha-1) \lg k)$ time on the completely connected graph, and in $O(\lg^4 k + \lg^2 k \lg n + (\alpha-1)(\lg^3 k + \lg k \lg n))$ time on the extended shuffle-exchange graph. These yield an average routing time of $O(\lg k)$ in the first case, and $O(\lg^3 k + \lg k \lg n)$ in the second case, for all $\alpha \geq \lg n$. These complexities indicate that the algorithm given in this paper is as fast as the algorithm given in [10] for unicast assignments, and with pipelining it is faster at least by a factor of $O(\lg n)$ on both topologies. Furthermore, for sparse assignments, i.e., when $k << n$, it is the first algorithm which has an average routing time of $O(\lg n)$ on a topology with $O(n)$ links.*

1 INTRODUCTION

The Benes network has received much attention in interconnection network literature because of its $O(n \lg n)$ cost and $O(\lg n)$ depth [1,8,9,10,7]. In a way, this network can be considered the forerunner of most multistage interconnection networks that have been extensively studied and used in some real parallel computer systems for interprocessor or processor-memory communications [12,6].

----
[0] * This work is supported in part by the National Science Foundation under Grant No: CCR-8708864, and in part by the Minta Martin Fund of the School of Engineering at the University of Maryland.

[1] All logarithms are in base 2 unless otherwise stated, and $\lg n$ denotes the logarithm of n in base 2.

One problem regarding the Benes network that remains to have a satisfactory solution is its routing. Many routing algorithms have been reported extending from the $O(n \lg n)$ time looping procedure of Waksman [14], and Opferman and Tsao-Wu [11] to the $O(\lg^2 n)$ time parallel algorithms of Lev et al [8], and Nassimi and Sahni [10]. Other routing algorithms for the Benes network include matching and edge-coloring schemes [4,2,3].

The looping algorithm was developed with the realization that there are two subnetworks in the center stage for the network's inputs to connect to its outputs, and with the added constraint that no two inputs can be connected to two outputs through the same subnetwork in the center stage unless those outputs belong to different switches in the last stage. One gets around this constraint simply by looping between the switches in the first stage and those in the last stage, and assigning the paths in an alternate fashion to the subnetworks in the center stage. Given that there are n inputs to route, it takes the looping algorithm $O(n)$ time to set the switches in a Benes network, and if the same algorithm is applied recursively to the center-stage subnetworks then all the switches in a recursively decomposed Benes network can be set in $O(n \lg n)$ time.

The routing time of the Benes network can be reduced in several ways. The most obvious approach is to use a binary tree to set the subnetworks in the center stage in parallel as was done in [3]. The root processor of the tree sets the switches in the first level, its children set the switches in the next level, and so on. It is easy to see that the time complexity of this parallel routing scheme is $O(n)$. The routing time can be reduced further by introducing parallelism into the setting of switches in each level. This can be done by dividing the outputs into equivalence classes such that two outputs will be in the same class if and only if they must be routed through the same center-stage subnetwork. Once the equivalence classes are decided, the switches in the first and last stages can be set in parallel. The parallel algorithm of Nassimi and Sahni [10] is based on this discovery, and its time complexity depends on the complexity of the parallel computer model and the number of processors available. Their algorithm takes $O(\lg^2 n)$ time on a completely interconnected n-processor computer, and $O(\lg^4 n)$ time on a shuffle-exchange interconnected n-processor computer. Lev et al [8] provided similar parallel algorithms in a more general framework by using edge-coloring schemes. Assuming a parallel computer with conflict free access between $O(n)$ processors and $O(n)$ memory elements, their algorithm also takes $O(\lg^2 n)$ time.

While these parallel algorithms are fast, their time complexities are still higher than the $O(\lg n)$ depth of the Beneš network. Furthermore, these algorithms can only route permutation assignments. In case of incomplete assignments where some inputs may remain idle, they cannot be used unless the idle inputs are given dummy outputs. This, however, takes additional time and may render these algorithms inefficient, especially in case of sparse assignments–assignments involving $O(k)$ pairs of inputs and outputs, where $k << n$.

In this paper, we present an efficient parallel algorithm for routing incomplete assignments on the Beneš network. We also pipeline this algorithm to attain a factor of $O(\lg n)$ speed up over the parallel algorithms of Nassimi and Sahni and Lev et al. Pipelining is made possible by routing the Beneš network stage by stage from left to right and overlapping the routing steps for consecutive stages. Unlike our routing algorithm, the cited parallel algorithms determine the settings from outer-stage switches toward inner-stage switches. One other difference is that, in our routing scheme, each of the first $\lg n - 1$ stages is provided with its own special routing module rather than using a single parallel computer to set all switches. Once these stages are set, the stages in the second half are then self-routed as suggested in [7].

Our parallel routing algorithm can run on any parallel computer whose processors are equipped with a constant number of $O(\lg n)$-bit registers and some simple arithmetic and logic circuitry that can compare $O(\lg n)$ bit numbers and perform some counting and decoding functions. We realize our routing schemes on two different topologies. We show that if every pair of processors are interconnected by a direct arc, then routing an assignment involving $O(k)$ pairs of inputs and outputs takes $O(\log^2 k + \lg n)$ time without pipelining and $O(\lg k)$ time with pipelining. We also establish that using a weaker topology, namely the extended shuffle-exchange graph (which will be defined in Section 4), leads to a routing algorithm with $O(\lg^4 k + \lg^2 k \lg n)$ time without pipelining and $O(\lg^3 k + \lg k \lg n)$ time with pipelining.

2   BASIC FACTS AND DEFINITIONS

An n-network is a directed acyclic graph with n distinguished source vertices, called inputs, n distinguished sink vertices, called outputs, and some internal vertices, called switches. An assignment for an n-network is a pairing of its inputs with its outputs such that each output appears in at most one pair. An assignment consisting of k pairs will be called a k-assignment. An assignment is called one-to-one or unicast if each input appears in at most one pair. A permutation assignment for an n-network is a unicast n-assignment. An

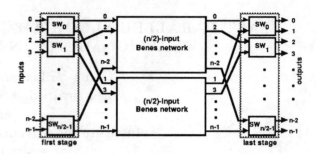

Figure 1: The recursive construction of an n-input Beneš network.

n-network is said to realize an assignment if, for each pair (a, b) in the assignment, a path can be formed from input a to output b by setting the switching nodes in the network with the constraint that the paths for no two pairs (a, b) and (c, d) overlap unless $a = c$. An n-network that can realize all unicast assignments is called a unicast n-network.

The well-known Beneš network is a unicast n-network that is constructed recursively as shown in Figure 1. Each of the first and the last stages consists of $n/2$ 2×2 switches, and the center stage consists of two $n/2$-input Beneš networks. Each 2×2 switch can be set in two ways: either through state where the two inputs are connected straight-through to the two outputs, or cross state where the two inputs are connected to opposite outputs. The inputs and outputs of the upper $n/2$-input Beneš network are numbered $0, 2, \ldots, n-2$, and the inputs and outputs of the lower $n/2$-input Beneš network are numbered $1, 3, \ldots, n-1$, from top to bottom. If the half-size Beneš networks in the center stage are recursively decomposed then one obtains a $(2 \lg n - 1)$-stage network consisting of 2×2 switches, assuming that n is a power of 2.

Paths in an n-network will be established by specifying some routing information at its inputs. It is assumed that each input holds its own routing information unless otherwise stated. It is also assumed that the routing information for each input is accompanied by some binary coded message that is to be routed from that input to the output specified in the routing information. A message and routing information combined together will be termed a packet. For an n-input Beneš network, the routing part of a packet, to be called the header, is assumed to have $\lg n + 1$ bits, and will be denoted as $(r_i, d^i_{\lg n-1}, \ldots, d^i_1, d^i_0)$ for a packet at input i. The bit r_i specifies whether input i is paired with some output. Input i is said to be busy if $r_i = 1$, and it is said to be idle if $r_i = 0$. The remaining bits, i.e., $(d^i_{\lg n-1}, \ldots, d^i_1, d^i_0)$, form the output address which specifies the binary representation of the output paired with input i with $d^i_{\lg n-1}$ being the most significant bit. Besides, a switch is said to be busy if both of its inputs are busy; it is said to be semi-busy if one of its inputs is busy and the other input is idle, and it is said to be idle if both of its inputs are idle.

In light of these facts, we make the notion of a unicast assignment more precise.

Definition 1: A unicast k-assignment for an n-network is a set $\{(i, (r_i, d^i_{\lg n-1}, \ldots, d^i_1, d^i_0)) : 0 \leq i \leq n-1\}$ such that exactly k r_i's are equal to 1, and $(d^i_{\lg n-1}, \ldots, d^i_1, d^i_0) \neq (d^j_{\lg n-1}, \ldots, d^j_1, d^j_0)$ whenever $i \neq j$ and $r_i = r_j = 1$. $\|$

3 THE ROUTING PRINCIPLE

In this section, we describe a routing principle which establishes that unicast assignments for the Beneš network can be recursively decomposed into half-sized unicast assignments stage by stage from left to right.

Notation: For $0 \leq i \leq n-1$, if $(b^i_{\lg n-1}, \ldots, b^i_1, b^i_0)$ is the binary representation of i, then $\bar{i}$ denotes the integer which has the binary representation $(b^i_{\lg n-1}, \ldots, b^i_1, \bar{b}^i_0)$, and[2] i and $\bar{i}$ are called a *dual pair* of integers. $\|$

Now let $H(n)$ denote a single stage n-network which comprises $n/2$ 2×2 switches, $SW_0, SW_1, \ldots, SW_{n/2-1}$.

Definition 2: Given a unicast assignment $\{(i, (r_i, d^i_{\lg n-1}, \ldots, d^i_1, d^i_0)) : 0 \leq i \leq n-1\}$ for $H(n)$, a sequence of switches $SW_{i_0}, SW_{i_1}, \ldots, SW_{i_{p-1}}$ in $H(n)$ is said to form a *chain* with respect to that assignment if, for all $0 \leq q \leq p-2$, one of the inputs of SW_{i_q} and one of the inputs of $SW_{i_{q+1}}$ are paired with a dual pair of outputs and have their connecting bits set to 1, i.e., there exist $x = 2i_q$ or $x = 2i_q + 1$ and $y = 2i_{q+1}$ or $2i_{q+1} + 1$ for which $(d^x_{\lg n-1}, \ldots, d^x_2, d^x_1, d^x_0) = (d^y_{\lg n-1}, \ldots, d^y_2, d^y_1, \bar{d}^y_0)$ and $r_x = r_y = 1$. Furthermore, $SW_{i_0}, SW_{i_1}, \ldots, SW_{i_{p-1}}$ is said to be a *closed chain* if the other input of $SW_{i_{p-1}}$ and the other input of SW_{i_0} are also paired with a dual pair of outputs. It is said to be an *open chain* otherwise. $\|$

The *size* of a chain is the number of switches in the chain. Given a unicast k-assignment for $H(n)$, the size of a chain can be as large as $\min\{\lfloor (k+2)/2 \rfloor, n/2\}$ and as small as 1, and the number of chains can be as large as $\min\{k, n/2\}$ (when each chain has size 1) and as small as 1 (when that chain is of size $\lfloor (k+2)/2 \rfloor$ or $\lfloor k/2 \rfloor$). Figure 2 shows a closed chain C_1 and an open chain C_2 with respect to a unicast 15-assignment for $H(n)$ where $n = 16$.

An open chain, say $SW_{i_0}, SW_{i_1}, \ldots, SW_{i_{p-1}}$, has two *end switches* (i.e., SW_{i_0} and $SW_{i_{p-1}}$). Based on the status of the end switches, two types of open chains are distinguished.

Definition 3: An open chain is said to be a *full open chain* if both of its end switches are busy or semi-busy, and it is said to be a *half open chain* if one of

---

[2] $\bar{b}$ denotes the binary complement of bit b.

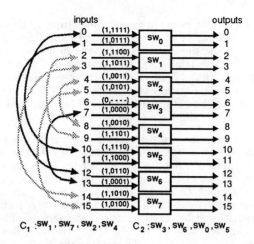

Figure 2: Two chains with respect to a unicast 15-assignment for $H(n)$ where $n = 16$.

its end switches is busy and the other end switch is semi-busy. $\|$

For example, C_2 given in Figure 2 is a half open chain since SW_5 is busy and SW_3 is semi-busy.

Theorem 1: Given $\{(i, (r_i, d^i_{\lg n-1}, \ldots, d^i_1, d^i_0)) : 0 \leq i \leq n-1\}$, a unicast k-assignment for $H(n)$, let $(r'_i, p^i_{\lg n-1}, \ldots, p^i_1, p^i_0)$ denote the header of the packet at output i of $H(n)$, $0 \leq i \leq n-1$. There exist settings for $SW_0, SW_1, \ldots, SW_{n/2-1}$ such that $\{(2i, (r'_{2i}, p^{2i}_{\lg n-1}, \ldots, p^{2i}_2, p^{2i}_1)) : 0 \leq i \leq n/2-1\}$ is a unicast k_0-assignment and $\{(2i+1, (r'_{2i+1}, p^{2i+1}_{\lg n-1}, \ldots, p^{2i+1}_2, p^{2i+1}_1)) : 0 \leq i \leq n/2-1\}$ is a unicast k_1-assignment, where $k_0 = \lceil k/2 \rceil$ and $k_1 = \lfloor k/2 \rfloor$.

Proof: With respect to the given unicast k-assignment and without loss of generality, suppose that there exist c chains, $1 \leq c \leq \min\{k, n/2\}$, and f of these c chains are full open chains, $0 \leq f \leq c$. From Definition 2, the switches of each chain can be set such that, given inputs i and j which have $r_i = r_j = 1$ and $(d^i_{\lg n-1}, \ldots, d^i_2, d^i_1, d^i_0) = (d^j_{\lg n-1}, \ldots, d^j_2, d^j_1, \bar{d}^j_0)$, one is routed to an even-numbered output and the other is routed to an odd-numbered output, and once a switch of the chain is set then the settings of the other switches of the chain are fixed. Moreover, different chains can be set mutually independently resulting in 2^c settings since each chain has two ways of settings. It is obvious that, with each of these 2^c settings in $H(n)$, $\{(2i, (r'_{2i}, p^{2i}_{\lg n-1}, \ldots, p^{2i}_2, p^{2i}_1)) : 0 \leq i \leq n/2-1\}$ is a unicast k_0-assignment and $\{(2i+1, (r'_{2i+1}, p^{2i+1}_{\lg n-1}, \ldots, p^{2i+1}_2, p^{2i+1}_1)) : 0 \leq i \leq n/2-1\}$ is a unicast k_1-assignment for some integers k_0 and k_1 with $k_0 + k_1 = k$. Thus, it suffices to show at least one of the 2^c settings will result in $k_0 = \lceil k/2 \rceil$ and $k_1 = \lfloor k/2 \rfloor$. The closed and full open chains, no matter how they are set, will increase k_0 and k_1 by the same integer since such chains have even numbers of busy inputs. However, in one setting of a half open chain (*Type-0 Setting*), k_0 will increase one more than k_1, and in its other setting (*Type-1 Setting*), k_0 will increase one less than k_1. Suppose that $\lceil f/2 \rceil$ of the f full open chains are in their *Type-0 Settings* and the other $\lfloor f/2 \rfloor$ full open chains are in their *Type-1 Settings*. Then, $k_0 = \lceil k/2 \rceil$ and $k_1 = \lfloor k/2 \rfloor$, and the statement follows. ‖

Given a unicast assignment for an n-input Beneš network, let the switches in the first stage be set such that the statement of Theorem 1 is satisfied. Then the two established half-sized assignments for the two center-stage $n/2$-input Beneš networks are also unicast and can be realized by these networks, respectively. Hence, any algorithm that satisfies the statement of Theorem 1 can be used recursively to set the switches in the first $\lg n - 1$ stages. Thereafter the packets can be routed on a self-routing basis through the last $\lg n$ stages to their final destinations as identified before [7].

4 THE PARALLEL ROUTING ALGORITHM

Following the discussion in the previous section, it is only necessary to describe a parallel algorithm for $H(n)$ so that the routing principle stated in Theorem 1 is satisfied. The parallel routing algorithm for the Beneš network then follows.

In the parallel algorithm for $H(n)$, it is assumed that there are n interconnected processors, $PR(0)$, $PR(1), \ldots, PR(n-1)$, and that a packet header $(r_i, d^i_{\lg n - 1}, \ldots, d^i_1, d^i_0)$ is initially input to $PR(i)$. $PR(i)$ and $PR(\bar{i})$ are called a *dual pair* of processors and will determine the setting of $SW_{\lfloor i/2 \rfloor}$ of $H(n)$, $0 \le i \le n - 1$. The time complexity of the parallel algorithm depends on the interconnection topology between these n processors. The algorithm will run on two connection topologies, namely the *completely connected graph* and the *extended shuffle-exchange* network.

A. The Routing Model

The parallel algorithm uses three kinds of macro functions: *move process*, *concentrate process* and *broadcast process* for exchanging packets between the n processors. A move process transfers packets from a subset of processors to a subset of processors in a one-to-one manner. A concentrate process transfers packets from a subset of processors to a single processor in a many-to-one manner. A broadcast process transfers packets from a single processor to a subset of processors in a one-to-many manner.

The time complexity to execute any of these processes depends on the topology which interconnects the processors. We will consider two topologies. The first is the completely connected graph in which there is a link between every two processors. It is obvious that each of the above three processes can be executed in $O(1)$ time over the completely connected graph.

The second topology we will use is called the *extended shuffle-exchange* network in which (a) each dual pair of processors is connected by a link, and (b) processors $PR(0), PR(1), \ldots, PR(n/2^m - 1)$ are $(n/2^m)$-shuffle connected for $m = 0, 1, \ldots, \lg n - 2$, (i.e., $PR(0), PR(1), \ldots, PR(n-1)$ is n-shuffle connected, $PR(0), PR(1), \ldots, PR(n/2-1)$ is $n/2$-shuffle connected, and so on), where the replicated links between the processors due to the shuffle connections are coalesced together. Essentially, the extended shuffle-exchange network is obtained by superposing Stone's perfect shuffle-exchange network [13,5] for n, $n/2$, $\ldots$, 4 inputs. Compared with the $O(n^2)$ links needed for the completely connected graph, it is easily verified that the extended shuffle-exchange network requires $O(n)$ links which is also the case with Stone's perfect shuffle-exchange network.

Any concentrate or broadcast process involving any subset of the processors can be completed over an extended shuffle-exchange network by passing $O(\lg n)$ times through the n-shuffle links between $PR(0), PR(1), \ldots, PR(n-1)$, and hence each of the concentrate and broadcast processes can be executed in $O(\lg n)$ time. To execute a move process, the extended shuffle-exchange graph takes three steps. Suppose that some k processors have packets (one each) which are to be to moved to some k processors, $2 \le k \le n$ ($k = 1$ is a trivial case). In the first step, the k packets are moved to any k of the first $2^{\lceil \lg k \rceil}$ processors by passing $\lg n$ times through the n-shuffle connection. In the second step, these k packets are sorted according to their destination addresses by passing $\lg^2 k'$ times through the k'-shuffle interconnection between $PR(0), PR(1), \ldots, PR(k'-1)$, where $k' = 2^{\lceil \lg k \rceil}$ [13,5]. In the third step, the sorted packets are moved to their final destinations by passing them $\lg n$ times through the n-shuffle connection. Therefore, a move process involving k packets can be executed over the extended shuffle-exchange graph in $O(\lg^2 k + \lg n)$ time. We note that, the extended shuffle-exchange network was intentially designed to contain a shuffle exchange network for each power of 2 number of inputs between 1 and n. This is needed since the value of k' varies between 1 and n.

B. The Parallel Routing Scheme

From the proof of Theorem 1 in Section 3, each chain has two settings (*Type-0* and *Type-1 Settings*), and the inputs of its switches can be partitioned into two *equivalence classes* such that the inputs in one equivalence class are connected to even-numbered outputs and the inputs in the other equivalence class are connected to odd-numbered outputs. Once the equivalence classes of inputs in a chain are established, the settings of switches in the chain are straightforward. The parallel algorithm that follows will use such equivalence classes to speed up the switch settings. The following proposition shows how to determine the pairs of inputs that are in the same equivalence class, and is similar to the observation given on p. 150 in [10].

Proposition: Let $\{(i, (r_i, d^i_{\lg n - 1}, \ldots, d^i_1, d^i_0)) : 0 \le i \le n - 1\}$ be a unicast assignment for $H(n)$. If

$(d^i_{\lg n-1}, \ldots, d^i_1, d^i_0) = (d^j_{\lg n-1}, \ldots, d^j_1, \bar{d}^j_0)$, $r_i = r_j = 1$ and $i \neq j$, then inputs i and $\bar{j}$ are in the same equivalence class and inputs j and $\bar{i}$ are in another equivalence class.

Proof: The proof is straightforward and is omitted. ||

Remark: This equivalence among the inputs will be noted by associating each input with an ordered *quadruple* in which the first element is a single bit used to indicate if this input belongs to a closed or open chain, the second element corresponds to this input, the third element points to an input that is in the same equivalence class as this input, and the fourth element, called the *representative* of this input, will be used to route this input. If i and j satisfy the hypothesis of Proposition 1, two ordered quadruples $\langle 1; i, \bar{j}; p_i \rangle$ and $\langle 1; j, \bar{i}; p_j \rangle$ will be established, where their first elements are initialized to 1 and their fourth elements are initialized to $p_i = \min\{i, \bar{j}\}$ and $p_j = \min\{j, \bar{i}\}$. If $r_i = 1$ and there is no $r_j = 1$ for which $(d^i_{\lg n-1}, \ldots, d^i_1, d^i_0) = (d^j_{\lg n-1}, \ldots, d^j_1, \bar{d}^j_0)$ (i.e., input i is paired with an output whose dual output is idle), an ordered quadruple $\langle 1; i, -1; i \rangle$ will be established, where -1 is used to denote that this input belongs to a busy end switch of an open chain, or an open chain of size 1. If $r_i = 0$ and $r_{\bar{i}} = 1$, an ordered quadruple $\langle 0; i, -2; i \rangle$ will be established, where -2 is used to denote that this input belongs to a semi-busy end switch of an open chain. ||

The parallel algorithm can be roughly divided into four phases. In the first phase, ordered quadruples as defined in Remark 1 are established. In the second phase, the representative in each quadruple is computed such that each input knows to which chain it belongs and all the inputs in the same equivalence class will agree on a common representative. In the third phase, each half open chain is assigned a *Type-0* or *Type-1 Setting* so that the statement of Theorem 1 (i.e., $k_0 = \lceil k/2 \rceil$ and $k_1 = \lfloor k/2 \rfloor$) is satisfied. In the fourth phase, each switch is set by using the representatives of its two inputs.

Using the unicast assignment given in Figure 2 as an example, the operation of each phase is outlined as follows.

The first phase applies Remark 1 to establish ordered quadruples, and can be further decomposed into three steps. At the first step, packet headers are input to the processors, and the idle inputs which belong to semi-busy switches have their quadruples established. At the second step, packet headers are moved to new processors so that each dual pair of processors can apply the steps in Remark 1 in parallel. At the third step, the remaining quadruples are established and moved to new processors which are specified by their second elements, and their first elements are changed to 0 if their third elements are -1. An illustration of these three steps is shown in Figure 3, where a column with 16 entries is used to express the data stored in the 16 processors at each step.

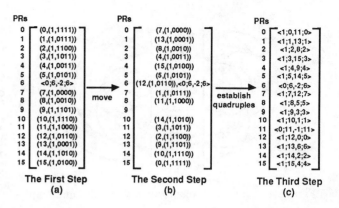

Figure 3: The three steps of the first phase by using the unicast 15-assignment given in Figure 2 as an example.

The second phase is an iterative procedure which computes the representative of each quadruple. This computation is the crux of the parallel algorithm, and it needs to be explained in detail.

A chain, depending upon its type, is decomposed into two sequences of quadruples in the first phase where the quadruples in each sequence belong to the same equivalence class. Let us use the quadruples in each sequence as nodes to form a directed graph in which a directed arc is established from a quadruple to another quadruple if and only if the third element of the former quadruple is equal to the second element of the latter quadruple. Then, a k-size closed chain will form two k-node *closed subchains* for which each node has an incoming arc and an outgoing arc, and a k-size open chain will form two k-node *open subchains* for which each source node has an outgoing arc, each sink node has an incoming arc, and each of the other nodes has an incoming arc and an outgoing arc. For example, Figure 4 explicitly depicts the four subchains obtained from the quadruples in Figure 3(c).

These subchains will be used to facilitate an understanding of the iterative procedure. If a quadruple is in a closed subchain in the m-th iteration, its first element is 1, its third element points to the second element of its 2^m-th successor and its last element points to the smallest input among the 2^m second elements of its first 2^m successors. If a quadruple is in an open subchain, the way its elements are updated depends on when it is known that this quadruple is in an open subchain. When this quadruple is at least 2^m far away from the sink quadruple in the m-th iteration, its elements are updated in the same way as if this quadruple is in a closed subchain. When this quadruple is at distance less than $2^m - 1$ away from the sink quadruple in the m-th iteration, it is recognized to be in an open subchain, and its first element is changed to 0, its third element is changed to -1 or -2 (the same as the third element of the sink quadruple) and its last element points to the second element of the sink quadruple. That is, for each open subchain,

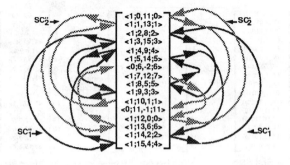

Figure 4: The subchains formed from the quadruples in Figure 3(c).

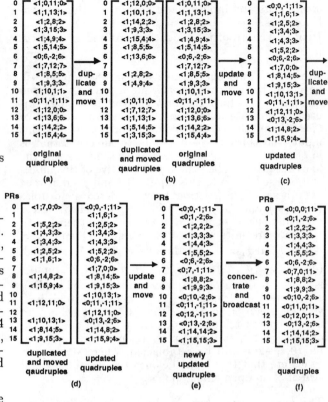

Figure 5: The second and third phase of the parallel algorithm that follows the first phase shown in Figure 3.

the updating information is exponentially propagated from the sink quadruple to the source quadruple. Using this updating procedure, after $\lg\lceil k\rceil$ iterations, any k quadruples that form a closed subchain will select the smallest input among their second elements as their common representative, and any k quadruples that form an open subchain will select the second element of the sink quadruple as their common representative. For example, the four subchains in Figure 4 need two iterations to determine their representatives, and inputs 2, 3, 11 and 6 will be selected as the representatives of the quadruples in SC_1', SC_1'', SC_2' and SC_2'', respectively.

Now we proceed to describe the second phase of the algorithm. Given a unicast k'-assignment, the second phase is assumed to consist of $\lg\lceil k\rceil$ iterations, where $k = \min\{\lfloor(k'+2)/2\rfloor, n/2\}$. This assumption is justified since there is no prior information about the exact sizes of the subchains and $\min\{\lfloor(k'+2)/2\rfloor, n/2\}$ is an upper bound for the sizes of the subchains with respect to the k'-assignment as stated in the previous section. Technically, each iteration can be further decomposed into three steps. In the first step, those quadruple whose first elements are 1 are duplicated and their copies are moved to new processors specified by their third elements. In the second step, each processor updates the quadruple(s) that it holds as follows: (1) when it holds only a quadruple, it "kills" the quadruple if the first element of the quadruple is 1; otherwise it keeps the quadruple intact; (2) when it holds $\langle 1;l,i;p_l\rangle$ and $\langle 1;i,j;p_i\rangle$, it replaces the first quadruple by $\langle 1;l,j;p_l'\rangle$ where $p_l' = \min\{p_i,p_l\}$ and kills the second quadruple; (3) when it holds $\langle 1;l,i;p_l\rangle$ and $\langle 0;i,*;j\rangle$ where $*$ is -2 or -1, it replaces the first quadruple by $\langle 1;l,*;i\rangle$ and keeps the second quadruple intact. In the third step, each updated quadruple is moved to a new processors specified by its second element if its first element remains 1, and its first element is changed to 0 if its third element is -1 or -2. For example, we illustrate the second phase in (a) to (e) of Figure 5, where the transitions from (a) to (b) and from (b) to (c) constitute the first iteration, and the transitions from (c) to (d) and from (d) to (e) constitute the second iteration. Note that, after the computation of the second phase, the two quadruples

held by $PR(i)$ and $PR(\bar{i})$ correspond to inputs i and $\bar{i}$ of $SW_{\lfloor i/2\rfloor}$. Besides, if the first elements of the two quadruples are 1, $SW_{\lfloor i/2\rfloor}$ belongs to a closed chain; otherwise, it belongs to an open chain. Furthermore, $SW_{\lfloor i/2\rfloor}$ belongs to a full open chain if the third elements are both -1 or -2, and to a half open chain if one of the third elements is -1 and the other is -2.

The third phase assigns types of settings to half open chains as discussed in the proof of Theorem 1, and can be decomposed into three parts. In the first part, for each half open chain, the representative of the quadruple that corresponds to the busy input of the semibusy end switch is concentrated to a specific processor, say $PR(0)$. In the second part, $PR(0)$ assigns Type-0 Setting to half of the concentrated representatives and Type-1 Setting to the other half, and broadcasts this information to all the processors. In the third part, each half open chain has its quadruples determine the type of settings it is assigned. For example, Figure 5(f) shows a result of the third phase, where the quadruples in $PR(0), PR(7), PR(11)$ and $PR(12)$ have their third elements change to 0 since these quadruples belong to half open chain C_2 which is assigned the Type-0 Setting.

Having decided the types of each chain and computed the representative of each quadruple, the fourth phase determines the settings of switches in $H(n)$. Each switch is set by a dual pair of processors which hold the two quadruples that correspond to the inputs of this switch, and the setting depends on the type of the chain to which this switch belongs.

Case 1: If $PR(i)$ holds $\langle 1; i, k; j \rangle$ and $PR(\bar{i})$ holds $\langle 1; \bar{i}, l; \bar{j} \rangle$, then $SW_{\lfloor i/2 \rfloor}$ is in a closed chain. Assuming that $SW_{\lfloor j/2 \rfloor}$ is set *through*, $SW_{\lfloor i/2 \rfloor}$ must be set *through* if $i - j$ is even and *cross* otherwise.

Case 2: If $PR(i)$ holds $\langle 0; i, p; j \rangle$ and $PR(\bar{i})$ holds $\langle 0; \bar{i}, p; l \rangle$ where $p = -1$ or $p = -2$, then $SW_{\lfloor i/2 \rfloor}$ is in a full open chain, and $SW_{\lfloor j/2 \rfloor}$ and $SW_{\lfloor l/2 \rfloor}$ are the end switches. Assuming that the smaller one of $SW_{\lfloor j/2 \rfloor}$ and $SW_{\lfloor l/2 \rfloor}$ is set *through*, when $j < l$, $SW_{\lfloor i/2 \rfloor}$ must be set *through* if $i - j$ is even and *cross* otherwise, and when $l < j$, $SW_{\lfloor i/2 \rfloor}$ must be *through* if $\bar{i} - l$ is even and *cross* otherwise.

Case 3: If $PR(i)$ holds $\langle 0; i, q; j \rangle$ and $PR(\bar{i})$ holds $\langle 0; \bar{i}, -2; l \rangle$ where $q = 0$ or 1, then $SW_{\lfloor i/2 \rfloor}$ is in a half open chain and $SW_{\lfloor j/2 \rfloor}$ is the semi-busy end switch. When $q = 0$ (the half open chain is assigned *Type-0 Setting*), $SW_{\lfloor i/2 \rfloor}$ must be set *through* if i is even and *cross* otherwise, and when $q = 1$ (the half open chain is assigned *Type-1 Setting*), $SW_{\lfloor i/2 \rfloor}$ must be set *through* if i is odd and *cross* otherwise.

For example, switches SW_0, SW_1, SW_4, SW_6 and SW_7 in Figure 2 are set *through* and switches SW_2, SW_3 and SW_5 are set *cross* by checking the final quadruples held in each dual pair of processors as shown in Figure 5(f).

C. The Parallel Algorithm

The following parallel routing algorithm formalizes the steps outlined in the previous subsection.

Step 1: Given $(i, (r_i, d^i_{\lg n - 1}, \ldots, d^i_1, d^i_0))$ input to $PR(i)$, $PR(i)$ establishes $\langle 0; i, -2; i \rangle$ if $r_i = 0$, $0 \le i \le n - 1$. Let k' be the number of r_i's whose value are 1, and let $k = \min\{\lfloor (k' + 2)/2 \rfloor, n/2\}$. Let m be a parameter initialized to 0.

Step 2: Move $(i, (r_i, d^i_{\lg n - 1}, \ldots, d^i_1, d^i_0))$ to $PR(x)$ if $r_i = 1$ and $(d^i_{\lg n - 1}, \ldots, d^i_1, d^i_0)$ is the binary representation of x, $0 \le i \le n - 1$.

Step 3: Given $P\bar{R}(x)$ holding (i, r_i) and $PR(\bar{x})$ holding (j, r_j), $PR(x)$ establishes $\langle 1; i, \bar{j}; p_i \rangle$ and $PR(\bar{x})$ establishes $\langle 1; j, \bar{i}; p_j \rangle$ where $p_i = \min\{i, \bar{j}\}$ and $p_j = \min\{j, \bar{i}\}$ if $r_i = r_j = 1$, or $PR(x)$ establishes $\langle 1; i, -1; i \rangle$ if $r_i = 1$ and $r_j = 0$, or $PR(\bar{x})$ establishes $\langle 1; j, -1; j \rangle$ if $r_i = 0$ and $r_j = 1$, $0 \le x \le n - 1$. Then, Move $\langle 1; i, j; p_i \rangle$ to $PR(i)$, and if the third element is "-1" then $PR(i)$ changes the first element to 0, $0 \le i \le n - 1$.

(Step 1, Step 2 and Step 3 constitute the first phase.)

Step 4: $m = m + 1$. If $m \le \lceil \lg k \rceil$ then go to Step 5, else go to Step 8.

Step 5: Duplicate $\langle 1; l, i; p_l \rangle$ and move a copy to $PR(i)$, $0 \le i \le n - 1$. (Those quadruples whose first elements are 1 are duplicated and moved to new processors specified by their third elements.)

Step 6: (The action of each processor depends on the quadruple(s) that it holds.)

(1) when $PR(i)$ holds only a quadruple: If the first element of the quadruple is 1 then $PR(i)$ kills the quadruple, else $PR(i)$ keeps the quadruple intact;

(2) when $PR(i)$ holds $\langle 1; l, i; p_l \rangle$ and $\langle 1; i, j; p_i \rangle$: $PR(i)$ replaces the first quadruple by $\langle 1; l, j; p'_l \rangle$ where $p'_l = \min\{p_i, p_l\}$ and kills the second quadruple;

(3) when $PR(i)$ holds $\langle 1; l, i; p_l \rangle$ and $\langle 0; i, *; j \rangle$ where $*$ is -2 or -1 : $PR(i)$ replaces the first quadruple by $\langle 1; l, *; i \rangle$ and keeps the second quadruple intact.

Step 7: Move $\langle 1; i, j; p_i \rangle$ to $PR(i)$, and if the third element is "-1" or "-2" then $PR(i)$ changes the first element to 0, $0 \le i \le n - 1$. Go to Step 4.

(Step 4, Step 5, Step 6 and Step 7 constitute the second phase.)

Step 8: If $PR(i)$ holds $\langle 0; i, -1; j \rangle$ and $PR(\bar{i})$ holds $\langle 0; \bar{i}, -2; \bar{i} \rangle$ (i.e., input i is the busy input of $SW_{\lfloor i/2 \rfloor}$ which is the semi-busy switch of a half open chain), then representative j of input i is concentrated $PR(0)$, $0 \le i \le n - 1$. Upon receiving the concentrated numbers, say $j_1, j_2, \ldots, j_f$, $PR(0)$ assigns 0 to the first $\lceil f/2 \rceil$ numbers and 1 to the last $\lfloor f/2 \rfloor$ numbers, and broadcasts the sequence, $(j_1, j_2, \ldots, j_f)$, to $PR(i)$, $0 \le i \le n - 1$. Then, the third element of the quadruple held in $PR(i)$ is changed from -1 to 0 if the fourth element is among the first $\lceil f/2 \rceil$ numbers of the broadcast sequence, and from -1 to 1 if the fourth element is among the last $\lfloor f/2 \rfloor$ numbers of the broadcast sequence, $0 \le i \le n - 1$.

(Step 8 constitutes the third phase.)

Step 9: Case 1: If $PR(i)$ holds $\langle 1; i, j; p_i \rangle$ and $PR(\bar{i})$ holds $\langle 1; \bar{i}, l; \bar{p}_i \rangle$, then $PR(i)$ sets $SW_{\lfloor i/2 \rfloor}$ *through* if $i - p_i$ is even and *cross* otherwise, $0 \le i \le n - 1$.

Case 2: If $PR(i)$ holds $\langle 0; i, p; j \rangle$ and $PR(\bar{i})$ holds $\langle 0; \bar{i}, p; l \rangle$ where $p = -1$ or $p = -2$, then they compare j and l. When $j < l$, $PR(i)$ sets $SW_{\lfloor i/2 \rfloor}$ *through* if $i - j$ is even and *cross* otherwise, and when $l < j$, $PR(\bar{i})$ sets $SW_{\lfloor i/2 \rfloor}$ *through* if $\bar{i} - l$ is even and *cross* otherwise, $0 \le i \le n - 1$.

Case 3: If $PR(i)$ holds $\langle 0; i, q; j \rangle$ and $PR(\bar{i})$ holds $\langle 0; \bar{i}, -2; l \rangle$ where $q = 0$ or $q = 1$, $PR(i)$ sets $SW_{\lfloor i/2 \rfloor}$ *through* if $i - q$ is even and *cross* otherwise, $0 \le i \le n - 1$.

(Step 9 constitutes the fourth phase.) ‖

That this algorithm is correct can easily be proved and is omitted for lack of space. In the algorithm, move processes dominate the time complexity, and each move process can be executed in $O(1)$ time if the processors are completely interconnected, and in $O(\lg^2 k + \lg n)$ time if the processors are extended shuffle-exchange interconnected, as shown in Subsection A. For a unicast k-assignment, since there are $O(\lg k)$ move processes in the algorithm, the switches in $H(n)$ can be set in $O(\lg k)$ time if the interprocessor topology is the complete graph, and in $O(\lg^3 k + \lg k \lg n)$ time if the interprocessor topology is extended shuffle-exchange graph. Furthermore, as discussed in the previous section, this algorithm can be recursively applied to set switches in the first half stages. In fact, only the first $\lfloor \lg k \rfloor$ stages would apply this algorithm for a unicast k-assignment since this algorithm decomposes an assignment into two half-sized assignments stage by stage as required in the statement of Theorem 1. Therefore, by using this parallel algorithm, the routing time for an n-input Beneš network to realize a unicast k-assignment becomes $O(\lg^2 k + \lg n)$ if the interprocessor connection topology is complete, and becomes $O(\lg^4 k + \lg^2 k \lg n)$ if the interprocessor connection topology is the extended shuffle-exchange network.

Also, unicast assignments can be pipelined over the stages of the Beneš network by using this parallel routing algorithm. That is, when the switch settings in a stage for a unicast assignment are finished, the switch settings in that stage for another unicast assignment can proceed. Pipelining will reduce the average routing time needed to realize a series of unicast assignments. Suppose that there are α consecutive unicast assignments to be realized over an n-input Beneš network. Without pipelining, the average routing time to realize an assignment is $O(\lg^2 k + \lg n)$ if the interprocessor connection topology is complete, and is $O(\lg^4 k + \lg^2 k \lg n)$ if the interprocessor connection topology is the extended shuffle-exchange network. With pipelining, the total routing time to realize these α assignments reduces to $O(\lg^2 k + \lg n + (\alpha - 1) \lg k)$ for completely connected network, and to $O(\lg^4 k + \lg^2 k \lg n + (\alpha - 1)(\lg^3 k + \lg k \lg n))$ for extended shuffle-exchange network. The average routing time to realize an assignment is reduced to $O(\lg k)$ in the first case, and to $O(\lg^3 k + \lg k \lg n)$ in the second case, for $\alpha \geq \lg n$.

5 CONCLUDING REMARKS

The paper presented a nontrivial extension of Nassimi and Sahni's parallel algorithm for routing unicast assignments in Beneš networks. Unlike the original algorithm of Nassimi and Sahni, the new algorithm can route all unicast assignments (including the permutation assignments) without using dummy outputs. In general, on a parallel processor with a completely connected or extended shuffle-exchange network, the time complexity of this algorithm increases polylogarithmically with the size of the assignments, and it matches the time complexity of Nassimi and Sahni's algorithm for permutation assignments.

References

[1] V. Beneš. *Mathematical Theory of Connecting Networks and Telephone Traffic.* Academic Press Publishing Company, New York, 1965.

[2] C. Cardot. Comments on a simple algorithm for the control of rearrangeable switching networks. *IEEE Transactions on Communications,* COM-34:395, April 1986.

[3] J. Carpinelli and A. Y. Oruç. Applications of edge-coloring algorithms to routing in parallel computers. In *Proceedings of 3rd International Conference on Supercomputing,* Santa Clara, CA, May 1988.

[4] F. R. Hwang. Control algorithms for rearrangeable Clos networks. *IEEE Transactions on Communications,* COM-31:952–954, August 1983.

[5] D. Knuth. *The Art of Computer Programming: Sorting and Searching.* Addison and Wesley, Reading, MA, 1973.

[6] J. Konicek and et al. The organization of the cedar system. In *International Conference on Parallel Processing,* pages 49–56, St. Charles, IL., August 1991.

[7] K. Y. Lee. A new Benes network control algorithm. *IEEE Transactions on Computers,* C-36:768–772, June 1987.

[8] G. F. Le, N. Pippenger, and L. Valiant. A fast parallel routing algorithm for routing in permutation networks. *IEEE Transactions on Computers,* C-30:93–100, February 1981.

[9] D. Nassimi and S. Sahni. A self-routing Benes network and parallel permutation algorithms. *IEEE Transactions on Computers,* C-30:332–340, May 1981.

[10] D. Nassimi and S. Sahni. Parallel algorithms to set up the Benes network. *IEEE Transactions on Computers,* C-31:148–154, February 1982.

[11] D. Opferman and N. Tsao-Wu. On a class of rearrangeable switching networks. *Bell Systems Technical Journal,* 50:1579–1618, May-June 1971.

[12] H. J. Siegel and et al. Using the multistage cube network topology in parallel supercomputers. *Proceedings of the IEEE,* 77:1932–1953, December 1989.

[13] H. Stone. Parallel processing with the perfect shuffle. *IEEE Transactions on Computers,* C-20:153–161, February 1971.

[14] A. Waksman. A permutation network. *Journal of the ACM,* 15:159–163, January 1968.

Optimal Routing Algorithms for Generalized de Bruijn Digraphs[*]

Guoping Liu and Kyungsook Y. Lee
Department of Mathematics and Computer Science
University of Denver
Denver, Colorado 80208

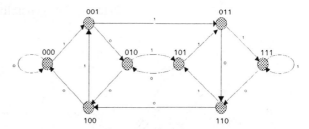

Figure 1: de Bruijn digraph B(2, 3).

Abstract -- *Generalized de Bruijn digraph, $G_B(d, n)$, was proposed independently by Imase and Itoh, and Reddy, Pradhan and Kuhl to include any number of nodes. Du and Hwang showed that $G_B(d, n)$ essentially retains all the nice properties of de Bruijn digraphs, and presented a $\lceil \log_d n \rceil$-step routing algorithm, where d is the indegree and outdegree, and n is the number of nodes in the graph.*

We present optimal routing algorithms for $G_B(d, n)$ and $G_I(d, n)$, a variation of $G_B(d, n)$ proposed by Imase and Itoh. We also present routing algorithms for $G_B(d, n)$ and $G_I(d, n)$ under faulty nodes or links.

Key words: *Generalized de Bruijn daigraphs, Optimal routing algorithms, Fault-tolerant routing, Average internode distance.*

1 Introduction

A de Bruijn digraph B(d, k) [3] consists of d^k nodes, where each node is represented as a k-vector in the d-ary number system. Node $u = (u_{k-1}, \cdots, u_0)$ has an edge to node $v = (v_{k-1}, \cdots, v_0)$ if and only if $v_i = u_{i-1}$ for $1 \le i \le k - 1$. Thus, both the indegree and outdegree of B(d, k) are d. The diameter of B(d, k) is k which is optimal. Due to an optimal diameter and a simple routing algorithm, de Bruijn digraphs have widely been used as a model for communication networks and processor networks. A de Bruijn digraph B(2, 3) is shown in Fig. 1.

de Bruijn digraph was generalized by Imase and Itoh [5] to $G_B(d, n)$ to include any number of nodes n. This generalization was also proposed by Reddy, Pradhan and Kuhl [8] independently. Later Imase and Itoh [6,7] proposed a variation of $G_B(d, n)$, denoted by $G_I(d, n)$, whose diameter is smaller than or equal to that of $G_B(d, n)$.

$G_B(d, n)$ and $G_I(d, n)$ have a low degree and optimal or nearly optimal diameters [5,6]. Routing algorithms were given by Du and Hwang [4], which set up a path of a fixed length $\lceil \log_d n \rceil$. This path length is equal to the diameter of $G_B(d, n)$ and equal to or greater than the diameter of $G_I(d, n)$. In this paper we propose optimal routing algorithms for $G_B(d, n)$ and $G_I(d, n)$, as well as fault-tolerant routing algorithms.

In Section 2, the definitions and the known routing algorithms of $G_B(d, n)$ and $G_I(d, n)$ are briefly reviewed. In Section 3, optimal routing algorithms for $G_B(d, n)$ and $G_I(d, n)$ are presented. A fault-tolerant routing algorithm for $G_B(d, n)$, which can be easily adapted for $G_I(d, n)$, is presented in Section 4. The average internode distances of $G_B(d, n)$ and $G_I(d, n)$ are discussed in Section 5. Conclusions are given in Section 6.

---

[*] This work is supported in part by the National Science Foundation under Grant No. MIP-9210605.

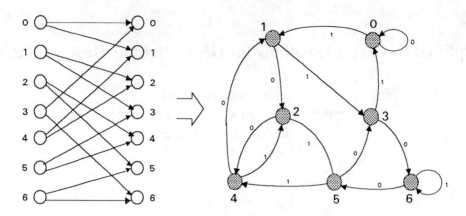

Figure 2: The generalized de Bruijn digraph $G_B(2, 7)$.

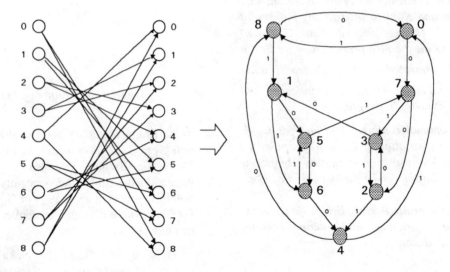

Figure 3: The generalized de Bruijn digraph $G_I(2, 9)$.

2 Generalized de Bruijn digraphs $G_B(d, n)$ and $G_I(d, n)$

2.1 Definitions

Label the n nodes of the generalized de Bruijn digraph $G_B(d, n)$ [5] as 0, 1, 2, $\cdots$, n-1. Node i has a directed link to node j if and only if

$$j = i \cdot d + r \pmod{n}, \quad 0 \le r \le d - 1.$$

The generalized de Bruijn digraph $G_B(2, 7)$ is shown in Fig. 2.

In $G_I(d, n)$ [6], a variation of $G_B(d, n)$, node i has a directed link to node j if and only if

$$j = d(n - 1 - i) + r \pmod{n}, \quad 0 \le r \le d - 1.$$

$G_I(2, 9)$ is shown in Fig. 3.

The diameter of $G_B(d, n)$ was shown to be $\lceil \log_d n \rceil$, which reaches the lower bound if $(d^m - 1)/(d - 1) \le n \le d^m$ [3,1]. The diameter of $G_I(d, n)$ is less than or equal to $\lceil \log_d n \rceil$. It equals m when $n = d^{m-b}(d^b + 1)$, where m is a natural number and $b \le m$ is an odd integer. In this case, the diameter of $G_I(d, n)$ reaches the lower bound. For example, the diameters of $G_I(2, 9)$ and $G_B(2, 9)$ are 3 and 4, respectively.

2.2 Routing algorithms

In the routing algorithms due to Du and Hwang [4], a control tag of length $m = \lceil \log_d n \rceil$, $t = (t_{m-1} \cdots t_0)$, is calculated based on the source node u and the destination node v. The control tag is then used to route a message from the source to the destination.

For given u and v in $G_B(d, n)$, let

$$t = v - u \cdot d^m \pmod{n}$$

be a d-ary number $(t_{m-1} \cdots t_0)$. Then there exists a path of length m from u to v:

$$u = x_m \to x_{m-1} \to \cdots \to x_0 = v$$

such that $x_i = dx_{i+1} + t_i \pmod{n}$ for $i = 0, \cdots, m-1$.

The length of the path generated by this algorithm is always m which is the diameter of the digraph. For example, $t = 110$ for $u = 6$ and $v = 5$ in $G_B(2, 7)$ of Fig. 2. The corresponding path is

$$u = 6 \xrightarrow{1} 6 \xrightarrow{1} 6 \xrightarrow{0} 5 = v.$$

For $u = 1$ and $v = 5$, $t = 100$. The corresponding path is

$$u = 1 \xrightarrow{1} 3 \xrightarrow{0} 6 \xrightarrow{0} 5 = v.$$

In $G_1(d, n)$, if m is even, let

$$t' = v - u \cdot d^m \pmod{n}.$$

If m is odd, let

$$t' = v + (u+1)d^m \pmod{n}.$$

Then there exists a path of length m from u to v:

$$u = x_m \to x_{m-1} \to \cdots \to x_0 = v$$

such that $x_i = d(n-1-x_{i+1}) + t_i \pmod{n}$, for $i = 0, \cdots, m-1$, where $t_i = t_i'$ for i even and $t_i = d - 1 - t_i'$ for i odd.

3 Optimal routing algorithms for $G_B(d, n)$ and $G_1(d, n)$

3.1 Optimal routing algorithm for $G_B(d, n)$

We present an optimal routing algorithm for $G_B(d, n)$ which generates a control tag t of length j, $j \le m$. The control tag corresponds to a shortest path from the source node u to the destination node v.

Algorithm A: Optimal routing algorithm for $G_B(d, n)$;

```
if u = v then
    j = 0
else begin
        j = 1;
        found = FALSE;
        while not found do
        begin
            t = (v - d^j u) (mod n);
            if t < d^j then found = TRUE else j = j + 1;
        end;
    end;
```

With this algorithm, for $u = 6$ and $v = 5$ in $G_B(2, 7)$, we have $j = 1$ and $t = (t_0) = (0)$. The corresponding path of length 1 is

$$u = 6 \xrightarrow{0} 5 = v.$$

For $u = 1$ and $v = 5$, $j = 2$ and $t = (t_1 t_0) = (01)$. The corresponding path of length 2 is

$$u = 1 \xrightarrow{0} 2 \xrightarrow{1} 5 = v.$$

Theorem 1: There exists a path of length j from a source node u to a destination node v, which is controlled by the control tag t of length j generated by Algorithm A.

Proof: Let $t = (t_{j-1} \cdots t_0)$. t_{j-1} is used to control the first step from u to node

$$u \cdot d + t_{j-1} \pmod{n}.$$

After j steps, it reaches the node

$$x = u \cdot d^j + t_{j-1} \cdot d^{j-1} + \cdots + t_1 \cdot d + t_0 \pmod{n}.$$

We need to show that $x = v$.

$v - x$

$$= v - (u \cdot d^j + t_{j-1} \cdot d^{j-1} + \cdots + t_1 \cdot d + t_0) \pmod{n}$$

$$= (v - u \cdot d^j) - (t_{j-1} \cdot d^{j-1} + \cdots + t_1 \cdot d + t_0) \pmod{n}$$

$$= (v - u \cdot d^j) - t \pmod{n}$$

$$= 0.$$

Thus, $x = v$. Q.E.D.

Theorem 2: The control tag t generated by Algorithm A leads to a shortest path.

Proof: Let i be the length of the shortest path from u to v. We have

$$v = u \cdot d^i + t'_{i-1} \cdot d^{i-1} + \cdots + t'_1 \cdot d + t'_0 \pmod{n}$$

where $0 \le t'_l \le d - 1$ for $l = 0, \cdots, i\text{-}1$. Let $t' = t'_{i-1} \cdot d^{i-1} + \cdots + t'_1 \cdot d + t'_0$. Clearly, $t' < d^i$ and $t' = v - u \cdot d^i \pmod{n}$. From Algorithm A, j is the smallest number such that $t = v - u \cdot d^j \pmod{n}$ and $t < d^j$. So, $j \le i$. Therefore, the control tag t generated by Algorithm A leads to a shortest path. Q.E.D.

Theorem 3: Let $m = \lceil \log_d n \rceil$ and $dist(u, v)$ be the distance from u to v. If $n < d^m$ and $dist(u, v) = m$, there are at most d shortest paths from u to v. Otherwise, the shortest path from u to v is unique.

Proof: Let $u = x_j \to x_{j-1} \to \cdots \to x_0 = v$ be a shortest path from u to v. Let $t = (t_{j-1} \cdots t_0)$ be the control tag corresponding to the shortest path. We have

$$x_{j-1} = u \cdot d + t_{j-1} \pmod{n}$$

and

$$v = u \cdot d^i + t'_{i-1} \cdot d^{i-1} + \cdots + t'_1 \cdot d + t'_0 \pmod{n}$$

$$= u \cdot d^j + t \pmod{n}.$$

Let $u = y_j \to y_{j-1} \to \cdots \to y_0 = v$ be another shortest path from u to v. Let $t' = (t'_{j-1} \cdots t'_0)$ be the corresponding control tag. So,

$$v = u \cdot d^j + t' \pmod{n}.$$

Therefore,

$$t = t' \pmod{n}.$$

Case 1: $j < m$. We have $t < n$ and $t' < n$. So, $t = t'$.

Case 2: $j = m$ and $n = d^m$. We also have $t < n$, $t' < n$ and $t = t'$.

Case 3: $j = m$ and $n < d^m$. Let $\bar{t} = t + i \cdot n$, where i is an integer. We have

$$u \cdot d^m + \bar{t} \pmod{n}$$

$$= u \cdot d^m + t + i \cdot n \pmod{n}$$

$$= u \cdot d^m + t \pmod{n}$$

$$= v.$$

If $\bar{t} < d^m$, it can be used as a control tag which leads to a path of length m from u to v. Since there are at most d different i's such that $0 \le t + i \cdot n < d^m$, there are at most d different paths of length m from u to v. Q.E.D.

If $n = d^k$, Algorithm A can be simplified as follows (where $u = u_{k-1} \cdots u_1 u_0$ and $v = v_{k-1} \cdots v_1 v_0$).

Algorithm A': Optimal routing algorithm for $G_B(d, n)$, $n = d^k$;

```
if u = v then
    j = 0
else  begin
        j = 1;
        while (j < k) and ((u_{k-j-1} ··· u_0) <> (v_{k-1} ··· v_j)) do
            j = j + 1;
        t = v_{j-1} ··· v_0;
    end;
```

3.2 Optimal routing algorithm for $G_I(d, n)$

A similar algorithm exists for $G_I(d, n)$, which generates a control tag t of length j which leads to a shortest path from u to v in $G_I(d, n)$. .

Algorithm B: Optimal routing algorithm for $G_I(d, n)$;

```
if u = v then
    j = 0
else begin
        j = 1;
        found = FALSE;
        while not found do
        begin
            if j is odd then
                t' = v + (u + 1)d^j (mod n)
            else
                t' = v - u · d^j (mod n);
            if t' < d^j then
            begin
                found = TRUE;
                for i = 1 to j do
                    if i odd then
                        t_i = d - 1 - t'_i
                    else
                        t_i = t'_i;
            end else
```

$$j = j + 1;$$
$$\text{end};$$
$$\text{end};$$

For example, to route a message from node $u = 6$ to node $v = 2$ in $G_I(2, 9)$, we have $j = 3$ and $t = (t_2 t_1 t_0) = (110)$. The corresponding path is

$$u = 6 \xrightarrow{\ 1\ } 5 \xrightarrow{\ 1\ } 7 \xrightarrow{\ 0\ } 2 = v.$$

For $u = 6$ and $v = 4$, $j = 1$ and $t = (t_0) = (0)$. The corresponding path is

$$u = 6 \xrightarrow{\ 0\ } 4 = v.$$

For $u = 6$ and $v = 8$, $j = 2$ and $t = (t_1 t_0) = (00)$. The corresponding path is

$$u = 6 \xrightarrow{\ 0\ } 4 \xrightarrow{\ 0\ } 8 = v.$$

4 Fault-tolerant routing in $G_B(d, n)$ and $G_I(d, n)$

In this section we present a fault-tolerant routing algorithm for $G_B(d, n)$ based on the depth-first search [2]. We assume that each node has the knowledge of all the faulty links in the graph. A faulty node can be treated as the case in which all links of the faulty node are faulty. Since the algorithm is a specialized case of the depth-first search, it can always find a path from a source node u to a destination node v if u is connected to v (i.e., if there is a directed path from u to v). In Algorithm C given below, GetShortestPath is either Algorithm A or Algorithm A' depending upon the size of the graph. Function GetPath in Algorithm C generates a control tag t and returns TRUE if there is a path from u to v. It visits each node at most once. So, the length of the path is less than or equal to n-1.

Algorithm C: Fault-tolerant routing in $G_B(d, n)$;

Function GetPath(u, v, t): Boolean;
begin
   GetShortestPath(u, v, t);
      /* where $t = (t_{j-1} t_{j-2} \cdots t_0)$ */

Let $u = x_j \rightarrow x_{j-1} \rightarrow \cdots \rightarrow x_0 = v$
   be the shortest path controlled by t;
if all the links on the path are not faulty then
   return TRUE   /* found a path */
else begin
   Let i be the largest number such that

      $x_{i-1} \in SeenList$ or link $x_i \rightarrow x_{i-1}$ is
      faulty;
   put $x_j, \cdots, x_i$ into $SeenList$;
   $done = $ FALSE;
   while (not $done$) and ($i <= j$) do
   begin
      $count = 1$;
      while (not $done$) and ($count <= d$) do
      begin
         $count = count + 1$;
         $t_i = t_i + 1 \pmod{d}$;
         if GetPath($x_i \cdot d + t_i \pmod{n}$, v,
            t') then
            $done = $ TRUE;
      end;
      if $done$ then   /* found a path from
                  x_i to v */

        $t = concatenate(t_{j-1} t_{j-2} \cdots t_i, t')$
      else
         $i = i + 1$;   /* Back tracking */
   end;

   return $done$;
   end;
end;

begin /* main */
   $SeenList = []$;   /* initialize the set to be empty */
   if GetPath(u, v, t) then
      control tag t is used to route the message
   else
      there is no path from u to v;
end.

For example, assume that link (1, 3) is faulty in $G_B(2, 7)$. To send a message from node 1 to node 6, by Algorithm C we have $t = (0100)$. The corresponding path is

$$1 \xrightarrow{\ 0\ } 2 \xrightarrow{\ 1\ } 5 \xrightarrow{\ 0\ } 3 \xrightarrow{\ 0\ } 6.$$

Theorem 4: Let u and v be two nodes in $G_B(d, n)$ with f faulty links. Let $l = \text{dist}(u, v)$ be the distance from u to v. The probability that an optimal path from u to v can be taken is

$$p_{opt} = \left(\frac{dn - f}{dn} \right)^l.$$

Proof: There are total dn links among which f links are faulty. We assume that these f faulty links are randomly distributed. The probability that a link is not faulty is $(dn - f)/(dn)$. There are l links on the optimal path from u to v. The probability that there are no faulty links on this path, p_{opt}, is

$$p_{opt} = \left(\frac{dn - f}{dn} \right)^l. \qquad \text{Q.E.D.}$$

For example, let $n = 128$, $d = 2$ and $f = 1$. The probability that a particular link is the faulty link is 0.004. The average internode distance is 5.46 (see Table 1). So, on average the probability that the optimal path can be taken is $(1 - 0.004)^{5.46} = 0.978$. If $f = 2$, the probability that the optimal path can be taken is 0.958. This shows that the probability of taking an optimal path is high when few links are faulty.

Algorithm C can easily be modified for $G_I(d, n)$; Algorithm B should be used for GetShortestPath, and $d(n - 1 - x_i) + t_i$ (mod n) should be used to generate the node address to which x_i is connected.

5 Average internode distances for $G_B(d, n)$ and $G_I(d, n)$

As mentioned before, generalized de Bruijn digraphs G_B and G_I are optimal or nearly optimal in diameter. In this section we examine their average internode distances. It is difficult to express the average internode distances for G_B and G_I explicitly. However, since Algorithms A and B always find the shortest path from any node u to any node v in G_B and G_I, respectively, we can use these algorithms to compute the average internode distances for G_B and G_I. Imase and Itoh [5] calculated the average internode distance of G_B by using adjacent matrices for up to $n = 100$. By using the new method we can calculate the average distances of $G_B(d, n)$ and $G_I(d, n)$ for larger n (e.g., $n = 1024$). For very large n, we present an approximate formula later in this section. The diameters and average internode distances of $G_B(2, n)$ and $G_I(2, n)$ are shown in Fig. 4 and Fig. 5, respectively. The average internode distances of $G_B(d, n)$ are shown in Fig. 6 for $d = 2$, 3 and 4. It can be seen that the average internode distances of $G_B(4, n)$ are about one half the average distances of $G_B(2, n)$.

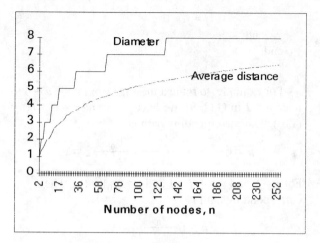

Fig. 4: Diameter and average internode distance of $G_B(2, n)$.

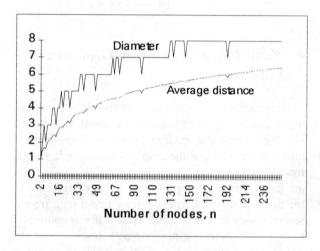

Fig. 5: Diameter and average internode distance of $G_I(2, n)$.

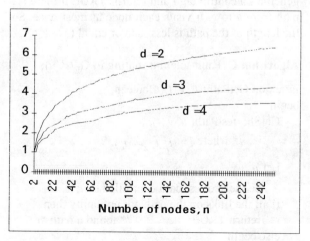

Fig. 6: Average internode distance of $G_B(d, n)$.

The average distances of $G_B(2, n)$ and $G_I(2, n)$ from Figures 4 and 5 are superimposed in Fig. 7. It can be seen that G_B and G_I have almost identical average internode distances except the cases of $n = d^{m-b}(d^b + 1)$, where G_I has slightly lower average internode distances. Recall that these are the cases when G_I has the optimal diameter, while G_B does not [6].

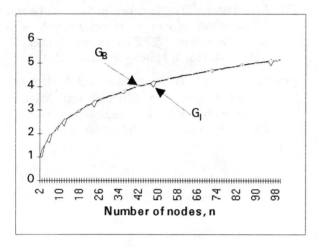

Fig. 7: Average internode distances of $G_B(2, n)$ and $G_I(2, n)$.

To estimate the average internode distance for $n = d^k$, we assume that a node sends a message to all other nodes in the graph with an equal probability. Consider any node pair $u = u_{k-1} \cdots u_1 u_0$ and $v = v_{k-1} \cdots v_1 v_0$. The probability that

$$u_{k-1} \cdots u_1 u_0 = v_{k-1} \cdots v_0 \qquad (1)$$

is $1/d^k$. We denote this probability by the function $p(d, k, k)$. In this case the source is the destination. The probability that

$$u_{k-2} \cdots u_1 u_0 = v_{k-1} \cdots v_1 \qquad (2)$$

is $1/d^{k-1}$. The probability that u and v satisfy (2) but not (1) is $1/d^{k-1}(1-1/d^k)$, which is denoted by the function $p(d, k, k\text{-}1)$. One step is enough to route a message from u to v in this case. Continuing this process, we can estimate the average internode distance of $G_B(d, d^k)$ and $G_I(d, d^k)$, AD_E, as

$$AD_E = \frac{d^k}{d^k - 1} \sum_{j=0}^{k} j \cdot p(d,k,k-j), \qquad (3)$$

where

$$p(d,k,i) = \frac{1}{d^k}, \qquad \text{if } i = k,$$

$$p(d,k,i) = \frac{1}{d^i}\left(1 - \sum_{m=i+1}^{k} p(d,k,m)\right), \qquad \text{if } 0 \le i < k.$$

The estimated average internode distances AD_E by (3) and the actual average internode distances AD for $G_B(2^k, 2)$ are compared in Table 1. It can be seen that the estimated values are very close to the actual values. The differences between AD_E and AD are less than 1% of AD.

Table 1: Estimated and actual average internode distances of $G_B(2^k, 2)$.

Size n	Estimated AD_E	Actual AD	$\dfrac{AD_E - AD}{AD}$
8	2.125	2.107	0.85%
16	2.860	2.833	0.95%
32	3.681	3.649	0.88%
64	4.565	4.532	0.73%
128	5.494	5.460	0.62%
256	6.451	6.417	0.53%
512	7.427	7.392	0.47%
1024	8.411	8.377	0.41%

6 Conclusion

The generalized de Bruijn digraphs $G_B(d, n)$ and $G_I(d, n)$ allow any number of nodes to be included in the graph, while preserving desirable properties of de Bruijn digraphs such as a small degree and diameter, and a small average internode distance.

We have presented optimal routing algorithms and fault-tolerant routing algorithms for $G_B(d, n)$ and $G_I(d, n)$. Fault-tolerant routing algorithms were obtained by use of the optimal routing algorithms. An approximate equation for the average internode distance of $G_B(d, d^k)$ and $G_I(d, d^k)$ has also been derived. The average internode distance obtained from the equation has been shown to be accurate within a 1% error range.

References

[1] W. G. Bridges, and S. Toueg, "On the Impossibility of Directed Moore Graphs," *Journal of Combinational Theory*, Series B 29, 1980, pp. 339-341.

[2] M.-S. Chen, K. G. Shin, "Depth-first Search Approach for Fault-tolerant Routing in Hypercube Multicomputers," *IEEE Trans. on Parallel and Distributed Systems*, Vol. 1, No. 2, April 1990, pp. 152-159.

[3] N. G. deBruijn, "A Combinatorial Problem," *Proc. Akademe Van Weteschappen*, 1946, Vol. 49, part 2, pp. 758-764.

[4] D. Z. Du and F. K. Hwang, "Generalized de Bruijn Digraphs," *Networks*, Vol. 18, 1988, pp. 27-38.

[5] M. Imase and M. Itoh, "Design to Minimize Diameter on Building-block Network," *IEEE Trans. on Computers*, Vol. C-30, No. 6, June 1981, pp. 439-442.

[6] M. Imase and M. Itoh, "A Design for Directed Graphs with Minimal Diameter," *IEEE Trans. on Computers*, Vol. C-32, No. 8, Aug. 1981, pp. 782-784.

[7] M. Imase, T. Soneka and K. Okada, "Connectivity of Regular Directed Graphs with Small Diameters," *IEEE Trans. on Computers*, Vol. C-34, No. 3, March 1985, pp. 267-273.

[8] S. M. Reddy, D. K. Pradhan and J. G. Kuhl, "Direct Graphs with Minimal and Maximal Connectivity," School of Engineering, Oakland University, Tech. Rep., July 1980.

A Class of Partially Adaptive Routing Algorithms for n_dimensional Meshes*

Younes M. Boura and Chita R. Das
Department of Electrical and Computer Engineering
The Pennsylvania State University
University Park, PA 16802
E-mail : {boura,das}@cmpe.psu.edu

Abstract

A simple model, called the direction restriction model, for developing partially adaptive routing algorithms for n_dimensional meshes is introduced in this paper. This model is based on dividing a system into two unidirectional networks that contain all physical channels of the system. Delivery of a message is broken into two phases. In the first phase, a message is routed adaptively to an intermediate node using one unidirectional network. In the second phase, the message is routed adaptively to its destination in the remaining network. Messages are routed by using the wormhole switching paradigm. All of the algorithms that are developed by using this model are proven to be deadlock free at no extra cost in terms of virtual channels. Routing algorithms are developed for 2_dimensional meshes, 3_dimensional meshes, and hypercubes. The effectiveness of these algorithms for different traffic patterns is evaluated via simulation. The ease of use and simplicity of the direction restriction model make it an attractive tool for developing partially adaptive routing algorithms for n_dimensional meshes.

1 Introduction

Recent research on interconnection networks for distributed and shared memory multiprocessor architectures has focused on direct networks. n_dimensional meshes represent a class of scalable direct interconnection networks that have become popular for building large scale multiprocessors. An n_dimensional mesh is a generic topology where each dimension could have an arbitrary number of elements. It is a generalization of the binary hypercube. Parallel machines that have adopted n_dimensional meshes as the underlying topology include MPP [1],

Tera computer system [2], Touchstone Delta system [3], K2 [4], DASH [5], ALEWIFE [6], iPSC [7], and N_CUBE [8].

A wormhole routing algorithm is designed with the objective of maximizing the utilization of channels in the network. It could be either deterministic or adaptive. Deadlock and livelock freedom are two important characteristics of routing algorithms that contribute to achieving such objective. One deterministic routing algorithm that is proven to be deadlock free for n_dimensional meshes is the *dimension-order* algorithm (*e-cube routing, xy routing*) [16]. This algorithm routes messages deterministically by traversing dimensions in a predetermined order. However, because of the inherent determinism in the algorithm, the maximum bandwidth of the network is not achievable. Some links in the network might be overused and others underused depending on the traffic pattern in the network. In order to distribute the traffic evenly among the channels and maximize the channel utilization in the network, a routing algorithm should have some degree of adaptiveness.

The concept of virtual channels was introduced in [9] for the prevention of deadlock in a variety of network architectures. Virtual channels are implemented by using either physical channels or buffers. In [10], virtual channels were used for designing deadlock-free fully adaptive routing algorithms. These routing algorithms require $(n + 1) \times 2^{n-1}$ and 2^{n-1} virtual channels for n_dimensional meshes with and without wraparound connections, respectively. This high virtual channel requirement for making routing algorithms fully adaptive motivated researchers to develop partially adaptive routing algorithms. *Planar adaptive routing* is one such scheme [15]. This algorithm routes a message adaptively along two dimensions (a plane) at a time until the message gets to its destination. Deadlock and livelock problems are solved by using 6

*This research was supported in part by the National Science Foundation under grant MIP-9104485.

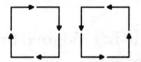

Figure 1. Possible turns in a 2D mesh.

and 3 virtual channels in n-dimensional meshes with and without wraparound connections, respectively.

In [12], the *turn model* was introduced for developing deadlock and livelock free partially adaptive routing algorithms. A turn is defined as a message's ability to move from one dimension to another. Figure 1 shows the possible turns that a message can make in a 2D mesh. The model is used to show that the removal of only a quarter of the turns results in deadlock and livelock free partially adaptive routing algorithms that do not require any virtual channels. No systematic procedure was introduced for finding the right turns that should be prohibited. As the number of dimensions increases, the number of possible turns and cycles increases as well and finding the right turns becomes more difficult. For example, in a 3D mesh, there are 24 possible turns that form 6 cycles. There are 4096 different ways to prohibit six turns (a quarter of 24), but only 176 prevent deadlock [13]. The process of determining the specific turns that should be prohibited to avoid deadlock becomes very complex as the number of dimensions increases.

In this paper, a simple model, called the *direction restriction model*, is introduced for developing partially adaptive routing algorithms for n-dimensional meshes. The model is based on dividing the routing process into two phases. In the first phase, a message is routed to an intermediate node by traversing dimensions in specified directions (positive or negative). In the second phase, a message is delivered to its destination by traversing dimensions in the opposite directions of the first phase. We call it *direction restriction model* because a message is restricted in each phase of the routing process to traverse dimensions in specified directions (positive or negative). Adaptiveness of the routing is attributed to the flexibility in choosing any available channel that gets the message closer to its destination in each phase. The routing is partially adaptive since a message is routed adaptively between the source node and an intermediate node in the first phase, and in the second phase, the message is routed adaptively between the intermediate node and the destination node.

The *direction restriction model*, like the *turn model*,

does not require any virtual channels for preventing deadlock. It is shown that there are 2^n possible adaptive routing schemes for an n-dimensional mesh. All the routing algorithms are deadlock-free. Routing in 2D meshes, 3D meshes, and n-cubes is discussed using this model. Different traffic patterns are simulated to analyze the effectiveness of the various routing alternatives. It is shown that these partially adaptive schemes could provide significant performance improvement compared to the deterministic routing algorithm for nonuniform traffic patterns.

Necessary notations and definitions are introduced in Section 2. Section 3 describes the model for developing partially adaptive routing algorithms. Section 4, deals with the development of routing algorithms for specific networks (2D mesh, 3D mesh, and hypercubes). Performance evaluation of the algorithms for different traffic patterns is discussed in Section 5. Finally, conclusions are drawn in Section 6.

2 Preliminaries

An n-dimensional mesh is defined formally as an interconnection structure that has $K_0 \times K_1 \times \ldots \times K_{n-1}$ nodes. n is the number of dimensions of the network, and K_i is the radix for dimension i. Each node is identified by an n-coordinate vector $(x_0, \ldots, x_{n-1})$, where $0 \leq x_i \leq K_i - 1$. Two nodes, $X(x_0, \ldots, x_{n-1})$ and $Y(y_0, \ldots, y_{n-1})$ are connected if and only if there exists an i such that $x_i = y_i \pm 1$, and $x_j = y_j$ for all $j \neq i$.

Definition 1 : An interconnection network is a strongly connected graph, $IN(N, C)$, where N represents the set of processing nodes and C represents the set of communication channels.

Definition 2 : A routing function, $R : C \times N \to C$, maps the current channel c_c and destination node n_d to the next channel c_n on the route from c_c to n_d. A channel is not allowed to route to itself, $c_c \neq c_n$ [9].

Definition 3 : A channel dependency graph, CDG, of an interconnection network IN and routing function R, is a directed graph, $CDG(C, E)$, where the set of vertices, C, represents the set of channels of IN, and the set of edges, E, represents the set of pairs of channels connected by R [9]. Thus,

$$E = \{(c_i, c_j) | R(c_i, n) = c_j \quad for \ some \ n \in N\}$$

Theorem 1 : A routing function R for an interconnection network is deadlock free, iff there are no cycles in CDG . In other words, R is deadlock free, iff CDG is acyclic [9].

Definition 4 : A message is said to move in the positive direction along dimension i , if it can move from node $X(x_0,\ldots,x_{n-1})$ to node $Y(y_0,\ldots,y_{n-1})$ where $x_i \leq y_i$.

Definition 5 : A message is said to move in the negative direction along dimension i, if it can move from node $X(x_0,\ldots,x_{n-1})$ to node $Y(y_0,\ldots,y_{n-1})$ where $x_i > y_i$.

Definition 6 : A network, IN, where messages move along any dimension in only one direction (positive or negative) is identified by a set of n integers as $IN_{(d_0,\ldots,d_{n-1})}$ where

$$d_i = \begin{cases} 0 & \text{if the direction is positive} \\ K_i - 1 & \text{if the direction is negative.} \end{cases}$$

As an example, Figure 2 shows two 4×4 2-dimensional meshes $IN_{(0,3)}$ and $IN_{(3,0)}$. In the first network, messages move in the positive direction along the zeroth dimension (x-axis), and in the negative direction along the first dimension (y-axis). In the second network, messages move in the negative direction along the zeroth dimension (x-axis) and in the positive direction along the first dimension (y-axis).

Definition 7 : The complement of d_i, $\overline{d_i}$, is defined as

$$\overline{d_i} = K_i - 1 - d_i.$$

Definition 8 : Two networks $IN_{(x_0,\ldots,x_{n-1})}$ and $IN_{(y_0,\ldots,y_{n-1})}$ are said to be complementary iff $x_i = \overline{y_i}$. Assuming that two unidirectional channels exist between any two connected nodes, any two complementary networks contain all the physical channels of the system. For example, the two networks in Figure 2 represent two complementary networks.

Definition 9 : The base K exclusive-or of two integers i and j, where $0 \leq i \leq K-1$ and $j = 0$ or $j = K-1$, is defined as follows :

$$i \oplus_K j = \begin{cases} i & \text{if } j = 0 \\ K-1-i & \text{if } j = K-1. \end{cases}$$

We use Λ to denote the digitwise base K_i exclusive-or of two strings of integers, $(x_0,\ldots,x_{n-1})$ and $(y_0,\ldots,y_{n-1})$, where $0 \leq x_i \leq K_i - 1$ and $y_i = 0$ or $y_i = K_i - 1$. Hence, if $(x_0,\ldots,x_{n-1})\Lambda(y_0,\ldots,y_{n-1}) = (z_0,\ldots,z_{n-1})$, then $z_i = x_i \oplus_{K_i} y_i$. For example, $(0,4,7,3,2) \Lambda (0,7,0,7,7) = (0,3,7,4,5)$.

3 The Direction Restriction Model

The *direction restriction model* is based on the concept of routing adaptively in two unidirectional networks that make up all of the channels in the system (complementary networks). The routing process is divided into two phases where in the first phase messages are forwarded to intermediate nodes in one

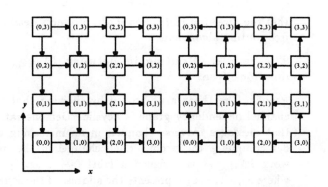

Figure 2. Two unidirectional networks in a 2D mesh a) $IN_{(0,3)}$ b) $IN_{(3,0)}$.

unidirectional network, and in the second phase messages are delivered to their destinations in the second unidirectional network. In each phase, the message routing is performed adaptively. Routing to an intermediate node gives rise to the partial adaptivity of the algorithm. Depending on the selection of the unidirectional networks, different algorithms are produced. The algorithms could be either minimal or nonminimal. We consider only minimal algorithms in this paper.

The following is a detailed description of the model. The description is presented in a procedure-like manner and shows the steps that need to be followed in order to develop partially adaptive routing algorithms.

1. Divide the network into two complementary networks. In an n-dimensional mesh, there are 2^{n-1} different pairs of complementary networks. (This is discussed with examples in section 4).

2. For a pair of complementary networks, identify one as the primary, and the other as the secondary network.

3. Messages can move among complementary networks in a unidirectional manner. For example, a message can only move from the primary to the secondary network.

4. Routing is accomplished in two phases.

   phase 1 : Route adaptively in the primary network by using any available links that get messages closer to their destinations.

   phase 2 : Route adaptively in the secondary network to deliver messages to their destinations.

Note that messages may use only one phase of the routing procedure to get to their destinations (either phase 1 or phase 2). The following theorems prove that the *direction restriction model* produces deadlock free routing algorithms.

Theorem 2: Routing in network $IN_{(0,...,0)}$ is deadlock free.

Proof : The theorem is proved by showing that the channel dependency graph is acyclic. Demonstrating that messages traverse channels in an increasing order completes the proof. Each channel in the network $IN_{(0,...,0)}$ is assigned a label $(s_0,...,s_{n-1},d)$, where $s_0,...,s_{n-1}$ represents the address of the source node of the channel and d is the dimension in which the channel exists. A message traverses dimension i (for $0 \leq i \leq n-1$) by going from node $(x_0,...,x_i,...,x_{n-1})$ to node $(x_0,...,y_i,...,x_{n-1})$ in network $IN_{(0,...,0)}$ iff $y_i = x_i + 1$. Since messages could move only in the positive direction

$$x_0,...,x_i,...,x_{n-1},i <$$
$$x_0,...,x_i+1,...,x_{n-1},i \qquad (1)$$

$$x_0,...,x_{i-l},...,x_{n-1},i <$$
$$x_0,...,x_{i-l}+1,...,x_{n-1},i-l,$$
$$for \quad 0 < l \leq i \qquad (2)$$

$$x_0,...,x_{i+l},...,x_{n-1},i <$$
$$x_0,...,x_{i+l}+1,...,x_{n-1},i+l,$$
$$for \quad 0 < l \leq n-1-i \qquad (3)$$

According to the above three inequalities and Theorem 1, a message can move in the positive direction along any dimension without introducing deadlock.
Q.E.D.

Theorem 3 : Routing in any unidirectional network $IN_{(d_0,...,d_{n-1})}$ is deadlock free.

Proof : The proof is based on demonstrating that $IN_{(d_0,...,d_{n-1})}$ and $IN_{(0,...,0)}$ are topologically equivalent by relabeling the nodes and the channels. The relabeling procedure and Theorem 1 complete the proof. The node relabeling procedure is accomplished by performing the base K_i exclusive-or operation $(s_0,...,s_{n-1})\Lambda(d_0,...,d_{n-1})$, where $(s_0,...,s_{n-1})$ is the address of a node in the network. This operation changes a node address along dimension i to $K_i - 1 - s_i$, if $d_i = K_i - 1$. Thus by transforming the coordinate system such that a permissible message transfer in the negative direction along dimension i in $IN_{(d_0,...,d_{n-1})}$ becomes a permissible message transfer in the positive direction, the resulting labeling produces a network that is topologically equivalent to $IN_{(0,...,0)}$. The channels of $IN_{(d_0,...,d_{n-1})}$ are

labeled using the newly generated addresses. Since $IN_{(d_0,...,d_{n-1})}$ and $IN_{(0,...,0)}$ are topologically equivalent, Theorem 1 leads to the conclusion that routing in $IN_{(d_0,...,d_{n-1})}$ is deadlock free. Q.E.D.

Theorem 4 : A routing algorithm produced by the *direction restriction model* is deadlock free.

Proof : Since messages can only move from a primary unidirectional network to a secondary unidirectional network, no cycles are created between the two networks. Theorem 3 indicates that no cycles are created in any of the two unidirectional networks. Therefore, a routing algorithm that satisfies the *direction restriction model* is deadlock free. Q.E.D.

A class of partially adaptive routing algorithms for n-dimensional meshes is defined by the *direction restriction model*. Since there are 2^{n-1} different pairs of complementary networks in an n-dimensional mesh, the number of these algorithms is equal to 2^n. Theorem 4 proves that all of these algorithms are deadlock free. In addition, the algorithms are also livelock free since they are minimal. We label a routing algorithm R by a string of digits that corresponds to the primary network used in the first phase of the routing process. For example, if a routing algorithm R uses $IN_{(d_0,...,d_{n-1})}$ as the primary network, then it is identified as $R_{(d_0,...,d_{n-1})}$.

4 Partially adaptive routing in n_dimensional meshes

In this section, we develop routing algorithms for 2D meshes, 3D meshes, and n-cubes.

4.1 A Partially adaptive routing in 2_dimensional meshes

In a 2_dimensional mesh, there are 2^2 different unidirectional networks, $IN_{(0,0)}$, $IN_{(0,K_1-1)}$, $IN_{(K_0-1,0)}$, and $IN_{(K_0-1,K_1-1)}$. According to definition 8, $IN_{(0,0)}$ and $IN_{(K_0-1,K_1-1)}$ are complementary networks, and so are $IN_{(K_0-1,0)}$ and $IN_{(0,K_1-1)}$. Therefore, four partially adaptive routing algorithms exist for a 2_dimensional mesh.

Corollary 1 : The partially adaptive routing algorithms $R_{(0,0)}$, $R_{(K_0-1,K_1-1)}$, $R_{(0,K_1-1)}$, and $R_{(K_0-1,0)}$ are deadlock free.

Proof: Corollary 1 follows from Theorem 4. Q.E.D.

Figures 3 and 4 show the labels of channels of two routing algorithms, $R_{(0,0)}$ and $R_{(0,3)}$ for a 4×4 mesh. Channels belonging to $IN_{(d_0,...,d_{n-1})}$ have a label equal to $(s_0,...,s_{n-1})\Lambda(d_0,...,d_{n-1}),dim,p$ where $(s_0,...,s_{n-1})$ is the address of the source node

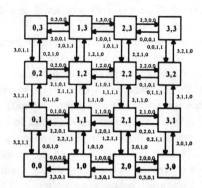

Figure 3. Labeling of channels for algorithm $R_{(0,0)}$ in a 2D mesh.

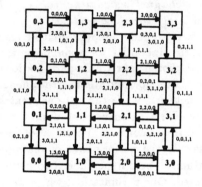

Figure 4. Labeling of channels for algorithm $R_{(0,3)}$ in a 2D mesh.

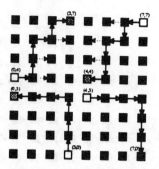

Figure 5. Routing examples of $R_{(0,0)}$ in an 8 × 8 mesh.

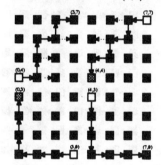

Figure 6. Routing examples of $R_{(7,7)}$ in an 8 × 8 mesh.

of the channel, *dim* is the dimension in which the channel resides, and $p = 0$ if the network $IN_{(d_0,...,d_{n-1})}$ is the primary network, otherwise $p = 1$. For example, in Figure 4 the label of the channel leaving *node* $(2, 2)$ and entering *node* $(1, 2)$ is calculated as follows. Since the routing algorithm is $R_{(0,3)}$, the primary unidirectional network is $IN_{(0,3)}$ and the secondary unidirectional network is $IN_{(3,0)}$. The channel of concern belongs to the secondary unidirectional network $IN_{(3,0)}$ because it resides in the zeroth dimension and its direction is negative. The source node of the channel is *node* $(2, 2)$. Therefore, $(s_0, s_1) = (2, 2)$, $(d_0, d_1) = (3, 0)$, $dim = 0$, and $p = 1$. The label becomes equal to $(2, 2)\Lambda(3, 0), 0, 1 = 3 - 2, 2, 0, 1 = 1, 2, 0, 1$. It is easily seen that messages traverse channels in an increasing order to reach their destinations (Theorem 3 and Theorem 4).

In the following, we describe the partially adaptive routing algorithms for an 8 × 8 2-dimensional mesh. The possible networks that could be used in the development of such algorithms are $IN_{(0,0)}$, $IN_{(7,7)}$, $IN_{(0,7)}$, and $IN_{(7,0)}$. $IN_{(0,0)}$ and $IN_{(7,7)}$ are two complementary networks that contain all the physi-

cal channels in the system. By making $IN_{(0,0)}$ the primary network and $IN_{(7,7)}$ the secondary network, we obtain one algorithm $R_{(0,0)}$. Similarly, by making $IN_{(7,7)}$ the primary network and $IN_{(0,0)}$ the secondary network, we obtain a different algorithm $R_{(7,7)}$. $IN_{(0,7)}$ and $IN_{(7,0)}$ represent a pair of complementary networks covering all the physical channels in the system. Hence, two routing algorithms are produced by using these two networks. If $IN_{(0,7)}$ is selected as the primary network and $IN_{(7,0)}$ is selected as the secondary network, the adaptive routing algorithm $R_{(0,7)}$ is produced. On the other hand, the selection of $IN_{(7,0)}$ as the primary network and $IN_{(0,7)}$ as the secondary network, results in a different routing algorithm $R_{(7,0)}$. Figures 5, 6, 7, and 8 show routing examples in the system according to algorithms $R_{(0,0)}$, $R_{(7,7)}$, $R_{(0,7)}$, and $R_{(7,0)}$ respectively. In the figures, source and destination nodes of messages are represented by white and shaded squares, respectively. Busy channels are indicated by dotted arrows.

4.2 Partially adaptive routing in 3-dimensional meshes

In a 3-dimensional mesh, there are 2^3 different unidirectional networks. Four of these networks, $IN_{(0,0,0)}$, $IN_{(0,0,K_2-1)}$, $IN_{(0,K_1-1,0)}$, and $IN_{(0,K_1-1,K_2-1)}$, are the complements of the others. Therefore, 8 partially adaptive routing algorithms exist for a 3-dimensional mesh. Our simulations indicate

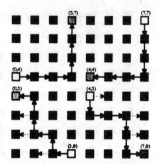

Figure 7. Routing examples of $R_{(0,7)}$ in an 8×8 mesh.

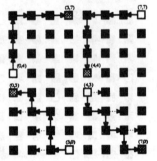

Figure 8. Routing examples of $R_{(7,0)}$ in an 8×8 mesh.

that four of these algorithms behave almost similarly to the other remaining four. Therefore, in the corollary below, we consider only half the algorithms.

Corollary 2 : The partially adaptive routing algorithms $R_{(0,0,0)}$, $R_{(0,0,K_2-1)}$, $R_{(0,K_1-1,0)}$, **and** $R_{(0,K_1-1,K_2-1)}$ **are deadlock free.**

Proof: Corollary 2 follows from Theorem 4. Q.E.D.

4.3 Partially adaptive routing in hypercubes

For an n-dimensional hypercube, 2^n different unidirectional networks could be identified. Half of those networks are the complements of the other half. Hence, 2^n different partially adaptive routing algorithms exist. Three of them are selected randomly here for the purpose of demonstrating the effectiveness of these routing strategies.

Corollary 3 : The partially adaptive routing algorithms $R_{(0,...,0)}$, $R_{(0,K_1-1,...,K_{n-2}-1,0)}$, **and**

$$R_{\underbrace{(0,...,0}_{\lceil \frac{n}{2} \rceil},\underbrace{K_{\lceil \frac{n}{2} \rceil}+1-1,...,K_{n-1}-1}_{\lfloor \frac{n}{2} \rfloor})} \text{ are deadlock}$$

free.

Proof: Corollary 3 follows from Theorem 4. Q.E.D.

5 Experimental results

Extensive simulations are conducted to analyze the performance of these newly generated routing al-

gorithms. The systems are an 8×8 2-dimensional mesh and a 6-dimensional hypercube. Two unidirectional channels are assumed to exist between two connected nodes in the two systems. All of the buffers in the routing hardware are assumed to be a single flit wide. Processors at each node generate messages at time intervals selected from a negative exponential distribution. Messages are assumed to be 20 flits long. Conflicts of requests for the same output channel by multiple messages are resolved by using a local first-come-first-served scheduling policy inside each router (*input channel scheduling policy*). When a message has a number of valid and available output channels, the first channel along the lowest dimension is selected (*output channel scheduling policy*). Messages are routed using *wormhole routing*.

Since partially adaptive routing algorithms manifest their superiority for nonuniform traffic patterns, we have chosen four traffic permutations to simulate. Three of these permutations create nonuniform traffic in the network, while the last one creates the uniform pattern. For a 2-dimensional mesh, the perfect shuffle (σ) and the uniform (u) permutations were simulated to quantify the performance of the newly developed algorithms. For a hypercube, two permutations were simulated. One permutation, which we denote as hotspot, (hs), is well known for creating hot spots in hypercubes [11]. In addition to (hs), the reverse-flip permutation, (rf), was simulated. These traffic patterns have been used in previous research [11,12]. Destination addresses generated by these traffic permutations are given below using $a_0, a_1, \ldots, a_{n-1}$ as the source node address.

$$u(a_0, a_1, \ldots, a_{n-1}) = \textit{any node in the system.} \quad (4)$$

$$\sigma(a_0, a_1, \ldots, a_{n-1}) = a_1, a_2, \ldots, a_{n-1}, a_0. \quad (5)$$

$$hs(a_0, a_1, \ldots, a_{n-1}) =$$
$$a_{\lceil \frac{n}{2} \rceil + 1}, \ldots, a_{n-1}, a_0, \ldots, a_{\lceil \frac{n}{2} \rceil}. \quad (6)$$

$$rf(a_0, a_1, \ldots, a_{n-1}) =$$
$$\overline{a_{n-1}}, \ldots, \overline{a_1}, \overline{a_0}. \quad (7)$$

Each simulation was run for a total of 30000 flit cycles. Output data were not collected in the first 10000 flit cycles in order to allow the system to stabilize. The 95% confidence interval of the results was estimated to be within 1% of the mean.

Figure 9 shows the latency of messages versus the channel utilization for the 2D mesh when the traffic pattern is uniform. The curves show that the partially adaptive algorithms saturate at a channel utilization rate that is less than that of the deterministic routing

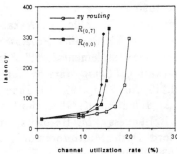

Figure 9. Performance of routing algorithms in an 8 × 8 mesh for *uniform* traffic pattern.

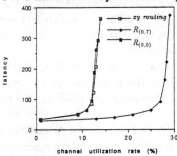

Figure 10. Performance of routing algorithms in an 8 × 8 mesh for *perfect shuffle* traffic pattern.

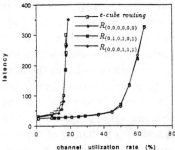

Figure 11. Performance of routing algorithms in a 6_cube for *hot spot* traffic pattern.

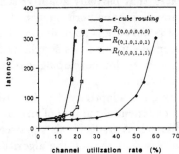

Figure 12. Performance of routing algorithms in a 6_cube for *reverse flip* traffic pattern.

algorithm (*xy routing*). Similar conclusions are drawn in the *turn model* [12]. However, when the traffic pattern is nonuniform, some partially adaptive routing strategies perform better than the deterministic *dimension order* scheme. In fact, the performance of a partially adaptive scheme is very much dependent on the traffic pattern in the network [12].

Figures 10, 11, and 12 show the performance of the algorithms for nonuniform traffic. The figures indicate that some partially adaptive routing algorithms saturate at a channel utilization rate that is two to three times larger than that of the deterministic algorithm. Moreover, the figures demonstrate that some traffic patterns are more suitable for some routing algorithms than others. For example, in the 6_dimensional hypercube, $R_{(0,1,0,1,0,1)}$ and $R_{(0,0,0,1,1,1)}$ outperformed $R_{(0,0,0,0,0,0)}$ for the *hot spot* traffic pattern. However, for the *reverse flip* traffic pattern, the situation was reversed and $R_{(0,0,0,0,0,0)}$ outperformed $R_{(0,1,0,1,0,1)}$ and $R_{(0,0,0,1,1,1)}$. This discussion confirms that the performance of these partially adaptive routing algorithms is dependent on the traffic pattern in the network. Given a traffic pattern, some partially adaptive routing algorithms perform much better than others.

The *dimension order routing* algorithm tries to distribute traffic evenly among the channels in the system. Even though, this deterministic algorithm does not achieve this evenness in traffic distribution [14], it accomplishes a better job than the partially adaptive routing algorithms for uniform traffic. The deterministic routing algorithm distributes traffic by using global information that is provided for free to the routing algorithm by the structure of the network. The reason for the poor performance of the partially adaptive algorithms is due to the fact that messages are routed by using only local information. Hence, what seems to be a good local routing decision is actually a not very good global routing decision. Therefore, more messages get blocked and the system saturate at a lower channel utilization rate. When the traffic in the network is nonuniform, the routing decisions made by the deterministic algorithm tend to overload some channels and underload others. The overloaded channels deteriorate the performance of the network due to the increase in the number of blocked messages. The partially adaptive schemes tend to distribute traffic much better for nonuniform traffic. The local routing decisions become much better from a global perspective as elaborated in the figures.

6   Conclusions

In this paper, a model was introduced for developing partially adaptive routing algorithms for a family of interconnection networks, n_dimensional meshes. The model is based on dividing a network into two disjoint unidirectional networks according to the directions in which messages can move along the different dimensions of the network. The routing process is divided into two phases, where messages are forwarded to intermediate nodes in the first network, and delivered to their destinations in the second network. Messages may use only one phase of the routing procedure depending on the locations of the source and destination nodes. The family of these partially adaptive routing algorithms include 2^n algorithms where n is the number of dimensions of the network. The paper illustrated that partially adaptive schemes are capable of providing good performance when the traffic pattern in the network is nonuniform. The attractiveness of this model is due to its simplicity and straightforward procedure for developing partially adaptive routing algorithms.

Providing full adaptivity in n-dimensional meshes requires an exponential number of virtual channels per physical channel. A tradeoff exists between the degree of adaptiveness that the routing algorithm possesses, and the number of virtual channels needed to prevent deadlock and livelock [10,12,16]. Therefore, routing algorithms that have some degree of adaptivity at a low cost are desirable. The model introduced in this paper provides such routing algorithms with no additional cost incurred in terms of virtual channels. The only additional cost is associated with the decision making logic at each routing node. This cost is inescapable for any adaptive routing algorithm.

Some partially adaptive routing algorithms are more suitable for some traffic patterns than others. This observation gives rise to exploring the possibility of building programmable routers that route messages using an algorithm that would minimize the latency and maximize the throughput of the network for the specific application at hand. The selection of an algorithm could be performed by examining a set of digits that identify one of the 2^n routing algorithms. Hence, applications could run faster by determining the dominant traffic pattern generated by them and deciding on the routing algorithm that suites them the most. This is one of the avenues that we will pursue in future.

7 References

[1] K. E. Batcher, "Design of a Massively Parallel Processor," *IEEE Trans. on Computers*, Vol. C-29, pp. 863-840, Sept. 1980.

[2] R. Alverson *et al.*, "The Tera Computer System," *Proc. 1990 int'l Conf. on Supercomputing*, pp. 1-6, June 1990.

[3] Intel Corporation, *A Touchstone DELTA System Description*, 1990.

[4] M. Annaratone, M. Fillo, K. Nakabayashi, and M. Viredaz, "The K2 Parallel Processor: Architecture and Hardware Implementation," *Proc. The 17th Annual International Symposium on Computer Architecture*, Vol. 18, No. 2, pp. 92-101, June 1990.

[5] D. Lenoski, J. laudon, K. Gharachorloo, W. Weber, A. Gupta, J. Hennessy, M. Horowitz, and M. Lam, "The Stanford Dash Multiprocessor," *IEEE Computer*, pp. 63-79, March 1992.

[6] A. Agarwal, B. Lim, D. Kranz, and J. Kubiatowicz, "APRIL: A Processor Architecture for Multiprocessing," *Proc. The 17th Annual International Symposium on Computer Architecture*, Vol. 18, No. 2, pp. 104-114, June 1990.

[7] G. Zorpetta, "Technology 1991: Minis and Mainframes," *IEEE Spectrum*, pp. 40-43, Jan. 1991.

[8] NCUBE Company, *NCUBE 6400 Processor Manual*, 1990.

[9] W. J. Dally and C. L. Seitz, "Deadlock-free message routing in multiprocessor interconnection networks," *IEEE Transcations on Computers*, Vol. C-36, pp. 547-553, May 1987.

[10] D. H. Linder and J. C. Harden, "An adaptive and fault-tolerant wormhole routing strategy for K-ary n-cubes," *IEEE Transcations on Computers*, Vol. C-40, pp. 178-186, Jan. 1991.

[11] S. Konstantinidou, "Adaptive, minimal routing in hypercubes," *Proc. The 6th MIT conference: Advanced Research in VLSI*, pp. 139-153, 1990.

[12] C. J. Glass and L. M. Ni, " The turn model for adaptive routing," *Proc. The 19th Annual International Symposium on Computer Architecture*, pp. 278-287, May 1992.

[13] C. J. Glass and L. M. Ni, " Adaptive routing in mesh-connected networks," *Proc. The 12th International Conference on Distributed Computing Systems*, pp. 12-19, June 1992.

[14] A. Agarwal, "Limits on Interconnection Network Performance," *IEEE Transactions on Parallel and Distributed Systems*, Vol. 2, No. 4, pp. 398-412, Oct. 1991.

[15] A. A. Chien and J. H. Kim, " Planar-adaptive Routing: Low-cost Adaptive Networks for Multiprocessors," *Proc. The 19th Annual International Symposium on Computer Architecture*, pp. 268-277, May 1992.

[16] H. Sullivan and T. R. Bashkow, " A large scale, homogeneous, fully distributed parallel machine," *Proc. The 4th Annual International Symposium on Computer Architecture*, pp. 105-117, May 1977.

SESSION 8C

SORTING/SEARCHING

GENERATION OF LONG SORTED RUNS ON A UNIDIRECTIONAL ARRAY

Yen-Chun Lin and Horng-Yi Lai

Dept. of Electronic Engineering, National Taiwan Institute of Technology

P.O. Box 90-100

Taipei 106, Taiwan, R.O.C.

y.lin@ieee.org   or   yclin@twnntit.bitnet

Abstract -- *A parallel algorithm is presented for generation of long sorted runs as the first phase of sorting a large file. It can generate runs of length about $2mp + 2$ on a unidirectional linear array of p processors each with a heap of size m. Internal computations can be completely overlapped with I/O, and almost only disk read time is required. Experiments with the algorithm and its variations have been conducted. The results show that all the versions can generate longer runs than a previous algorithm run on a bidirectional linear array.*

1. INTRODUCTION

Sorting is very important because it has practical use and is of great theoretical interest. Generation of sorted sequences, called runs, is usually the first phase of sorting a large file. The second phase is external merge sorting of the runs. On sequential machines, using a heap of size m can produce runs of length about $2m$ [3], rather than just m as by other algorithms. This can greatly reduce the sorting time of a file that requires external merge sorting, as fewer runs may require fewer passes over all data. The algorithm has two other advantages: Data I/O and computations can be performed simultaneously, and sorting can be done in one pass over the data if the data have a certain ordering [8].

A parallel algorithm has been proposed to achieve these advantages [6]. The algorithm was then improved to be simpler and to generate longer runs [4, 5]. For ease of presentation, we will, at times, call the improved version the Lin algorithm. Specifically, the Lin algorithm runs on a bidirectional linear array of processing elements (PEs), in which each PE can send and receive messages with its two neighbors simultaneously. It can generate runs of length about twice the total amount of memory in the array. Because I/O can be performed in parallel with internal computations, perfect overlapping [4, 5, 7] can be achieved. Perfect overlapping is achieved when the computations are completely overlapped with I/O, and input and output are mostly overlapped, reducing the elapsed time to almost only input time. Specifically, perfect overlapping requires that the time to input a block of data from the disk be equal to (or greater than, if the equivalence is not possible) the time required to process

the data, and the time to output a block of data to the disk be equal to (or greater than) the time required to produce the data. A linear array is the simplest to implement among the various distributed-memory parallel computers, such as mesh-connected and cube-connected computers. An algorithm for the linear array can easily be simulated on mesh-connected, cube-connected, or even shared-memory parallel computers.

In this paper, an algorithm is proposed to run on a unidirectional linear array, in which data are required to move in only a single direction. Compared with the Lin algorithm, the new algorithm has two merits. First, the required hardware to support the algorithm is simpler. Second, when perfect overlapping is achieved, the average run length obtained can be larger. The algorithm uses the idea of feeding back output values to circumvent the inherent limitation of a unidirectional array. Experiments with the algorithm have been conducted to verify that the advantages can be achieved in a real environment. Specifically, the algorithm can produce runs of length about $2mp + 2$ on a linear array of p PEs each with a heap of size m. In addition, it can completely sort an arbitrarily large file in one pass providing no item in the file has $mp + 1$ larger items before it. Based on the algorithm, many variations are devised and implemented to obtain longer runs.

2. BRIEF REVIEW OF THE SEQUENTIAL ALGORITHM

The sequential algorithm to produce runs of length about twice the size of the heap used is briefly reviewed. We describe a variant of that presented in [3]. Sorting items in ascending order is considered in this paper; however, sorting in descending order can be performed similarly. Initially, a min heap is constructed from inputs. A min heap is a complete binary tree in which the item of each non-leaf node is not larger than the items of its children. Hereafter a min heap will be called a heap. In each following step, the root item of the heap will be output and be replaced by an input item. If the input is larger than its smaller child, it is sifted down by exchanging it with the smaller child until it is not larger than its smaller child or becomes a leaf. However, if the input is smaller than the last output, then it belongs to the next run and is considered to be larger than all current-run items when compared for exchange. Thus, a next-run item will be sifted down to be a descendent of current-run items if the heap contains current-run items. When the whole heap contains only next-run items, a new run is ready to be output.

Using a heap of size m for random items, the first run produced will have about $1.718m$ items; the second, about $1.952m$ items; and the following runs, except for the last two, will have about $2m$ items. The penultimate and the last runs respectively have about $5m/3$ and $m/3$ items [3].

3. MODEL

We consider generating runs from a file of n items on a unidirectional linear array of p PEs as depicted in Fig. 1. Each PE has two links to communicate with its neighboring PEs, and can compute, receive from its west neighbor, and send to its east neighbor simultaneously. Unless otherwise stated, assume that each PE can manage a heap of m items, and each PE has one buffer for each inter-PE link. It is also assumed that $n \gg 2(m + 1)p$. The first PE can receive from the host processor, and the pth PE can send to the host. The transfer rate between the host and either of its neighboring PE is assumed to be equal to that between neighboring PEs. The coordination of neighboring processors is done by message passing.

The host processor can input from and output to separate disks simultaneously. Assume that reading a block from the input disk and writing a block to the output disk take the same amount of time. To avoid possible confusion, in this paper, data transfer between the host and disks is called external I/O, and data transfer between the host and either of its neighboring PE as well as that between neighboring PEs is called inter-PE transfer.

The input file and sorted runs are stored on disks in blocks. Thus, the host must perform deblocking of input blocks from the input disk for item-by-item transfer to the linear array and perform blocking of the sorted runs from the array for writing to the output disk. As the external I/O is usually the bottleneck of a computing system, double buffering is used for the external I/O to improve the performance. We will further assume that the heap of size m managed by a PE guarantees perfect overlapping. As will be seen, the computing time of each step is only $O(\log m)$; thus, perfect overlapping can be achieved if the external I/O time of a block of b items is not smaller than either the computing time of b steps, which is $O(b \log m)$, or the inter-PE transfer time of b items.

4. BASIC ALGORITHM FOR RUNS GENERATION

Heap restructuring takes most of the computing time in the array. We will use the same heap restructuring technique as that used in [4-6] to maintain a heap in each PE. After introducing the heap restructuring, an overview of our algorithm will be presented, which is then followed by a more detailed description.

4.1. Heap restructuring

The heap in each PE can easily be implemented with the array data structure, e.g., $a[1..m]$. In this case, the parent of node $a[i]$ is node $a[\lfloor i/2 \rfloor]$, $2 \le i \le m$. Every item in a PE is either a current-run item, which belongs to the same run as the item being sent by the array back to the host, or a next-run item. A next-run item is regarded as larger than a current-run item. In each PE, the first next-run item to reside in the heap is placed at node $a[m]$, the second next-run item is placed at node $a[m-1]$, and so on. Therefore, the locations occupied by next-run items are contiguous, and the next-run item with the smallest index will be called the leading next-run item. The current-run item adjacent to the leading next-run item will be called the trailing current-run item. The lower part of the heap occupied by next-run items will be called the next-run part, and the part of the heap that contains current-run items will be called the current-run part. For example, Fig. 2 shows a heap containing both the current-run and the next-run parts. The leading next-run item is 4, and the trailing current-run item is 25.

A pointer pointing to the trailing current-run item (hereafter called the TCR pointer) can be used for the heap management. When the heap contains only current-run items, the TCR pointer is m, or points to $a[m]$. If the heap contains only next-run items, the TCR pointer is 0. When the heap contains both the current-run part and the next-run part, the TCR pointer is between 1 and $m-1$.

When the root item in a heap (hereafter called the root) has been changed, the SIFT-DOWN operation [4-6] is used to restructure the modified heap to maintain the heap property, as described in the following:

(1) If the heap contains current-run items, the root is checked to see if it should be sifted down to its proper location in the current-run part. Specifically, if the root is larger than the smaller of its children, the positions of the root and the smaller child will be exchanged. The comparison and exchange are repeated until the original root is not larger than either of its children or it becomes a leaf of the current-run part. (2) If the heap contains next-run items, the leading next-run item is checked to see if it should be sifted down to its proper location in the next-run part. Specifically, if the leading next-run item is larger than the smaller of its children, the positions of the leading next-run item and the smaller child will be exchanged. The comparison and exchange are repeated until the original leading next-run item is not larger than

either of its children or it becomes a leaf of the next-run part.

The application of the SIFT-DOWN operation maintains the heap property of both the current-run part and the next-run part and takes only O(log m) time [5, 6].

We use two examples to illustrate the heap restructuring performed by the SIFT-DOWN operation for two possible situations. The first example is for the situation when the leading next-run item is not a new one. Fig. 3 illustrates the heap restructuring after the root has been changed. Fig. 3(a) depicts that the root of the heap shown in Fig. 2 is being replaced by a new item 30, and Fig. 3(b) shows the result of applying the SIFT-DOWN operation to the disordered heap. The item 30 has been sifted down to be a leaf of the current-run part.

The second example is for the situation when the leading next-run item is a new one. Fig. 4(a) shows that two items in Fig. 2 are being modified. First, the trailing current-run item, 25, is copied to the root. Second, a new next-run item, 14, is moved to replace 25. Fig. 4(b) shows the result of applying the SIFT-DOWN operation to the disordered heap. Note that the new leading next-run item, 1, and its children constitute a heap.

4.2. Overview of The Algorithm

The operations of PEs can be considered as executing an infinite loop. Each iteration of the loop is a step. In a step, each PE receives a data item from the west neighbor, sends an item to the east, and does some computations.

The last PE regards an item as a current-run item if the item is not smaller than the last output item sent from the array back to the host (hereafter called the last output); otherwise, the item belongs to the next run. Ideally, every item in the array smaller than the last output is a next-run item, and all other items are current-run items.

However, the first $p - 1$ PEs do not know what the last output is. To provide a remedy, the host will send the last output back to the array periodically so that the first $p - 1$ PEs know the last output. Actually, only the key value of the last output is required. The key value sent back to the array will be called the feedback value. The first feedback value is taken from the first output item of the array. The period between when two feedback values are sent to the array, to be called the feedback period, will be measured in number of inputs to the array, or equivalently, in number of outputs to the host.

In the first $p - 1$ PEs, an item is called a current-run item if it is not smaller than the feedback value. Because the array sends a new output to the host at the step when the host sends the feedback value to the array, the feedback value is outdated unless many items with the same key value as the feedback value are to be output.

Thus, a current-run item in the first $p - 1$ PEs may not be a current-run item in the last PE. An item will be called a pseudo-current-run item if it is a current-run item in the first $p - 1$ PEs and becomes a next-run item when sent to the last PE. Such an item will still be called a current-run item when it is in the first $p - 1$ PEs, because there is no way to predict whether a current-run item in the first $p - 1$ PEs will become a next-run item in the last PE. In contrast, a next-run item in the first $p - 1$ PEs is smaller than the feedback value. Before a new run is output, such an item will still be a next-run item in the last PE.

Each PE always sends out the smallest current-run item it contains unless all its items are next-run items. Thus, a PE can receive a next-run item only after all heaps to its west are full of next-run items. However, there is one exception to this: the last PE may receive pseudo-current-run items before all heaps to its west are full of next-run items. If an item to be sent out belongs to the next run, it is the smallest next-run item in the PE.

When the last PE contains only next-run items, a new run is ready to be output to the host at the next step. Ideally, immediately after the first item of the new run is output, all next-run items in the array should then become current-run items, and all feedback values should be updated. Because there is some delay before a new feedback value is sent to the first $p - 1$ PEs to enforce the necessary changes, it may happen that a small next-run item in the $p - 1$ PEs is sent to the last PE and becomes a next-run item there. Such a premature next-run item will shorten the current run.

To provide a remedy, the ith PE ($1 \leq i \leq p - 1$) can count how many next-run items have been sent out, and when the number of next-run items sent out equals a preset value, the PE can reset the number of next-run items sent out to zero and update the feedback value with the root item. So, an item smaller than the new feedback value will not be sent out to be a premature next-run item. As a result of our experiment, when the heap size m is between 63 and 1023 and $p = 8$, using a preset value of $m(p - i) - 2p$ for the ith PE helps obtain satisfactory run lengths. If the preset value is too large, it has no effect at all. However, if the value is too small, the generated runs may be short as a result of the following two possible situations. First, when the last PE still contains many current-run items, but the first PE has updated the feedback value with the root item, a current-run item sent to the first PE is not sent out early enough to be a real current-run item because the item is larger than the feedback value and all the next-run items in the first PE. Second, when the last two PEs each contain current-run items, but the first PE has updated the feedback value with the root item, a small next-run item sent to the first PE is kept in the first PE because the item is smaller than the feedback value, and will be sent out of the array following the next run. In contrast, if the first PE has not updated the feedback value, the small next-run item can be sent to the $(p - 1)$th PE and will belong to the next run.

The host can identify an output item of the array as the first item of a new run if it is smaller than the preceding output. The host will then send a new feedback value to the array. More detailed operations will be described in the following subsections.

4.3. Operations of the First $p - 1$ PEs

A step of these PEs can be divided into two stages for ease of description. In the first stage, data transfers with the neighbors are performed in parallel with the SIFT-DOWN operation.

In the second stage, the input item is checked to see if it is a feedback value. If the input is a feedback value and is smaller than the old feedback value, meaning that a new run is being generated, then (1) the number of next-run items sent out is reset to zero, and (2) if the TCR pointer shows that the heap contains only next-run items, then the TCR pointer is reset to indicate that the heap contains only current-run items. No matter whether the new feedback value is smaller than the original feedback value or not, the new feedback value is used to replace the old feedback value in the PE and is copied to the output buffer.

If the input is not a feedback value, both the root in the PE and the input should be moved to proper locations, for data transfer and heap restructuring at the next step. The basic idea is that the smaller of the two items will be sent to the east, and the larger will be kept in the heap. Specifically, the PEs operate in one of four different modes depending on whether the root and the input belong to the current run or the next run.

The first mode is employed when both the input and the root belong to the current run. If the root is smaller than the input, then the root is moved to the output buffer, and the input becomes the new root. (Fig. 3 illustrates such a situation and the result of applying the SIFT-DOWN operation at the next step.) Otherwise, the input is moved to the output buffer.

The second mode is selected when the input and the root, respectively, belong to the current run and the next run. The input is moved to the output buffer.

The third mode is activated when the input and the root, respectively, belong to the next run and the current run. The root is moved to the output buffer. Then, the trailing current-run item is moved to the root node, and the input is moved to the node where the new root comes from, i.e., the previous trailing current-run node, as the leading next-run item. The TCR pointer is therefore decreased by one. (Fig. 4 illustrates such a situation and the result of applying the SIFT-DOWN operation at the next step.)

The fourth mode occurs when both the input and the root belong to the next run. Initially, this mode performs the same operations as in the first mode. Subsequently, the number of next-run items sent out is then increased by one. If the number equals the preset

value, the number is reset to zero, the feedback value is updated with the root item, and the TCR pointer is reset to indicate that the heap contains only current-run items.

We can see that most of the computing time is spent in performing the SIFT-DOWN operation. Thus, the computing time of each step is only $O(\log m)$.

4.4. Operations of the last PE

For this PE, the operations performed in the first stage are concurrent data transfers and heap restructuring, and are the same as those performed by the other PEs in the first stage.

In the second stage, the input is checked to see if it is a feedback value. If so, the feedback value is copied to the output buffer; otherwise, the PE also operates in one of four different modes depending on whether the root and the input belong to the current run or the next run. Specifically, the first two modes are exactly the same as those of the other PEs.

The third mode performs most of the operations of the third mode of the other PEs. The root is moved to the output buffer. Then, the trailing current-run item is moved to the root node, and the input is moved to the node where the new root comes from as the leading next-run item. The TCR pointer is then decreased by one.

The fourth mode begins by doing the same operations as in the first mode. Then, because the next output will be the first item of a new run, the TCR pointer is reset to indicate that the heap contains only current-run items.

4.5. Initial and Final Steps

The above description of the algorithm focuses on PE operations already in progress. Usually, an algorithm includes initial data preloading steps and final output steps that require special processing. Fortunately, we can use an ingenious method to make the initial and final steps of the PEs the same as those described previously.

While the host is reading initial inputs from the disk, each PE should be setting up the environment for processing the initial inputs: assigning the least possible value $(-\infty)$ to be the initial feedback value, to the output buffer, and to every location of the heap; and setting the TCR pointer to indicate that the heap contains only current-run items. In this way, the $(m + 1)p$ items of value $-\infty$ in the output buffers and heaps will be the first $(m + 1)p$ outputs to the host while initial inputs are being sent to the array as current-run items.

As for orderly extraction of the final outputs, $(m + 1)p$ largest possible items should follow the real input items. Thus, the last $(m + 1)p$ data items can be output while the largest possible items are being input.

5. PROCESSING TIME AND AVERAGE RUN LENGTH

The total number of steps required to generate the

runs can be obtained by adding the first $(m + 1)p$ steps to output $(m + 1)p$ garbage items and the n steps to output n data items, totaling $n + (m + 1)p$ steps, then multiplying the sum by $(1 + 1/F)$, where F is the feedback period. Under the situation of perfect overlapping, the array can process a block of data no slower than the external I/O of a block of data, so, when $1 >> 1/F$, the total processing time is dominated by the external I/O time of n items. Thus, the total processing time is O(n). Furthermore, most of the input time and output time can also be overlapped.

Although the array can hold $(m + 1)p$ different items in the heaps and buffers, the average run length cannot achieve $2(m + 1)p$ for several reasons. The first cause of shorter runs is attributed to pseudo-current-run items. Pseudo-current-run items become next-run items in the last PE such that there may be real current-run items in the first $p - 1$ PEs after the last PE has contained only next-run items. That is, there may be current-run items that are not output before the next run is output, making the current run shorter than in the ideal case that no heap contains current-run items when the next run is ready to be output. When the next run is to be output, a heap in the first $p - 1$ PEs containing current-run items usually contains next-run items. In this case, these next-run items will still be next-run items when the new run is being output. Thus, the new run will be shorter than in the ideal case that no heap initially contains next-run items when the new run is being output.

The second cause of shorter runs is that each input to the array cannot be compared immediately with the last output for the generation of the new output item at the next step. In the worst case, at the end of a step, there are $p - 1$ items in the output buffers of the first $p - 1$ PEs that may be the new output if they are in the last PE; that is, $p - 1$ items are not used for generating the new output, and a total of $(m + 1)p - (p - 1) = mp + 1$ locations is used for generating runs. Therefore, $2mp + 2$ may be a better guess than $2(m + 1)p$ for the average run length, and we will say that the guessed average run length is $2mp + 2$, unless otherwise stated.

Two minor causes of shorter runs are related to the preset number of next-run items that can be sent out by each of the first $p - 1$ PEs. They have been considered in Sec. 4.2.

6. VARIATIONS OF THE BASIC ALGORITHM

Because pseudo-current run items hinder the generation of long runs, we should reduce the probability of having pseudo-current-run items in order to increase the run lengths. To achieve this, two possible approaches are considered, as follows.

An easy way to decrease the number of pseudo-current run items is to decrease the feedback period to avoid too much difference between the feedback value

saved in the array and the ever-changing output to the host. However, decreasing the feedback period also increases the overhead and thus the processing time. A compromise must be arranged to obtain satisfactory average run length and processing time.

A totally different approach to increasing the run lengths is by preventing the heap in the last PE (hereafter called the last heap) from containing only next-run items while there are real current-run items in the other PEs. Using a larger last heap than those in the other PEs helps fill the last heap up with next-run items only after the first $p - 1$ PEs contain no current-run items.

For the generation of long runs as fast as possible, variations of the basic algorithm are derived from the above approaches. These variations can be classified into two categories. The first category is related to decreasing the feedback period dynamically. Initially, a larger feedback period is used to avoid too much feedback overhead. Later, when the last heap contains next-run items, the feedback period is decreased. The feedback period can be decreased more than once. We will consider two such variations.

The first variation decreases the feedback period by 20 twice when the number of current-run items in the last heap is $3p$ and $2p$, respectively. This variation will be called the 2-period-decreases variation. The other variation decreases the feedback period by 20 three times when the number of current-run items in the last heap is $4p$, $3p$, and $2p$, respectively, and will be called the 3-period-decreases variation.

The second category of variations uses a last heap whose size is larger than the other heap size, m. Assume that m is the size of a full binary tree. Consequently, when the last heap size increases from m to $2m + 1$, the number of levels of the heap will only be increased by one. Thus, we will use a last heap of size $2m + 1$ as a variation, which will be called the $2m + 1$ variation. For comparison, we will also use $m + p$ and $m + 2p$ as the sizes of the last heap, and the resulting variations will be called the $m + p$ variation and the $m + 2p$ variation, respectively.

7. EXPERIMENTAL RESULTS

We have implemented the basic algorithm and its variations in the Occam 2 language [1] for execution on a ring of nine transputers [2]. A transputer has four links to communicate with its neighbors in parallel with internal computations. Transferring data from one transputer to a neighboring transputer can be 1.5 times faster when there are only one-way transfers between the two transputers than when there are two-way transfers [2].

One of the transputers is used as the host, and the other eight transputers are used as a linear array; thus, $p = 8$. Each input file has 400,000 random items, and, for ease of measuring, 400,000 items are loaded into the host before transferring data to the array and measuring the processing time. Experiments are conducted with heap size $m = 63$. We compare the average length of the generated runs with the guessed average run length. As mentioned earlier, the guessed average run length is $2mp + 2$, except for the larger last heap category of variations. Let d be the difference between the last heap size and m. The guessed average run length for the larger last heap category is $2mp + 2d + 2$. Because the expected length of each of the first two and the last two runs is theoretically smaller than the other runs, we do not count these four runs when calculating the average run length.

Fig. 5 shows the average run lengths of the various versions of our algorithm in percentage of the guessed average run length, where all run lengths are the average for ten input files. All versions have a satisfactory average run length when the feedback period is 51, but only the $2m + 1$ and the $m + 2p$ variations have a satisfactory average run length throughout the various feedback periods used.

Fig. 6 shows percentage improvement in processing times of various versions of the presented algorithm over the Lin algorithm. All versions are faster than the Lin algorithm; this must be due to both faster one-way transfers and fewer operations in every version of our algorithm. The $2m + 1$ variation of all these versions is slower than the others because it takes longer to restructure the last heap; however, as the feedback period increases, the improvement of the running time of the $2m + 1$ variation is greater than that of the other versions. Figures 5 and 6 confirm that as the feedback period increases, the run lengths tend to be shorter, but the processing times improve.

So far, to achieve satisfactory average run length and processing time, the $m + p$ variation, when the feedback period is around 67, and the $m + 2p$ variation, when the feedback period is around 83, are better than the others. On the other hand, because when the feedback period increases, the improvement of the processing time of the $2m + 1$ variation is greater than that of the other versions, further investigations are needed to see if the $2m + 1$ variation can be better than the $m + p$ and the $m + 2p$ versions. Our experiments show that before the average run length of the $2m + 1$ variation begins to decrease to less than 98% of the guessed average run length, the processing time fails to be better than those of the $m + p$ and the $m + 2p$ variations. Thus, the $m + p$ and the $m + 2p$ variations are better than the others when the running time and the average run length are considered.

From our experiments, it seems that although the guessed average run length $2mp + 2$ is a little smaller than $2(m + 2)p$, which is the expected average run length of the Lin algorithm, some versions of our new algorithm can be more than 25% faster than the Lin algorithm. In reality, data cannot be totally preloaded into the host as in our experiment, and disk I/O must be considered. Thus, to make a fair comparison between the

two algorithms, we should consider the situation when external I/O is considered and perfect overlapping is achieved. The external I/O time should be essentially the same regardless of the algorithm used; thus, we should compare the average run lengths of the two algorithms when the processing times are the same.

As it appears from our experiments, the average run length obtained by some versions of our algorithm can be larger than that obtained by the Lin algorithm when their processing times are the same. In particular, when $m > 2p - 2$, the guessed average run length of the $2m + 1$ variation, $2mp + 2m + 4$, is larger than $2(m + 2)p$. Our experiments also show that the $2m + 1$ variation can generate longer runs than the Lin algorithm and is 23% faster than the Lin algorithm when the feedback period is about 160. Therefore, by using larger heaps, the $2m + 1$ variation can generate even longer runs than the Lin algorithm when the two algorithms take the same running time on the array.

We have also verified by experiment that the algorithm can sort an arbitrarily large file in one pass providing no item in the file has $mp + 1$ larger items before it.

8. MORE VARIATIONS OF OUR ALGORITHM

Many more variations are possible to achieve longer runs and faster processing. As examples, three more variations are as follows. First, we can increase the feedback value kept in each of the first $p - 1$ PEs by one in each step. The feedback value will be closer to the ever-changing last output. If most of the key values are not continuous, we can even increase the feedback value by i, where $i > 1$ and is the average difference between every pair of nearest key values. This should reduce the probability of having pseudo-current-run items. Second, we can add p to the original feedback value before the host sends out the feedback value; the feedback value with p added will be called the modified feedback value. This will also reduce the probability of having pseudo-current-run items, especially when all key values are different. For example, if a data item sent to the array following a feedback value is between the original feedback value and the modified feedback value, it may become a pseudo-current-run item if the feedback value is not modified; however, because the data item is smaller than the modified feedback value and thus it will never be a pseudo-current-run item if the feedback value is modified and all key values are different. If most of the key values are not continuous, we can even add ip to the original feedback value, where $i > 1$ and is the average difference between every pair of nearest key values. Third, we can always send a new feedback value together with each real data item to quickly reflect the new ever-changing last output.

Note that some of the variations may be applied together. For example, the second and the third of the above three variations, as well as that using a larger last heap, can be combined. Thus, although we have presented some very satisfactory variations, there are so many possible modifications that extensive experiments are required to find the best.

However, this may not be the end in the search for the best algorithm. Because the heap size and the size of an item affect the degree of the overlapping of heap restructuring and inter-PE transfers, it may occur that a certain variation is the best for some heap size and item size but is not the best for other heap sizes or item sizes.

9. CONCLUSION

We have presented a parallel algorithm and its variations for execution on a unidirectional linear array of PEs to generate long runs as the first phase to sort a large file. They can achieve perfect overlapping and their relative advantages have been investigated by experiments. Our experiments show that the algorithm can generate longer runs than the previously proposed Lin algorithm for execution on a bidirectional linear array. Thus, faster sorting can be achieved by first using a variation of the algorithm presented here and then using the merging algorithm run on a linear array presented in [4].

ACKNOWLEDGEMENT

This research was supported in part by the National Science Council of the R.O.C. under contract NSC-82-0408-E011-132.

REFERENCES

[1] Inmos, *Occam 2 Reference Manual*, Prentice Hall, Englewood Cliffs, NJ (1988).

[2] Inmos, *The Transputer Databook, Third Edition*, Almondsbury, Bristol, UK (1992).

[3] D.E. Knuth, *The Art of Computer Programming* Vol. 3: Sorting and Searching, Addison-Wesley, Reading, MA, (1973).

[4] Y.C. Lin, "Perfectly overlapped sorting on a linear array," *Proc. 1992 Int. Conf. on Parallel Processing*, St. Charles, IL, (Aug. 1992), pp. III-285-288.

[5] Y.C. Lin, "Perfectly overlapped generation of long runs for sorting large files," *J. of Paral. and Distrib. Comput.*, to appear.

[6] Y.C. Lin and Y.H. Cheng, "Fast generation of long sorted runs for sorting a large file," *Proc. Int. Conf. on Application Specific Array Processors*, Barcelona, Spain, (1991), pp. 445-456.

[7] B. Salzberg, "Merging sorted runs using large main memory," *Acta Inform.*, Vol. 27, (1989), pp. 195-215.

[8] R. Sedgewick, *Algorithms*, Addison-Wesley, Reading, MA (1988).

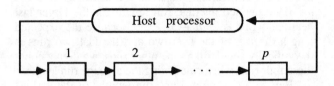

Fig. 1. A unidirectional linear array of *p* PEs connected to the host processor.

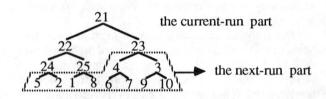

Fig. 2. A heap with the current-run and the next-run parts.

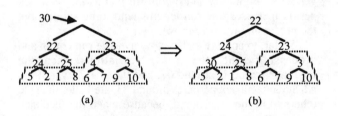

Fig. 3. Heap restructuring after the root has been replaced.

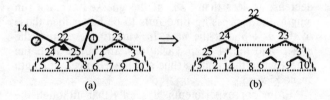

Fig. 4. Heap restructuring after two items have been replaced.

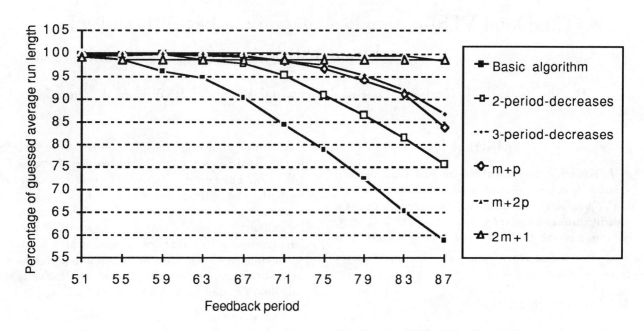

Fig. 5. Average run lengths of the various versions.

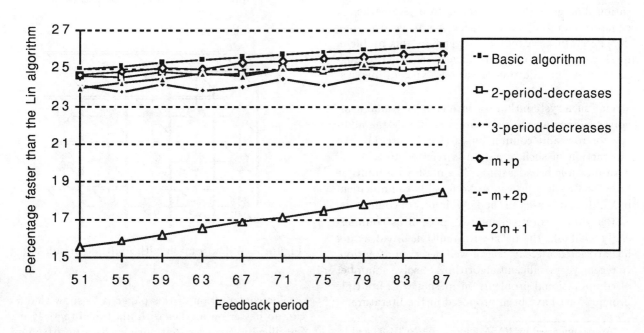

Fig. 6. Percentage improvement in processing times of various versions of our algorithm over the Lin algorithm.

Time- and VLSI-optimal sorting on meshes with multiple broadcasting *

D. Bhagavathi[†] H. Gurla[†] S. Olariu[†] J. Schwing[†] W. Shen[†] L. Wilson[†] J. Zhang[‡]

Abstract

In this work, we present a time- and VLSI-optimal sorting algorithm for meshes with multiple broadcasting. Specifically, we show that for every choice of a positive integer constant c, m items ($n^{\frac{1}{2}+\frac{1}{2c}} \leq m \leq n$) stored in the first $\lceil \frac{m}{\sqrt{n}} \rceil$ columns of a mesh with multiple broadcasting of size $\sqrt{n} \times \sqrt{n}$ can be sorted in $O(\frac{m}{\sqrt{n}})$ time.

1 Introduction

The mesh-connected computer architecture has emerged as a natural choice for solving a large number of computational tasks in image processing, computational geometry, and computer vision. The regular structure and simple interconnection topology make the mesh particularly well suited for VLSI implementation. However, due to its large communication diameter, the mesh tends to be slow when it comes to handling data transfer operations over long distances. In an attempt to overcome this problem, mesh-connected computers have recently been augmented by the addition of various types of bus systems.

One such system that is commercially available involves enhancing the mesh architecture by the addition of row and column buses (see Figure 1). An abstraction of such a system is referred to as mesh with multiple broadcasting. The mesh with multiple broadcasting has proven to be feasible to implement in VLSI, and is used in the DAP family of computers.

Being of theoretical interest as well as commercially available, the mesh with multiple broadcasting has attracted a great deal of well-deserved attention. In recent years, efficient algorithms to solve a number of computational problems on meshes with multiple broadcasting have been proposed in the literature.

It is well known that n data items can be sorted in $O(\sqrt{n})$ time on a mesh-connected machine of size $\sqrt{n} \times \sqrt{n}$. Furthermore, this result is easily shown to be both time-optimal and VLSI-optimal. Recently, a number of sorting algorithms have been proposed for enhanced meshes [4]. An easy information transfer argument shows that for meshes of area n, even when enhanced with multiple broadcasting or with a dynamically reconfigurable bus system, $\Omega(\sqrt{n})$ is a time lower bound for sorting n items.

This somewhat counter-intuitive result motivated us to look at the following problem. Given a mesh or enhanced mesh of size $\sqrt{n} \times \sqrt{n}$ and the goal is to sort m ($\sqrt{n} \leq m \leq n$) items stored in the first $\lceil \frac{m}{\sqrt{n}} \rceil$ columns of the machine. How fast can this task be performed? It is easy to show that the $\Omega(\sqrt{n})$ is a time lower bound for the unenhanced mesh. Clearly, no algorithm can be VLSI-optimal in this case.

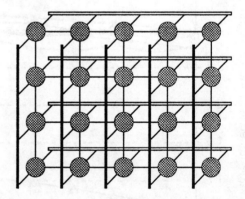

Figure 1: A 4 × 5 mesh with multiple broadcasting

The contribution of this paper is to show that we can do better on meshes with multiple broadcasting. Specifically, we show that once we fix a positive integer constant c, we can sort m items ($n^{\frac{1}{2}+\frac{1}{2c}} \leq m \leq n$) in $O(\frac{m}{\sqrt{n}})$ time. We show that this is time-optimal for this architecture. It is also easy to see that this achieves the VLSI lower bound in the word model.

*Work supported by NASA under grant NCC1-99 and by the NSF grant CCR-8909996

[†]Department of Computer Science, Old Dominion University, Norfolk, VA 23529

[‡]Department of Mathematics and Computer Science, Elizabeth City State University, Elizabeth City, NC 27909

2 The Lower Bound

The purpose of this section is to show that every algorithm that sorts m ($m \leq n$) items given in the first $\lceil \frac{m}{\sqrt{n}} \rceil$ columns of a mesh with multiple broadcasting must take $\Omega(\frac{m}{\sqrt{n}})$ time.

Our argument is of *information transfer* type. Consider the submesh $\mathcal{M}$ consisting of processors $P(r,s)$ with $1 \leq r \leq \frac{m}{2\sqrt{n}}$ and $1 \leq s \leq \frac{m}{\sqrt{n}}$. The input will be constructed in such a way that every element initially input into $\mathcal{M}$ will find its final position in the sorted order outside of $\mathcal{M}$. To see that this is possible, note that $m \leq n$ guarantees that the number of elements in $\mathcal{M}$ satisfies:

$$\frac{m}{\sqrt{n}} * \frac{m}{2\sqrt{n}} = \frac{m^2}{2n} \leq \frac{m}{2}.$$

Since at most $O(\frac{m}{\sqrt{n}})$ items can leave or enter $\mathcal{M}$ in $O(1)$ time, it follows that any algorithm that correctly sorts the input data must take at least $\Omega(\frac{m}{\sqrt{n}})$ time. Thus, we have the following result.

Theorem 2.1. Every algorithm that sorts m ($m \leq n$) items in the first $\frac{m}{\sqrt{n}}$ columns of a mesh with multiple broadcasting of size $\sqrt{n} \times \sqrt{n}$ must take $\Omega(\frac{m}{\sqrt{n}})$ time.

In addition to time-lower bounds for algorithms solving a given problem, one is often interested in designing algorithms that feature a good VLSI performance [6]. One of the most used metrics is the product AT^2, where A is the area of the chip and T is the time taken for the problem. In [6] it was shown that the VLSI lower bound for sorting m elements is $\Omega(m^2)$. If we use n processors (i.e $A = n$), the VLSI lower bound translates into $\Omega(\frac{m}{\sqrt{n}})$ which is the same as the time lower bound just proved. This means that a sorting algorithm running in $O(\frac{m}{\sqrt{n}})$ time on a mesh with multiple broadcasting of size $\sqrt{n} \times \sqrt{n}$ is both time- and VLSI-optimal.

3 The Algorithm

We state the following propositions which will be used later. For the details refer to [2].

Proposition 3.1. [5] Let $S_1 = (a_1, a_2, \ldots, a_r)$ and $S_2 = (b_1, b_2, \ldots, b_s)$, with $r + s = \sqrt{n}$, be sorted sequences stored in the first row of a mesh with multiple broadcasting of size $\sqrt{n} \times \sqrt{n}$, with $P(1,i)$ holding a_i ($1 \leq i \leq r$) and $P(1, r+i)$ holding b_i ($1 \leq i \leq s$). The two sequences can be merged into a sorted sequence in $O(1)$ time. $\square$

Proposition 3.2. A sequence consisting of k equal-sized sorted subsequences stored in the first row of a mesh with multiple broadcasting of size $\sqrt{n} \times \sqrt{n}$ can be sorted in $O(\log k)$ time. $\square$

Proposition 3.3. Given a $\sqrt{n} \times \sqrt{n}$ mesh with multiple broadcasting, with input elements stored in the first x columns in sorted row-major order, the data can be moved into a sorted column-major order in the first x columns, in $O(x)$ time. $\square$

Our algorithm implements the well-known bucket sort strategy. Throughout, we assume a mesh with multiple broadcasting $\mathcal{R}$ of size $\sqrt{n} \times \sqrt{n}$. We also assume that the processors in the first column of the mesh also serve as I/O ports.

Fix an arbitrary positive integer constant c. The input is assumed to be a set S of m items

$$n^{\frac{1}{2}+\frac{1}{2c}} \leq m \leq n \qquad (1)$$

from a totally ordered universe[1] stored in the first $\frac{m}{\sqrt{n}}$ columns of $\mathcal{R}$. The goal is to sort these items in column-major order, so that they can be output from the mesh in $O(\frac{m}{\sqrt{n}})$ time. We propose to show that with the above assumptions the entire task of sorting can be performed in $O(\frac{m}{\sqrt{n}})$ time. Thus, from our discussion in Section 2, we can conclude that our algorithm is both time- and VLSI-optimal.

To make the presentation more transparent and easier to follow we refer to the submesh consisting of the first $\frac{m}{\sqrt{n}}$ columns of $\mathcal{R}$ as $\mathcal{M}$. (In other words, $\mathcal{M}$ is the submesh that initially contains the input). Further, a *slice* of size k of the input consists of the items stored in k consecutive rows of $\mathcal{M}$.

We will first present an outline of our algorithm and then proceed to the details. Starting with slices of size $\frac{m}{\sqrt{n}}$ sorted in row major order, we use bucket sort to merge consecutive $\frac{m}{\sqrt{n}}$ of these into slices of size $(\frac{m}{\sqrt{n}})^2$ sorted in row major order. Using the same strategy, these slices are again merged into larger slices sorted in row major order. We proceed with the merging process until we have one slice of size $\sqrt{n}$, sorted in row major order. Finally, employing the data movement discussed in Proposition 3.3, the data is converted into column-major order.

We proceed to show that the merging of $\frac{m}{\sqrt{n}}$ consecutive sorted slices of size $(\frac{m}{\sqrt{n}})^i$ into sorted slices of size $(\frac{m}{\sqrt{n}})^{i+1}$ requires $O(\frac{n}{\sqrt{n}})$ time. It is convenient to view the original mesh $\mathcal{R}$ as consisting of submeshes $R_{j,k}$ of size $(\frac{m}{\sqrt{n}})^{i+1} \times (\frac{m}{\sqrt{n}})^{i+1}$ with $R_{j,k}$ involving processors $P(r,s)$ such that $(j-1)(\frac{m}{\sqrt{n}})^{i+1} < r \leq j(\frac{m}{\sqrt{n}})^{i+1}$ and $(k-1)(\frac{m}{\sqrt{n}})^{i+1} < s \leq k(\frac{m}{\sqrt{n}})^{i+1}$.

We refer to submeshes $R_{k,k}$ as *diagonal*. The diagonal submeshes can be viewed as independent meshes, since the same task can be performed, in

---

[1] We assume $O(1)$ time comparisons among the items in the universe

parallel, in all of them without broadcasting conflict. The algorithm begins by moving the data items in every $R_{k,1}$ to the diagonal submesh $R_{k,k}$. This can be accomplished column by column in $O(\frac{m}{\sqrt{n}})$ time. We now present the details of the processing that takes place in parallel in every diagonal submesh $R_{k,k}$.

The rightmost item in every row of $R_{k,k}$ will be referred to as the *leader* of that row (see Figure 2). To begin, the sequence of leaders $q_1, q_2, \ldots, q_{(\frac{m}{\sqrt{n}})^i+1}$ in $R_{k,k}$ is sorted in increasing order. Note that by virtue of our grouping, the sequence of leaders consists of $\frac{m}{\sqrt{n}}$ sorted subsequences, and so, by Proposition 3.2, the sequence of leaders can be sorted in $O(\log \frac{m}{\sqrt{n}})$ time. Let this sorted sequence be $a_1, a_2, \ldots, a_{(\frac{m}{\sqrt{n}})^i+1}$. For convenience, we assign $a_0 = -\infty$.

Next, in preparation for bucket sort, we define a set of $(\frac{m}{\sqrt{n}})^i$ buckets $B_1, B_2, \ldots, B_{(\frac{m}{\sqrt{n}})^i}$, such that for every j $(1 \le j \le (\frac{m}{\sqrt{n}})^i)$,

$$B_j = \{x \mid a_{\frac{(j-1)m}{\sqrt{n}}} < x \le a_{\frac{jm}{\sqrt{n}}}\} \qquad (2)$$

By definition, the leaders $a_{\frac{(j-1)m}{\sqrt{n}}+1}$ through $a_{\frac{jm}{\sqrt{n}}}$ belong to bucket B_j. This observation motivates us to call a row in $R_{k,k}$ *regular* with respect to bucket B_j if its leader belongs to B_j. Similarly, a row of $R_{k,k}$ is said to be *special* with respect to bucket B_j if its leader belongs to a bucket B_t with $t > j$, while the leader of the previous row belongs to a bucket B_s with $s \le j$. To handle the boundary case, we also say that a row is *special* with respect to B_j, if it is the first row in a slice and its leader belongs to B_t with $t > j$. Note that, all items in B_j must be in either regular rows or special rows.

Let us make a crucial observation.

Observation 3.4. With respect to every bucket B_j, there exist $\frac{m}{\sqrt{n}}$ regular rows and at most $\frac{m}{\sqrt{n}}$ special rows in $R_{k,k}$. □

In order to process each of the $(\frac{m}{\sqrt{n}})^i$ buckets individually, we view the mesh $R_{k,k}$ as consisting of submeshes $T_1, T_2, \ldots, T_{(\frac{m}{\sqrt{n}})^i}$ of size $(\frac{m}{\sqrt{n}})^{i+1} \times \frac{m}{\sqrt{n}}$. Each submesh T_l is dedicated to bucket B_l, in order to accumulate and process the elements belonging to that bucket.

In $O(\frac{m}{\sqrt{n}})$ time, we replicate the contents of T_1 in T_j $(2 \le j \le (\frac{m}{\sqrt{n}})^i)$. Using simple data movements, in each of the submeshes T_l, the values of $a_{(l-1)\frac{m}{\sqrt{n}}}$ and $a_{l\frac{m}{\sqrt{n}}}$ are broadcast to all the elements in it in $O(\frac{m}{\sqrt{n}})$ time so that all the elements that belong to B_l mark themselves. All the unmarked elements change their values to ∞.

Now the mesh $R_{k,k}$ is viewed as consisting of sub-

meshes $Q_{l,j}$ $(1 \le l \le \lceil \frac{m}{\sqrt{n}} \rceil^i, 1 \le j \le \lceil \frac{m}{\sqrt{n}} \rceil^i)$, of size $(\frac{m}{\sqrt{n}}) \times (\frac{m}{\sqrt{n}})$. The processor $P(r, s)$ is in $Q_{l,j}$ if $(l-1)\frac{m}{\sqrt{n}} < r \le l\frac{m}{\sqrt{n}}$ and $(l-1)\frac{m}{\sqrt{n}} < s \le l\frac{m}{\sqrt{n}}$. The objective of this step is to move all the elements belonging to bucket B_j in submesh T_j into submesh $Q_{j,j}$. Let q_k be the leader of a regular row. The rank r of this row is given by $r = v \bmod \frac{m}{\sqrt{n}}$, where $q_k = a_v$. Hence, in the order of their ranks, each of the regular rows is moved to the rth row of $Q_{j,j}$ $(j = \lceil \frac{v}{m/\sqrt{n}} \rceil)$, where r is its rank. Thus, all the regular rows with respect to B_j can be moved into the submesh $Q_{j,j}$ in $O(\frac{m}{\sqrt{n}})$ time.

A special row u in T_j with respect to bucket B_j is assigned a rank s, $s = \lceil \frac{u}{(\frac{m}{\sqrt{n}})^i} \rceil$. Note that no two special rows can have the same rank. In the order of their ranks, special rows are moved to the rows corresponding to their rank in $Q_{j,j}$. As the number of special rows is at most $\frac{m}{\sqrt{n}}$, the time taken to broadcast all the special rows to $Q_{j,j}$ is $O(\frac{m}{\sqrt{n}})$.

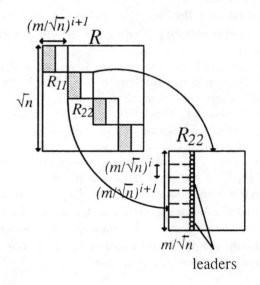

Figure 2: Illustration of diagonal submeshes and leaders

Now each processor in $Q_{j,j}$ holds at most one element from a regular row and one from a special row. We would now like to sort the elements in each of the submeshes $Q_{j,j}$, in an overlaid row-major order. In case the number of elements in $Q_{j,j}$ is less than or equal to $\frac{m^2}{n}$, after sorting, the elements can be placed one per processor. If the number of elements exceeds $\frac{m^2}{n}$, the first $\frac{m^2}{n}$ of them are said to belong to generation-1 and the remaining elements are said to belong to generation-2. We would like to place the elements belonging to generation-1 one per processor in row-major order and overlay this with those from generation-2, also in the same order. This is done as follows. Using optimal sorting algorithm for meshes, sort the elements from regular rows in $Q_{j,j}$ in $O(\frac{m}{\sqrt{n}})$ time and repeat the same for the elements from special rows. Merging the two sorted sequences thus obtained can be accomplished in another $O(\frac{m}{\sqrt{n}})$ time.

Now in each $Q_{j,j}$, all the elements know their ranks in bucket B_j. Our next goal is to compute the final rank of each of the elements in $R_{k,k}$. Before we give the details of this operation, we let $S_1, S_2, \ldots, S_{\frac{m}{\sqrt{n}}}$ be the sorted slices of size $(\frac{m}{\sqrt{n}})^i$ in $R_{k,k}$. Let m_j be the largest element in bucket B_j. In parallel, using simple data movement, each m_j is broadcast to all the processors in T_j in $O(\frac{m}{\sqrt{n}})$ time. Next, we determine the rank of m_j in each of the S_l's as follows: in every S_l we identify the smallest item (if any) strictly larger than m_j. Clearly, this can be done in at most $O(\frac{m}{\sqrt{n}})$ time, since every processor only has to compare m_j with the item it holds and with the item held by its predecessor. Now the rank of m_j among the items in $R_{k,k}$ is obtained by simply adding up the ranks of m_j in all the S_l's. Once these ranks are known, in at most $O(\frac{m}{\sqrt{n}})$ time they are broadcast to the first row of $Q_{j,j}$, where their sum is computed in $O(\log \frac{m}{\sqrt{n}})$ time. Observe that once m_j knows its rank in $R_{k,k}$, every item in bucket B_j finds its rank in $R_{k,k}$ by using its rank in the bucket, the size of the bucket and the rank of m_j in $O(1)$ time. Consequently we have proved the following result.

Lemma 3.5. The rank in $R_{k,k}$ of every element in every bucket can be determined in $O(\frac{m}{\sqrt{n}})$ time. $\square$

Finally, we need to move all the elements into the first $\frac{m}{\sqrt{n}}$ columns of $R_{k,k}$ in row major order. In $O(1)$ time, each element determines its final position from its rank r as follows. The row number x is given by $\lceil \frac{r}{(\frac{m}{\sqrt{n}})} \rceil$ and the column number y by $((r-1) \bmod \frac{m}{\sqrt{n}}) + 1$. In every submesh T_j, each element belonging to generation-1 is moved to the row x it belongs to (after sorting) by broadcasting the $\frac{m}{\sqrt{n}}$ rows of $Q_{j,j}$, one at a time. This takes $O(\frac{m}{\sqrt{n}})$ time. Now that every row of $R_{k,k}$ contains at most $\frac{m}{\sqrt{n}}$ elements. Knowing the columns they belong to, in another $\frac{m}{\sqrt{n}}$ time all the elements can be broadcasted to their positions along the row buses. This is repeated for the generation-2 elements. In parallel, every diagonal submesh $R_{k,k}$ moves back its data into the first $\frac{m}{\sqrt{n}}$ columns of submesh $R_{k,1}$. Thus, in an overall time of $O(\frac{m}{\sqrt{n}})$, all the elements are moved

to the first $\frac{m}{\sqrt{n}}$ columns of $\mathcal{R}$. Now $\mathcal{R}$ contains slices of size $(\frac{m}{\sqrt{n}})^{i+1}$, each sorted in row major order.

Lemma 3.6. Merging $\frac{m}{\sqrt{n}}$ consecutive sorted slices of size $(\frac{m}{\sqrt{n}})^i$ into sorted slices of size $(\frac{m}{\sqrt{n}})^{i+1}$ can be done in $O(\frac{m}{\sqrt{n}})$ time. $\square$

Let $T(i+1)$ be the worst-case running time of the basic step described above. It is easy to confirm that the recurrence describing the behavior of $T(i+1)$ is

$$\begin{cases} T(i+1) = T(i) + O(\frac{m}{\sqrt{n}}) & \text{for } i \geq 1 \\ T(1) = \frac{m}{\sqrt{n}} \end{cases}$$

The algorithm terminates after t iterations, when $(\frac{m}{\sqrt{n}})^{t+1} = \sqrt{n}$ i.e, $t + 1 = \frac{\log \sqrt{n}}{\log \frac{m}{\sqrt{n}}}$. $\quad$ (3)

By virtue of (1), (3) yields
$$t = c. \qquad (4)$$

Thus the total running time of our algorithm is
$$T(t+1) = O(c\frac{m}{\sqrt{n}}) \qquad (5)$$

Theorem 3.7. For every choice of a positive integer constant c, m items ($n^{\frac{1}{2}+\frac{1}{2c}} \leq m \leq n$) stored in the first $\lceil \frac{m}{\sqrt{n}} \rceil$ columns of a mesh with multiple broadcasting of size $\sqrt{n} \times \sqrt{n}$ can be sorted in $O(\frac{m}{\sqrt{n}})$ time. Furthermore, this is both time and VLSI-optimal. $\square$

References

[1] A. Aggarwal, Optimal bounds for finding maximum on array of processors with k global buses, *IEEE Trans. on Computers*, C-35, 1986, 62-64.

[2] D.Bhagavathi *et al.*, Time- and VLSI-optimal sorting on meshes with multiple broadcasting, *TR-92-29*, December 1992, Old Dominion University.

[3] D. Bhagavathi, S. Olariu, W. Shen, and L. Wilson, A Time-Optimal Multiple Search Algorithm on Enhanced Meshes, with Applications, *Proc. Journal of Parallel and Distributed Computing*, to appear.

[4] V. P. Kumar and D. I. Reisis, Image computations on meshes with multiple broadcast, *IEEE Trans. on Pattern Analysis and Machine Intelligence*, vol. 11, no. 11, 1989, 1194-1201.

[5] S. Olariu, J. L. Schwing, and J. Zhang, Time-Optimal Sorting and Applications on $n \times n$ Enhanced Meshes, *Proc. IEEE Internat. Conf. on Computer Systems and Software Engineering*, The Hague, May 1992.

[6] C. D. Thompson, The VLSI complexity of sorting, *IEEE Trans. on Computers*, C-32, 1983, 1171-1184.

A COMPARISON BASED PARALLEL SORTING ALGORITHM*

Laxmikant V. Kalé
Department of Computer Science
University of Illinois
Urbana, IL 61801
E-mail: kale@cs.uiuc.edu

Sanjeev Krishnan
Department of Computer Science
University of Illinois
Urbana, IL 61801
E-mail: sanjeev@cs.uiuc.edu

Abstract

We present a fast comparison based parallel sorting algorithm that can handle arbitrary key types. Data movement is the major portion of sorting time for most algorithms in the literature. Our algorithm is parameterized so that it can be tuned to control data movement time, especially for large data sets. Parallel histograms are used to partition the key set exactly. The algorithm is architecture independent, and has been implemented in the CHARM portable parallel programming system, allowing it to be efficiently run on virtually any MIMD computer. Performance results for sorting different data sets are presented.

1 Introduction

Sorting is one of the most basic algorithms in computer science. As a parallel application, sorting it is challenging because of the extent of communication it requires. Essentially, almost all data items must move from the processor they were originally on to some other processor. Moreover, in a network of processors, the average number of hops travelled by each data item is of the order of the network diameter.

The input to a sorting algorithm is a collection of records. Each record has a designated *key* field and possibly multiple data fields. Applications of sorting in different contexts may involve a variety of data types as keys. The keys may be integers, floating point number, long strings of characters, or records of arbitrary structure. Comparison based sorting methods only assume the existence of an operation which compares two keys and determines if one is smaller, larger or equal to the other, in some metric. Thus they are more general than methods such as radix sort which depend on knowledge of internal structure of the keys and its relationship with the underlying ordering.

In this paper we present a new comparison based parallel sorting algorithm, analyze its complexity, present performance results, and compare it with previous work.

2 The algorithm

We assume that there are initially n keys distributed among p processors such that each processor has approxi-

mately n/p keys[1]. At the end of sorting, the data must be approximately equally distributed among the processors, and for $i = 1$ to $n - 1$, all keys on processor $i - 1$ should be less than any key on processor i. In other words, processor 0 must have the smallest n/p keys, processor 1 must have the next n/p keys, and so on. The data within each processor must be in sorted order.

2.1 Overview of algorithm

The basic structure of the algorithm is similar to sample sort [4, 5, 6], load balanced sort [1], hyperquicksort [10] and binsort [11], in that in each phase, it finds $k - 1$ "splitter" keys that partition the linear order of keys into k equal partitions. These keys are found by an initial local sort on each processor followed by repeated iterations of global histogramming (Section 2.2). Each of the k partitions is then sent to the appropriate set of p/k processors such that the ith processor partition gets the ith data partition (Section 2.3). The next recursive phase of the algorithm can then run independently in all k processor partitions. Figure 1 gives a high level view of the algorithm.

k is the parameter that controls data movement. When k is equal to p, there is only one phase, and every key moves exactly once. At the other extreme, when k is 2, the algorithm essentially finds the median of the key set, then each processor sends all keys less than the median to the first $p/2$ processors and all keys greater than the median to the last $p/2$ processors, thus resulting in $log_k(p)$ phases of data movement.

2.2 Data Partitioning with Histograms

The object of this step is to find the $k - 1$ splitter keys, (partition boundaries) defined as the keys having ranks $n/k, 2*n/k \dots (k - 1)*n/k$ in the global order of keys. This step of the algorithm consists of a series of iterations consisting of upward passes (called *reductions*) and broadcasts along a logarithmic spanning tree. Starting with an initial set of splitter keys, each iteration refines the splitter keys till all partitions (as defined by the splitter keys) have approximately equal number of keys.

At the beginning of this step, each processor has a sorted key set. The first set of splitter keys can be found in var-

---

*This research was supported in part by the National Science Foundation grants CCR-90-07195 and CCR-91-06608.

[1]If the initial load distribution is unbalanced, the performance of the algorithm may degrade in proportion to the degree of imbalance. If so, we can use a load distribution phase similar to that in Section 2.3.

Perform Local Sort on each processor.

for (phase = 1 to $\log_2 p$)

    do /* This is the Histogramming step */

        Generate histogram probes and broadcast them.

        On each processor find key counts for probes.

        Send counts up spanning tree to root.

        At root processor, use new set of counts to refine

           current best values of partition boundaries.

    while (key counts for each partition are unequal)

    At root, generate k quintuples per subtree (section 2.3).

    Send quintuples down spanning tree. At each internal
node, split each quintuple among subtrees.

    On each processor, use k quintuples to find what data
to send to which processors.

    On each processor, send keys to other processors,
then merge keys received from other processors.

    Reconfigure tree into k separate spanning trees,
one for each partition.

endfor

Figure 1: High level description of algorithm. n is the total number of keys, p is the number of processors, k is the number of partitions

ious ways (see Section 6). In our current implementation we find the keys that equally divide the key set of the processor at the spanning tree root. If the root processor's distribution is representative of the global distribution, the splitter keys found by considering only the root would be reasonably good.

The number of probes (histogram boundaries) m may be more than the number of partition boundaries (number of splitter keys) $(k-1)$. Making m larger helps us refine the splitter keys faster, but increases the size of messages going up and down the spanning tree. We use $m = 3 * (k-1)$ as a heuristic, with $k-1$ of the m values being the best guesses for the splitter keys, another $k-1$ being slightly lesser than each of the first $k-1$, and the last $k-1$ being slightly greater than each of the first $k-1$.

The m probes are arranged in sorted order in a single message and broadcast to all processors using the spanning tree. Each processor then counts the number of keys less than or equal to each of the m probes. (A simple binary search is used). The array of m counts is sent up to the root by a reduction pass, with combining at internal nodes of the spanning tree, so that all message sizes are m. At internal nodes the counts received from subtrees are stored for use later during the data movement step.

The root maintains the current best known lower and higher values for each of the splitter keys. These values are updated using the new set of probes and their corresponding counts as follows : if any of the counts that came up the tree is nearer to the desired partition boundary (e.g.

$n/2$ for the median) than the current best known count, then that count is replaced with this new count and its corresponding probe value. If the partition counts as specified by the current splitter keys are not equal (within some user specified tolerance factor), the root calculates a new set of splitter keys. Each of the $k-1$ splitter key values is found by proportional linear interpolation[2] between the current best lower and higher values and their corresponding current best lower and higher counts, using the count for the desired partition boundary.

The root then starts the next histogram refinement iteration. If the partition counts are equal, the splitter keys have been successfully found, and the data movement step can be started.

2.3 Data Movement

Once the root has determined the splitter keys that partition the key set, we can efficiently move keys to their destination partitions as described below.

We initially have n/k keys in each partition distributed (possibly unequally) among p processors, which have to be transferred to a set of p/k processors such that each of the p/k processors ends up with approximately the same number of keys. Thus each processor must know whom to send to, as well as how many keys to send. We also want to ensure that each key moves exactly once per phase, hence it is not acceptable for a representative processor in each partition to collect all the keys for that partition and then distribute it among the processors of that partition.

The root knows at the beginning how many keys each partition has, but does not know which processors have those keys. However, this information is available at the internal nodes of the spanning tree. Hence in one downward pass, it is possible to tell each processor how many keys to send to which processors.

The entities being passed down the spanning tree are *constant sized* (5 integers) *quintuples* of the form *(startproc, startdata, middledata, endproc, enddata)* which indicates that processor *startproc* is supposed to receive *startdata* keys, processor *endproc* receives *enddata* keys, and the processors in between receive *middledata* keys each. There is a quintuple for each partition, hence the messages consist of k quintuples.

When a processor i at an internal node of the spanning tree receives a quintuple for a particular partition x, it means that all the keys in the subtree rooted at i belonging to partition x must be sent to the processors specified in the quintuple. Processor i has the latest histogram counts for its subtrees (which came from each of its subtrees during the last histogram refinement), hence it has information about how many keys each subtree has in each partition. This information is used to derive the quintuples for the subtrees from the quintuple that came from the parent.

---

[2]Interpolation is not strictly a comparison operation. However, an interpolation function can be provided for many cases where a comparison function is available. In Section 3 we prove that this step can be accomplished solely with a comparison function

At the root the global quintuple for the first partition would be $(0, n/p, n/p, p/k - 1, n/p)$. This is divided among its subtrees depending on how many keys each subtree has in that partition. Thus each internal node of the spanning tree receives a quintuple per partition from its parent and divides it among its children.

Finally, each processor has a quintuple per partition, indicating how many keys it has to send to which other processors for that partition. Now all sends and receives of keys for all partitions occur in parallel. Each processor sends the part of its local key set belonging to a partition to one or more processors in the set of p/k processors corresponding to the destination of that partition.

Note that in both data partitioning and data movement steps, the upward and downward passes along the spanning tree involve constant sized messages (i.e. the size is independent of the number of keys or number of processors) at all heights in the spanning tree, because of combining at internal nodes.

3 Complexity Analysis

We prove a bound on the number of histogramming iterations required, assuming no interpolation function is available. This bound is independent of the number of partitions because histogramming for all partitions is done together.

We consider the case where the number of partitions k is 2 , which means that we want to find the median of the key set. We maintain an upper and a lower bound on the value of the median. One of these bounds is refined in each histogramming iteration. We first prove that each refinement iteration decreases the number of keys between the bounds by a constant multiplicative factor c. Let n_i be the number of keys between the bounds at the ith iteration. Let m_i be the value to be guessed for the median at the ith iteration. Let j be the processor that has the most number of keys between the bounds. Processor j must have at least $n_i^j = \frac{n_i}{p}$ keys between the bounds. We choose m_i as the median of these n_i^j keys. Now after we get a global histogram, m_i may turn out to be either lesser or greater than the median we seek, so m_i becomes the new lower or upper bound, respectively. In either case, we have decreased the keys between the bounds by at least $\frac{n_i}{2*p}$. Thus in the worst case,

$$n_{i+1} = n_i - \frac{n_i}{2*p}$$
$$n_{i+1} = n_i/c \text{ where } c > 1.$$

Using this result we can prove a bound on the number of refinements. Let $n_0 = n$ be the initial number of keys, then

$$n_i = n_0/c^i .$$

If t is the total number of iterations required, then at the end of t iterations, the number of keys between the upper and lower bounds is 1, hence $n_t = 1$. Thus

$$1 = n_0/c^t , \text{ hence}$$
$$\mathbf{t = O(log(n))}.$$

It must be emphasized, however, that in practice, the number of histogram iterations required is much smaller than this worst case (depending on the data distribution, see Table 2). This is because we have an interpolation function, and we have more than one probe point, which allows us to converge towards the median much faster.

Using the above result, the total time taken by the algorithm is :

local sort time + $\lceil log_k(p) \rceil \{O(log(n))$(time per histogram) + data movement and merging time }

4 Implementation and results

We have implemented our algorithm in the CHARM portable parallel programming system [7]. CHARM supports C with a few extensions for creation of tasks and message passing. CHARM has a message driven model of execution, allowing overlap of computation and communication. Our implementation can run without change on nonshared as well as shared memory machines.

4.1 Basic results

Table 1 gives performance results for the nCUBE/2 and Intel iPSC/860 systems. The data set consists of 2^{23} integers formed by averaging four sets of random numbers as in the NAS Integer Sort Benchmark [3]. All random numbers for these measurements were generated using the C library function lrand48. Measurements for this table were taken with $k = 8$ and the tolerance factor as 1% (which means that the *final* counts of keys per processor can differ by only 1% of the total number of keys). The timings in all tables do not include startup, data generation and correctness checking times. From the table we can see good speedups as the number of processors increases. The results for the iPSC/860 compare well with other results reported in the literature [3, 2]. Considering the fact that parallel sorting is inherently a communication intensive application, these results demonstrate that our algorithm successfully reduces communication.

Table 1: Histogram Sort Basic Timings on the nCUBE/2 and iPSC/860. The keys are integers obtained by averaging 4 sets of random integers.

Number of Processors	Number of Keys	nCUBE/2 (s)	iPSC/860 (s)
64	2^{23}	12.30	3.87
128	2^{23}	6.87	2.66
128	2^{24}	-	5.04
256	2^{23}	3.93	-
512	2^{23}	2.46	-
1024	2^{23}	2.00	-
1024	2^{26}	9.14	-

4.2 Effect of data distribution

The performance of our algorithm depends, to some extent on the probability distribution of data in the space of possible key values. In general, uniform, random distributions are easier to sort as compared to non-uniform distributions having significant amounts of data concentrated in small value ranges.

Entropy [9, 8] has been suggested as a metric of distribution. Informally, the entropy of a key set corresponds to the number of "unique" bits in the key. However, entropy suffers from the drawback that low entropies (which mean that some bits are effectively unused) need not necessarily mean non-uniform distributions, and conversely, uniform distributions need not have high entropies. Consider a distribution consisting of equal numbers of all integers whose *least* significant 16 bits are 0. This distribution is uniform, but has entropy of 16 bits. Consider another distribution consisting of all integers whose *most* significant 16 bits are 0. This also has an entropy of 16 bits, but is highly nonuniform, in that all the data is concentrated in $1/2^{16}th$ of the data space. For most algorithms, this non-uniformity makes a difference in performance, still entropy usually will not distinguish between these distributions. Thus data distribution cannot be characterized by a single metric such as entropy.

Table 2 gives performance results for different distributions. These timings were taken on the nCUBE/2 with 256 processors, with number of partitions (k) as 8, tolerance factor 1% and a data set of 2^{23} integers. Distribution D1 is a uniform random distribution. Distribution D2 was generated by averaging four sets of random numbers. Distribution D3 has entropy 25.95 and was generated by performing a bitwise AND operation on two sets of random integers [9]. Distribution D4 has only 4 distinct values for the most significant 16 bits, while the least significant 16 bits have uniformly distributed random values. Distribution D4 has entropy 10.78 and was generated by doing an AND on four sets of random integers.

Although the table shows widely differing distributions, (from uniform, random to highly non-uniform), the histogramming time increases only slightly, from about 4 % to 10 % of the total execution time. This demonstrates that histogramming is an efficient method for obtaining the pattern of data distribution. This is because algorithm makes effective use of short, constant sized messages moving along the spanning tree.

Table 2: Timings as a function of distribution for sorting 2^{23} integers on the nCUBE/2 with 256 processors. There are 3 phases, because $k = 8$.

Distri-bution	Histogram iterations per phase	Histogram-ming Time (%)	Total Time (s)
D1	3+3+2	4.4	4.00
D2	4+3+2	5.0	3.93
D3	11+8+3	10.1	4.23
D4	12+5+2	10.4	4.28
D5	15+3+1	8.2	5.20

Table 3 gives a breakup of the times among the various steps. These timings were taken on the nCUBE/2, with

a data distribution as for Table 1. It can be seen that the time spent in communication, which is included completely in the sum of the histogramming and data movement times, (15 % on the average for the nCUBE), is not the major portion of the total execution time. (For the iPSC/860 about 45 % of the time was spent in communication). Moreover, data movement time decreases as a fraction of total time when k increases. In general, the optimum value of k depends on the number of processors. For small k, each data movement step takes less time because there are fewer messages, but there more phases. The opposite happens for large k. For 1024 processors, $k = 8$ was observed to be optimal.

Table 3: Breakup of timings for sorting 2^{23} integers on the nCUBE/2. Hist and Move are the times for histogramming and data movement, respectively.

Processors, Partitions, Phases	Local Sort (%)	Hist (%)	Move (%)	Total Time (s)
128, 4, 4	43.8	2.2	12.6	7.27
128, 8, 3	46.4	2.1	10.4	6.87
128, 16, 2	49.5	2.7	8.1	6.44
128, 128, 1	46.8	7.9	7.9	6.81

5 Previous work

The fast parallel sorting algorithms reported in the literature have been mostly non comparison-based ones, such as those based on radix-sort [2, 9]. Even with lexicographically-ordered data (such as names), radix sort based methods are inefficient or impossible to use, if the length of the keys is large and variable (and possibly unbounded). In addition, for non lexicographically-ordered data, such methods cannot be used at all.

For example, consider a set of customer records maintained by a consumer-service company. They wish to sort such records in decreasing order of importance/attention the company wishes to accord to customers (for a sophisticated mass mailing, say). Given two customer records, one can use a subroutine based on heuristics that decides which customer is worthy of more attention. However, the heuristic knowledge embodied in such a routine cannot always be extracted and quantified to yield a single numeric metric of importance. In such a situation, a comparison based sort can be used, while a radix sort cannot be used.

Moreover, for large data sets, most of the time for radix sort is taken up by data movement. For 32 bit keys, it is infeasible to use a 32 bit radix (that would involve 2^{32} buckets), hence at least two data moves involving all processors are required. In our algorithm, we can set k equal to p, so that only one move of data is required. Even if we set $k = \sqrt{p}$ so that two moves are required, the second move involves only $\sqrt{p}$ processors, hence will have less cost.

Load balanced sort [1] has the same high level steps as our algorithm. However, our data partitioning step avoids the transpose operation, and moreover, can use more probes than partitions, for faster convergence. Load balanced sort is a special case of our algorithm, with $k = p$, hence our algorithm has more flexibility in face of differing communication parameters. Finally, our algorithm does not depend on the topology of the underlying machine.

Our algorithm has the same high level steps as sample sort [4, 5, 6], with some important differences :

- Data Partitioning in our algorithm is exact; the time (number of histogram iterations) taken depends on the input distribution : for uniform distributions, one or two iterations are enough, hence an almost exact partition can be found quickly. For non-uniform cases the time taken depends on the tolerance (deviation from exact partitioning) specified by the user. Sample sort does not optimize the uniform case.

- Sample sort is not scalable, because the size of the messages that need to be communicated is $samplesize * O(p)$, which is usually $64 * p$ keys that could be large [4]. In our algorithm, all messages going up and down the spanning tree have size $O(k)$, which in the worst case is $5 * k$ integers, and the height of the spanning tree is $O(log(p))$. No keys are moved except in the actual data movement step.

The other salient features of our algorithm that differentiate it from most, if not all of the algorithms reported in the literature are :
- The number of partitions may be less than the number of processors, allowing the algorithm to run in more than one recursive phase, if that is faster.
- A novel method using histograms is used to find the partitions.
- All messages except those actually carrying keys are constant sized (independent of number of processors). Thus reductions take very little time.
- Key movement is reduced by requiring keys to move exactly once per phase.
- The algorithm does not depend on any particular architecture.

6 Discussion and Conclusions

We have described a comparison-based parallel sorting algorithm, and demonstrated its performance on a large parallel machine. The algorithm can be tuned to widely varying combinations of number of processors, data-sizes, and machine communication parameters by varying k, the number of partitions in each phase (and thus, the number of phases).

Further refinements to the sorting algorithm that we believe will improve its performance include
- A possible trade off involves doing away with the initial sorting phase, at the cost of spending more time on producing the histogram for a given probe with unsorted keys.
- Since the optimal value of k depends on the number of processors, we can allow k to dynamically vary from phase to phase, depending on the number of processors in a phase.
- The initial probe keys for the first histogramming iteration in each phase can be generated by more sophisticated methods. For example, each processor can find its local set

of splitter keys, which can be combined heuristically in a reduction, so that a more accurate picture of data distribution is obtained.
- The message driven nature of CHARM allows us to overlap communication and computation effectively. Thus the histogram phase itself can be pipelined by segmenting the set of probe keys and the corresponding histogram into separate pieces that can be broadcast and reduced concurrently with each other. The data movement step is currently overlapped with the merging step, and we could also overlap the histogramming and data movement steps.

Finally, we expect the algorithm to perform very well on MIMD computers with faster processors, because the time for local computations, which is the major fraction of the total time, will be reduced.

References

[1] B. Abali, F. Ozguner, and A. Bataineh. Load balanced sort on hypercube multiprocessors. In *Proc. Fifth Distributed Memory Computing Conference*, Apr. 1990.

[2] M. Baber. An implementation of the radix sorting algorithm on the Touchstone Delta Prototype. In *Proc. Sixth Distributed Memory Computing Conference*, May 1991.

[3] D. Bailey, E. Barszcz, L. Dagum, and H. Simon. NAS parallel benchmark results. In *Proc. Supercomputing*, Nov. 1992.

[4] G. Blelloch et al. A comparison of sorting algorithms for the Connection Machine CM-2. In *Proc. Symposium on Parallel Algorithms and Architectures*, July 1991.

[5] W. Fraser and A. McKellar. Samplesort : A sampling approach to minimal storage tree sorting. *Journal of the Association for Computing Machinery*, 17(3), July 1970.

[6] J. Huang and Y. Chow. Parallel sorting and data partitioning by sampling. In *Proc. Seventh International Computer Software and Applications Conference*, Nov. 1983.

[7] L. Kale. The Chare Kernel parallel programming language and system. In *Proc. International Conference on Parallel Processing*, Aug. 1990.

[8] C. Shannon and W. Weaver. *The Mathematical Theory of Communication*. University of Illinois Press, Urbana, 1949.

[9] K. Thearling and S. Smith. An improved supercomputer sorting benchmark. In *Proc. Supercomputing*, Nov. 1992.

[10] B. Wagar. Hyperquicksort: A fast sorting algorithm for hypercubes. In *Proc. Second Conference on Hypercube Multiprocessors*, Sept. 1986.

[11] Y. Won and S. Sahni. A balanced bin sort for hypercube multicomputers. *Journal of Supercomputing*, 2:435–448, 1988.

SnakeSort: A Family of Simple Optimal Randomized Sorting Algorithms

David T. Blackston *   Abhiram Ranade

Department of EECS
University of California at Berkeley
Berkeley, CA 94720
Email : davidb@cs.berkeley.edu, ranade@cs.berkeley.edu

Abstract – We present a family of exceedingly simple sorting algorithms called Snakesort. Snakesort is a natural generalization of the simple parallel bubble sort algorithm, also known as odd-even transposition sort. Instances of Snakesort can be implemented on any interconnection network so long as it has a Hamiltonian Path. We present two Snakesort algorithms for the hypercube and one for the two dimensional mesh. These algorithms are asymptotically optimal for the respective networks and have very small constant factors. The hypercube algorithms also give linear speedup over the best sequential algorithm. The main attraction of these algorithms is their extreme simplicity. Indeed, it seems unlikely that one could even conceive of simpler sorting algorithms for the hypercube and mesh.

1 Introduction

We present a family of exceedingly simple parallel sorting algorithms called Snakesort. Snakesort can be implemented on any interconnection network so long as it has a Hamiltonian path. Snakesort is a natural generalization of the simple bubble-sort algorithm (also called *odd-even transposition* sort[3]). Snakesort starts with a single preprocessing step (N denotes number of processors):

Step 0: Fix a Hamiltonian path, and number processors $0, \ldots, N-1$ according to their order along the path (Henceforth called the "Snake").

On all networks, the algorithm has the following structure:

Step 1: Locally sort the keys held by each processor.

---
*Research supported by an ONR Fellowship.

Step 2: Repeat the following *Merge-Split* step sufficiently many times:

2.1 Pick a matching from the edges in the network.

2.1 Let (u, v) be an edge in the matching, with $u < v$ according to the numbering along the Snake. Let each processor have k keys. For all (u, v), in parallel, first merge the $2k$ keys held by the two processors into a single sorted order, and then move the smallest k elements into processor u, and the largest k into processor v.

The odd-even transposition sort is obtained when the network is a linear array (and $k = 1$). N Merge-Split steps suffice, with the matching selected for the odd step consisting of processors $(2i, 2i + 1)$, and for the even step consisting of processors $(2i - 1, 2i)$ [3]. In the case of a richer graph, the hope is that by picking matchings that involve processors far apart on the Snake, we may be able to reduce the number of Merge-Split steps required considerably. This idea is similar to that of Shellsort.

We present two instances of Snakesort for hypercubes called Normal Sort (section 3) and Gray Sort (section 4), and an instance called Fast Shearsort which runs on meshes (section 5). All these algorithms sort correctly very quickly with high probability assuming that the input is a random permutation of the Nk keys, with k sufficiently large. In particular, Normal Sort will sort correctly in time $O(k \log k) + k(\log N + 1)$ for $k = \Omega(N^2 \log N)$. Gray Sort is a modification of Normal Sort that will sort correctly in time $O(k \log k) + k(2 \log N)$ but improves the bound on k to $\Omega(N \log N)$. It is easily seen from these bounds that both algorithms give linear speedup over the best sequential algorithms. Fast Shearsort will sort in time $O(k \log k) + k(2\sqrt{N} + o(\sqrt{N}))$ with high probability for $k = \omega(\sqrt{N} \log N)$, and this is also optimal to within tiny constant factors for the mesh. For all three algorithms it is possible to show that if k is substantially smaller than the bounds given then the algorithms will fail to sort quickly with constant probability. It can also be shown that if the initial distribution of keys is not uniform, then a simple randomization scheme can be used that will allow the algorithms to succeed with the same probability as if the initial distribution were uniform. Finally, we note that Normal Sort is in fact a normal algorithm, and can be implemented on shuffle-exchanges, CCCs etc. using well known transformations[3].

The main attraction of SnakeSort is extreme simplicity. Indeed, it seems unlikely that one could even conceive a simpler sorting algorithm for hypercubes than Normal Sort, or a simpler algorithm for meshes than Fast Shearsort. Our third algorithm, Gray Sort, is a slightly embellished version of Normal Sort, and is also interesting because it uses an unusual recursive formulation of Binary Reflected Gray Codes.

Our second motivation for studying Snakesort is its potential for use with irregular networks (e.g. local area networks of workstations), or even faulty networks. The only requirement in generating an instance of Snakesort for a network is that it contain a Hamiltonian path. Given a (possibly irregular) network, how fast can Snakesort run on it with the right choice of the Snake and matchings? Can we easily find the snake and the matchings? We have provided answers to these questions for hypercubes and meshes, but it is possible that one might be able to answer these questions in general using graph properties such as expansion.

1.1 Previous Work

There has been enormous work on sorting, starting with the pioneering work by Batcher who showed how to sort N keys on N processors in time $O(\log^2 N)$ (non-work-optimal), Ajtai Komlos and Szemeredi[1] (also see Leighton[2]) who gave a $O(\log N)$ time work-optimal algorithm. For the case in which each processor has a large number of keys (which is what we consider here), Leighton's Columnsort algorithm[2] can be used to derive work (deterministic) optimal sorting algorithms for hypercubic networks. Our (randomized) algorithms are considerably simpler, and have slightly smaller constant factors.

Much work has also been done on sorting on meshes, for example the work of Schnorr and Shamir[5], and Leighton, Makedon and Tollis[4], who give optimal algorithms. Our mesh algorithm in fact can be thought of as a (optimal, extremely abbreviated) version of the (non-optimal) Shearsort algorithm of [5]. Our algorithm is considerably simpler than the optimal algorithms in [4, 5].

2 Preliminaries

Our first two algorithms use an $N = 2^n$ node hypercube where each node holds k keys. Each node of the hypercube is labelled using an n-bit string, and nodes differing in a single bit are connected by an edge. For the Snake, we use the Binary Reflected Gray Code on n bits. This is defined recursively as follows. Let $G_n(x)$ denote the Gray code number of a processor labelled x in a 2^n node hypercube. We have $G_1(0) = 0$ and $G_1(1) = 1$. Further, $G_n(0x) = G_{n-1}(x)$, and $G_n(1x) = 2^n - 1 - G_{n-1}(x)$, where x is an $n-1$ bit string. If u is an n bit string, define $succ_n(u)$ to be the n bit string such that $G_n(u) = G_n(succ_n(u)) - 1$.

For convenience, we define the *lower* subcube to be the subcube consisting of those processors with Gray code numbers in the range $[0, \frac{N}{2} - 1]$, and the *upper* subcube to be those processors with numbers in the range $[\frac{N}{2}, N - 1]$.

Our analysis uses a randomized version of the 0-1 Sorting Lemma[3]. The standard version of the 0-1 Sorting Lemma states that if an oblivious algorithm sorts all zero-one input arrays correctly, then it will sort any input array, where an algorithm is said to be *oblivious* if the only operation it can perform is to compare two elements of the input and switch them if they are out of order, and, in addition, all the comparisons are specified in advance.

Lemma 1 *If, for every $0 \leq p \leq 1$, an oblivious algorithm sorts N keys with probability at least $1 - \delta$ in the case that each key is independently set to 1 with probability p and 0 otherwise, then with probability at least $1 - (N + 1)\delta$ the algorithm sorts any random permutation.*

The proof can be found in the full version of this paper.

The primary mathematical tool used in the analyses of our algorithms is the following lemma, which can be proven using Chernoff bound techniques.

Lemma 2 *Suppose* $X_1, \ldots, X_n$ *are independent Bernoulli random variables with parameter p. Let $X = \sum_i X_i$, with $E[X] = np$. Then for any β between 0 and 1*

$$Pr[|X - E[X]| \geq \beta n] \leq 2e^{-\frac{\beta^2 n}{3}}$$

3   Normal Sort

The Snake for Normal Sort is the Binary Reflected Gray Code. The algorithm has three phases:

1. Each processor does a local sort of its keys.

2. (Merge Phase) For i running from n to 1, processors that are connected along dimension i perform a Merge-Split.

3. (Mop-Up Step) For all x with $G_n(x)$ odd, do a Merge-Split between x and $succ_n(x)$.

Note that the merge phase is a normal algorithm[3], and the mop-up step can also be viewed as a normal algorithm by executing it one dimension at a time. With this change, Normal Sort can be ported to shuffle-exchanges, Butterflies, CCCs etc.

We analyze Normal Sort using the randomized version of the 0-1 Sorting Lemma which allows us to limit our analysis to inputs in which keys are either 0s or 1s. We need a few definitions. We say a subcube of the hypercube is *clean* if its processors contain all ones or all zeroes, and *dirty* if there are both zeroes and ones present. We also define a subcube to be *near-clean* if there are fewer than k ones or k zeroes present over the entire subcube. We state the following lemma without proof.

Lemma 3 (Near-Clean Lemma) *Suppose Normal Sort is run on an N node hypercube with k keys per node. The algorithm works correctly if the hypercube is near-clean initially.*

We next consider the case in which the input to the sorting algorithm is *balanced*, i.e. the input is zero-one and there exists numbers $0 \leq \alpha \leq \beta \leq 1$ such that the number of zeroes in each processor is between αk and βk, and $\beta - \alpha \leq \frac{1}{N}$.

Lemma 4 (Balanced Hypercube Lemma)
Suppose we run Normal Sort on an N node hypercube with k keys per node. The algorithm will sort correctly if the input is balanced.

(Sketch) We prove the lemma by induction on N. If $N = 1$ then after the local sort the input is sorted, and we are done. Suppose $N > 1$, and consider the effect of the first Merge-Split step. There are two cases to consider.

Suppose $\alpha \leq \beta < \frac{1}{2}$ or $\frac{1}{2} < \alpha \leq \beta$. In this case, it can be shown that after the first Merge-Split step one of the subcubes is clean, and the other is balanced. The induction hypothesis now holds.

If $\alpha \leq \frac{1}{2} \leq \beta$ then it can be shown that after the first Merge-Split step, both subcubes are near-clean. The Near-Clean Lemma now implies that after the Merge Phase both subcubes will be sorted. The Mop-Up Step now suffices to complete the sort. □

Theorem 1 *Suppose we run Normal Sort on an N node hypercube with k keys per node. The algorithm will sort a random permutation correctly using $\log N + 1$ Merge-Split steps with probability at least $1 - 2N(Nk+1)e^{-\frac{k}{12N^2}}$. If $k = \Omega(N^2 \log N)$, then the algorithm fails with polynomially small probability.*

(Sketch) Fix a probability p, and suppose that each input is 1 with probability p and 0 otherwise. We apply Lemma 2 to show that in this case the probability that the input is balanced is at least $1 - 2Ne^{-\frac{k}{12N^2}}$. The randomized 0-1 lemma now implies that the algorithm will sort a random permutation with probability at least $1 - 2N(Nk+1)e^{-\frac{k}{12N^2}}$, and the bound on k is easily observed. □

4   Gray Sort

As in the case of Normal Sort, the Snake for Gray Sort is also the Binary Reflected Gray Code. Let H be a 2^n node hypercube, and H_0 and H_1 be the subcubes with least significant bit 0 and 1 respectively. Gray Sort is defined recursively for a 2^n node hypercube as follows.

- If $n = 0$ do a local sort.

- If $n > 0$ then do the following.

  1. In parallel, run Gray Sort on H_0 and H_1.

  2. For all x with $G_n(x)$ even, do a Merge-Split between x and $succ_n(x)$.

3. For all x with $G_n(x)$ odd, do a Merge-Split between x and $succ_n(x)$.

This algorithm is very similar to Normal Sort; in fact it is easily seen that the algorithm consists of the Merge-Split steps of Normal Sort alternating with new Merge-Split steps that act as Mop-Up steps. These new Merge-Split steps allow us to improve the bound on k as per the following theorem which we state without proof.

Theorem 2 *If Gray Sort is run on an N node hypercube with k keys per node, then a random permutation is sorted correctly using $2 \log N$ Merge-Split steps with probability at least $1 - 4N(Nk + 1)e^{-\frac{k}{6N}}$. If $k = \Omega(N \log N)$ then this probability can be made polynomially close to one.*

5 Fast Shearsort

Consider a $\sqrt{N} \times \sqrt{N}$ mesh of processors, each with k keys. The sorted order will be the snake-like order used in Shearsort [3] and illustrated below.

Figure 1: The Snake for Fast Shearsort

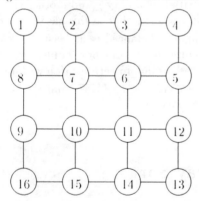

When $k = 1$, it has been shown that sorting requires $3\sqrt{N} - o(\sqrt{N})$ Merge-Split steps. For large k we will show that with very high probability $2\sqrt{N} + o(\sqrt{N})$ Merge-Split steps suffice for sorting. As in the case of the hypercube algorithm, the sorting algorithm for the mesh is quite simple. There are 4 phases.

- Each processor does a local sort.

- A complete parallel bubble sort is run along the columns.

- A complete parallel bubble sort is run along the rows.

- Parallel bubble sort is run over the entire sorted order of the mesh until the sort is complete.

The second and third phases each require $\sqrt{N}$ Merge-Split steps, and it can be shown that if k is large enough, then the final round of parallel bubble sort along the rows will only require $o(\sqrt{N})$ Merge-Split steps, allowing us to acheive the required bound. The following theorem, stated without proof, summarizes our results for this algorithm.

Theorem 3 *If Fast Shearsort is run on an $\sqrt{N} \times \sqrt{N}$ mesh with $k = \omega(\sqrt{N} \log N)$ keys per node, then the algorithm will require $2\sqrt{N} + o(\sqrt{N})$ Merge-split steps and sort correctly with probability that goes polynomially to one.*

References

[1] M. Ajtai, J. Komlos, and E. Szemeredi. An O(n log n) sorting network. In *Proceedings of STOC 83*, pages 1–10, 1983.

[2] F. T. Leighton. Tight bounds on the complexity of parallel sorting. *IEEE Transactions on Computers*, C–34(4):344–354, April 1985.

[3] F. T. Leighton. *Introduction to parallel algorithms and architectures*. Morgan-Kaufman, 1991.

[4] F. T. Leighton, F. Makedon, and I. Tollis. A $2N-2$ step algorithm for routing in an $N \times N$ mesh. In *Proceedings of the ACM Symposium on Parallel Algorithms and Architectures*, pages 328–335, June 1989.

[5] C. P. Schnorr and A. Shamir. An optimal sorting algorithms for mesh connected computers. In *Proceedings of the ACM Annual Symposium on Theory of Computing*, pages 255–263, 1986.

MERGING MULTIPLE LISTS IN $O(\log n)$ TIME

Zhaofang Wen

Department of Computer Science
University of Minnesota at Duluth
Duluth, MN 55812, USA
e-mail: zwen1@uc.d.umn.edu

Abstract– *The problem of merging k $(k \geq 2)$ sorted lists is considered. We give an optimal parallel algorithm which takes $O(\frac{n \log k}{p} + \log n)$ time using p processors on the CREW PRAM, where n is the total size of the input lists. The algorithm can be seen as a unified algorithm for parallel sorting and merging. To obtain this algorithm, interesting techniques for designing algorithms are also discussed, including a new processor assignment strategy.*

1 Introduction

The problem of merging arises naturally in several applications like database management systems. Because of its fundamental importance in computer science [9], much effort has been devoted to finding efficient parallel algorithms for the problem (see [1, 2, 3, 6, 7, 8, 10, 11, 12] to name just a few). Most study the problem of merging two sorted lists. For example, cost-optimal parallel algorithms for merging two sorted lists can be found in [1, 8]. In particular, the 2-way merge algorithm in [8] runs in $O(\log n)$ time using $O(\frac{n}{\log n})$ processors on the EREW PRAM. Parallel algorithms for merging multiple sorted lists have also been proposed under various assumptions [7, 11, 12]. These multi-way parallel algorithms are not cost-optimal. To merge k $(k \geq 2)$ lists, we can repeatedly merge the lists two at a time using an existing parallel 2-way merge algorithm (e.g, the one in [8]), until a single sorted list is obtained. Such a method can lead to a cost-optimal parallel k-way merge algorithm with time complexity $O(\log k \times \log n)$. Our interest is in finding faster and also cost-optimal parallel algorithm to merge k sorted lists. So far, no such algorithm has been reported.

In this paper, we give a cost-optimal parallel algorithm to $O(\frac{n \log k}{p} + \log n)$ time using p processors on the CREW PRAM, where n is the total size of the input lists. This algorithm achieves $O(\log n)$ time when $p = \frac{n \log k}{\log n}$. When $k = n$, it is an optimal sorting algorithm. Therefore, it can be seen as a unified algorithm for parallel sorting and merging. To obtain this cost-optimal algorithm, interesting techniques for designing parallel algorithms are discussed, including a new processor assignment

strategy.

2 Preliminaries

To simplify our presentation, we borrow some terminologies from [4, 8]. An interval $[a, b]$ *intersects* with (contains) an item e if $a \leq e \leq b$. Let $A = a_1, a_2, ..., a_l$ and $B = b_1, b_2, ..., b_m$ be two sorted lists, and let f be an item. We define an item f to be *ranked* in B, if we know the item b_j of B such that $b_{j-1} \leq f < b_j$ (if necessary, we let $b_{j-1} = -\infty$ or $b_j = \infty$). We say that f is *straddled* by the b_{j-1} and b_j; and we define the *rank* of f in B to be $j - 1$. We define A to be *ranked* in B (denoted $A \rightarrow B$) if each item of A is ranked in B. Define A and B to be *cross-ranked* (denoted $A \times B$) if both $A \rightarrow B$ and $B \rightarrow A$. We use $A \bigcup B$ to denote the sorted merged list of all items in A or B.

Let L and J be sorted lists. We say that L is a C-cover (for some constant C) of J if between each two adjacent items in $(-\infty, L, \infty)$ there are at most C items from J (where $(-\infty, L, \infty)$ denotes the list consisting of $-\infty$, followed by L, followed by ∞).

Let L be a sorted list of m elements. and t be an integer. $SAMPLE_t(L)$ denote a sorted list consisting of every t-th item of L. That is, $SAMPLE_t(L)$ consists of the t-th item of L followed by the $(2t)$-th item of L, followed by the $(3t)$-th item of L and so on. The size of $SAMPLE_t(L)$ is $\lfloor \frac{m}{t} \rfloor$.

3 Algorithm

In this section, we present an optimal parallel algorithm for the problem of merging k sorted lists. One way to solve the problem is by applying repeatedly an existing optimal parallel 2-way merge algorithm to merge the lists, two at a time, until a single sorted list is left. If we use the existing 2-way merge algorithms, we can merge only $\frac{k}{2}$ pairs of lists in the first round; and merge $\frac{k}{4}$ pairs of lists in the second round after the first round is finished and so on. Each round takes logarithmic time, As mentioned in the Section 1, this method will not lead to an $O(\log n)$ time optimal solution, because each round takes more than $O(1)$ time.

To obtain a faster algorithm, we have to pipeline the merges in different rounds. That is, the second round of merge will begin before the first round ends; and the third round will begin before the second round is finished and so on. However, this can not be done if we use the existing optimal parallel 2-way merge algorithms because they give output only when they are finished. Therefore, the first thing we need is a new 2-way merge algorithm.

In the following, we first give a new 2-way merge algorithm which increasingly (twice as before) generates sorted samples of the final output list. It stops when the size of the sample matches that of the expected output. With this 2-way merge algorithm, we then extend a technique of Cole [4] to obtain a parallel k-way merge algorithm. In order to implement optimally our k-way merge algorithm using p processors, we also need a new processor assignment technique, which will be presented in the next section.

Let L_1 and L_2 be two sorted lists. Assume $|L_1| \geq |L_2|$. Denote $n_1 = |L_1|$ and $n_2 = |L_2|$; also denote $U_s = SAMPLE_{\lceil \frac{n_1}{2^s} \rceil}(L_1)$ and $V_s = SAMPLE_{\lceil \frac{n_2}{2^s} \rceil}(L_2)$. Our parallel 2-way merge algorithm is as the following:

Algorithm 2-merge

Form U_0 and V_0;

for stage $s = 0$ to $\lceil \log n_1 \rceil$ do

    (1) Compute $U_s \cup V_s$;

    (2) Form U_{s+1}, and V_{s+1};

    (3) Compute $(U_s \cup V_s) \to U_{s+1}$ and $(U_s \cup V_s) \to V_{s+1}$;

end.

This algorithm merges the input lists in stages. In each stage it produces a sample of final output list (thus enabling the merges in the next round of the k-way to be performed in a pipelined fashion). We will show later that each stage can be done in $O(1)$ time using $|U_s| + |V_s|$ processors.

With the above 2-way merge algorithm, we are now ready to give the k-way merge algorithm. Let $L_1, L_2, ..., L_k$ be the sorted lists to be merged. Denote $n_1 = |L_1|$, $n_2 = |L_2|$, ..., and $n_k = |L_k|$. Assume $n_1 > n_2 > ... > n_k$. The computation is guided by a perfect binary tree with $\lceil \log k \rceil + 1$ levels (the root of the tree is at level 0, and the leaves are at level $\lceil \log k \rceil$). The input lists are placed at the leaves of the tree. In particular, from left to right, the j-th input list L_j is placed at the j-th leaf of the binary tree, where $1 \leq j \leq k$. The task is to compute, for each node v, the sorted list $L(v)$ that consists of all the items initially at the leaves of the subtree rooted at v.

The algorithm for constructing $L(v)$ for each node v of the tree proceeds in stages. To compute $L(v)$ for each node v, we maintain $U(v)$, a sorted subset of $L(v)$. Specifically, at stage s, we will compute $U_s(v)$. As s increases, so does $|U_s(v)|$; and $U_s(v)$ becames a more and more accurate approximation of $L(v)$. Also, after constructing $U_s(v)$ in

stage s, we will create $SU_{s+1}(v)$, a sample of $U_s(v)$. $SU_{s+1}(v)$ will be useful for constructing $U_{s+1}(p(v))$ in stage $s + 1$, where $p(v)$ is the parent of v.

We say v is *full* in stage s if $|U_s(v)| = |L(v)|$. We say that a node v is *active* in stage s if one of the following is true: (i) $0 < |U_s(v)| \leq |L(v)|$; (ii) v has become full for no more than three stages; and (iii) $|SU_s(lchild(v))| + |SU_s(rchild(v))| > 0$, where $lchild(v)$ and $rchild(v)$ are the left and right children of v.

Initially, at stage 0, $SU_0(v)$ is empty for every node v. At an intermediate stage s, if v is active, three steps will be performed for node v. Assume, at the beginning of stage s, we already have $U_{s-1}(v) \to SU_s(lchild(v))$ and $U_{s-1}(v) \to SU_s(rchild(v))$ for any non-leaf node v. The details of our k-way merge algorithm is given below.

Algorithm k-merge

$SU_0(v) = \Phi$, for all node v in the binary tree;

for stage $s = 0$ to $\lceil \log n_1 \rceil + 3(\lceil \log k \rceil + 1)$ do the following for each active node v

    (1) Compute $U_s(v)$. There are two cases.

        case 1.1 v is the j-th leaf (from left to right) where $1 \leq j \leq k$.
        $U_s(v) = SAMPLE_{\lceil \frac{n_j}{2^s} \rceil}(L_j)$

        case 1.2 v is not a leaf.
        $U_s(v) = SU_s(lchild(v)) \cup SU_s(rchild(v))$

    (2) Form $SU_{s+1}(v)$. There are four cases.

        case 2.1 v is active and not full
        $SU_{s+1}(v) = SAMPLE_4(U_s(v))$

        case 2.2 the first stage after v became full
        $SU_{s+1}(v) = SAMPLE_4(U_s(v))$

        case 2.3 the second stage after v became full
        $SU_{s+1}(v) = SAMPLE_2(U_s(v))$

        case 2.4 the third stage after v became full
        $SU_{s+1}(v) = SAMPLE_1(U_s(v))$

    (3) Compute $U_s(v) \to SU_{s+1}(lchild(v))$ and $U_s(v) \to SU_{s+1}(rchild(v))$

end.

Because of the sampling strategy in Step (2), a node v becomes full three stages after both of its children became full. Therefore, at the end of stage $\lceil \log n_1 \rceil + 3(\lceil \log k \rceil + 1)$ the root of the tree becomes full; and $U(root)$ contains the final output of the algorithm.

We now look at the behavior of the algorithm.

Lemma 3.1 *For any node v, $s \geq 0$, and $a, b \in (-\infty, SU_s(v), \infty)$, if interval $[a, b]$ intersects with $l + 1$ items in $(-\infty, SU_s(v), \infty)$, then it intersects with at most $2l + 1$ items in $SU_{s+1}(v)$ for all $l \geq 1$.*

Proof. (sketch) According the way samples are taken, the claim can be shown by induction on stage number. □

Taking $l = 1$ in the previous lemma, we have,

Lemma 3.2 $SU_s(v)$ is a 3-cover of SU_{s+1}. □

4 Processor Assignment

In the previous section, we only assume that processors are always in place whenever they are needed to do the merging of samples at different nodes of the tree. In this section, we present the processor assignment strategy that supports our algorithm in the previous section.

Assume $P_1, P_2, ..., P_p$ are the processors to be used by the algorithm. For simplicity of discussion, we assume $p \geq 3k$, where k is the number of input lists.

When we say that the processors are assigned to the tree nodes to perform the steps (1)–(3) in stage s of Algorithm k-merge, the following conditions are true.

1. Each processor knows which node v it is assigned to; it also knows the lists (and their sizes) to be processed at node v during stage s, including $SU_s(lchild(v))$, and $SU_s(rchild(v))$ etc.

2. The group of processors assigned to the same node v are indexed consecutively; and each processor knows the size of the group as well as its relative position within group.

In addition to these two obvious standards, we also wish to have an assignment scheme which can distribute the work load evenly among all processors.

In the rest of this section, we will define and then show how to implement a processor assignment scheme which satisfies all the standards listed above. Besides, as we will show, our processor assignment can be implemented without increasing the overall complexity of our k-way merge algorithm.

Consider stage s. At the beginning of stage s, define $SIZE_s(v) = |SU_s(lchild(v)) + |SU_s(rchild(v))|$; and define

$$w_s(v) = \begin{cases} 2^s & v \text{ is an active leaf} \\ SIZE_s(v) & v \text{ is an active non-leaf} \\ 0 & v \text{ is not active} \end{cases}$$

$$W_s = \sum_{\text{all } v} w_s(v)$$

Notice that $O(W_s)$ is the total amount of work to be performed by Algorithm k-merge in stage s.

Let $p' = p - 2k$. Define

$$p_s(v) = \lceil \frac{p' \times w_s(v)}{W_s} \rceil$$

We assign $p_s(v)$ processors to each node v. The total number of processors assigned in the way is at most

$$\sum_{\text{all } v} p_s(v) = \sum_{\text{all } v} \lceil \frac{p' \times w_s(v)}{W_s} \rceil$$

$$\leq 2k + \frac{p'}{W_s} \times \sum_{\text{all } v} w_s(v) = 2k + p' = p$$

With this processor assignment, the time required to finish the work at node v in stage s is $O(\frac{w_s(v)}{p_s(v)}) = O(\frac{w_s(v)}{\lceil \frac{p' \times w_s(v)}{W_s} \rceil}) = O(\frac{W_s}{p'}) = O(\frac{W_s}{p})$, which is the same for every node; in other words, the work load is evenly distributed among the p processors.

We now discuss how to assign processors, $P_1, P_2, ..., P_p$, on to the active nodes in each stage s. The assignment is done in two phases. We first decide, for each processor, which level of the binary tree the processor is to be assigned to; we then further assign the processor on to a node at that level. It happens that the implementations of these two phases are based on the same technique. So we only give the details for the first phase as follows.

Define $t_{s,i}$ to be the number of processors to be assigned to level i ($0 \leq i \leq \lceil \log k \rceil$) in the first phase such that

$$t_{s,i} = l_i + \frac{p'}{W_s} \times \sum_{v \text{ at level } i} w_s(v)$$

where l_i is the number of nodes at level i. Notice that

$$\sum_{i=0}^{\log k} t_{s,i} = p' + 2k = p$$

This means that, in phase 1, we will not need more processors than we actually have for the algorithm. Compute

$$q_{s,i} = \sum_{j=0}^{i} t_{s,i} \quad (0 \leq i \leq \lceil \log k \rceil)$$

We assign to level i ($1 \leq i \leq \lceil \log k \rceil$) the processors indexed from ($q_{s,i-1} + 1$) to ($q_{s,i}$), where we assume $q_{s,-1} = 0$.

In order to inform each of the processors the level number it is assigned to, we can use a vector $A_s[1..p]$ to keep the information. In particular, compute $A_s[1..p]$ such that

$$A_s[j] = \begin{cases} 0 & 1 \leq j \leq q_0 \\ q_{s,i} & q_{s,i-1} + 1 \leq j \leq q_{s,i} \end{cases}$$

(Note: $q_{s,i}$ ($0 \leq i \leq \lceil \log k \rceil$) and $A_s[1..p]$ can be computed by applying the optimal parallel algorithm for "prefix-sum" computation [5].)

Processor P_j $(1 \le j \le p)$ can read $A_s[j]$ to know which level it is assigned to.

Using exactly the same technique, we can further assign the processors at each level to the individual nodes at that level. We will need another vector $A'_s[1..p]$ to keep this information, i.e. $A'_s[j]$ $(1 \le j \le p)$ stores the information of the node to which processor P_j is assigned in stage s. Notice that the processor assignment thus obtained satisfies the two conditions stated at the beginning of this section.

There is one issue yet to be discussed, i.e. the time complexity of implementing the above processor assignment strategy. We know that quantities $w_s(v)$, W_s, and $t_{s,i}$ $(0 \le i \le \lceil \log k \rceil)$ can not be computed in $O(1)$ time. In order to use them in stage s, we have to precompute them before the merging process actually begins. Even if we have $w_s(v)$, W_s and $t_{s,i}$ $(0 \le i \le \lceil \log k \rceil)$ at the beginning of stage s, the computations of $q_{s,i}$ $(0 \le i \le \lceil \log k \rceil)$, and $A_s[1..p]$ still need more than constant time. So they also have to be precomputed. Altogether, we need to precompute, the following: $w_s(v)$, W_s, $t_{s,i}$, $q_{s,i}$, $A_s[1..p]$, and $A'_s[1..p]$, for every node v, $s = 0, 1, ..., \lceil \log n_1 \rceil + 3(\lceil \log k \rceil + 1)$, and $(0 \le i \le \lceil \log k \rceil)$. These quantities can be computed by applying the optimal parallel algorithm [5] for prefix-sum computation and thus can be done in $O(\log p + \lceil \log n_1 \rceil + (\lceil \log k \rceil))$ time using $O(p)$ processors on the EREW PRAM.

Therefore, processor assignment for all stages can be done in advance without increasing the overall time and processor complexities of the algorithm.

5 Complexity

In this section, we give the time complexity of the algorithm. The following lemma is crucial to the analysis.

Lemma 5.1 *Let* J_1, J_2, J_3, K_1, K_2, *and* K_3 *be sorted lists. If the following (input) conditions are true:* **(a)** J_1 *is a C-cover of* J_2, *and* K_1 *is a C-cover of* K_2, *where C is a constant;* **(b)** $J_1 \times K_1$ *;* **(c)** $(J_1 \cup K_1) \rightarrow J_2$ *and* $(J_1 \cup K_1) \rightarrow K_2$ *; then* $(J_2 \cup K_2)$ *(i.e.* $J_2 \times K_2$*) can be computed in* $O(1)$ *time using* $|J_1| + |K_1|$ *processors on the CREW PRAM.*

If in addition to conditions (a)-(c), the following conditions are also true, **(d)** J_2 *is a C-cover of* J_3, *and* K_2 *is a C-cover of* K_3. **(e)** $J_2 \rightarrow J_3$ *and* $K_2 \rightarrow K_3$*; then* $(J_2 \cup K_2) \rightarrow J_3$ *and* $(J_2 \cup K_2) \rightarrow K_3$ *can also be computed in* $O(1)$ *time using* $|J_1| + |K_1|$ *processors on the CREW PRAM.* □

With this lemma and the processor assignment strategy, we have the following results.

Theorem 5.1 *Using p processors on the CREW PRAM, Algorithm* 2-merge *takes* $O(\log n_1 + \frac{n_1 + n_2}{p})$ *time.* □

Theorem 5.2 k *sorted lists can be merged in* $O(\frac{n \log k}{p} + \log n)$ *time using p processors on the CREW PRAM, where n is the total size of the input lists. This result is cost optimal.*

Proof. (sketch) By the processor assignment scheme, the work load is evenly distributed among all processors. Thus, in every stage, the work at different nodes can be done using about the same amount of time. The total work performed by the algorithm is $O(n \log k)$; the time complexity follows. □

Acknowledgement Thanks to an anonymous referee for careful reading of the paper whose comments have helped improve the presentation.

References

[1] S. G. Akl and N. Santoro. "optimal parallel merging and sorting without memory conflict". *IEEE Trans. Comp.*, pages 1367–1369, 1987.

[2] K. E. Batcher. "sorting networks and their applications". In *Proc. of the AFIPS Spring Joint Comp. Conf. 32*, pages 307–314, 1968.

[3] A. Borodin and J. E. Hopcroft. "routing, merging and sorting on parallel models of comparison". *J. Comp. and Sys. Sci.*, 30:130–145, 1985.

[4] R. Cole. "parallel merge sort". *SIAM J. on Comput.*, pages 770–785, August 1988.

[5] R. Cole and U. Vishkin. "approximate parallel scheduling. part 1: the basic technique with applications to optimal parallel list ranking in logarithmic time". *SIAM J. on Comput.*, pages 128–142, 1988.

[6] E. Dekel and I. Ozsvath. "parallel external sorting". *J. of Parallel and Dist. Comput.*, 6:623–635, 1989.

[7] J. Y. Fu and F. C. Lin. "optimal parallel external merging under hardware constraints". In *Proc. of 1991 ICPP*, pages III–70–III74, 1991.

[8] T. Hagerup and C. Rub. "optimal merging and sorting on the EREW PRAM". *IPL*, 33:181–185, 1989.

[9] D. E. Knuth. *The Art of Computer Programming, Vol. 1, Fundamental Algorithms*. Second Edition, Addison-Wesley, Reading, Mass., 1973.

[10] Y. Shiloach and U. Vishkin. "finding the maximum, merging and sorting in a parallel computation". *J of Alg.*, 2:88–102, 1981.

[11] P. Valduriez and G. Gardarin. "join and semijoin algorithms for multiprocessors database machines". *ACM Trans. on DataBase Sys.*, 9(1):133–161, 1984.

[12] P. J. Varman, S. D. Scheufler, B. R. Iyer, and G. R. Ricard. "merging multiple lists on hierarchical-memory multiprocessors". *J. of Parallel and Dist. Comput.*, 12:171–177, 1991.

On the Bit-Level Complexity of Bitonic Sorting Networks*

Majed Z. Al-Hajery and Kenneth E. Batcher
Department of Mathematics and Computer Science
Kent State University
Kent, OH 44240-0001

Abstract – *Bitonic sorting networks can be implemented with a bit-level cost complexity of $O(N \log^2 N)$ and a bit-level time complexity of $O(\log^2 N)$ using comparators with bit-level $O(1)$ time and cost complexities. Items to be sorted are pipelined (worm-hole routed) bit-serially most-significant-bit first through the network. The cost complexity can be reduced to $O(N \log N)$ by recirculating items of length $O(\log N)$ through $\log N$ stages.*

Introduction

The sorting problem of N elements over a well defined topology network has been extensively studied by a number of researchers in the last few decades [1, 4, 5]. For the most part, two methods are used to measure the complexity of parallel sorting networks:

1. Number of comparators(Cost): This is equal to the number of constant comparators the network contains.

2. Time: For any input a, the network time is the number of comparators a has to visit before a exits the network.

A Bit-Serial Bitonic Sorting Network(BBSN) will be presented here. This network has a time complexity of $O(\log^2 N)$ and cost complexity of $O(N \log N)$ both in bit level. The cost and time complexities of this network compare excellently with other networks.

Stone [5] has shown an implementation of the Bitonic sorter with a single stage perfect shuffle network. Ironically, the large reduction in the cost complexity factor increases the time complexity in bit level since sorting N keys in bit level requires sorting N sequences of bits. Assigning this task to a single stage perfect shuffle network increases the time complexity by a factor of $\log N$. This is due to either:

1. The time and cost complexities of the comparator must be $O(\log N)$, or

2. Each bit slice must be sorted independently if a comparator of $O(1)$ time and cost complexities is used.

For those two reasons, many papers in the literature quote a bit level time and cost complexities for the bitonic sorter of $O(\log^3 N)$ and $O(N \log^3 N)$ respectively. Recently, Cam and Fortes [3] have compared their four permutation network to such implementation of bitonic sorting networks. Also, Chien and Oruç [2] have quoted such complexities for the bitonic sorting network. Our implementation of the bit level bitonic sorting network reduces the cost complexity by a factor of $O(\log)$ and still preserves the well known time complexity of an $O(\log^2 N)$. The bit level comparator in this network has time and cost complexities of $O(1)$.

Bitonic Sorter

The Bitonic Sort network requires $O(\log^2 N)$ time steps if $N/2$ comparators are used to perform their operations (compare and exchange) in parallel. Each comparator (I-comparator or D-comparator Figure 1) takes two numbers as inputs, where they are exchanged , if necessary, so they remain in an increasing or decreasing order according to the type of the comparator. The following definition and theorem are essential fundamentals in a Bitonic Sorting Network:

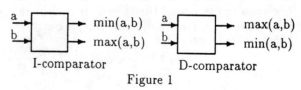

I-comparator   D-comparator

Figure 1

Definition: 1 *For any sequence of Bitonic numbers $\{a_1, a_2, a_3, \cdots, a_n\}$ one of two conditions must be met;*

*This material is based upon work supported by the National Science Foundation under grant MPP-9004127.

1. *There exists an index* i, $1 \leq i \leq n$, *such that*

$$a_1 \leq \cdots \leq a_{i-1} \leq a_i \geq a_{i+1} \geq \cdots \geq a_n \quad or$$

2. *The list can be cyclically shifted until the first condition is met.*

The first compare-exchange stage in the network (see Figure 2) divides the bitonic list into two bitonic lists. Every element in the first list is smaller than any element in the second list. In the next stage, the same process is applied to divide the two lists into four Bitonic lists ...etc. After $\log N$ stages, the list is sorted in an increasing order.

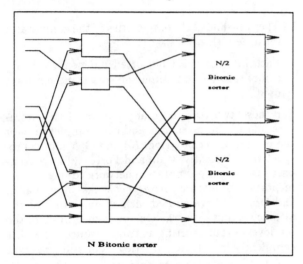

Figure 2

Theorem 1 (Batcher) *Let* $\{a_1, a_2, a_3, \cdots, a_{2n}\}$ *be bitonic. If* $D = \{d_1, d_2, \cdots, d_i, \cdots, d_n\}$ *where* $d_i = min(a_i, a_{n+i})$ *and* $E = \{e_1, e_2, \cdots, e_i, \cdots, e_n\}$ *where* $e_i = max(a_i, a_{n+i})$ *for* $1 \leq i \leq n$, *then*

1. *Both* D *and* E *are bitonic and*

2. $max\{d_1, \cdots d_i \cdots, d_n\} \leq min\{e_1, \cdots e_i \cdots, e_n\}$

Proof: A cyclic shift on the Bitonic sequence $a_1 \leq \cdots \leq a_k \geq \cdots \geq a_{2n}$, causes the two subsequences D and E to be cyclically shifted, that by itself does not affect the bitonic property of both D and E sequences.

Let k be the index of the largest element in the sequence (see Figure 3.a), if $k = 2n$ (or $k = 1$) then D consists of the first (second) half of the list and E consist of the second (first) half. It is clear that the properties 1 and 2 still hold. But, if $k \neq 1$ and $k \neq 2n$, then k either :

case 1: $k \geq n$ or

case 2: $k < n$

We will prove the first case, and the second case could be proven similarly. For the smallest i where $k \leq i \leq 2n$ such that $a_{i-n} > a_i$ then

$$a_j \in D \; \forall \begin{cases} 1 \leq j \leq i - (n+1) & Increasing \\ i \leq j \leq 2n & Decreasing \end{cases} \quad (1)$$

$$a_j \in E \; \forall \begin{cases} n+1 \leq j \leq k & Increasing \\ k+1 \leq j \leq i-1 & Decreasing \\ i-n \leq j \leq n & Increasing \end{cases} \quad (2)$$

Comparing every segment in (1) with every segment in (2), we conclude that every $a_i \in D$ is smaller than any $a_j \in E$. Both (1) and (2) consist of segments that make up bitonic lists D and E respectively (see figure 3.b).

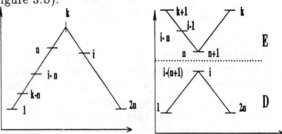

(a) Major indices in one bitonic list

(b) Dividing (a) into two bitonic lists D and E

Figure 3

Our proof for *case 1* could be violated only if the condition $a_{i-n} > a_i$ never occurs for any i where $k \leq i \leq 2n$. As a result of this condition, D is the first half of the bitonic sequence and E is the second half. Both D and E are bitonic and every $a_i \in D$ is smaller than any $a_j \in E \square$

When the bitonic sorter was first introduced, it was laid out as a sorting network that consisted of $\log N(\log N - 1)/2$ stages. Each stage consisted of $N/2$ comparators. In the remainder of this section we show an implementation of the bitonic sorter over a $\log N$ stage network. Such implementation reduces the cost complexity by a $\log N$ factor.

Theorem 2 *A network consisting of $\log N$ stages, each stage consisting of $N/2$ comparators, can sort a list of N elements in time complexity of $O(\log^2 N)$.*

Proof: Sorting N elements using bitonic merges, can be viewed as first, dividing the elements into $N/2$ groups of size 2, since any two elements are bitonic. After a single compare-exchange stage, the pairs could be sorted iteratively ascending and descending. Next, all groups of size 2^2 elements are bitonic as a result of the previous step. Two compare-exchange stages are necessary for sorting the 2^2 elements iteratively ascending and descending (Theorem 1). The same process is applied to groups of size $2^3, 2^4, \cdots$, *and* $2^{\log N}$. It is clear that elements are compared in the list that differ in their indices by bit 0 to form a list of $N/2$ sorted groups of size 2. Next, to form $N/4$ sorted groups of size 2^2 within the list, the two parallel comparisons are done on elements that differ in their indices by bit

1 then bit 0. Finally, a sorted list is formed by performing the $\log N$ parallel comparisons on elements that differ in their indices by bit $\log N - 1, \cdots, 1$, and 0.

Given a $\log N$ stage network connected by perfect shuffle, we could sort the N elements by sequentially sorting groups size $2, 2^2, \cdots, and\, 2^{\log N}$ within the list. Since the $\log N$ stage perfect shuffle network has the capabilities of performing the function

$$F(i) = (\log N - i)shuffle; comp; i(shuffle; comp;)$$

which does i parallel comparisons between elements in the list that differ in their indices by bit $i - 1$, $i - 2, \cdots$, and 0. A list with $2^{\log N - i}$ sorted groups size 2^i (where $0 \le i < \log N$) enters the network with the first $\log N - i$ stages set to straight (only shuffling). The remaining i stages compare normally resulting in a list with sorted groups size 2^{i+1}. To sort the $2^{\log N - (i+1)}$ groups size 2^{i+1} list into $2^{\log N - (i+2)}$ groups size 2^{i+2} list. the output of the network is fed back into the input of the network. This feedback is applied until $i = \log N$. The total time consumed is: number of passes through the network $\times$ number of stages in the network $= \log^2 N \square$

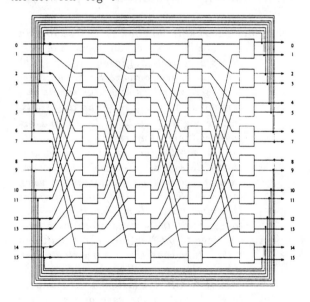

Figure 4. 4 Stage Bitonic Sorting Network (N=16).

Now, we present Algorithm 1 simulating the network presented in Theorem 2 that sorts N numbers in $O(\log N^2)$. A construction of this network is also shown in Figure 4.

Algorithm 1

Comment: The shuffle function performs a perfect shuffle on its parameter input vector.

The Compare-Exchange function takes i-vector as an input and compares each a_{2*i-1} and a_{2*i} i-vector elements and exchanges them, if necessary, according to the comparator type. Each comparator behaves according to the mask. For example, if the mask is "0", the comparator behaves as an I-comparator, else it behaves as a D-comparator.

Input : unsorted i-vector$=[I_1, I_2, \cdots, I_N]$
Output : sorted i-vector$=[I_{r_1}, I_{r_2}, \cdots, I_{r_N}]$

---

for $i \leftarrow 1$ **to** $\log N - 1$ **do**
    $mask \leftarrow 0, 1, 0, 1, \cdots, 0, 1$
    /*align elements*/
    for $j \leftarrow i$ **to** $\log N - 1$ **do**
        $shuffle(i\text{-}vector)$
    /*sort groups of size i*/
    for $j \leftarrow 1$ **to** i **do**
        $shuffle(i\text{-}vector)$
        $compare\text{-}exchange(i\text{-}vector)$
        $shuffle(mask)$
    /*set all as I-comparators*/
    $mask \leftarrow mask \oplus mask$
/*sort the group of size N*/
for $i \leftarrow 1$ **to** $\log N - 1$ **do**
    $shuffle(i\text{-}vector)$
    $compare\text{-}exchange(i\text{-}vector)$

Bit-Serial Bitonic Sorting Network

This network receives its input words in parallel with each word being presented one bit every time unit. The bits are processed most significant bit first as they arrive through the network input lines. This process is also applied within the network through pipelining, until the output is produced.

Each comparator uses the fact that the comparison of the binary representations of two natural numbers depends solely on the bits in the most significant position where the representations differ. Two numbers, a and b, are input bit-serially most-significant-bit first. As long as corresponding bits of a and b agree the comparator remains in the Neutral state and outputs their common value (whether $a > b$ or $a < b$ is not known at this time but it doesn't matter since their bits agree). When corresponding bits of a and b first disagree the comparator locks up in the Straight state or the Cross state and outputs the bits of $max(a, b)$ on one line and the bits of $min(a, b)$ on the other line.

The network is obtained by interconnecting $\log N$ stages through perfect shuffle links. The output of the last stage is connected to the first stage's perfect shuffle links. Each stage consists of $N/2$ comparators. Each comparator receives two bits as inputs, compares those two bits and exchanges them if necessary, then produces the two bits as output. Each of the output bits may be one of another comparator's two inputs. Figure 5 demonstrates such a network.

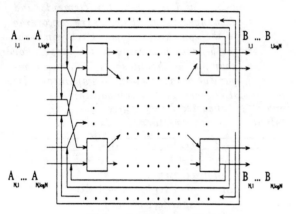

Figure 5. Layout of BBSN.

The comparators in the Bit-Serial Bitonic sorting Network differ from those described in section 2 in that status states are established for the former and always set to one of three states (as in the I-comparator):

1. Neutral :When through its input lines the comparator receives two alike bits, those bits are produced unchanged to the corresponding comparator output lines.

2. Straight: When the first upper input is "0" and the lower input line is "1", then these two bits as well as the remaining bits of the two input integers, are produced on the output lines unchanged.

3. Cross: When the first upper input is "1" and the lower input line is "0", then these two bits and the remaining bits of the two input integers. are produced on the output lines exchanged.

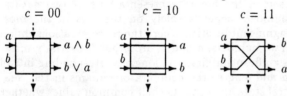

Figure 6: The three status states function for the I-comparator.[a]

In the case of the D-comparator, the output bits in the "Straight" state are exchanged, and in the "Cross" state, are unchanged, rather than exchanged. The comparator states are implemented using a two bit mask for each comparator 00-Neutral, 10-Straight and 11-Cross. The function of each I-comparator type, with regard to its state mask c, is illustrated in Figure 6. In the case of the

[a]The symbols $\oplus, \wedge, \vee$ and $|$ are *xor, and, or* and *concatenation* of two bits respectively.

D-comparator, the output lines labeled a and b are exchanged.

The comparator state mask is reset to Neutral state if the two input bits are the first bits of the input integers. Although we used two bits for the state mask, it is conceivable to use only one bit state mask (0-Neutral, and 1-Set [Straight or Cross])for each comparator, providing we preserve the state of the comparator as soon as the first two input bits differ. If the reset condition rises as indicated above, then the state mask will reset to 0-Neutral.

Theorem 3 *For any $\log N$ stage network of comparators M, with N input lines that can sort any N numbers in $O(\log^2 N)$ time, there exists a network M', topologicaly equivalent to M, that can sort any N sequences of binary bits; each of maximum size $\log N$ in $O(\log^2 N)$ time.*

Proof: Let M be the network indicated in Theorem 2. Now, let M' be a network similar to M, except M' uses the new type of (Bit-Serial) comparators. Hence, M' is topologicaly equivalent to M and inspection of Algorithm 2, shows that M' sorts the N sequences in $O(\log^2 N)$ time □

Algorithm 2 *[Bit serial Bitonic Sorting Network]*

Comment: The shuffle function takes i-vectors an $N \times \log N$ table as an input and performs a perfect shuffle on all the $\log N$ columns. The Compare-Exchange function takes i-vectors and n as inputs and compares each $a_{j,(2*i)-1}$ and $a_{j,2*i}$ bits, (where $\log N \geq j \geq n \geq 1$) and exchanges them, if necessary, according to the comparator type and the comparator status state mask "S-mask". Each comparator behaves as I-comparator when the T-mask = 0, else, it behaves as D-comparator. Reset function is used to reset all the comparators in a single stage to 00-Neutral. The c_shift and r_shift functions shift every element in the i-vectors one cell to the right. In the case of c_shift, the last element is moved to the front. In the case of r_shift, the last element is output.
The Load function shifts the i-vectors one cell to the right without the wrap around, and also loads the N input bits in the first column cells of i-vectors.

Input : *unsorted N streams of $\log N$ binary bits*
Output : *sorted N streams of $\log N$ binary bits*

/* *initialization* */

$i \leftarrow \log N - 1$

$V \leftarrow \{0, 1, 0, 1, \cdots, 0, 1\}$

$T\text{-}mask \xleftarrow{all} 0$ /*reset comparators type mask*/

$T\text{-}mask \xleftarrow{column \log N} V$

$S\text{-}mask \xleftarrow{all} 00$ /*reset status state mask*/

/*fill the network pipe */

for $l \leftarrow 1$ to $\log N$ **do**
    load(I-vectors)
    shuffle(I-vectors)

/* processing within the network pipe */

for $l \leftarrow 1$ to $\log N - 1$ **do**
    $m \leftarrow V$
    If $l = \log N - 1$ **then**
      $m \leftarrow V \oplus V$
    for $j \leftarrow 1$ to $\log N$ **do**
      compare-exchange(I-vectors,i+1)
      shuffle(I-vectors)
      c-shift(I-vectors)
      If $j \geq i$ **then**
        If $j = i$ **then**
          $i = i - 1$
        $T\text{-}mask \xleftarrow{column\ j} m$
      shuffle(m)
      reset(j)

/* empty the network pipe */

for $l \leftarrow 1$ to $\log N$ **do**
    compare-exchange(I-vectors,1)
    r-shift(I-vectors)
    shuffle(I-vectors)

It is clear from the algorithm that it takes $O(\log N)$ steps to fill the network pipe, $O(\log^2 N)$ steps to process within the network pipe, and $O(\log N)$ steps to flush the pipe. The total time complexity = $O(\log N) + O(\log^2 N) + O(\log N) = O(\log^2 N)$. Since the network is composed of $O(\log N)$ stages and each stage consists of $O(N)$ comparators, then the total cost complexity = $O(N) \times O(\log N) = O(N \log N)$.

As an example consider the input vector [6 5 3 1 7 2 4 0], where $N = 8$. The binary representation(most significant bit first) of the input vector and the transformation it goes through after each step is shown below:

$$
\begin{bmatrix} 011 \\ 101 \\ 110 \\ 100 \\ 111 \\ 010 \\ 001 \\ 000 \end{bmatrix}
\begin{bmatrix} 011 \\ 101 \\ 111 \\ 100 \\ 110 \\ 011 \\ 000 \\ 000 \end{bmatrix}
\begin{bmatrix} 011 \\ 110 \\ 101 \\ 111 \\ 111 \\ 000 \\ 000 \\ 000 \end{bmatrix}
\begin{bmatrix} 011 \\ 111 \\ 110 \\ 000 \\ 101 \\ 000 \\ 111 \\ 000 \end{bmatrix}
\begin{bmatrix} 101 \\ 010 \\ 111 \\ 100 \\ 011 \\ 111 \\ 000 \\ 000 \end{bmatrix}
$$

$$
\begin{bmatrix} 010 \\ 101 \\ 101 \\ 111 \\ 111 \\ 000 \\ 010 \\ 000 \end{bmatrix}
\begin{bmatrix} 100 \\ 010 \\ 011 \\ 101 \\ 111 \\ 001 \\ 110 \\ 000 \end{bmatrix}
\begin{bmatrix} 011 \\ 110 \\ 000 \\ 101 \\ 100 \\ 011 \\ 111 \\ 000 \end{bmatrix}
\begin{bmatrix} 001 \\ 101 \\ 111 \\ 010 \\ 000 \\ 101 \\ 110 \\ 010 \end{bmatrix}
\begin{bmatrix} 100 \\ 110 \\ 110 \\ 000 \\ 111 \\ 001 \\ 001 \\ 011 \end{bmatrix}
$$

$$
\begin{bmatrix} 100 \\ 000 \\ 110 \\ 110 \\ 101 \\ 001 \\ 011 \\ 011 \end{bmatrix}
\begin{bmatrix} 100 \\ 000 \\ 010 \\ 110 \\ 101 \\ 001 \\ 111 \\ 011 \end{bmatrix}
\begin{bmatrix} 000 \\ 100 \\ 010 \\ 110 \\ 001 \\ 101 \\ 011 \\ 111 \end{bmatrix}
$$

Conclusions

This paper presented a bit-serial bitonic sorting network. This network has a time complexity of $O(\log^2 N)$ and a cost complexity of $O(N \log N)$. Keys are sorted in a pipeline fashion (worm-hole routed) bit-serially most-significant-bit first through the network. The reduction in cost complexity of this network is achieved by recirculating the keys of length $\log N$ through $\log N$ stages. The bit level comparator in this network has time and cost complexities of $O(1)$.

References

[1] K. Batcher, Sorting networks and their applications, in *Proc. AFIPS Spring Joint Computer Conference*, 1968, vol. 32, pp. 307-314.

[2] M. V. Chien and A. Y. Oruç, Adaptive Binary Sorting Schemes and Associated Interconnection Networks, in *Proceedings of the 1992 International Conference on Parallel Processing*, pp. I-289 - I-293, 1992.

[3] Hasan Cam and Jose A.B. Fortes , Fault-Tolerant Self-Routing Permutation Networks, in *Proceedings of the 1992 International Conference on Parallel Processing*, pp. I-243 - I-247, 1992.

[4] D. Knuth, The Art of Computer Programming, *vol. 3: Sorting and Searching*. Reading, MA: Adison-Wesly, 1973.

[5] H. S. Stone, Parallel Processing with the Perfect Shuffle, *IEEE Transactions on Computers*, vol. C-20, February 1971, pp. 153-161.

SESSION 9C

GRAPH ALGORITHMS (II)

MULTICOLORING FOR FAST SPARSE MATRIX-VECTOR MULTIPLICATION IN SOLVING PDE PROBLEMS

H.C. Wang and Kai Hwang
University of Southern California
Los Angeles, CA 90089-2562
{hcwang,kaihwang}@aloha.usc.edu

Abstract: A new multicoloring technique is proposed for parallel sparse matrix-vector multiplication, which dominates the computing cost of iterative PDE (partial differential equation) solvers. The new technique enables parallel solution of grid-structured nonsymmetric PDE problems on shared-memory multiprocessors through resolving memory access conflicts by multiple processors. The coloring scheme is formulated as an algebraic mapping which can be implemented with low overhead. The proposed multicoloring scheme has been tested on an Alliant FX/80 for solving 2D and 3D PDE problems using CGNR method. Compared to the results reported in [8] for matrix-vector multiplication operations, which were obtained on an identical Alliant system, our results show an improvement factor of more than 30 times in terms of Mflops. This proves that multicoloring is indeed effective for parallel scientific computations.

Keywords: Parallel processing, matrix-vector multiplication, conjugate gradient methods, multicoloring, partial differential equations.

1 Introduction

In this paper, we present a new multicoloring technique which allows the parallelization of *generalized conjugate gradient* (GCG) methods on shared-memory multiprocessor systems. These methods are used to solve nonsymmetric systems of equations and are characterized by the need to perform vector multiplication by a matrix and its transpose. Such operations dominate the overall computing cost of the PDE algorithms.

With most storage formats for sparse matrices that have been adopted in numerical packages for scientific computations, the matrix-vector multiplications form a duality of reduction-extension operations. It is illustrated that without proper coloring, the extension operation in a parallel processing environment either does not show much improvement in performance over serial execution,   or cause the

algorithms to diverge due to simultaneous updates to the same memory location by several processors. Both must be avoided. To achieve the goals, a multicoloring technique for grid-structured PDE problems is developed. In this approach, coloring is formulated as a mapping from the coordinates of a grid point to a certain color. The sufficient and necessary conditions for a mapping to generate a valid coloring scheme are derived. The minimum number of colors needed for different discretization stencils is also established.

Numerical experiments conducted on a shared-memory Alliant FX/80 multiprocessor system for 2D and 3D PDE problems clearly demonstrate the improvement in performance. Indeed, the results obtained for extension type of operation show a significant gain in speed over those reported in the past. With the use of the proposed coloring scheme, these operations also exhibit better scalability with respect to machine and problem sizes.

The fundamental idea of coloring is to decouple the connections among grid points. It has been used extensively to improve parallel processing efficiency. The best known coloring technique is the two-coloring developed for the parallel processing of Gauss-Seidel and SOR methods. Diagonal coloring has been used to allow a wave-front sweeping along diagonals. Coloring has also been used with irregular grid structures [5].

Another use of coloring is to form long vectors in order to reduce the startup cost of pipeline processing on some vector processors. This class of methods is exemplified by the *continuous coloring* scheme proposed by Poole and Ortega in [7]. The main objective of these schemes is to obtain a matrix the diagonal blocks of which are all diagonal. Such a matrix is said to possess a DDB (diagonal diagonal blocks) structure [8].

The methods have been successfully used to vectorize the preconditioning phase of conjugate-gradient methods, in particular, preconditioning techniques based on incomplete (block) factorization of the coefficient matrix.   These multicoloring techniques share some similarity with the method described in this paper. However, the objectives are different, and so the outcome also differs. For instance, the continuous coloring method does not allow any point connected to a point P to have the same color as P.

This is inadequate for concurrent processing in the context of extension type of operations. In fact, we introduce the concept of the *turf* of a point P as an area within which no other point is allowed to have the same color as P, and show that the area needs to be extended beyond those points directly connected to pint P. Moreover, the number of colors used in Poole's continuous coloring scheme varies with the number of grid points along each coordinate direction and requires trial-and-error to determine a suitable coloring. Our method enumerates the valid coloring schemes, thereby eliminating the potentially tedious process.

2 Storage of Sparse Matrices

The manipulation of sparse matrices is quite different from that of dense matrices. Special data structures are needed for the storage of such matrices to conserve memory space. Several commonly used formats are described below. The corresponding code for matrix-vector multiplications is then specified. These operations are required in GCG methods.

Suppose A is a sparse $N \times N$ real matrix. N is usually large, effective storage of the matrix is essential. Irrespective of the format used, the basic information required is the value of each nonzero element and its row and column indices in the matrix. Two storage schemes are described below:

- Diagonal format. For matrices having a banded structure with a small bandwidth or matrices with nonzero elements clustered along fixed diagonals, the nonzero elements can be stored in a vector A of dimension $N \times Ndiag$. The value of $Ndiag$ depends on the characteristic of the problem under consideration. A small vector IA of dimension $Ndiag$ stores the offset of each element relative to the main diagonal.

- Rowwise format. A vector A of size NZ (the total number of nonzero elements) is used to store the nonzero elements of the matrix row by row. A second vector JA of the same length as A stores the column index of each nonzero element. A third vector ISA of size $N+1$ indicates where in vector A each row of the sparse matrix starts,

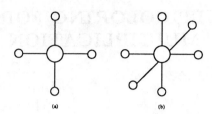

Figure 1: (a) 5-star and (b) 7-star stencils used for discretizing 2D and 3D differential operators.

with the exception that $ISA(N + 1)$ points to the first free memory location at the end of the matrix.

Of course, diagonal and rowwise formats are not the only schemes used for the storage of sparse matrices. Several other formats have also been used adopted, including ITPACK format used in ITPACK and ELLPACK. Columnwise format, which is used in Harwell-Boeing Sparse Matrix Collection [1] is a counterpart of rowwise format. Jagged row and column formats, which store a matrix by row or by column according to the number of nonzero elements in each row or column, have been used to improve vector processing efficiency [5].

We limit the discussion to a system of linear equations arising from the discretization of a PDE-like problem defined over a regular region. For a 2D problem, the grid size is $m \times n$ with a 5-star discretization stencil; for a 3D problem, a grid of size $m \times n \times p$ is used with a 7-star stencil. See Figure 1. These problems will be referred to as the *model problems* with $N = mn$ and $N = mnp$, respectively. The resulting coefficient matrix A is sparse.

A difference between the model problems considered and typical PDE problems is that the coefficients in the stencil are randomly generated positive numbers between 0 and 1. A ramification is that the corresponding matrices are neither M-matrix nor diagonally dominant. Therefore, the algorithms used tend to converge more slowly. Nevertheless, the model problems capture the essence of nonsymmetric PDE problems for which GCG methods have to be used.

Two such methods are *CG method applied to normal equations* (CGNR) and *biconjugate-gradient* (BCG) methods. These methods have been shown by examples in [6] to be suitable for solving different types of linear systems. Detailed analysis of these methods can be found in [2]. One important property of CGNR method is that it converges monotonically for any nonsingular coefficient matrix. Algorithm 1 shows the computation steps of CGNR for solving the linear system $Ax = f$. It does so by applying CG method to the normal equation $A^T A x = A^T f$.

Algorithm 1 (CGNR method)
1. Initialization: Choose x_0 as an initial guess and compute $r_0 = f - A x_0$ and $p_0 = A^T r_0$.
2. **For** $k = 0, 1, 2, ...$ until convergence **do**

- Compute $\alpha_k = (A^T r_k, A^T r_k)/(A p_k, A p_k)$.
- Update solution vector $x_{k+1} = x_k + \alpha_k p_k$.
- Compute residual $r_{k+1} = r_k - \alpha_k A p_k$.
- Compute $\beta_k = (A^T r_{k+1}, A^T r_{k+1})/(A^T r_k, A^T r_k)$.
- Update direction vector $p_{k+1} = A^T r_{k+1} + \beta_k p_k$.

Both CGNR and BCG involve the multiplication of vectors by both A and A^T. It has been suggested that performing both multiplications in the same algorithm can be inefficient because of the different properties of the two operations. Suppose v is an $N \times 1$ column vector. Av is a *reduction* operation in which two vectors are combined to generate a single scalar quantity. On the contrary, operation $A^T v$ is an *extension* operation whereby the action of a scalar quantity can affect the values of several others.

More accurately, the two operations form a duality. That is, the reduction and extension nature of the operations can be reversed, depending on the storage scheme used. For instance, if columnwise storage is used, the operation $A^T v$ is a reduction operation and Av is an extension operation. In contrast, Av is a reduction and $A^T v$ an extension if rowwise format is used.

The basic operation in the matrix-vector multiplication $u = Av$ is the following:

$$u_i = \sum_{j=1}^{NZ_i} A_{ij} v_j, \qquad (1)$$

where NZ_i is the number of nonzero elements in row i of A. Similarly, the basic step in $u = A^T v$ is

$$u_i = \sum_{j=1}^{NZ_i} (A^T)_{ij} v_j = \sum_{j=1}^{NZ_i} A_{ji} v_j. \qquad (2)$$

It is convenient to interchange the subscripts in Eq. (2) and rewrite it as

$$u_j = \sum_{i=1}^{NZ_j} A_{ij} v_i. \qquad (3)$$

In the sequel, we describe the two types of multiplications for rowwise storage format in the following pseudo code. Computations for the other storage schemes can be similarly formulated.

- **Operation $u = Av$:**

  For i = 1 to N **do**
    For j = ISA(i) to ISA(i+1) − 1 **do**
      u(i) = u(i) + A(j) * v(JA(j))
    Enddo
  Enddo

- **Operation $u = A^T v$:**

  For i = 1 to N **do**
    For j = ISA(i) to ISA(i+1) − 1 **do**
      u(JA(j)) = u(JA(j)) + A(j) * v(i)
    Enddo
  Enddo

3 Multicoloring on 2D Grids

Consider the operation $A^T v$ for rowwise storage format. The code consists of an outer loop indexed by i and an inner loop indexed by j. At compile time, it is impossible to know the value of $JA(j)$ for a particular combination of i and j. Therefore, the code is generally not optimized for parallel execution by a compiler. However, since vector v is read-only, the access order to its elements is irrelevant. Moreover, for the model problem it is known that each $JA(j)$ corresponds to one of the neighbors that the grid point indexed by i is adjacent to. Hence, the values of $JA(j)$ are all different for a given i. Consequently, it is permissible to carry out concurrent update operations in the inner loop. This will be referred to as *concurrent inner* (CI) mode of operation. Many systems provide compiler directives for this purpose.

On the other hand, different combinations of i and j may lead to the same value of $JA(j)$. For the model PDE problem, this aliasing problem can occur at adjacent grid points. Thus, if two iterations in the outer loop are allowed to proceed concurrently, they will create a *write-after-write* (WAW) hazard condition. This mode of execution is termed *concurrent outer* (CO) mode. The hazard is likely to deteriorate the performance of the GCG algorithms. In Table 1 and Figure 2, we show the execution times of the Av operation and the relative residual norm for the first few iterations when CGNR method is used to solve a 2D model problem.

Table 1: Sequential and parallel execution times (in ms) for the operation $u = A^T v$ using rowwise storage format.

Grid size	Seq.	CI	CO	6-coloring
15×15	14.8	4.5	1.2	2.3
63×63	275.6	71.8	12.0	17.2
255×255	4512.6	1183.0	207.3	292.5

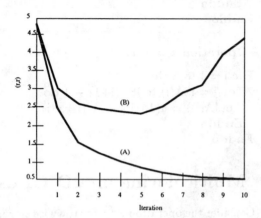

Figure 2: Value of (r, r) for the first 10 CGNR iterations in sequential and CI mode (curve A) and CO mode (curve B) of execution.

The sequential execution time is obtained on an Alliant FX/80 detached processor running in scalar mode, and the parallel execution time is obtained on a 4-processor cluster with vector processing enabled. The curve for the sequential and CI mode shows a monotonic decrease in the residual norm whereas that for the CO mode shows a slower convergence at first and then diverges. The hazard condition observed for rowwise format also exists for diagonal storage format.

To improve parallel performance while avoiding the hazard conditions, a coloring technique is devised. Define the range R_{idx} of an operation on index idx as the set of points whose values are affected by the operation. From the perspective of the outer loop in the code for 2D extension operation, the range consists of the five grid points in the neighborhood of the point corresponding to index i. In this case, the range of a grid point coincides with the discretization stencil centered at it.

To circumvent the concurrent write conflicts, it is important to avoid simultaneous operations at grid points indexed by different i's whose ranges intersect with each other. In other words, two grid points indexed by i_1 and i_2 can be updated simul-

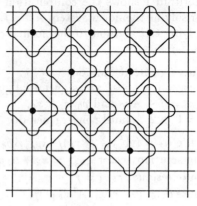

(a) A 6-coloring scheme

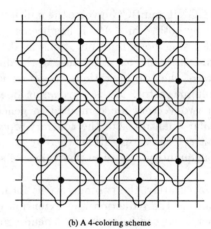

(b) A 4-coloring scheme

Figure 3: 6- and 4-coloring schemes. Each diagram shows the ranges of grid points in the same color.

taneously only if $R_{i_1} \cap R_{i_2} = \phi$.

Since the range of a grid point is a 5-star stencil centered at point (r, s), a two-coloring scheme does not work. Suppose two points P at (r_1, s_1) and Q at (r_2, s_2) have the same color. Then an examination of their ranges indicates that the coordinates must satisfy the condition

$$|r_1 - r_2| + |s_1 - s_2| \geq 3 \qquad (4)$$

in order for P and Q to be updated concurrently. Graphically, a rhombic region surrounding point P can be drawn, which will be referred to as the turf of point P and denoted $\mathcal{R}_p$. Equation 4 then requires that no other point in $\mathcal{R}_p$ is allowed to have the same color as P. The turf associated with a point always contains the range of the point in an extension operation.

Clearly, a 9-coloring, in which points separated by a distance of 3 along each coordinate have the same color, satisfies Eq. (4). More compact 6-coloring

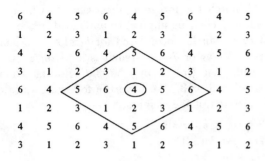

Figure 4: Color assignment on a 9×8 grid using a 6-coloring scheme. The rhombic area represents the turf of the circled point.

and 5-coloring schemes are also allowed. With 6-coloring (3a), points in the same color are separated by 3 along x direction and by 2 along y direction. Figure 3b shows that a 4-coloring scheme, in which points at a distance of 4 along one direction and 1 along the other are assigned the same color, leads to write interference and is not allowed. This can be clearly seen from the diagram. A proof can also be furnished by verifying that the condition embodied in Eq. (4) is violated.

Figure 4 shows the color of each grid point on a 9×8 grid using a 6-coloring scheme. Note that the colors of grid points in the turf are all different from the one circled. The turf associated with point P (circled) is also indicated in the figure.

Using the 6-coloring, we obtain the timing results shown in the last column of Table 1. It is interesting to compare the results with those of the CI and CO modes. As the problem size is increased, the performance of the 6-coloring scheme improves steadily relative to that of the concurrent outer mode. This is so because the number of grid points in each color is larger and concurrent processing is more efficient. CI mode shows little improvement in speedup because of the limited amount of parallelism.

4 Multicoloring on 3D Grids

For 3D grids using 7-star discretization stencils, we expect seven colors to be sufficient to avoid the concurrent write conflicts. The geometric rendition useful for 2D grids is difficult to depict for a 3D grid. In the following, an algebraic approach to coloring is described. We use a seven color scheme as an illustration and later generalize to other numbers of colors.

We formulate a coloring scheme as a linear mapping M denoted by a 3-tuple $\langle m_1, m_2, m_3 \rangle$. M as-

signs color c to a grid point at (r, s, t) according to the following formula:

$$c = (m_1 r + m_2 s + m_3 t) \bmod 7, \ 1 \leq m_1, m_2, m_3 \leq 6. \tag{5}$$

The integers m_1, m_2, m_3 are the *weights* of the mapping.

Conceptually the determination of valid coloring can be envisioned as searching a state space comprising various combinations of the weights. For 7-coloring, there is a total of $6^3 = 216$ possible states. An exhaustive search for valid combinations is tedious. Since our concern is to group grid points by different colors and color assignment is invariant with regard to coordinates, the number of distinct mappings is much smaller. In fact, if the coordinate-invariance property is taken into account, the number is reduced to 56. In general, for a b-coloring scheme, a simple analysis shows that $(b-1)b(b+1)/6$ distinct combinations are obtained after considering the invariance property. For large b, exhaustive search is still time-consuming. Therefore, it is desirable to determine valid mappings systematically. To this end, we first introduce the concept of equivalent mappings.

Definition 1 Two mappings are equivalent if two grid points assigned the same color c_1 by one mapping are assigned the same color c_2 by the other mapping.

Formally, mappings M_1 and M_2 are equivalent if for any $i \neq j$, $M_1(r_i, s_i, t_i) = c_1$, $M_1(r_j, s_j, t_j) = c_1$, and $M_2(r_i, s_i, t_i) = c_2$, then $M_2(r_j, s_j, t_j) = c_2$.

Based on the above definitions, we obtain

Theorem 1 Suppose mapping M_1 is defined by the 3-tuple $\langle m_1, m_2, m_3 \rangle$ and M_2 by $\langle Sm_1 \bmod 7, Sm_2 \bmod 7, Sm_3 \bmod 7 \rangle$ with S an integer, $1 \leq S \leq 6$. Then mappings M_1 and M_2 are equivalent.

Detailed proof of the theorem can be found in [9]. An immediate result of the theorem is given below:

Corollary 1 All the mappings $\langle Sm_1, Sm_2, Sm_3 \rangle$ for $S = 1, 2, 3, 4, 5, 6$ are equivalent.

For instance, mappings $\langle 1, 1, 2 \rangle$, $\langle 2, 2, 4 \rangle$, $\langle 3, 3, 6 \rangle$, $\langle 4, 4, 1 \rangle$, $\langle 5, 5, 3 \rangle$, and $\langle 6, 6, 5 \rangle$ are all equivalent, and we say $\langle 1, 1, 2 \rangle$ *covers* the other equivalent mappings. By the equivalence theorem, the number of unique mappings can be cut down to 22. Compared to the original 216 possible combinations, the reduction in complexity is significant, which makes the search for legitimate coloring schemes much less expensive. The set comprising the unique mappings is called the *minimum cover set*.

In the 3-tuple notation, the minimum cover set for 7-coloring consists of the following mappings: $\langle 1, 1, 1 \rangle$, $\langle 1, 1, 1 \rangle$, $\langle 1, 1, 2 \rangle$, $\langle 1, 1, 3 \rangle$, $\langle 1, 1, 4 \rangle$, $\langle 1, 1, 5 \rangle$, $\langle 1, 1, 6 \rangle$, $\langle 1, 2, 3 \rangle$, $\langle 1, 2, 4 \rangle$, $\langle 1, 2, 5 \rangle$, $\langle 1, 2, 6 \rangle$, $\langle 1, 3, 4 \rangle$, $\langle 1, 3, 5 \rangle$, $\langle 1, 3, 6 \rangle$, $\langle 1, 4, 5 \rangle$, $\langle 1, 4, 6 \rangle$, $\langle 1, 5, 6 \rangle$, $\langle 2, 3, 4 \rangle$, $\langle 2, 3, 6 \rangle$, $\langle 2, 5, 6 \rangle$, $\langle 3, 4, 5 \rangle$, $\langle 3, 4, 6 \rangle$, and $\langle 3, 5, 6 \rangle$.

Not all these 22 mappings will lead to colorings which prevent concurrent write conflicts. Thus, the final step in the coloring process is to determine valid mappings in the minimum cover set. Similar to 2D cases, the turf of a grid point P at (r, s, t) is a rhomboid with a side length of 3 surrounding P; i.e., it is defined by the following inequality:

$$\mathcal{R}_p = \{Q : |\triangle r| + |\triangle s| + |\triangle t| \leq 2\}, \quad (6)$$

where $\triangle$ stands for the distance along each direction between point P and another point Q in the 3D grid. Any other grid point in $\mathcal{R}_p$ is not allowed to have the same color as P.

From the definition of color mapping, the distance between two grid points in the same color satisfies the condition

$$(m_1 \triangle r + m_2 \triangle s + m_3 \triangle t) \bmod 7 = 0. \quad (7)$$

Hence, the goal is to find values for m_1, m_2, m_3 such that any point $Q \in \mathcal{R}_p$, $Q \neq P$ will have a different color than P. From Eq. (6), a point in $\mathcal{R}_p$ satisfies one of the following conditions:

1. One of $|\triangle r|$, $|\triangle s|$, $|\triangle t|$ is 2, the others are 0;
2. One of $|\triangle r|$, $|\triangle s|$, $|\triangle t|$ is 1, the others are 0;
3. Two of $|\triangle r|$, $|\triangle s|$, $|\triangle t|$ are 1, the other is 0.

In the first two cases, since $1 \leq m_1, m_2, m_3 \leq 6$, the condition specified by Eq. (7) is always violated, hence Q and P will always have different colors. In the last case, Eq. (7) is satisfied only if

$$(m_i \leq m_j) \bmod 7 = 0 \text{ for } i \neq j. \quad (8)$$

Because $1 \leq m_i \leq 6$, the condition Eq. (8) holds if

$$m_i = m_j \text{ or } m_i + m_j = 7 \text{ for } i \neq j. \quad (9)$$

A direct consequence of Eq. (9) is that any coloring scheme with two or more identical weights will give rise to WAW conflicts and should be eliminated from the minimum cover set. Likewise, mappings with two weights that add up to 7 are also prohibited. Based on these criteria, only five coloring schemes, $\langle 1, 2, 3 \rangle$, $\langle 1, 2, 4 \rangle$, $\langle 1, 3, 5 \rangle$, $\langle 2, 3, 6 \rangle$, and $\langle 3, 5, 6 \rangle$, and their equivalent mappings will produce valid 7-coloring.

Although the foregoing discussion has been focused on 7-coloring applied to a 3D grid, more general rules can be formulated for b-coloring scheme when a 5-star or 7-star stencil is used on a 2D or 3D grid. Let $\langle m_1, ..., m_d \rangle$ be the coloring scheme where $d = 2$ or 3. We have the following rules on the selection of the weights:

1. $0 \leq m_i \leq b$ for all i.
2. $2m_i \neq b$ for all i.
3. $m_i + m_j \neq b$, $1 \leq i, j \leq d$.
4. $m_i \neq m_j$, $1 \leq i, j \leq d$.

Based on the rules, it is straightforward to determine valid choices for the weights. Moreover, using these rules, it can be shown that both 4-coloring for 2D grids and 6-coloring for 3D grids are not valid for avoiding WAW hazards. For instance, if 6-coloring is applied to a 3D grid, the possible choices for any of the three weights are 1 through 5. Value 3 is disallowed since it violates rule 2. The pair 1 and 5 cannot appear together since rule 3 will be violated, neither can 2 and 4. Also, rule 4 disallows any number to be chosen more than once. Thus, it is impossible to form any legitimate 3-tuple, and hence 6-coloring is not feasible. Thus, 5 and 7 are the minimum numbers of colors needed for 2D and 3D grids, respectively. The results are consistent with intuitive reasoning.

The above rules reflect the similarity in the way turfs for 5-and 7-star stencils are constructed. For different stencils, such as the 9-point stencil used for 2D differential operators with mixed differential terms, these rules will need to be modified. This and other important issues are more fully addressed in [9].

5 Experimental Results

Numerical experiments have been conducted on an Alliant FX/80 with multiple vector processors. The machine has a memory hierarchy consisting of cache and main memory, both of which are shared among processors. The problems tested are the model problems described in Section 2. Only the performance data for matrix-vector multiplications are collected and expressed in Mflops (millions of floating-point operations per second) in the ensuing discussions. 5-star and 7-star stencils are used for 2D and 3D grids, respectively.

The total of number nonzero elements in the coefficient matrix obtained from discretization is $5mn - 2(m + n)$ on an $m \times n$ 2D grid and $7mnp - 2(mn +$

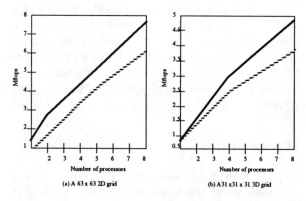

Figure 5: Mflops achieved for different numbers of processors.

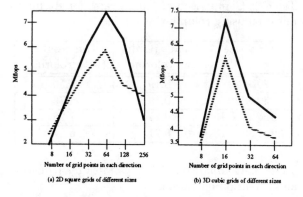

Figure 6: Mflops achieved for different grid sizes using 8 processors.

$mp + np$) on an $m \times n \times p$ 3D grid. For each of these elements, two arithmetic operations, one multiplication and one addition, are performed. Therefore, the number of floating-point operations is $10mn - 4(m + n)$ and $14mnp - 4(mn + mp + np)$ for 2D and 3D model problems. This information together with the execution time allows us to estimate the Mflops rate.

Define the *degree of parallelism* (DOP) as the number of grid points that can be operated upon simultaneously without causing instability to the algorithms used. In the extension operation, the DOP is equal to the number of grid points divided by b when a b-coloring scheme is used. DOP represents the software or algorithmic parallelism. When the algorithm is executed on an Alliant FX/80, the software parallelism is split to match hardware parallelism which is a combination of vector processing within each processor and concurrent processing across multiple processors. Color assignment is performed only once at the beginning with the information stored for subsequent use.

In the following, we investigate several factors that may affect the performance of matrix-vector multiplication operations, such as problem size and machine size. In general, with the use of more processors, performance should improve. But certain hardware constraints may limit the gain derived from grid size increase. Comparison with previous results is presented where appropriate.

In Figure 5 we show the measured Mflops versus the number of processors for grid sizes of 63×63 and $31 \times 31 \times 31$, respectively. Vector processing is enabled in all cases. In each figure, solid curves represent results for diagonal storage format and dashed curves correspond to those for rowwise format.

Four different machine sizes are compared with the number of processors equal to 1, 2, 4, and 8, respectively. It is clear that with the increase in hardware parallelism, the Mflops rate increases monotonically. It also indicates that coloring has generated enough parallelism to utilize the available computing resources.

Figure 6 shows the Mflops rate obtained in performing the extension operation using 8 processors for different grid sizes. As before, vector processing is enabled. When the grid size is small, there is a mismatch between hardware and software parallelism. Hence the Mflops rate is moderate. As the grid size is increased, the Mflops rate improves rapidly for both storage formats. However, when the grid size is increased beyond some threshold values, the Mflops drops sharply. The effect can be attributed to the fact that the storage space requirement of large grids exceeds the cache capacity. This phenomenon is referred to as *cache saturation*.

When cache saturation takes place, data have to be loaded from main memory frequently, prompting the replacement of those already residing in the cache. This data thrashing slows down processing speed considerably. Indeed, at the onset of cache saturation, the time spent on memory access may dominate the overall execution time. A possible way to overcome cache saturation is to apply strip mining or tiling [10] technique to improve the reuse of a data item once it is brought into the cache instead of reloading the same data from main memory.

We also compare the results with those reported in [8] for Av operation using columnwise storage. The multiplication, as we noted before, is an extension operation. Table 2 lists the Mflops rates from the experiments. The results in [8] were obtained on an Alliant FX/80 using 8 processors. The same environment is used in our experiments. Clearly,

Table 2: Comparison of Mflops rates for Av multiplication using columnwise storage.

Grid size	Saad's results (no coloring)	Our results (with coloring)
20×20	0.19	3.50
30×30	0.19	4.78
$20 \times 20 \times 10$	0.21	7.28
$30 \times 30 \times 10$	0.21	6.27

Table 3: Comparison of Mflops rates for reduction and extension operations.

Grid size	Av	$A^T v$
$7 \times 7 \times 7$	3.97	3.53
$15 \times 15 \times 15$	7.37	6.14
$31 \times 31 \times 31$	5.16	4.12
$63 \times 63 \times 63$	4.44	3.70

coloring has significantly improved the performance and shows better scalability with respect to grid size before cache saturation sets in.

Finally, we compare the performance results of reduction and extension operations for the same storage format. Table 3 shows the Mflops rates for the operations Av and $A^T v$ using diagonal storage format. The results were obtained for a 3D grid with different sizes using 8 processors with vector processing. In all cases, the reduction operation outperforms the extension operation by 10% to 20%. The difference may be ascribed to the higher DOP of reduction operation and the need for extra memory accesses to retrieve the coloring information in the extension operation. A similar observation has been made for rowwise storage format.

6 Conclusions

We have characterized matrix-vector multiplications as reduction and extension operations depending on the storage scheme used. We showed that for extension type of operation, concurrent writes can cause the numerical algorithm to diverge. A coloring technique has been developed to avoid hazard conditions and improve the performance of extension operations. The geometric rendition through the concepts of ranges and turfs gives a clear illustration of the hazard conditions and their avoidance.

By recasting coloring in an algebraic mapping setting, a systematic approach has been derived for assigning colors to the grids points in a straight-

forward manner. Using this method, we have been able to obtain performance results far superior to those without coloring. Another remarkable feature of the coloring scheme is that it incurs little overhead compared to other parallelization techniques. The property ensures high quality of parallel computation [3, 4] with the proposed coloring technique.

References

[1] I.S. Duff, R.G. Grime, and J.G. Lewis. Sparse matrix test problems. *ACM Trans. Math. Software*, 15:1–14, 1989.

[2] H.C. Elman. *Iterative Methods for Large, Sparse, Nonsymmetric Systems of Linear Equations*. PhD thesis, Yale University, 1982.

[3] K. Hwang. *Advanced Computer Architecture: Parallelism, Scalability, Programmability*, pages 105–113. McGraw-Hill, New York, 1993.

[4] R.B. Lee. Empirical results on the speedup, efficiency, redundancy, and quality of parallel computations. In *Proc. Int. Conf. on Parallel Processing*, pages 91–96, August 1980.

[5] R. Melhem and K. Ramarao. Multicolor reordering of sparse matrices resulting from irregular grids. *ACM Trans. Math. Software*, 14(2):117–138, 1988.

[6] N.M. Nachtigal, S.C. Reddy, and L.N. Trefethen. How fast are nonsymmetric matrix iterations? *SIAM J. Matrix Anal. Appl.*, 13(3):778–795, 1992.

[7] E.L. Poole and J.M. Ortega. Multicolor ICCG methods for vector computers. *SIAM J. Numer. Anal.*, 24:1394–1418, 1987.

[8] Y. Saad. Krylov subspace methods on supercomputers. *SIAM J. Sci. Stat. Comp.*, 10(6):1200–1232, 1989.

[9] H.C. Wang. *Parallelization of Iterative PDE Solvers on Shared-Memory Multiprocessors*. PhD thesis, University of Southern California, 1992.

[10] M.J. Wolfe. Automatic vectorization, data dependence, and optimizations for parallel computers. In Hwang and DeGroot, editors, *Parallel Processing for Supercomputing and Artificial Intelligence*, chapter 11. McGraw-Hill, New York, 1989.

Efficient Parallel Shortest Path Algorithms for Banded Matrices

Y. Han*   Y. Igarashi**

*Department of Computer Science
University of Hong Kong
Pokfulam Road, Hong Kong

**Department of Computer Science
Gunma University
Kiryu, 376 Japan

Abstract

We present efficient parallel shortest path algorithms for an $n \times n$ banded matrix of bandwidth b. Our algorithm computes all pair shortest distances within the band in time $O(nb^2/p + I(b) \log b \log(n/b))$ on the PRAM using p processors, where $I(b)$ is $\log b$ on the EREW PRAM, $\log \log b$ on the CRCW PRAM, and a constant on the randomized CRCW PRAM. It computes all pair shortest distances in time $O(n^2 b/p + I(b) \log b \log(n/b))$ using p processors.

1. Introduction

In this paper we consider the problem of computing shortest paths for graphs whose underlying matrix is a banded matrix. The input is an $n \times n$ matrix A with bandwidth b, $\lfloor b/2 \rfloor$ diagonals on either side of the main diagonal. Each a_{ij} within the band is the weight of the arc from vertex i to vertex j. Entries outside the band are ∞'s. We consider two problems. One is the problem of computing all pair shortest distances within the band. The output is also a banded matrix B with bandwidth b, where each b_{ij} within the band represents the shortest distance from i to j. The other problem is the problem of computing all pair shortest distances. The output is a matrix giving all pair shortest distances.

Recently Allison *et al.* showed a fast sequential shortest path algorithm for banded matrices of bandwidth b [ADY]. Their algorithm computes all shortest distances within the band in $O(nb^2)$ time and computes all pair shortest distances in $O(n^2 b)$ time. This represents a substantial saving compared with the general case where the best algorithm has time complexity slightly less than $O(n^3)$ [Fr].

Unfortunately, the algorithm given by Allison *et al.* is a highly sequential algorithm. Their algorithm is an incremental algorithm which adds one vertex at a time to the graph. Therefore their algorithm requires $O(n)$ time when implemented on a PRAM even if the number of available processors is unlimited.

In this paper we use a new smoothing technique to design shortest path algorithms for banded matrices. The technique allows us to smooth shortest paths in the process of contracting them. Parallel algorithms designed using this technique exhibit excellent speedup compared to their sequential counterparts. Our algorithms have time complexity $O(nb^2/p + I(b) \log b \log(n/b))$ for computing all pair shortest distances within the band, and time complexity $O(n^2 b/p + I(b) \log b \log(n/b))$ for computing all pair shortest distances. Our algorithms show that time complexity smaller than $O(I(n) \log n)$ can be achieved when b is small and enough processors are available. Our algorithms can also be used for detecting a negative cycle in the input graph.

2. Path Contraction and Smoothing

Let A be the input matrix. A is an $n \times n$ matrix with bandwidth b. The rows and columns of A are numbered from 1 to n. Let $k = \lceil b/2 \rceil$. Assume that n is a multiple of k. Let A_{ij}, $1 \leq i, j \leq n/k$, be a $k \times k$ submatrix containing elements in rows $(i-1)k+1, ..., ik$ and columns $(j-1)k+1, ..., jk$ of A. We use $[i]$ to denote any vertex u such that $(i-1)k+1 \leq u \leq ik$. A path is a sequence of vertices. A path can be denoted by $[i_1][i_2]...[i_j]$. A path $[i_1][i_2]...[i_j]$ is a loop if $i_1 = i_j$. It is a simple loop anchored at i_1 if $i_t \neq i_1, 2 \leq t < j$. The transitive closure of a matrix M is denoted by M^*. The transitive closure gives all pair shortest distances. For an $n \times n$ matrix M, the transitive closure M^* can be computed in time $O(n^3/p + I(n) \log n)$ using p processors [HPR].

Because the input matrix is a banded matrix, all entries outside the band are ∞'s. Therefore, a shortest path $[i_1][i_2]...[i_j]$ has the property that $i_k - 1 \leq i_{k+1} \leq i_k + 1$.

The computation of shortest distances can be done by matrix multiplication over the semiring $(A, \min, +)$, which may be viewed as a process of path contraction. For example, the shortest path from vertex $[i]$ to vertex $[i]$ may have the form $[i][i-1]...[1][0][1]...[i-1][i]$, the distance of this shortest path can be computed by matrix multiplication

$$A_{i,i-1}A_{i-1,i-2}...A_{1,0}A_{0,1}...A_{i-2,i-1}A_{i-1,i}.$$

This matrix multiplication can be viewed as path contraction. After $A_{i,i-1}A_{i-1,i-2}$ is computed, an arc $[i][i-2]$ labeled with the new weight obtained from the computation can be added to the input graph. Now the shortest path becomes $[i][i-2]...[1][0][1]...[i-1][i]$. The path is contracted and its length is decremented by 1.

We use path smoothing techniques to solve the shortest path problems. A loop $[i_0][i_1]...[i_j]$ has amplitude d if $\max\{|i_t - i_0| \mid 1 \leq t < j\} \leq d$. Such a loop is smoothed if it is contracted to a single arc $[i_0][i_j]$. When the loop represents a shortest path the weight on the contracted arc $[i_0][i_j]$ gives the shortest distance.

3. Computing Shortest Paths

We first consider the problem of computing all pair shortest distances within the band.

Assume that $n/k = 2^a - 1$ for an integer a. There are a total of $2^a - 1$ submatrices $A_{ii}, 1 \leq i \leq 2^a - 1$, on the main diagonal of the input matrix A. The idea of path smoothing can be used for computing the shortest distances. In order to compute all shortest distances for all entries within the band, we use two stages.

First stage: For each i, let j be the largest integer such that $i/2^j$ is odd. Smooth loops of the form $[i]...[i]$ with amplitude $2^j - 1$.

Second stage: Smooth all loops and compute all pair shortest distances within the band.

For example, when first stage finishes, loops of the form $[2]...[2]$, $[6]...[6]$, $[10]...[10]$ with amplitude 1 are smoothed, loops of the form $[4]...[4]$ with amplitude 3 are smoothed, and loops of the form $[8]...[8]$ with amplitude 7 are smoothed.

The first stage can be accomplished by the following algorithm.

Procedure Band-1.1
  for all i, $1 \leq i \leq 2^a - 1$, **do in parallel**
    $A_{ii} := A_{ii}^*$;
  for $t := 1$ to $a - 1$ **do**
  /* $2^a - 1 = n/k$. */
  begin
    for all i, $1 \leq i \leq 2^a - 1$, $i \bmod 2^t = 0$, **do in parallel**
    begin
      $A_{ii} := \min\{A_{ii},$
$(\prod_{j=i}^{i+2^{t-1}-1}(A_{j,j+1}A_{j+1,j+1}))(\prod_{j=i+2^{t-1}}^{i+i}(A_{j,j-1}A_{j-1,j-1})),$

$(\prod_{j=i}^{i-2^{t-1}+1}(A_{j,j-1}A_{j-1,j-1}))(\prod_{j=i-2^{t-1}}^{i-1}(A_{j,j+1}A_{j+1,j+1}))\};$

      $A_{ii} := A_{ii}^*$;
  end
end

In the h-th iteration of the loop indexed by t in Band-1.1, if 2^h divides i, then loops of the form $[i]...[i]$ with amplitude $2^h - 1$ are smoothed. Therefore, procedure Band-1.1 accomplishes the tasks of the first stage.

Theorem 1: Procedure Band-1.1 smoothes loops of the form $[i]...[i]$ with amplitude $2^h - 1$, where h is the largest integer such that $i/2^h$ is odd.

Proof: By induction on h. Before the execution of the loop indexed by t in procedure Band-1.1, instruction $A_{ii} := A_{ii}^*$ is executed. Therefore loops of the form $[i]...[i]$ with amplitude 0 have been contracted to a single arc, that is, they have been smoothed. Assume that, after the h-th iteration, loops of the form $[i]...[i]$ with amplitude $2^h - 1$, where 2^h divides i, have been smoothed. Consider the case of $h + 1$. After the h-th iteration, a simple loop of the form $[i]...[i]$ with amplitude $2^{h+1} - 1$, where 2^{h+1} divides i, has been contracted to a loop of the forms $[i][i+1][\delta_{i+1}][i+2][\delta_{i+2}]...[i+2^j-1]...[i+1][\delta_{i+1}][i]$ and $[i][i-1][\delta_{i-1}][i-2][\delta_{i-2}]...[i-2^j-1]...[i-1][\delta_{i-1}][i]$ by the principal of path smoothing, where δ_s is either s or empty ϵ. Instruction

$A_{ii} := \min\{A_{ii},$
$(\prod_{j=i}^{i+2^{t-1}-1}(A_{j,j+1}A_{j+1,j+1}))(\prod_{j=i+2^{t-1}}^{i+1}(A_{j,j-1}A_{j-1,j-1})),$
$(\prod_{j=i}^{i-2^{t-1}+1}(A_{j,j-1}A_{j-1,j-1}))(\prod_{j=i-2^{t-1}}^{i-1}(A_{j,j+1}A_{j+1,j+1}))\}$

in the $h + 1$ iteration smoothes such a simple loop. And then instruction $A_{ii} := A_{ii}^*$ smoothes non-simple loops. $\square$

Theorem 2: The time complexity of Band-1.1 is

$$O\left(\frac{nb^2 \log \frac{n}{b}}{p} + I(b) \log n \log \frac{n}{b}\right).$$

Proof: $O(\log(n/b))$ iterations of the loop indexed by t in Band-1.1 is executed. In the h-th iteration, $\lfloor n/(2^h k)\rfloor$ parallel matrix products and matrix transitive closures are computed, each product contains $O(2^h)$ matrix multiplications. Each matrix multiplication takes $O(b^3/p + I(b))$ time, and each matrix transitive closure takes $O(b^3/p + I(b) \log b)$ time. Therefore, the h-th iteration takes $O((nb^2)/p + I(b) \log b + I(b)h)$ time. The time complexity of the algorithm is

$$O\left(\frac{nb^2 \log \frac{n}{b}}{p} + I(b) \log b \log \frac{n}{b} + I(b) \log^2 \frac{n}{b}\right)$$

$$= O\left(\frac{nb^2 \log\frac{n}{b}}{p} + I(b)\log n \log\frac{n}{b}\right).$$

□

The computation of matrix products in Band-1.1 can be improved. We note that after the first instruction of computing the matrix transitive closure is executed, matrices A_{ii}, $i \bmod 2 \neq 0$, are fixed and will not be modified in the remaining executions of Band-1.1. After the first iteration of the loop indexed by t, matrices A_{ii}, $i \bmod 4 \neq 0$, are fixed and will not be modified. And so on. Therefore, we can compute certain matrix products and save them to be used in the later iterations. The modified procedure Band-1.2 follows.

Procedure Band-1.2
  for all i, $1 \leq i \leq 2^a - 1$, **do in parallel**
    $A_{ii} := A_{ii}^*$;
  for $t := 1$ to $a - 1$ **do**
  /* $2^a - 1 = n/k$. */
  begin
    for all i, $1 \leq i \leq 2^a - 1$, $i \bmod 2^t = 0$, **do in parallel**
    begin
      if $t \neq 1$ **then**
      begin
        $A_{i,i+2^{t-1}} := A_{i,i+2^{t-2}}A_{i+2^{t-2},i+2^{t-2}}A_{i+2^{t-2},i+2^{t-1}}$;

        $A_{i+2^{t-1},i} := A_{i+2^{t-1},i+2^{t-2}}A_{i+2^{t-2},i+2^{t-2}}A_{i+2^{t-2},i}$;

        $A_{i,i-2^{t-1}} := A_{i,i-2^{t-2}}A_{i-2^{t-2},i-2^{t-2}}A_{i-2^{t-2},i-2^{t-1}}$;

        $A_{i-2^{t-1},i} := A_{i-2^{t-1},i-2^{t-2}}A_{i-2^{t-2},i-2^{t-2}}A_{i-2^{t-2},i}$;

      end
      $A_{ii} := \min\{A_{ii},\ A_{i,i+2^{t-1}}A_{i+2^{t-1},i+2^{t-1}}A_{i+2^{t-1},i},$
  $A_{i,i-2^{t-1}}A_{i-2^{t-1},i-2^{t-1}}A_{i-2^{t-1},i}\}$;
      $A_{ii} := A_{ii}^*$;
    end
  end

Theorem 3: The time complexity of Band-1.2 is

$$O\left(\frac{nb^2}{p} + I(b)\log b \log\frac{n}{b}\right).$$

Proof: In the h-th iteration, $\lfloor n/(2^h k)\rfloor$ parallel matrix products and matrix transitive closures are computed, each product contains a constant number of matrix multiplications. Each matrix multiplication takes $O(b^3/p + I(b))$ time, and each matrix transitive closure takes $O(b^3/p + I(b)\log b)$ time. Therefore, the h-th iteration takes $O((nb^2)/(2^h p) + I(b)\log b)$ time. The time complexity of the algorithm is

$$O\left(\frac{nb^2}{p} + I(b)\log b \log\frac{n}{b}\right).$$

□

The second stage is to smooth loops $[i]...[i]$ with any amplitude and to compute all pair shortest distances within the band. When the first stage finishes, loops of the form $[2^{a-1}]...[2^{a-1}]$ with any amplitude are smoothed, where $2^a - 1 = n/k$. This results in two graphs G_1, G_2 from the input graph, G_1 contains vertices in $[1], [2], ..., [2^{a-1}]$, G_2 contains vertices in $[2^{a-1}], ..., [2^a - 1]$. Each original loop of the form $[i]...[i]$ is now a loop in one of the resulting graphs. To continue the smoothing process, we need to work on these two resulting graphs independently. Let us consider G_1. A simple loop of the form $[2^{a-2}]...[2^{a-2}]$ should have already been smoothed in the first stage if it does not contain a vertex in $[2^{a-1}]$. If it does, then the loop is now in the form $[i][\delta_i][i+1][\delta_{i+1}]...[2^{a-1}]...[i][\delta_i]$, where $i = 2^{a-2}$ and δ_j is either j or ϵ. We can use matrix multiplication to smooth such a loop. After all loops of the form $[2^{a-2}]...[2^{a-2}]$ are smoothed, G_1 is divided into two graphs, one contains vertices in $[1], [2], ..., [2^{a-2}]$, the other contains vertices in $[2^{a-2}], ..., [2^{a-1}]$. G_2 can be processed in a similar way. Thus we can continue the dividing process recursively.

After we smoothed all loops, we have to compute shortest distances for all other entries within the band. The lower left triangle of $A_{i,i+1}$ is updated with $A_{ii}A_{i,i+1}$ and upper right triangle of $A_{i,i-1}$ is updated with $A_{ii}A_{i,i-1}$.

We now give procedure Band-2 below for the second stage.

Procedure Band-2
  for $t := a - 2$ downto 0 **do**
  /* $2^a - 1 = n/k$. */
  begin
    for all i, $1 \leq i \leq 2^a - 1$, $i \bmod 2^t = 0$ and $i \bmod 2^{t+1} \neq 0$, **do in parallel**
    begin
      if $i \neq 2^a - 2^t$ **then** $A_{ii} := \min\{A_{ii},$
  $A_{i,i+2^t}A_{i+2^t,i+2^t}A_{i+2^t,i}\}$;
      if $i \neq 2^t$ **then** $A_{ii} := \min\{A_{ii},$
  $A_{i,i-2^t}A_{i-2^t,i-2^t}A_{i-2^t,i}\}$;
      $A_{ii} := A_{ii}^*$;
    end
  end
  /* Fill all entries within the band. */
  for all i, $1 \leq i \leq 2^a - 2$, **do in parallel**
  begin
    $A_{i,i+1} := A_{ii}A_{i,i+1}$;
    $A_{i+1,i} := A_{i+1,i+1}A_{ii}$;
  end

Band-2 actually fills certain entries outside the band.

Theorem 4: Band-2 computes all pair shortest distances within the band in time

$$O\left(\frac{nb^2}{p} + I(b)\log b \log\frac{n}{b}\right).$$

Proof: Each iteration of the loop indexed by t in Band-2 divides the graph into two graphs. Thus after $O(\log(n/b))$ iterations a remaining graph contains only vertices in $[i]$ for some i. In the iteration with $t = h$ all loops of the form $[i]...[i]$ are smoothed, where $i \bmod 2^h = 0$ and $i \bmod 2^{h+1} \neq 0$. Therefore, when Band-2 finishes all loops are smoothed. Instructions $A_{i,i+1} := A_{ii}A_{i,i+1}$ and $A_{i+1,i} := A_{i+1,i+1}A_{ii}$ are used to fill all remaining entries within the band. They have the side effect of filling some entries outside the band.

In the iteration with $t = h$, $O(n/(2^h b))$ matrix products and transitive closures are computed. The time complexity for the iteration is $O(nb^2/(2^h p) + I(b)\log b)$. Since there are $O(\log(n/b))$ iterations, the time complexity of Band-2 is

$$O\left(\frac{nb^2}{p} + I(b)\log b \log\frac{n}{b}\right).$$

$\square$

We now show how to compute all pair shortest distances. After all pair shortest distances are computed within the band, the all pair shortest distances outside the band can be computed progressively from the main diagonal toward the top right corner and the bottom left corner. For convenience we add n/k dummy vertices and number them from $n + 1$ to $n + n/k$. Arcs incident with dummy vertices are labeled with weight ∞. There are now 2^a submatrices on the main diagonal. The algorithm for computing the all pair shortest distances is given below.

Procedure Band-3
for $t := 1$ to $a - 1$ **do**
/* $2^a - 1 = n/k$. */
 begin
  for all i, $1 \leq i \leq 2^a$, $i \bmod 2^t = 0$ and $i \bmod 2^{t+1} \neq 0$,
do in parallel
  begin
  for all j, k, $i - 2^t + 1 \leq j \leq i$, $i + 1 \leq k \leq i + 2^t$,
do in parallel
   begin
    $A_{jk} := A_{ji}A_{i,i+1}A_{i+1,k};$
    $A_{kj} := A_{k,i+1}A_{i+1,i}A_{ij};$
   end

Theorem 5: Band-3 correctly computes all pair shortest distances.

Proof: After h-th iteration of the loop indexed t in Band-3, shortest distances for arcs of the form $[j][k]$

have been correctly computed, where $(i - 1)2^h + 1 \leq j, k \leq i2^h$, $1 \leq i \leq (n/k + 1)/2^h$. In the $h + 1$-th iteration Band-3 computes shortest distances for pairs of the forms $([j], [k])$ and $([k], [j])$, where $i - 2^t + 1 \leq j \leq i$ and $i + 1 \leq k \leq i + 2^t$, $i \bmod 2^{h+1} = 0$ and $i \bmod 2^{h+2} \neq 0$. The shortest paths for pair $([j], [k])$ must go through arc $([i], [i+1])$ and the shortest paths for pair $([k], [j])$ mush go through arc $([i+1], [i])$. Therefore, Band-3 correctly computes all pair shortest distances. $\square$

Theorem 6: The time complexity of Band-3 is $O(n^2b/p + I(b)\log(n/b))$.

Proof: There are $O(\log(n/b))$ iterations in Band-3. Iteration h takes $O(nb^2 2^h/p + I(b))$ time. Thus the whole algorithm has time complexity $O(n^2b/p + I(b)\log(n/b))$. $\square$

References

[ADY] L. Allison, T. I. Dix and C. N. Yee. Shortest path and closure algorithms for banded matrices. Information Processing Letters 40, 317-322(December 1991).

[Fr] M. L. Fredman. New bounds on the complexity of the shortest path problem. *SIAM J. Comput.*, Vol. 5, No. 1, 83-89(March 1976).

[HPR] Y. Han, V. Pan, J. Reif. Efficient parallel algorithms for computing all pair shortest paths in directed graphs. To appear in Proc. 1992 ACM Symposium on Parallel Algorithms and Architectures, San Diego, June-July 1992.

PARALLEL IMPLEMENTATIONS OF A SCALABLE CONSISTENT LABELING TECHNIQUE ON DISTRIBUTED MEMORY MULTI-PROCESSOR SYSTEMS

Wei-Ming Lin and Zhenhong Lu

Dept. of Electrical and Computer Engineering

Mississippi State University

Mississippi State, MS 39762

wlin@ee.msstate.edu

Abstract

An efficient implementation of a consistent labeling technique can lead to a fast solution in solving all kinds of tree search problems by reducing the search space from a potentially exponential size to a much smaller size. In this paper a scalable parallel algorithm is presented in achieving labeling consistency. This algorithm is optimal in execution time as well as in processor-time product. It is suitable for implementation on most distributed memory multiprocessor system.

1 Introduction

Consistent labeling techniques have found their extensive applications in the areas of pattern recognition, artificial intelligence and computer vision [5, 13, 1, 11]. Constraint satisfaction problems in artificial intelligence and various levels of scene labeling problems in pattern recognition and computer vision involve a tree search process to search for one or all available solutions from a search space of potentially exponential size. An application of a consistent labeling technique preceding the search process can eliminate all infeasible labeling assignments by checking up to a certain level of local consistency, in contrast to the global consistency to be satisfied during the final tree search process. A search space of exponential size can be dramatically reduced with a small polynomial time spent in implementing such a preprocessing technique.

There have been several sequential consistent labeling algorithms proposed in the literature [13, 9, 12, 8]. Optimal performance can be reached with the ones presented in [12, 8]. Based on these sequential algorithms, several parallel algorithms [10, 14, 6, 3, 2] have been designed to show various speedup on various platforms with multiple processors. Most of these work lead to a theoretically optimal speedup; however, optimality in processor-time product has not been obtained in any of these results. This usually means poor utilization in the worst case. None of these are designed for efficient implementation on a distributed memory system, either due to the inherently tight data sharing among processors or some data structure costly in time to maintain when implemented in a distributed fashion.

In this paper, we present a modified version of our previously presented parallel algorithm in [8], also an optimal one in execution time as well as in processor-time product, suitable for implementations both on distributed or shared memory multiprocessor systems and easily scalable. To be able to predict and correctly evaluate the overall performance of an algorithm run on a distributed system, the underlying communication structure, the provided communication mechanisms and the time ratio between a unit communication step and a unit computation step in the target system have to be taken into consideration. Our general parallel algorithm is capable of balancing the communication overhead against the necessary computation work for implementations on various sys-

tems with its inherent flexibility. Performance with speedups close to the linear speedup theoretically claimed is demonstrated with a series of tests on the iPSC/860.

2 Consistent Labeling

A typical tree search problem involves labeling a set of n objects, $\{o_1, o_2, \ldots, o_n\}$ with a set of m labels, $\{l_1, l_2, \ldots, l_m\}$ by satisfying some compatibility constraints. The definition of a consistent labeling technique can be found in several papers in the literature [13, 6]. A concise definition of this technique is given in this section. Each object o_i has a set of distinct label indices, Λ_i. The labels corresponding to the indices in this set can be assigned to the object. The compatibility constraints between any two objects, o_i and o_j, are represented by a set of pairs of compatible label indices Ω_{ij}, and $\Omega_{ij} \subseteq \Lambda_i \times \Lambda_j$. A pair of label indices in Ω_{ij}, (p,q), represents that assigning label l_p to object o_i is compatible with assigning label l_q to object o_j. A typical constraint to be satisfied in a tree search problem would be that each assignment has to be compatible with any other assignment for any other object to be feasible. The relaxed constraint to be satisfied in a consistent labeling process is that, for each assignment to be feasible, there has to be at least one assignment for any other object that is compatible with this assignment.

The purpose of the labeling process is to find an n-tuple $L = (L_1, \ldots, L_n)$ such that $\forall i$, $L_i \subseteq \Lambda_i$, and object o_i is assigned the set of labels L_i. A labeling L is said to be <u>consistent</u> with respect to the relaxed constraint if

$$\forall i, j, \quad \forall p \in L_i, \quad (\{p\} \times L_j) \cap \Omega_{ij} \neq \{ \} \tag{1}$$

where the $\times$ denotes the Cartesian product of two sets. The consistent labeling technique is to obtain a labeling L satisfying the consistency criteria defined in equation 1.

The above definition can also be represented using a matrix notation. The compatibility constraints between object o_i and o_j, Ω_{ij}, can be represented by an m-by-m matrix Ω_{ij}^x. $\Omega_{ij}^x[p,q]$ is 1 if assigning label l_p to object o_i is compatible with assigning label l_q to object o_j, 0 otherwise. In the matrix notation, $1 \leq i, j \leq n$, $1 \leq p, q \leq m$, $\Omega_{ij}^x[p,q] = 1(0)$ if $(p,q) \in (\notin)\Omega_{ij}$. The initial label sets, Λ_i's, and the labeling status can be represented by nm boolean variables, v_{ip}, $1 \leq i \leq n, 1 \leq p \leq m$. The labeling process becomes a search process for suitable value for each of the v_{ip}'s. The final (as well as intermediate) labeling can be represented as, for $1 \leq i \leq n$, $1 \leq p \leq m$, $v_{ip} = 1(0)$ if $p \in (\notin)L_i$, and a labeling is <u>consistent</u> if, for all i, p,

$$v_{ip} \leq \prod_{j=1}^{n} \left[\sum_{q=1}^{m} (v_{ip} * v_{jq} * \Omega_{ij}^x[p,q]) \right]. \tag{2}$$

The "$*$" operation represents the logical AND operation, and the summation ($\sum$) and the product ($\prod$) operations are also logical OR and AND operations. This inequality relation corresponds to the definition of consistency in equation 1.

3 Sequential Algorithms

In this section we briefly introduce the idea shared by several similar sequential algorithms previously introduced [12, 8], all of which lead to optimal sequential execution time. More details of this algorithm can be found in [8].

The sequential algorithm is presented in Algorithm 3. An

```
     { Initialization }
· 1.  for i = 1 to n do
  2.    for p = 1 to m do
  3.      v_{ip} ← 1;
  4.      for j = 1 to n do
  5.        N_{ip}[j] ← 0;
  6.        for q = 1 to m do
  7.          if (Ω_{ij}^x[p, q] = 1)  then  N_{ip}[j] ← N_{ip}[j] + 1
  8.        end;
  9.        if (N_{ip}[j] = 0) then do
 10.          pushstack(< i, p >, S); v_{ip} ← 0;
 11.          exit j-for-loop
 12.    end ;
     { Iteration }
 13.  while ((< j, q > ← popstack(S)) ≠ nil) do
 14.    for i = 1 to n do
 15.      for p = 1 to m do
 16.        if ((v_{ip} = 1) AND (Ω_{ij}^x[p, q] = 1))  then do
 17.          N_{ip}[j] ← N_{ip}[j] − 1;
 18.          if (N_{ip}[j] = 0)  then do
 19.            pushstack(< i, p >, S); v_{ip} ← 0;
 20.  end
```

Algorithm 1: An Optimal Sequential Algorithm

ordered pair $< i, p >$ is used to denote the assignment of label l_p to object o_i. Throughout this paper, such a pair is referred to an assignment. This assignment of label l_p to object o_i is said to be feasible if $\forall j \; \exists q$ such that $\Omega_{ij}^x[p, q] = 1$ according to the consistency criteria in equation 1. The counter variable $N_{ip}[j]$, $1 \le i \le n$, $1 \le p \le m$, $1 \le j \le n$, represents the number of 1's in the p-th row in the updated Ω_{ij}^x, which is also the current number of compatible labels that can be assigned to object o_j when object o_i is assigned with label l_p. Each $< i, p >$ assignment has n such counter variables, $N_{ip}[1], N_{ip}[2], \ldots, N_{ip}[n]$. During the initialization procedure (lines 1-12 in Algorithm 3), all the n $N_{ip}[j]$ variables of each assignment $< i, p >$ are initialized and the feasibility of each assignment is also determined. A stack S is used to store infeasible assignments for further update operation.

Since there are at most nm assignments that can be placed on the stack, the number of iterations is upper bounded by nm. It is easy to verify that Algorithm 3 runs in $O(n^2 m^2)$ time with each time unit corresponding to the time for a simple arithmetic/logic operation. This represents a factor of $O(nm)$ faster than the well-known sequential approach [13] in which, in each iteration, $O(nm)$ time is spent for each assignment in checking if it becomes infeasible by inspecting all its corresponding nm Ω entries.

4 Known Parallel Approaches

In [14] several parallel algorithms are proposed based on the sequential algorithms proposed in [9, 12]. PAC-1, PAC-3 and PAC-4 lead to solutions optimal in execution time on a shared memory platform. PAC-1 and PAC-3 are based on non-optimal sequential algorithms; therefore, both of them require more than minimal number of processors, nm, necessary to

arrive at the optimal parallel solution of $O(nm)$ time. Each of them uses $n^2 m^2$ processors to achieve its optimal speedup. PAC-4 is based on a similar optimal sequential algorithm described in section 3 to achieve optimal speedup on a shared memory system; however, due to an inefficient approach in exploiting parallelism, it still requires $O(n^2 m^2)$ processors, more than minimally required number of processors. This leads to a non-optimal processor-time product, neither can they be effectively implemented on a distributed-memory system.

In [2], several algorithms are proposed to parallelize the sequential algorithms in [9, 12] to exploit parallelism in a way similar to the one we have in [8]. Linear speedup is achieved on a shared memory system; however, the algorithms proposed cannot be effectively run on a distributed memory system due to the tight data sharing among processors and data structure costly to maintain when run in a distributed fashion. The basic computation step in these algorithms involves set operations (eg. include, union, delete, etc.) which are not usually considered a unit-time operation.

Some special purpose parallel architectures for solving the consistent labeling problem have also been proposed. The designs in DRA2, DRA3 [6] and DRA5 by Gu [4] using array type of hardware are based on the traditional sequential approach. Thus, in theory, optimality in execution time cannot be reached. These designs do not lead to scalable implementations.

5 Parallel Implementations

Two scalable parallel algorithms are presented in this section along with the implementations on the iPSC/860.

5.1 A Scalable Parallel Algorithm

With a fixed number of processing elements (PEs), say P ($P \le nm$), we are able to design a scalable implementation by enabling each PE to process $\frac{nm}{P}$ distinct v_{ip} variables. An external memory module is attached to each PE to store all the relevant data and all the PEs are connected through an interconnection network, such as bus, mesh, hypercube, etc. The data stored in each of the P memory modules includes $\frac{nm}{P}$ v_{ip} variables. In addition, for each v_{ip}, the corresponding nm $\Omega_{ij}^x[p, q]$ variables ($1 \le j \le n$, $1 \le q \le m$) and n $N_{ip}[j]$ counter variables ($1 \le j \le n$) are also stored in the memory module.

A simple parallel algorithm abstracted from the one we have in [8] is described in Algorithm 2. The procedure Initialize() performed by each PE, for each v_{ip} it is responsible, initializes the n counter variables for this assignment and returns the Id (Identification) of an infeasible assignment as $< i, p >$, if there is any. During each iteration, the interconnection network, either with centralized or distributed control depending on the underlying communication architecture, selects one non-nil assignment Id from the ones sent to the network and broadcasts it to every PE. This operation is performed by the procedure Id-select-broadcast(i, p) in Algorithm 2, and an Id (represented by $< j, q >$ in line 6) is received by each PE. The algorithm stops when each of the Ids sent to the network is a nil Id. The procedure Update(j, q) performs the update operation necessary for each assignment in each PE, same as the operations performed in lines 13-20 in Algorithm 3.

It can be easily derived that the time to accomplish the Initialization part is $O(\frac{n^2 m^2}{p})T_{comp}$, where T_{comp} corresponds to the time required for a basic arithmetic operation, either an update operation on a counter variable or a simple compare operation on an Ω entry. The execution time for each iteration can be represented by $T_{comm}^s(1) + O(\frac{nm}{P})T_{comp}$, where $T_{comm}^s(1)$ denotes the communication time for select/broadcasting one Id throughout the network. Thus, with the maximal number of iterations equal to nm, the overall execution time can be represented as:

$$T_1 = O(nmT^s_{comm}(1)) + O(\frac{n^2m^2}{P}T_{comp}). \qquad (3)$$

The performance can be evaluated by considering the follow-

```
    { Initialization }
1.  parallel do (in all PEs)
2.      < i, p > ←  Initialize()
3.  parallel end ;
    { Iteration }
4.  repeat
5.      parallel do (in all PEs)
6.          < j, q > ←  Id-select-broadcast(i,p);
7.          if < j, q >= nil then stop;
8.          else < i, p > ←  Update(j,q)
9.      parallel end
10. forever
```

Algorithm 2: A Scalable Parallel Algorithm

ing two factors: 1. the relation between nm and P, and 2. the relation between $T^s_{comm}(1)$ and T_{comp}. With a theoretical assumption on the second factor, $T^s_{comm}(1) = D \cdot T_{comp}$ in which D depends on the configuration of the interconnection network, linear speedup can be reached in various types of networks with certain condition on the first factor satisfied. In theory, the D here is usually the value of the diameter of the network (or with a small constant factor) with the assumption that adjacent-PE unit data transfer consumes the same amount of time as a unit computation step, i.e. T_{comp}. In [8] we show that with a shared bus as the interconnection network, under the assumption of $T^s_{comm}(1) = T_{comp}$ (i.e. $D = 1$), optimal speedup can be reached if $nm > P$. We also show that with a linear array connection as the interconnection network, under the assumption of $T^s_{comm}(1) = P \cdot T_{comp}$ (i.e. $D = P$), optimal speedup can be reached if $nm > P^2$. The condition for a hypercube network becomes $nm > P \log P$ with the assumption becoming $T^s_{comm}(1) = \log P \cdot T_{comp}$ (i.e. $D = \log P$). These results hold only when the corresponding assumption on the second factor is not impractical. Unfortunately, for most of the real-life parallel machines, we have much larger D values than the ideal ones, i.e. $T^s_{comm}(1) >> D \cdot T_{comp}$. In this case, $T^s_{comm}(1)$ becomes a much more dominant factor in deciding the overall performance. Unless the balance between computation and communication work is further adjusted, linear speedup can only exist in theory.

5.2 A Modified Parallel Algorithm

The problem presented in the previous section in excessive communication overhead can be relieved by notifying each processor with more than one infeasible assignment Ids, if available; during each iteration. It is found that in practice, especially in early iterations, most partitions of assignments in PEs each leads to multiple infeasible assignments. This approach allows a better balance between computation and communication times by having more computation work to be performed during each iteration for a slightly increased amount of communication work. In addition to this, total number of iterations can be expected to be smaller; therefore, the overall execution time is further reduced due to the fact that a certain degree of redundant communication and computation work performed in the previous approach is eliminated as well.

It would be an optimal scheme if, for each iteration, each processor is capable of effectively appending all its currently found infeasible Ids to the message to be broadcast to all the PEs, if not considering the overhead involved in each PE in this append operation and the necessary dynamic space allocation for the unknown message length. This overhead can sometimes offset the intended saving in communication time. We choose to balance the communication and computation times using a fixed-size message concatenation scheme, in which each

PE contributes no more than a fixed number of infeasible Ids for the computation (update) work in the current iteration. The modified algorithm is described in Algorithm 3. The

```
    { Initialization }
1.  parallel do (in all PEs)
2.      X ←  Initialize()
3.  parallel end ;
    { Iteration }
4.  repeat
5.      parallel do (in all PEs)
6.          Y ←  Id-concatenate-broadcast(X);
7.          if Y = nil then stop;
8.          else X ←  Update(Y)
9.      parallel end
10. forever
```

Algorithm 3: A Modified Scalable Parallel Algorithm

X in line 2 denotes an array of predetermined size x with each element contains an available infeasible Id, from the result of Initialize() or Update() procedure. The Y in line 6 denotes the message of all X concatenated together with its size fixed as $y = xP$. This approach not only regularizes the message passing process among processors but eliminates the extra overhead in dynamic space allocation as well. Theoretically, this has the same worst case performance as that of the Algorithm 2. A simple timing analysis for the average case performance is conducted as follows. The initialization part takes about the same amount of time as the one required in Algorithm 2. Assume that during each iteration an average number of $c_1 x$ Ids, where $0 \leq c_1 \leq 1$, are present in each X; therefore, an average number of $c_1 x P$ infeasible Ids can be expected in Y. Also, we assume that the total number of infeasible assignment Ids to be found throughout the process is $c_2 nm$, where $0 \leq c_2 \leq 1$. The computation time in each iteration becomes $c_1 x P \cdot \frac{nm}{P} T_{comp} = c_1 xnmT_{comp}$, and the average number of iterations can be approximated by $\frac{c_2 nm}{c_1 x P}$. With $T^c_{comm}(y)$ denoting the time in performing the communication step in line 6 in Algorithm 3, the overall execution time for the iteration part becomes

$$T_2(x) = \frac{c_2 nm}{c_1 x P}T^c_{comm}(xP) + \frac{c_2 n^2 m^2}{P}T_{comp}. \qquad (4)$$

By taking the parameter c_2 into consideration in equation 3, the average execution time required for the approach in section 5.1 becomes

$$T_1 = c_2 nmT^s_{comm}(1) + \frac{c_2 n^2 m^2}{P}T_{comp}. \qquad (5)$$

The new approach in Algorithm 3 outperforms the previous approach in Algorithm 2 when

$$T^s_{comm}(1) > \frac{T^c_{comm}(xP)}{c_1 x P}. \qquad (6)$$

The above condition is usually satisfied by most of the distributed memory systems available, especially when an efficient concatenate mechanism is available and c_1 is not too small. It would have been a strictly increasing performance function with larger x if the following two scenarios are not taken into consideration: 1. c_1 becomes smaller with a larger x, and 2. computation overhead involved in handling the message during each iteration increases with a larger x. This implies that a saturation point in performance can be expected with a value of x depending on nature of the input (c_1) and the efficiency of the concatenation mechanism used ($T^s_{comm}(1)/T^c_{comm}(xP)$).

5.3 Implementations and Results

In this section the implementations of the Algorithm 3 on the iPSC/860 system are presented. Our machine has 32 nodes interconnected by a five-dimensional hypercube network in which a modified "wormhole" routing hardware is used to provide efficient multi-hop message routing between non-adjacent processors.

On this system there are several approaches available to implement the necessary Id-broadcast operation. Explicit

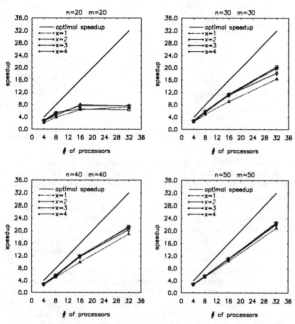

Figure 1: Experimental Timing Results

message passing mechanisms, **send** and **receive**, can be used to implement the 'Id-select-broadcast' operation in Algorithm 2. A system library call **gcol** (global concatenate) can be adapted to effectively carry out the operation 'Id-concatenate-broadcast' in Algorithm 3. It is found that the unit communication times for both algorithms measured on this system, $T^s_{comm}(1)$ and $T^c_{comm}(xP)$, easily satisfy the condition in equation 6 for most x values chosen. Thus, we only show the

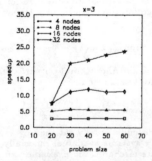

Figure 2: Speedups vs. Problem Size with $x = 3$

results in running the Algorithm 3 on this system. Shown in Figure 1 are the speedups from a series of tests run on inputs with different input sizes (n, m) each with a message of different sizes (x). As expected, a saturation point in performance is reached with a value of x depending on the input size. From the figures shown, $x = 3$ leads to a smallest $T^c_{comm}(xP)/c_1 xP$ value in average for these cases. The speedups as a function of problem size run on various configurations of different numbers of processors are shown in Figure 2 with the message size fixed at $x = 3$.

The implementation results just shown demonstrate a parallelization approach with speedups up to 70-80% of linear

speedups. We expect to see even better numbers when a shared memory version of our algorithm is run on a shared memory platform.

6 Conclusion

We have given two parallel algorithms easily implementable on various distributed memory systems and other shared-memory systems without major modifications. Linear speedup is reached from the theoretical point of view. Implementations with speedups close to linear speedup are presented using a novel but simple approach intended to balance the computation work against the originally time-dominating communication work. One of the potential future extension of this work is to develop a complete parallel search algorithm that utilizes the consistent labeling technique presented in this paper as a preprocessing technique. We expect to see various applications of this search algorithm in solving problems in different fields.

References

[1] A. Bundy, "Catalogue of Artificial Intelligence Tools" in *Symbolic Computation Artificial Intelligence*, L. Bolc, A. Bundy, P. Hayes and J. Siekmann, eds. Berlin: Springer-Verlag, 1984.

[2] J. M. Conrad, D. P. Agrawal and D. R. Bahler, "Scalable Parallel Arc Consistency Algorithms for Shared Memory Computers", *Int'l Parallel Processing Symposium*, 1992.

[3] P. R. Cooper and M. J. Swain, "Parallelism and Domain Dependence in Constraint Satisfaction", *Technical Report TR 255*, Dept. of Computer Science, Univ. of Rochester, 1988.

[4] J. Gu, "An Optimal Lookahead Processor to Prune Search Space", *3rd Symposium on the Frontiers of Massively Computation*, 1990.

[5] A. Guzman, "Computer Recognition of Three Dimensional Objects in a Visual Scene", *PhD thesis*, MIT, 1968.

[6] J. Gu, W. Wang and T. C. Henderson, "A Parallel Architecture for Discrete Relaxation Algorithm", *IEEE Trans. on PAMI*, vol. PAMI-9, no. 6, pp. 816-831, Nov. 1987.

[7] W.-M. Lin, "Mapping Image Algorithms onto Fixed Size Window Architectures," *PhD thesis*, USC, 1991.

[8] Wei-Ming Lin and V. K. Prasanna Kumar, "Parallel Algorithms and Architectures for Discrete Relaxation Technique," *IEEE Conference on Computer Vision and Pattern Recognition*, 1991.

[9] A. K. Mackworth and E. C. Freuder, "The Complexity of some Polynomial Network Consistency Algorithms for Constraint Satisfaction Problems," *Artificial Intelligence*, Jan. 1985.

[10] J. T. Mccall, J. T. Tront, F. G. Gray, R. M. Haralick and W. M. Mccormack, "Parallel Computer Architectures and Problem Solving Strategies for the Consistent Labeling Problem", *IEEE Trans. on Computer*, vol. C-34, pp. 973-980, Nov. 1985.

[11] G. Medioni and R. Nevatia, "Matching Images Using Linear Features", *IEEE Trans. on PAMI*, vol. PAMI-6, no. 6, pp. 675-785, Nov. 1984.

[12] R. Mohr and T. C. Henderson, "Arc and Path Consistency Revisited," *Artificial Intelligence*, Mar. 1986.

[13] A. Rosenfeld, R. A. Hummel and S. W. Zucker, "Scene Labeling by Relaxation Operations", *IEEE Trans. on Syst., Man, Cyber.*, vol. SMC-6, pp. 420-433, June 1976.

[14] A. Samal and T. Henderson, "Parallel Consistent Labeling Algorithms", *International Journal of Parallel Programming*, vol. 16, no. 5, Oct. 1987.

Maximally Fault Tolerant Directed Network Graph With Sublogarithmic Diameter For Arbitrary Number of Nodes

Pradip K Srimani

Department of Computer Science
Colorado State University
Ft. Collins, CO 80523

Abstract

An efficient fault tolerant communication network is an integral part of designing today's parallel processing systems. In this paper we propose a new family of almost regular directed network graphs based on permutation of groups of elements for an arbitrary number of nodes. The graph is shown to have sublogarithmic diameter and almost linear number of edges and is proved to be maximally fault tolerant.

1 Introduction

Recently there has been a spurt of research on Caley graphs, symmetric graphs defined on permutation of distinct symbols [1]. Most important of them are the so-called star graphs which seem to enjoy most of the desirable properties of the hypercubes at considerably less cost; they accommodate more nodes with less interconnection hardware and less communication delay. Only very recently Corbett [4] has proposed a family of directed network graphs based on permutation of elements, called the *rotator graphs*. These rotator graphs are also Caley graphs except that the generators are not closed under inverse and hence the graph is directed (communication channels are unidirectional). These rotator graphs are significant in the sense that they are the only known directed graphs based on permutation of groups of symbols. But they have a serious drawback that they can be defined for N nodes only when $N = n!$ for some integer n; they cannot be defined for an arbitrary number of nodes. This incremental extensibility is a very essential and desirable property in real life applications of a topology in designing computer networks. A few of the network graphs described in the literature [3, 7, 8] are incrementally extensible; but all of them are symmetric graphs. No family of directed network graphs is known to be incrementally extensible.

Our purpose in the present paper is to propose a new incrementally extensible directed network topology that can be defined for an arbitrary number of nodes. The design philosophy basically involves appropriate interconnection of different sized rotator graphs of different sizes. If N is the given number of nodes, $n! < N < (n+1)!$, the proposed graph is a superset of several rotator graphs of size less than or equal to n; we call it a super rotator graph.

2 Basic Concepts

In this section we briefly introduce the *rotator graphs*, discuss relevant properties and introduce a few new concepts that will be needed to describe the new topology and to study its properties. Graph theoretic terms not defined here can be found in [5] and a detailed treatment of the rotator graphs can be found in [4].

A rotator graph R_n, of order (dimension) n, is defined to be a directed graph $G = (V, E)$, where V is the set of $n!$ vertices (nodes), each representing a distinct permutation of n distinct symbols, and E is the set of directed edges such that there is an edge from one permutation (node) v to another permutation (node) u iff u can be reached from v by rotating its first ℓ symbols one place left ($2 \le \ell \le n$). These rotator graphs are Caley Graphs [2] except that the generators (different values of ℓ give the distinct generators) are not closed under inverse operation and hence the graph is a directed graph. For example, in R_3, the node abc has outgoing edges to nodes bca and bac as well as has incoming edges from nodes bac and cab. We denote the nodes as permutations of English alphabets; for example, the identity permutation is denoted by $I = (abc...z)$ (z is the last symbol in the string, not necessarily the 26th letter). It has been shown in [4] that: (1) R_n is $(n-1)$ regular in the sense that each node has both an in-degree and an out-degree of $n-1$ (hence, the number of edges is $n!(n-1)$), (2) the diameter of R_n is given by $\mathcal{D}(R_n) = n-1$, and (3) R_n is vertex symmetric for all values of n like other Caley graphs [1].

It is also to be noted that the rotator graphs are hierarchical in the sense that R_n can be decomposed into n number of R_{n-1}'s. In R_n we use V_x to denote the set of nodes (permutations) that end with the symbol "x"; obviously V_x is a rotator graph of dimension $n-1$. Similarly, we use V_α to denote the set of nodes that end with α where α represents a sequence of symbols. V_α is a rotator graph of dimension $n - |\alpha|$ if V_α is a subgraph of R_n.

Definition 1 *Consider any two mutually disjoint subgraphs V_x and V_y of a rotator graph R_n. The nodes of V_x that are directly connected to some node of V_y by outgoing edges are called the type I gateway nodes of V_x with respect to V_y; we denote this set of nodes by $G^I_{x,y}$. The nodes of V_x that are directly connected to some node of V_y by incoming edges are called the type II gateway nodes of V_x with respect to V_y; we denote this set of nodes by $G^{II}_{x,y}$.*

Definition 2 *A directed graph G is called* **strongly connected** *iff for an arbitrary pair of vertices u and v, there exists a directed path from u to v in G.*

Definition 3 *A directed graph G is called strongly k-connected if it remains strongly connected after removal of an arbitrary set of k or less nodes. This k is called the* **measure of strong connectedness** *ξ of the graph G.*

Remarks:

- A strongly 0-connected graph is simply a strongly connected directed graph.

- The rotator graph R_n is strongly $(n-2)$-connected, i.e., $\xi(R_n) = n-2$, for all $n \geq 3$.

- The measure of strong connectedness of a directed graph defines the node fault tolerance

of the graph. A directed graph G, $\xi(G) = k$ remains strongly connected when an arbitrary set of k or less nodes are faulty.

Definition 4 *Any positive integer N, $n! \leq N < (n+1)!$, can be expressed in its mixed-radix form as $< a_n, a_{n-1}, \cdots, a_1 >$, where*

$$N = a_n.n! + a_{n-1}.(n-1)! + ... + a_1.1!$$

and $0 \leq a_i \leq i$ for $i = 1, \cdots, n-1$, and $0 < a_n \leq n$.

For example, $110 = < 4, 2, 1, 0 >$, since $4.4! + 2.3! + 1.2! + 0.1! = 96 + 12 + 2 = 110$.

In order to design the proposed super rotator graphs we need two types of connections between rotator graphs of different dimensions. We define them as follows.

Definition 5 *Given m copies of R_k where $m < k$, we say that these m copies are joined by type A connections when they are connected by the directed edges as if they were subgraphs of the larger R_{k+1}.*

Remark: There are exactly $(k-1)!$ directed type-A edges from each R_k to each of the other R_k's; similarly there are exactly $(k-1)!$ directed type-A edges from each of the other R_k's to a specific R_k.

Definition 6 *When m copies of R_k, $m \leq k$, are joined by the type A connections, the resulting graph is called a class $C_k(m)$.*

Definition 7 *Given two rotator graphs R_m and R_n, $m < n$, they are said to be joined by type B connections if outgoing edges are added from each node w in R_m to $|n-m|$ different nodes of R_n (type B successors of w in R_n) and incoming edges are added to each node w in R_m from $|n-m|$ different nodes of R_n (type B predecessors of w in R_n). It is required that type B successors of an arbitrary pair of nodes in R_m are mutually disjoint and so are their type B predecessors; but type B successors and type B predecessors of nodes may overlap.*

Vertex Numbering: It is well known [6] that all the $n!$ permutations of n distinct symbols can be uniquely numbered from 0 through $n! - 1$. We use this scheme to number the vertices of any rotator graph R_n; we also extend this scheme to number the vertices of a class $C_k(m)$. The class $C_k(m)$ has $m.k!$ nodes; the nodes of C_k^1 are numbered from 0 to $k! - 1$, the nodes of C_k^2 are numbered from $k!$ to $2k! - 1$ and so on.

3 Topology for Super Rotator Graphs

The basic idea behind the design of the super rotator graph of N nodes, when $n! < N < (n=1)!$, is to express N as sum of several factorials, build smaller rotator graphs of appropriate dimensions, and then add appropriate type A and type B edges to connect those smaller graphs. The following algorithm builds the super rotator graph for any given N, $n! < N < (n=1)!$.

The Algorithm

Step 1: [Build the smaller rotator subgraphs]

Compute the mixed radix representation of $N = < c_n, c_{n-1}, \cdots, c_1 >$ and construct c_i copies of R_i for all i, $1 \le i \le n$ (note $c_n \ne 0$).

Step 2: [Label the nodes]

- Choose $n + 1$ symbols to label the nodes (permutations). We use $n + 1$ consecutive English letters starting with "a".

- For $i = n$ to 1 do the following (fix the i-th symbol for the nodes):

  - if $c_i \ne 0$ then label each of the c_i copies of R_i as $V_{\alpha_j \beta}$ where $\beta = symbol(i + 1) symbol(i + 2) \cdots symbol(n)$, and α_j, $1 \le j \le c_i$, are chosen in alphabetic order from the set of symbols that are yet to be allocated to the "symbol" array.

  - Set $symbol(i)$ to be equal to the next available English letter in alphabetic order.

Step 3: [Provide type A connections among rotator subgraphs to form classes]

- For each i, $1 \le i \le n$, join the c_i components of R_i's by type A connection as defined earlier to get the different classes C_i (note that this does not connect the rotator subgraphs of different dimensions).

- Each class C_i has c_i number of components C_i^ℓ, $1 \le \ell \le c_i$ each of which is a rotator graph of dimension i. The vertices in C_i are numbered from 0 to $c_i . i! - 1$ by using the vertex numbering scheme as described before (the vertices of C_i^1 are numbered from 0 to $i! - 1$, those of C_i^2 are numbered from $i!$ to $2i! - 1$ and so on).

Step 4: [Construct the super rotator graph in steps by providing the type B connections]

Find the minimum i such that $c_i \ne 0$ and then set $j = i$ and set $SR_j = C_i$ (SR_j denotes the super rotator graph with $\sum_{k=1}^{j} c_k k!$ nodes). while $i \le n$ do if $c_i \ne 0$ then

- Establish type B connections between SR_j and C_i. Each node in SR_j is assigned $(i-j)$ type B successors as well as $(i - j)$ type B predecessors in the leader L_i of the class C_i. This is easily done by using the node numberings in both the graphs SR_j and C_i (e.g., outgoing edges are introduced from node "0" of SR_j to nodes "0" through "i-j-1" of L_i as well as incoming edges are introduced from nodes "$i! - 1$" to "$i! - (i-j)$" of L_i to node "0" of SR_j, outgoing edges are introduced from node "1" of SR_j to nodes "i-j" through "2(i-j)-1" of L_i as well as incoming edges are introduced from nodes "$i! - (i - j) - 1$" to "$i! - 2(i - j)$" of L_i to node "0" of SR_j, and so on).

- Renumber the nodes of SR_j by adding $c_i i!$ to each node number.

- Set $j = i$ and set SR_j to be the composite graph generated in the previous steps. Note that SR_j has now $\sum_{k=1}^{j} c_k k!$ nodes and they are numbered from 0 to $\sum_{k+1}^{j} c_k k! - 1$.

$i = i + 1$

Return SR_n as the desired super rotator graph of N vertices.

Example 1: Let $N = 23$. Then N can be expressed as $N = < 3, 2, 1 >$ or $c_3 = 3$, $c_2 = 2$, and $c_1 = 1$. We have 3 non-null classes. See Figure 1. The class C_3 have 3 components, e.g., V_a, V_b and V_c each of which is a rotator graph of dimension 3; the vertices are numbered from 0 to 17. Also, $symbol(3) = d$. The class C_2 has two components V_{ad} and V_{bd} each of which is a rotator graph of dimension 2; the vertices are numbered from 0 to 3. As before, the class C_1 is a single node (a rotator graph of dimension 1) and since $symbol(2) = c$, this single node is labeled as the permutation "bacd" and is numbered 0. Type A connections are provided in each class as shown in the figure. In the first iteration of step 4 of the design algorithm, type B connections are provided to nodes of C_1 and C_2 by adding directed edges from node "0" of C_1 to node "0" of C_2 and from node "1" of C_2 to node "0" of C_1 and we get SR_2.

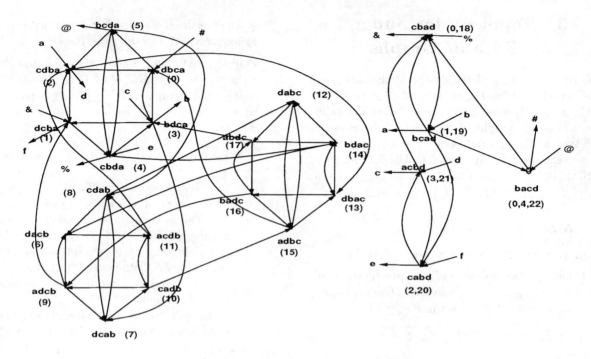

Figure 1: Example Graph with $N = 23$ nodes

Nodes of SR_2 are renumbered (actually the nodes of C_1 only need be renumbered; the node "*bacd*" is renumbered as 4). Next, C_3 and SR_2 are joined by type B connections to get the desired super rotator graph $SR_3(23)$ (nodes "0" through "4" of SR_2 are connected to "0" through "4" of C_3 as well as nodes "5" through "1" of C_3 are connected to "0" through "4" of SR_2).

4 Conclusion

We have proposed a new class of directed network graphs that can be effectively used in designing the communication architecture for distributed processing systems. The design of the graphs is based on the theory of Caley graphs. Recently almost regular symmetric graphs have been proposed for arbitrary number of nodes [7, 8], but regular directed graphs are known only for N nodes where N is a factorial of an integer [4]. As far as we know, this is the first attempt in designing almost regular directed graphs for an arbitrary number of nodes. The proposed family of graphs has many interesting properties; we plan to reprt them shortly.

References

[1] S. B. Akers and B. Krishnamurthy. A group-theoretic model for symmetric interconnection networks. *IEEE Transactions on Computers*, 38(4):555–566, April 1989.

[2] F. Annexstein, M. Baumslag, and A. L. Rosenberg. Group action graphs and parallel architectures. *SIAM Journal on Computing*, 19(3):544–569, March 1990.

[3] B. W. Arden and H. Lee. A regular network for multiprocessor systems. *IEEE Transactions on Computers*, 31(1):60–69, January 1982.

[4] P. F. Corbett. Rotator graphs: an efficient topology for point-to-point multiprovcessor networks. *IEEE Transactions on Parallel and Distributed Systems*, 3(5):622–626, September 1992.

[5] F. Harary. *Graph Theory*. Addison-Wesley, Reading, MA, 1972.

[6] D. E. Knuth. *The Art of Computer Programming, Volume II*. Addison-Wesley, 1972.

[7] S. Sur and P. K. Srimani. Super star: a new optimally fault tolerant network architecture. In *Proceedings of the International Conference on Distributed Computing Systems (ICDCS-11)*, pages 590–597, Texas, 1991.

[8] S. Sur and P. K. Srimani. Incrementally extensible hypercube (IEH) graphs. In *Proceedings of the International Conference on Computers and Communication (IPCCC-92)*, pages 1–7, Phoenix, Arizona, April 1992.

SESSION 10C

RECONFIGURABLE ARCHITECTURE AND DATABASE APPLICATIONS

FAST ARITHMETIC ON RECONFIGURABLE MESHES*

Heonchul Park, Viktor K. Prasanna and Ju-wook Jang

Department of Electrical Engineering-Systems

University of Southern California

Los Angeles, CA 90089-2562

Abstract–*We show an $O(1)$ time algorithm for computing the N-point Discrete Fourier Transform (DFT) on an $N \log N \times N \log N$ reconfigurable mesh, where each point is represented using $\Theta(\log N)$ bits. Based on the DFT algorithm, we derive $O(1)$ time algorithms for N-bit multiplication, N-bit modular operations, and computing the quotient of two N-bit numbers on an $N \times N$ bit model of the reconfigurable mesh. The proposed algorithms lead to AT^2 optimal designs in the bit model of VLSI for $1 \leq T \leq \sqrt{N}$. All these algorithms run on the MRN, RMESH, and PARBUS models of the reconfigurable mesh, and employ optimal size of the mesh.*

1 Introduction

The complexity of the computations in VLSI has been measured in terms of the AT^2 product, where A is the VLSI layout area of the design and T is the computation time. The lower bound for the computation of N-point Discrete Fourier Transform (DFT) is $AT^2 = \Omega(N^2 \log^2 N)$[1] in the bit model of VLSI. Several basic arithmetic operations such as multiplication, modular operations, and division are known to be closely related to the computation of the DFT. Area-time tradeoffs satisfying $AT^2 = O(N^2)$ are known for performing these arithmetic operations in the bit model of VLSI [1, 14].

VLSI architectures for performing the above arithmetic operations including these for DFT have been studied extensively during the past decade. Computation of the DFT, multiplication of two N-bit numbers, modular operations and integer division have been considered on various architectures [1, 4, 14, 19, 21]. However, the fastest known time bounds on these architectures are limited to $O(\log N)$ [19].

Recently, the reconfigurable mesh has received a lot of attention. The reconfigurable mesh is a two

*This research was supported in part by NSF under grant IRI-9145810 and in part by Defense Advanced Research Project Agency under contract F-49620-89-C-0126.

[1]Unless specified, all logarithms in this paper are to base 2.

dimensional mesh of Processing Elements (PEs) connected by reconfigurable buses[15]. In this model, the connections between the PEs can be dynamically rearranged during the execution of algorithms. Several algorithms on the reconfigurable mesh are known [3, 7, 8, 15, 17, 22].

In this paper, we show algorithms for fast arithmetic operations on the reconfigurable mesh. First, we show a design which computes the N-point DFT in constant time on an $N \log N \times N \log N$ bit model of the reconfigurable mesh. Based on this, we present $O(1)$ time algorithms for multiplication, modular operations, and integer division using $N \times N$ bit model of reconfigurable mesh. These result in AT^2 optimal designs for these problems.

The organization of this paper is as follows. In the next section, we briefly introduce the reconfigurable mesh. In Section 3, $O(1)$ time algorithm for computing DFT is introduced. Several $O(1)$ time algorithms for performing arithmetic operations on an $N \times N$ reconfigurable mesh are shown in Section 4. Concluding remarks are made in Section 5.

2 Reconfigurable Mesh

The reconfigurable mesh architecture used in this paper is based on the architecture defined in [15]. The $N \times N$ bit model of the reconfigurable mesh consists of an $N \times N$ array of PEs connected to a grid-shaped reconfigurable broadcast bus, where each PE has locally controllable bus switches. The switches allow the broadcast bus to be divided into *subbuses*, providing smaller reconfigurable meshes. For a given set of switch settings, a *subbus* is a maximal connected subset of the PEs. In this paper, we use the exclusive write model which allows only one PE to broadcast to a *subbus* shared by multiple PEs at any given time. We assume that the value broadcast consists of $O(1)$ bits and takes $\Theta(1)$ time. The bus can carry one of *1-signal* or *0-signal* at any time. Each PE can perform arithmetic and logic operations on $O(1)$ bits in unit time. The size of the local storage in each PE is $O(1)$

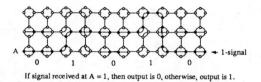

If signal received at A = 1, then output is 0, otherwise, output is 1.

Figure 1: A 3×10 reconfigurable mesh for computing EXOR of 5 input bits

bits. The array operates in synchronous mode. Other than the buses and the switches, the reconfigurable mesh is similar to the standard mesh in that it has $\Theta(N^2)$ area, under the assumption that a PE and its switches, and a link between adjacent switches occupy unit area in the bit model of VLSI. A 3×10 reconfigurable mesh for computing the EXOR of 5 bits in constant time is shown in Figure 1.

Each PE has four I/O ports (N, E, W, and S). The internal connection among the four ports of a PE can be configured during the execution of algorithms. Figure 2 shows some possible connection patterns. We use the notation {SW, EN} to denote the configuration in which S port is connected to W port while N port is connected to E port.

After the definition of reconfigurable mesh in [15], other models have been defined [3, 16]. These models restrict the allowed connection patterns. The most general and powerful one is the PARBUS model [22] which allows any combination of 4-port connections in each PE. Notice that the PARBUS model is same as our model. In [3], the MRN model has been introduced. The connection patterns allowed in this model are shown in Figure 2. In MRN, the number of possible connection patterns for each PE is 10. In [16], RMESH is introduced, which doesn't allow {NS, EW}, {NE, SW}, and {NW, SE} connections allowed in MRN. However, RMESH allows {NEWS}, {NEW, S}, {NES, W}, {NWS, E}, and {N, EWS} connections. Thus, the total number of connection patterns allowed in each PE of RMESH is 12. This corresponds to having switches on mesh links only. In addition, a variation of the reconfigurable mesh is introduced in [13] and has been denoted as REBSIS. The REBSIS of size $N \times N$ consists of N^2 word-level switches $S(w, d)$, where w denotes the length of a word. A switch setting can be specified using d bits. In the known solutions on REBSIS [13], each $S(w, d)$ consists of two types of bit-level switches and each switch has 4 at most 5 I/O ports and $d \leq 2$.

In [9], it has been shown that the word model of the reconfigurable mesh of size $N \times N$ with w bits per word (having an area of $N^2 w^2$) and the REBSIS of

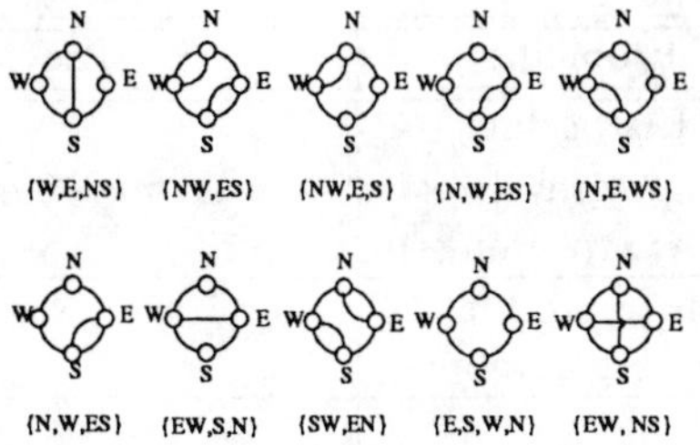

Figure 2: Connection patterns allowed in MRN

size $N \times N$ consisting of $S(w, 2)$ can be simulated with $O(1)$ time overhead per step using the bit model of the reconfigurable mesh of size $Nw \times Nw$, $\log N \leq w \leq N$. Further, for simulating the REBSIS of size $N \times N$ consisting of $S(w, 2)$, only $O(N^2 w)$ PEs are needed in the bit model of the reconfigurable mesh. In addition, it is shown that the bit model of reconfigurable mesh can improve several known results. Table 1 summarizes these results.

Throughout this paper, reconfigurable mesh denotes the PARBUS (bit) model unless noted otherwise. However, the algorithms derived in this paper can be simulated on an MRN of size $N \times N$ without slowdown, since the proposed algorithms employ a subset of the connection patterns shown in Figure 2. Also, it is shown that an $N \times N$ MRN can be simulated using $N/p \times N/p$ MRN with slowdown by a factor of $4p^2$ [3]. Based on this, sometimes constant factors are not considered in the size of the reconfigurable mesh.

The reconfigurable mesh algorithms in this paper can also be simulated on an RMESH (of corresponding size), since we do not employ the {EW,NS} connection, and {NW,ES} and {SW,EN} connections in the MRN model can be simulated using 4 PEs in the RMESH model. In addition, any MRN algorithm can be simulated on a 2-layer RMESH without asymptotic loss in time [19].

3 Computing the DFT on the Reconfigurable Mesh

The number of multiplications needed for computing the DFT can be reduced by the decomposition technique which leads to the FFT algorithm [6]. The main idea of this technique is to expand the dimension of the index space of the input points using a radix-representation. Consider the following one dimensional DFT computation:

$$X(m) = \sum_k x_0(k) \cdot W_N(k, m) \qquad (3.1)$$

Problems	Previous results	Improved results
Histogram (h grey levels of an $N \times N$ image)	$T = O(min\{\sqrt{h} + \log^*(N/h), N\})$[8]	$T = O(\sqrt{h})$ where $h \leq N$
Inner product (N w-bit numbers)	$A = O(Nw^3)$, $T = O(\log N)$ [12]	$A = O(w^2 N \log^2 N)$, $T = O(1)$
Radix sort (N w-bit numbers)	$A = O(N^{3/2}\log^2 N)$, $T = O(N^{1/4}w)$ if $w \leq \log N$ [11]	$A = N^{1/2}Bw \times N^{1/2}Bw$, $T = O((\log N/2\log B)^c N^{1/2}/B)$ where $w \leq N/2B\log B$ and c is a constant

Table 1: A summary of improved results using the bit model of the reconfigurable mesh

where $0 \leq k, m \leq N - 1$, $W_N(k,m) = \omega^{(km/N)}$, and ω is the N-th root of unity. The computation of DFT shown in equation (3.1) can be expressed as a d-dimensional DFT by using a simple change of radix. The d-dimensional DFT computation can be performed by computing n^{d-1} n-point DFTs d times and rotations of $N = n^d$ data items.

Computing the DFT can be considered as a matrix-vector multiplication which consists of sequences of inner product operations. First, we show that the computation of inner product and matrix-vector multiplication can be performed in constant time on the reconfigurable mesh. Computing the DFT in $O(1)$ time follows this observation.

Lemma 1 *Given two k-bit numbers, generating the k $(2k-1)$-bit partial products for multiplication of these two numbers can be performed in constant time on a $k \times 2k$ reconfigurable mesh.*

Proof: Load the k-bit number into the PEs in the topmost row such that PEs in columns k to $(2k-1)$ have one input bit. Set the rest of the PEs in the top row to have a '0'. The PEs in the top row broadcast their bits to all the PEs along the anti-diagonal using {NW, SE} connection such that the bit at the j-th PE in the topmost row is in the $(j-l)$-th PE in row l, $0 \leq l \leq k - 1$. Load the other k-bit number of PEs in the leftmost column such that each PE has one bit. Broadcast the i-th bit to the PEs in the i-th row using {WE, N, S} connection. Each PE performs an AND operation. □

We use the following result related to addition of numbers:

Lemma 2 *[8] Given q k-bit numbers, these numbers can be added in constant time on an $2q \times 2qk$ reconfigurable mesh.*

Theorem 1 *Given an $s \times q$ matrix and a $q \times 1$ vector, where each element of the vector consists of k bits,* the matrix-vector multiplication can be performed in constant time on a $2ksq \times 2k^2q$ reconfigurable mesh, assuming sq (q) elements of the matrix (vector) are available in the PEs in the leftmost column (topmost row) in row major order.

Proof: Let the leftmost column be partitioned into sq disjoint subcolumns of size $2k$. Input the i-th element of the input matrix in row major order into the bottom k PEs of the i-th subcolumn. Consider the topmost row to be made of q disjoint subrows of size $2k^2$. Input the $q \times 1$ vector to the PEs in the topmost row such that an element of the input vector is available in the leftmost k PEs of each subrow. Denote $B(i)$ as the i-th block of size $2kq \times 2k^2q$, $0 \leq i \leq s - 1$. Also, denote $S(j)$ as the j-th sub-block of size $2k \times 2k^2q$ in a block, $0 \leq j \leq q - 1$. Figure 3 shows this arrangement.

The i-th element of the output is computed in the i-th block as follows. Broadcast the elements of the input vector such that each block has a copy of the vector in the PEs in its topmost row. In each block, broadcast the vector using {NS,E,W} connection such that $S(j)$ has the j-th element of the vector in the PEs in its topmost row. Notice that $S(j)$ has an element of the j-th column of the $s \times q$ matrix. Multiplication of two k-bit numbers can be performed in constant time on $2k \times 2k^2$ reconfigurable mesh using Lemmas 1 and 2. Thus, q multiplications of two k-bit numbers can be performed in constant time within each block. In each sub-block, the result of the multiplication is moved to the leftmost column PEs. Using Lemma 2, addition of q $(2k-1)$-bit numbers can be performed in constant time in each block. Since the matrix-vector multiplication is composed of s inner products, it can be performed in constant time on a $2ksq \times 2k^2q$ reconfigurable mesh. □

Computing the N-point DFT can be considered as four sets of computations of $N^{\frac{3}{4}}$ $N^{\frac{1}{4}}$-point DFTs and $N^{\frac{1}{4}}$-shuffle on N data between successive com-

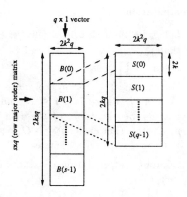

Figure 3: Reconfigurable mesh for performing matrix-vector multiplication

putations. The computation of $N^{\frac{1}{4}}$-point DFT is a matrix-vector multiplication of $N^{\frac{1}{4}} \times N^{\frac{1}{4}}$ matrix and a $N^{\frac{1}{4}} \times 1$ vector. Thus, using Theorem 1, computing the $N^{\frac{1}{4}}$-point DFT with $\Theta(\log N)$ bits per input element can be performed in constant time on a $N^{1/2} \log N \times N^{1/4} \log^2 N$ reconfigurable mesh.

Theorem 2 *The computation of N-point DFT can be performed in constant time on an $N \log N \times N \log N$ reconfigurable mesh, assuming each input point is represented using $\Theta(\log N)$ bits.*

Proof: Divide the $N \log N \times N \log N$ mesh into $N^{1/2}$ blocks of size $N^{1/2} \log N \times N \log N$. Denote $B(i)$ as the i-th block, $0 \leq i \leq N^{1/2} - 1$. Also, denote $S(j)$ as the j-th sub-block of size $N^{1/2} \log N \times N^{\frac{3}{4}} \log N$, $0 \leq j \leq N^{1/4} - 1$ within each block. Each sub-block computes the $N^{1/4}$-point DFT as its input. Note that the size of the sub-block is larger than what is needed to apply Theorem 1.

Load the input vector of $N \log N$-bit numbers into the PEs in the leftmost column. Move the data within each $B(j)$ such that $S(k)$ has the $(jN^{1/2} + k)$-th block of $N^{1/4} \log N$ bits in its topmost row, $0 \leq j \leq N^{1/2} - 1$, $0 \leq k \leq N^{1/4} - 1$. Load $\sqrt{N} (\log N)$-bit numbers corresponding to the coefficient matrix of $N^{1/4} \times N^{1/4}$ DFT into the PEs in the leftmost column. Broadcast these data such that each sub-block has these data in its PEs in the leftmost column. Using Theorem 1, simultaneously compute $N^{3/4} N^{1/4}$-point DFTs. Collect the intermediate results ($N \log N$-bit numbers) in the leftmost column of the reconfigurable mesh. Shuffle these N data and move them into the PEs in the leftmost column for the next iteration. Repeating the above four times, the result is available in the PEs in the leftmost column. □

Thus, the DFT can be computed in $O(1)$ time and the layout area of the reconfigurable mesh is

$O(N^2 \log^2 N)$. Thus, the AT^2 product of the proposed design matches the $\Omega(N^2 \log^2 N)$ lower bound for computing the DFT in the two dimensional grid model of VLSI.

4 Arithmetic Operations on Reconfigurable Mesh

Based on the algorithm for computing the DFT in Section 3, we present several fast algorithms for basic arithmetic operations on the reconfigurable mesh. First, an $O(1)$ time multiplication algorithm is shown. This is followed by algorithms for modular operations and integer division.

4.1 Multiplication

Given two N-bit binary numbers, X and Y, the multiplication problem is to compute the $(2N-1)$-bit product. Previously, this problem has been solved in $O(1)$ time using a reconfigurable mesh of size $N^2 \times N^2$ [22]. However, it is not optimal in the bit model of VLSI, since the known AT^2 bound for the multiplication of two N-bit numbers is $\Omega(N^2)$ [4].

Our approach to perform integer multiplication is well known [4]: The N-bit numbers X and Y are subdivided into n terms of r bits each such that $X = \sum_{j=0}^{n-1} x_j 2^{jr}$ and $Y = \sum_{j=0}^{n-1} y_j 2^{jr}$, where $0 \leq x_j, y_j < 2^r$ and $x_j = y_j = 0$ for $j \geq n/2$. The multiplication of two N-bit numbers can be considered as a convolution of two sequences of length n where $c_i = \sum_{h+k \equiv i \bmod n} x_h y_k$ for $0 \leq i \leq n - 1$ and $XY = \sum c_i 2^{ir}$. The convolution can be performed using the DFT. Thus, the algorithm for multiplication as follows:

1. Compute two n-point DFTs on numbers X and Y.

2. Perform pointwise multiplication of the resulting coefficients.

3. Compute n-point inverse DFT of the n coefficients in Step 2.

Choosing $n = N/\log N$ and $r = \log N$, Step 1 can be performed in $O(1)$ time on an $N \times N$ reconfigurable mesh. Steps 2 and 3 can also be performed in constant time using $N \times N$ mesh. It leads to the following:

Theorem 3 *Multiplication of two N-bit numbers can be performed in constant time on an $N \times N$ reconfigurable mesh.*

An alternative proof of this theorem can also be found in [10].

4.2 Modular Operations

Let X be an integer in the range $[0, M)$, $M = 2^N$. The mod m conversion problem is to compute X modulo a given integer m, where $m \leq M/2$. Several fast algorithms for modular operations such as mod m conversion, mod m addition and mod m multiplication follow from Theorem 3. Let $|X|_M$ denotes $X \bmod M$. In [1], the following algorithm is shown to perform mod m conversion.

1. Input is X, m, and k. Output is $X \bmod m$.

2. Let $k = \lfloor M/m \rfloor$. For a given X, k, and m, compute $|Xk|_M$ using N-bit binary multiplier. Let T denote the N-bit integer part of the output of the multiplication.

3. Compute $X_1 = |X - mT|_M$ using an N-bit binary multiplier and an N-bit adder.

4. Compute $X_2 = |X_1 - m|_M$ using an N-bit adder.

5. If the most significant bit of X_2 is 1, then output X_1. Otherwise, output X_2.

Using Theorem 3 in the above mod m conversion algorithm, we obtain the following:

Corollary 1 *Given X and m where $X \in [0, M)$, $M = 2^N$, and $m \leq M/2$, $X \bmod m$ can be computed in $O(1)$ time on an $N \times N$ reconfigurable mesh.*

In a similar spirit, we can show:

Corollary 2 *$XY \bmod m$ operation can be performed in $O(1)$ time on $N \times N$ reconfigurable mesh, where $X, Y \in [0, M)$, $M = 2^N$, and $m \leq M/2$.*

4.3 Division

The division problem is to compute the (most significant) N-bit quotient of Y/X, given two N-bit (non-negative) integers X and Y as inputs. This problem has been studied in [5] using (second order) Newton approximation which takes $O(\log N)$ iterations, where each iteration requires an N-bit integer multiplication. Note that the most significant 2^i bits are computed at the end of the i-th iteration of the Newton approximation. It results in $O(\log^2 N)$ time solution with $O(M(N))$ area, where $M(N)$ denotes the area for multiplying two N-bit numbers in $O(\log N)$ time. $O(\log N)$ time algorithm for the above problem has been shown in [2] using the Chinese remainder theorem which results in a circuit of size $O(N^4 \log^3 N)$. VLSI architectures satisfying $AT^2 = O(N^2)$ where $T = O(\log^{1+\epsilon} N)$, $0 < \epsilon \leq 1$, have been shown in [14] which employs a modified design of [2] and the integer divider. Recently, $O(\log N)$

- Precomputation

  1. Let p_i be the i-th prime. Choose r primes such that $p_1 < p_2 < \cdots < p_r$ and $\prod_{j=1}^{r} p_j > 2^{N^2}$. Let $M = \prod_{i=1}^{r} p_j$.

  2. For each a, $0 \leq a < p_j$, compute Nr tables of $a^i \bmod p_j$, where $1 \leq i \leq N$ and $1 \leq j \leq r$. Let T_{ij} be the table of the i-th powers mod p_j. Notice that each entry is $O(\log N)$-bit wide. Each table has $O(p_r)$ entries.

  3. Compute interpolation constants $0 \leq u_j < M$ such that
  $$u_j \equiv \begin{cases} 1 \bmod p_j, & i \neq j \text{ for } 1 \leq i \leq r \\ 0 \bmod p_i, & \end{cases}$$
  Note that u_js are $O(N^2)$-bit numbers.

- Algorithm for powering {input is x, $0 \leq x \leq 2^N - 1$: output is x^i, for $1 \leq i \leq N$ }

  1. Compute $z_j = x \bmod p_j$ for $1 \leq j \leq r$.

  2. Read $y_{ij} = z_j^i \bmod p_j$ from the tables (in Step 2 of the Precomputation phase) for $1 \leq i \leq N$ and $1 \leq j \leq r$.

  3. Compute $y_i = \sum_{j=1}^{r} y_{ij} u_j$, for $1 \leq i \leq N$.

  4. Output $x^i = y_i \bmod M$ for $1 \leq i \leq N$.

Figure 4: An algorithm for powering using the Chinese remainder theorem

time algorithm has been shown in the CRCW PRAM model [20] using Newton approximation and using DFT. The integer division method shown in [2] computes the inverse of an N-bit number X by adding the first N powers of $U = 1 - X$ and truncating the N^2-bit result to its leading N-bits. Each power of U is computed simultaneously using the Chinese remainder theorem and all the powers are added. The algorithm for powering [2] is shown in Figure 4:

Based on this algorithm, we can show:

Lemma 3 *Given an N-bit integer X, X^i in N^2-bit precision can be computed in $O(1)$ time, simultaneously for $1 \leq i \leq N$, on an $N^4 \log^2 N \times N^5$ reconfigurable mesh.*

Proof: Load the N-bit data into the leftmost N PEs in the top row of the mesh. Divide the mesh into N blocks of size $N^4 \log^2 N \times N^4$ and denote these as $B(p)$, $0 \leq p \leq N - 1$. Also, within $B(p)$, let $S(p, q)$ denote the q-th sub-block of size $N^2 \log^2 N \times N^4$, $0 \leq q \leq N^2 - 1$. The arrangement of these blocks is shown in Figure 5. Broadcast the N-bit integer to the leftmost N PEs in the top row of $S(0, q)$, $0 \leq q \leq N^2 - 1$.

Notice that in the algorithm for powering in Figure

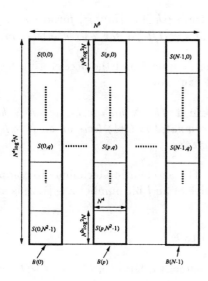

Figure 5: Arrangement of sub-meshes for powering

4, $r < N^2$ and $p_r = O(N^2 \log N)$. Step 1 can be performed in constant time using $S(0, q)$, $0 \leq q \leq N^2-1$, using Corollary 1. Step 2 can be performed by table look-up for each i and j. For each y_{ij}, the table requires $O(p_r)$ entries of $O(\log N)$ bits each. Each y_{ij} is computed in $S(i, j)$ after row broadcast from $B(0)$. The table T_{ij} is stored in $S(i, j)$. In each $S(i, j)$, the k-th row stores the entry for $k^i \bmod p_j$, where $0 \leq k \leq p_j - 1$. Table look-up using z_j can be performed in constant time. Step 3 computes N outputs. Each output consists of $O(N^2)$ terms and each term requires a multiplication of an $O(\log N)$-bit integer by an $O(N^2)$-bit integer. Each block performs summation of N^2 terms after each sub-block computes the product of y_{ij} and u_j. This multiplication can be performed in constant time using Theorem 3 on a $N^2 \times N^2$ mesh. Addition of N^2 terms can be performed within each block in constant time using Lemma 2. Finally for each i, $1 \leq i \leq N$, Step 4 can also be performed in constant time within $B(i)$ using Corollary 1. Since the total number of the blocks available is N, all the N powers of X can be computed in constant time on an $N^4 \log^2 N \times N^5$ mesh. $\square$

In Lemma 3, the powers are computed to N^2 bits of precision. However, it is not necessary to compute the powers with N^2 bits of precision in computing the reciprocal [14]. Consider the factorization

$$1+u+u^2+\cdots+u^{N-1} = (1+u+u^2+\cdots+u^{s-1}) \times (1+u^s+u^{2s}+\cdots+u^{(s-1)s})\cdots(1+u^{s^{m-1}}+\cdots+u^{(s-1)s^{m-1}})$$

$$(4.3.2)$$

where $s = N^{1/m}$, m is a fixed integer. Note that there are m factors in this factorization. In [20], it

```
function Modpower(x, r, s, ε)
(* x = η_{r-1}η_{r-2}···η_0 is an r bit number, r is power of 2,
s is a positive integer, and 0 < ε ≤ 1/2 *)
begin
if r, s ≤ 4
    Modpower ← x^s mod 2^r + 1
else
begin
    case r ≥ s^2 do (* case-1 *)
    begin
        l ← ½√(r/s);
        k ← 2√(rs);
        (* divide x into k equal blocks of l bits each *)
        (* form a vector (x_0, x_1, ···, x_{k-1})
            where x_i = η_{r-il-1}···η_{r-(i+1)l} *)
        (x_0, x_1, ···, x_{k-1}) ← (x_0, x_1 2, ···, x_{k-1} 2^{k-1}) mod 2^k+1;
        (x_0, x_1, ···, x_{k-1}) ←
        DFT_k(x_0, x_1, ···, x_{k-1}) mod Z_{2^k+1};
        pardo x_i ← Modpower(x_i, k, s, ε)
        (x_0, ···, x_{k-1}) ← DFT_k^{-1}(x_0, ···, x_{k-1}) mod Z_{2^k+1};
        (x_0, ···, x_{k-1}) ←
        (x_0, x_1 2^{-1}, ···, x_{k-1} 2^{-(k-1)}) mod 2^k + 1;
        Modpower ← (x_0, x_1 2l, ···, x_{k-1}(2^l)^{k-1});
    end
    case r < s^2 do (* case-2 *)
    begin
        α ← log_r s; ε ← α/⌈α/ε⌉; s ← r^ε;
        for i = 1 to ⌈α/ε⌉ do Modpower(x, r, s, ε)
    end
end
end
```

Figure 6: An algorithm for computing $x^s \bmod (2^r+1)$

is shown that the inverse of X with N-bit precision can be obtained by computing each factor with Ns-bit precision, multiplying these m factors, and truncating the result to N bits [20].

The number of PEs used to compute the powers can be reduced by using Lemma 3 in the modular power algorithm [20] and our technique to compute the DFT. The modular power algorithm which computes $x^s \bmod (2^r + 1)$ for a given r-bit number x is shown in Figure 6.

In [20], it is shown that case-1 of the algorithm is initiated at most $\log(1/\epsilon)$ times recursively, then case-2 is initiated, which requires the s-th power of an s^2 bit numbers, assuming $\epsilon = 1/m$ for a fixed integer m. case-2 initiates case-1 $O(1/\epsilon)$ times. Thus, the total number of recursions of DFT is $O(\frac{1}{\epsilon} \log \frac{1}{\epsilon})$.

Lemma 4 *Given an N-bit integer X and i, $1 \leq i \leq N$, X^i with Ns bits of precision can be computed in constant time on an $N^{1+\frac{1}{m}} \times N^{1+\frac{1}{m}}$ reconfigurable* mesh, where $s = N^{\frac{1}{m}}$ and m is a fixed integer.

Proof: Consider computing [the i-th power of an N-bit number] mod $2^N + 1$. **case-1** consists of computing the DFT of k numbers of l bits each (each number is represented using k-bit precision, $k \geq l$). This can be performed in constant time on an $N^{1+\frac{1}{m}} \times N^{1+\frac{1}{m}}$ mesh, since the N-bit number can be considered as k k-bit numbers such that $k^2 = O(Ns) = O(N^{1+\frac{1}{m}})$. Denote k_j, A_{DFT}^j as the length of k and the available mesh size to compute DFT_k at the j-th call of Mod-power, respectively. Also, denote A_{power}^j as the size of the mesh required to compute the power of a k_j bit number using Lemma 3. Then,

$$A_{DFT}^j = \frac{(Ns)^2}{k_j^2 \prod_{i=1}^{j} k_i} = \frac{(Ns)^2}{4^{j+1-(1/2)^j} N^{1+(1/2)^j} s^{j+1}},$$

In the above, we have used $k_j = 4^{1-(1/2)^j} N^{(1/2)^j} s$ as in [20]. Also, we know that $A_{power}^j < (k_j)^8 \log^2 k_j$ from Lemma 3 and $A_{DFT}^j > (k_j)^9$ for $j \geq 4$. This implies that Lemma 3 can be used to compute the powers after 4 iterations of computing DFTs. Thus, computing the power of X takes $O(m \log m)$ computations of DFTs for a fixed integer m, for $m \geq 32$. Thus, all the computations can be performed in constant time on an $N^{1+\frac{1}{m}} \times N^{1+\frac{1}{m}}$ mesh. $\square$

Lemma 5 *Given an N-bit integer X, the inverse of X with N bits of precision can be computed in constant time on an $N^{1+\frac{1.5}{m}} \times N^{1+\frac{1.5}{m}}$ reconfigurable mesh, where m is a fixed integer.*

Proof: Using Lemma 4, compute $N^{\frac{1}{m}}$ powers of $(1 - X)$, simultaneously, in constant time on an $N^{1/m}(N^{1+\frac{1}{m}} \times N^{1+\frac{1}{m}})$ mesh. Summation of $N^{\frac{1}{m}}$ numbers of Ns bits can be performed in constant time on an $N^{\frac{1}{m}}(N^{1+\frac{1}{m}} \times N^{1+\frac{1}{m}})$ mesh using Lemma 2. Multiplication of two $N^{1+\frac{1}{m}}$-bit numbers can be performed in constant time on an $N^{1+\frac{1}{m}} \times N^{1+\frac{1}{m}}$ mesh using Theorem 3. Multiplying the m terms in equation (4. 3. 2) takes $\log m$ steps. Thus, computing the N-bit inverse of an N-bit number can performed in constant time on a mesh of size $N^{1+\frac{1.5}{m}} \times N^{1+\frac{1.5}{m}}$ for a fixed integer m. $\square$

Suppose x is an l-bit number, where $x \in [1, 2)$. Let $u = 1 - x$. If u has l_1 zeros immediately to the right of the fixed point, where $0 \leq l_1 \leq l$, then $u^{\lceil l/l_1 \rceil} < 2^{-l}$. It means that addition of the first $\lceil l/l_1 \rceil$ consecutive powers of u leads to the inverse of x [14]. Combining this fact and Lemma 5 leads to:

Lemma 6 *If an l-bit number $x \in [1, 2)$ has $l_1 - 1$ zeros immediately to the right of the fixed point, the l-bit inverse of x can be computed in time $T = O(1)$ and*

size of the mesh $A = (ls^{1.5})^2$, for any s, $2 \leq s \leq l/l_1$ such that $s = (l/l_1)^{\frac{1}{m}}$ for a fixed integer m.

Using Lemma 6 and the accelerating technique in [14], we can show:

Theorem 4 *The N-bit inverse of an N-bit number can be computed in $O(1)$ time on an $N \times N$ reconfigurable mesh.*

Proof: The accelerating technique to compute the l-bit inverse of an l-bit number $x \in [1, 2)$ is as follows:
begin
$z_1 \leftarrow x$
for $i = 1$ to k
  begin
    $x_i \leftarrow$ leftmost l_i bits of z_i (* $l_1 < \cdots < l_k = l$ *)
    $v_i \leftarrow (l_i + 1)$-bit inverse of x_i
    $z_{i+1} \leftarrow z_i v_i$
  end
$v \leftarrow v_1 v_2 \cdots v_k$
end

A number $x \in [1, 2)$ can be represented as $x = x_1 + 2^{-l_1} w$ where x_1 is an l_1-bit number and w is an $(l - l_1)$-bit number. Let v_1 be an l_1-bit approximation to x_1. Then $v_1 x = 1 + \mu + v_1 w 2^{-l_1}$, where $\mu < 2^{-l_1}$, and $v_1 x$ has at least $l_1 - 1$ consecutive 0's immediately to the right of the fixed point. If it is successively applied, then we can obtain $1/x \cong v_1 v_2 \cdots v_k$.

In this technique, let A_i denote the size of the mesh, and T_i be the time taken for the i-th iteration of the above, for $i = 1, 2, \cdots, J$, and J is an integer to be chosen. Choose, $l_i = N/(c^{1+\delta} s_i^{1.5})$, $s_i^{1.5} = (N/c^{1+\delta})^{1/(2 \log^{\delta(i-1)} N)}$ such that $A_i T_i^2 = O(N^2)$ and $A_i = (l_i s_i^{1.5})^2$ from Lemma 6, for a fixed integer c, and a fixed constant δ, $0 < \delta < 1/2$, and $1 \leq i \leq J$. The parameter J is chosen as the smallest value of i for which $s_i > 2$ and is readily found to be $\Theta(1/\delta)$. Then, each iteration can be performed in constant time on an $N \times N$ mesh. The last value of l_J is approximately $l_J = \frac{N}{2c^{1+\delta}}$, since $s_J \simeq 2$. If the Newton approximator is employed after the computation of the l_J-bit result, then we can compute the N-bit inverse of the x in $O((1 + \delta) \log(2c))$ time using the constant time multiplication algorithm on an $N \times N$ mesh. All the computations are performed using $O(N)$ bits. Thus, all the computations can be performed in constant time on an $N \times N$ reconfigurable mesh. $\square$

5 Conclusion

We have proposed an algorithm which can compute the N-point DFT in constant time using a

$N \log N \times N \log N$ reconfigurable mesh. This algorithm is optimal in the AT^2 sense. Based on this, we have shown $O(1)$ time algorithms for multiplication, modular computations, and integer division using $N \times N$ reconfigurable mesh. In addition, these algorithms lead to AT^2 optimal designs for $1 \leq T \leq \sqrt{N}$ in the 2-dimensional grid model of VLSI. The set of switch connection patterns used by our algorithms is a subset of those allowed in the MRN model. Also, {EW, SN} connection is not used in all our algorithms, which implies that the proposed algorithms can be executed on the RMESH model. Thus, our algorithms run on these models as well, while maintaining AT^2 optimality.

References

[1] F. Barsi, "Mod m arithemetic in binary systems," *Info. Processing Letters,* pp. 303-309, 1991.

[2] P. W. Beame, S. A. Cook and H. J. Hoover, "Log depth circuits for division and related problems," *SIAM J. Comput.,* pp. 994-1003, 1986.

[3] Y. Ben-Asher, D. Gordon, A. Schuster, "Optimal simulations in reconfigurable arrays," *Tech. Rep.* No. 716, Dept. of Computer Science, Technion-Israel Institute of Technology, 1992.

[4] R. P. Brent and H. T. Kung, "The chip complexity of binary arithmetic," *J. ACM,* pp. 521-534, 1981.

[5] S. A. Cook, *On the minimum computation time of functions,* Ph. D. Thesis, Harvard Univ., 1966.

[6] J. W. Cooley and J. W. Tuckey, "An algorithm for the machine calculation of complex fourier series," *Math. Comput.,* 1965.

[7] J. Jang and V. K. Prasanna, "An optimal sorting algorithm on reconfigurable mesh," *International Parallel Processing Symp. (IPPS),* pp. 130-137, 1992.

[8] J. Jang, H. Park and V. K. Prasanna, "A fast algorithm for computing a histogram on reconfigurable mesh," *Symp. on the Frontiers of Massively Parallel Computation,* 1992.

[9] J. Jang, H. Park and V. K. Prasanna, "A bit model of the reconfigurable mesh," *Manuscript,* Dept. of EE-Systems, USC, January 1993.

[10] J. Jang, H. Park and V. K. Prasanna, "An optimal multiplication algorithm on reconfigurable mesh," *IEEE Symp. on Parallel and Distributed Processing,* Dec. 1992.

[11] R. Lin, "Reconfigurable buses with shift switching-VLSI radix sort," *ICPP,* 1992.

[12] R. Lin and S. Olariu, "Computing the inner product on reconfigurable buses with shift switching," *Joint Conf. on Vector and Parallel Processing,* 1992.

[13] R. Lin and S. Olariu, "Reconfigurable buses with shift switching-Architectures and application," *International Pheonix Conf. on Computers and Communications,* 1993.

[14] K. Mehlhorn and F. P. Preparata, "Area-time optimal division for $T = \Omega((\log n)^{1+\epsilon})$," *Symp. on Theoretical Aspects of Comp. Sci.,* pp. 341-352, 1986.

[15] R. Miller, V. K. Prasanna Kumar, D. I. Reisis and Q. F. Stout, "Meshes with reconfigurable buses," *MIT Conf. on Advanced Research in VLSI,* 1988.

[16] M. Nigam and S. Sahni, "Sorting n numbers on $n \times n$ reconfigurable meshes with buses," *International Parallel Processing Symp. (IPPS),* 1993.

[17] S. Olariu, J. L. Schwing and J. Zhang, "Fast computer vision algorithms for reconfigurable meshes," *International Parallel Processing Symp. (IPPS),* 1992.

[18] H. Park and V. K. Prasanna, "A Class of Optimal VLSI Architectures for Computing Discrete Fourier Transform," *ICPP,* 1992.

[19] H. Park, *Area efficient VLSI architectures for image computations,* Ph. D. Thesis, Dept. of EE-Systems, University of Southern California, 1993.

[20] N. Shankar and V. Ramachandran, "Efficient parallel circuits and algorithms for division," *Info. Processing Letters,* pp. 307-313, 1988.

[21] C. D. Thompson, "Fourier transforms in VLSI," *IEEE Trans. on Comp.,* pp. 1047-1057, 1983.

[22] B. F. Wang, G. H. Chen and H. Li, "Configurational computation: A new computation method on processor arrays with reconfigurable bus systems," *ICPP,* 1991.

List Ranking and Graph Algorithms on the Reconfigurable Multiple Bus Machine

(*Extended Abstract*)

C. P. Subbaraman   Jerry L. Trahan   R. Vaidyanathan*

Department of Electrical & Computer Engineering
Louisiana State University, Baton Rouge, LA 70803–5901

Abstract

The Reconfigurable Multiple Bus Machine (RMBM) is a model of parallel computation based on reconfigurable buses. We present constant time algorithms for list ranking, integer sorting and a number of fundamental graph problems on the RMBM. The algorithms are more efficient in terms of processors than the corresponding PARBS algorithms. The algorithms demonstrate some of the potential for computation available in the ability to manipulate communication paths as a vital part of computation.

1   Introduction

Recently, the processor array with a reconfigurable bus system (PARBS) has drawn considerable attention and several interesting results have been obtained for it (see [2, 4, 7] for additional references). An $n \times n$ PARBS consists of an array of processors, each of which may fuse together two or more of its ports to create a bus through the processor. As a result, even processors that are not neighbors in the array may communicate directly. Other models related to the PARBS have been proposed, such as the Reconfigurable Mesh [2], Polymorphic Torus [1] and Bus Automaton [3]. Computationally, these models are not significantly different from the PARBS. A drawback of the PARBS and related models is that the number of buses that the model can generate depends on the number of processors in it. Consequently, for computations that require a high communication bandwidth, the PARBS uses a large number of processors solely for reconfiguring the buses. In other words, several PARBS algorithms use processors to perform tasks that can be accomplished by simple switches; as a result, these algorithms are inefficient.

In this paper, we develop constant time algorithms for ranking a linked-list and some fundamental graph problems on a model called the Reconfigurable Multiple Bus Machine (RMBM) proposed by Thiruchelvan *et al.* [5]. Like the PARBS, the RMBM is based on reconfigurable

buses. Unlike the PARBS, however, the number of processors and and the number of buses in the RMBM are independent quantities. This allows some RMBM algorithms to be more efficient (in terms of number of processors used) than corresponding PARBS algorithms.

The algorithms we present illustrate this advantage of the RMBM over the PARBS. Olariu *et al.* [4] proposed the best constant time PARBS algorithm for list ranking. This algorithm uses an $(n^\epsilon(n+1)+2) \times 3n$ processor 2-dimensional PARBS to rank a list of n elements in constant time, for any constant $0 < \epsilon < 1$. The proposed RMBM algorithm uses only $n^{1+\epsilon}$ processors and $n^{1+\epsilon}$ buses for the same problem. The number of switches needed for both algorithms is $\Theta(n^{2+\epsilon})$, for any $0 < \epsilon < 1$. We note here that the new algorithm is fundamentally different from that of Olariu *et al.* [4]. The RMBM list ranking algorithm can also apply to sorting n $O(\log n)$-bit integers (integer sorting) in constant time on an RMBM with $n^{1+\epsilon}$ processors and $n^{1+\epsilon}$ buses, for any constant $0 < \epsilon < 1$. Wang *et al.* [7] proposed constant time PARBS algorithms for fundamental graph problems such as finding the transitive closure of an undirected graph, finding all connected components of an undirected graph, finding all bridges and cut vertices of a connected undirected graph, etc. Their transitive closure and connected components algorithms use an $n \times n \times n$ (or $n^2 \times n^2$) PARBS. The proposed RMBM algorithms to solve these problems use only $n^{2+\epsilon}$ processors. The number of switches used by our algorithms is $n^{3+2\epsilon}$, for any constant $0 < \epsilon < 1$, however.

2   The Reconfigurable Multiple Bus Machine

Thiruchelvan *et al.* [5] proposed several versions of the Reconfigurable Multiple Bus Machine (RMBM). In this paper, we use one of these versions called the E-RMBM; we therefore describe only this restricted form of the RMBM.

The E-RMBM consists of P processors (numbered $0, 1, \cdots, P-1$), B buses (numbered $0, 1, \cdots, B-1$), P "fuse lines" and PB sets of switches, $Q_{i,j} = \{c_{i,j,0}, c_{i,j,1}, s_{i,j,0}, s_{i,j,1}, s_{i,j,2}, f_{i,j}\}$, for $0 \leq i < P$ and $0 \leq j < B$. The buses provide communication paths

*Supported in part by the Summer Stipend Program Grant from the Council on Research, Louisiana State University, Baton Rouge, LA 70803.

between processors, and the switches and fuse lines manipulate these communication paths. The switches in $Q_{i,j}$ correspond to processor i and bus j and are controlled only by processor i. Each processor has its own local memory; there is no shared memory. Each processor has a write port (port 0) and a read port (port 1). For all $0 \leq i < P$, $0 \leq j < B$ and $0 \leq k < 2$, processor i can connect its port k to bus j by setting the switch $c_{i,j,k}$. For $0 \leq \ell < 3$, switches $s_{i,j,\ell}$ (segment switches) are located on bus j. In fact, the three segment switches $s_{i,j,0}$, $s_{i,j,1}$ and $s_{i,j,2}$ are placed before $c_{i,j,0}$, between $c_{i,j,0}$ and $c_{i,j,1}$, and after $c_{i,j,1}$, respectively. When set, $s_{i,j,\ell}$ segments the bus at the point where it is located. Each processor has a "fuse line" associated with it, placed perpendicular to the buses. A fuse switch $f_{i,j}$ is located at the intersection of bus j and the fuse line of processor i. In fact, $f_{i,j}$ is located on bus j, after the segment switch $s_{i,j,2}$. When set, $f_{i,j}$ fuses the fuse line of processor i to bus j. The effect of setting two fuse switches $f_{i,j}$ and $f_{i,j'}$ is to fuse buses j and j'. Let E-RMBM(P, B) denote an E-RMBM with P processors and B buses.

The read port of a processor has a buffer that holds the value read from the bus to which it is connected. The processors operate synchronously and can perform one of the following operations in constant time: set or reset a switch, access a local memory word or the read port buffer, write on a bus, or execute an instruction from its instruction set. If the read port of a processor is connected to a bus, then at each step, if the bus has been written on, the value written to the bus is written into the read port buffer. A new value written on the bus destroys the old contents of the buffer. When the bus to which a read port is connected is not written on, the contents of the buffer of that port remain unchanged.

Several processors may read from a set of fused buses (or bus segments) concurrently, and several processors may attempt to write on a set of fused buses (or bus segments) concurrently as long as all such processors are writing the same value. Thus the E-RMBM operates under the Common Concurrent Read Concurrent Write (Common CRCW) rule. Though all algorithms in this paper run on the Common CRCW E-RMBM, the Collision rule can also be used to resolve write conflicts.

3   List Ranking on the E-RMBM

The list ranking problem can be defined as follows. The inputs are ordered sets $L = \langle a_0, a_1, \cdots, a_{n-1} \rangle$ and $P = \langle x_0, x_1, \cdots, x_{n-1} \rangle$. Each pointer $x_i \in \{0, 1, \cdots, n-1, \text{NULL}\}$ is associated with element a_i. If $x_i = j \neq \text{NULL}$, then the element following a_i in the list is a_j. If $x_i = \text{NULL}$, then a_i is the last element of the list. The list ranking problem is finding, for each a_i, the value $rank(i) \in \{0, 1, \cdots, n-1\}$ that is the number of elements before a_i in the list.

For any constant $0 < \epsilon < 1$, the proposed algorithm proceeds in $\lceil \frac{1}{\epsilon} \rceil$ iterations as follows. For simplicity of

description, we assume n^ϵ to be an integer. We also assume that each processor holds the values of n, $\lceil \frac{1}{\epsilon} \rceil$ and n^ϵ. In iteration 0 (the first iteration), the E-RMBM partitions the n input elements of L into $n^{1-\epsilon}$ groups G_i^0 (where $0 \leq i < n^{1-\epsilon}$), with each group having n^ϵ elements that are contiguous in the input list. Next, the E-RMBM finds the rank of each element within its group. In iteration 1, the E-RMBM considers the $n^{1-\epsilon}$ groups G_i^0 as atomic elements and partitions them into $n^{1-2\epsilon}$ groups G_i^1 (where $0 \leq i < n^{1-2\epsilon}$), each containing n^ϵ contiguous groups from iteration 0. In iteration 1, the E-RMBM determines the ranks of the elements within each group G_i^1. This easily translates into the ranks of the $n^{2\epsilon}$ input elements in each group G_i^1. In general, in iteration t (for $0 \leq t < \lceil \frac{1}{\epsilon} \rceil$), the E-RMBM first generates the groups G_i^t (where $0 \leq i < n^{1-(t+1)\epsilon}$), each of which contains n^ϵ contiguous groups G_j^{t-1} of the previous iteration. Next it ranks the G_j^{t-1}'s within each group G_i^t of the current iteration. Once again, it is easy to now find the ranks of the $n^{(t+1)\epsilon}$ input elements in each group G_i^t. At the end of iteration $\lceil \frac{1}{\epsilon} \rceil - 1$, all elements of L are in one group $G_0^{\lceil \frac{1}{\epsilon} \rceil}$ and each of these elements have been ranked with respect to the entire list.

The proposed E-RMBM algorithm for list ranking uses $n^{1+\epsilon}$ processors and $n^{1+\epsilon}$ buses, for any constant $0 < \epsilon < 1$. Partition the processors and buses into n clusters, each with n^ϵ processors or buses. Let $p_{i,j}$ and $b_{i,j}$ denote the processor and bus respectively, in the ith cluster and the jth position within a cluster, where $0 \leq i < n$ and $0 \leq j < n^\epsilon$. Assume without loss of generality that each processor $p_{i,j}$ initially holds element a_i and its pointer x_i. If the input is distributed in any other manner, it is easy to relocate the elements to the appropriate clusters and if necessary broadcast this information to all processors within a cluster. This can be done in constant time, as there are sufficient buses.

We now give a pseudo-code description of the list ranking algorithm, then we prove it correct and analyze its time complexity. In the following pseudo-code, a variable of the form $var(i, j)$ is local to processor $p_{i,j}$ and a variable of the form $var(i)$ is held by each processor in cluster i.

Algorithm List_Ranking
/* Initially, all switches are reset */
  $rank(i) \longleftarrow 0$ /* Initially each input element is
            in a separate group */
  $last(i) \longleftarrow 1$ /* Initially each input element is
           the last element of its group */
  /* find the first element of the list */
  {**Execute in parallel by each processor** $p_{i,j}$}
    set read port buffer to 1
    connect read port to $b_{i,j}$
    if ($x_i \neq \text{NULL}$) then write a 0 on $b_{x_i,j}$
    $first(i) \longleftarrow$ value in read port buffer
    /* The buffer of $p_{i,j}$ is unchanged iff it has
          the first element of list */

```
/* Iterate 1/ε times */
{Execute in parallel by each processor p_{i,j}}
for iter ⟵ 0 to 1/ε − 1 do
    if (last(i) = 1) then
            fuse b_{i,j} and b_{x_i,(j+1 (mod n^ε))}
    else fuse b_{i,j} and b_{x_i,j}
    connect read port to b_{i,j}
    initialize read port buffer to 0
    if (first(i) = 1) then write 1 on b_{i,0}
    last(i) ⟵ (last(i) and p_{i,n^ε −1} reads a 1)
    /* For each cluster of processors, exactly one
                            processor reads a 1 */
    if (p_{i,j} reads a 1) then subrank(i) ⟵ j
    rank(i) ⟵ rank(i) + subrank(i) × (n^{iter×ε})
    end
end
```

From the pseudo-code, it is clear that the algorithm requires $\Theta(\frac{1}{\epsilon})$ time. We devote the remainder of this section to establishing the algorithm's correctness.

We first show that the algorithm contains no concurrent writes. To find the first element of the list, each processor $p_{i,j}$ (associated with input element a_i) writes on bus $b_{x_i,j}$ (if $x_i \neq$ NULL). Since the input is assumed to be a linked list, each of the x_i's is different, and so each processor writes on a distinct bus. During each iteration only one processor writes. Therefore, there are no concurrent writes in the algorithm.

The first part of the algorithm involves finding the first element of the list. An element a_i is the first element of the list iff there is no i' such that $x_{i'} = i$. In this part of the algorithm, for each $0 \leq j < n^\epsilon$, processor $p_{i,j}$ reads the 0 written by processor $p_{i',j}$ iff $x_{i'} = i$. If there is no i such that $x_{i'} = i$, then, for each j, $p_{i,j}$ does not read any value; that is, its read port buffer is unchanged from the initialized value 1. Processors in all other processor clusters read 0's.

For the following discussion, let $rank(i,t)$, $last(i,t)$ and $subrank(i,t)$ denote the values of the variables $rank(i)$, $last(i)$ and $subrank(i)$, respectively, at the beginning of iteration t (or the end of iteration $t-1$), where $0 \leq t \leq \frac{1}{\epsilon}$.

Lemma 1 *For $0 \leq t \leq \frac{1}{\epsilon}$, at the beginning of iteration t*
   (i)   *$rank(i,t) = r_i \pmod{n^{t\epsilon}}$, where r_i is the actual rank of a_i.*
   (ii)   *$last(i,t) = 1$ iff $rank(i,t) = n^{t\epsilon} - 1$.*

Proof: We proceed by induction on t. For $t = 0$, the lemma clearly holds, as initially $rank(i) = 0$ and $last(i) = 1$, for all $0 \leq i < n$. During each iteration it is clear that the 1 written (henceforth referred to as the signal) traverses each cluster of buses exactly once, as the list traverses each of the n nodes exactly once. Moreover, the signal traverses exactly one bus of each cluster of buses, as no pair of buses within a cluster is fused. Consider iteration t. For any arbitrary element a_i, let its actual rank be r_i. Therefore, $rank(i,t) = r_i \pmod{n^{t\epsilon}}$, by the

induction hypothesis. Also, of the r_i elements that precede a_i in the list, exactly $\lfloor \frac{r_i}{n^{t\epsilon}} \rfloor$ have a rank of $n^{t\epsilon} - 1$ at the start of iteration t; that is, exactly $\lfloor \frac{r_i}{n^{t\epsilon}} \rfloor$ input elements $a_{i'}$ that precede a_i in the list have $last(i',t) = 1$. Consider now the signal emanating from some bus $b_{y,z}$. If $last(y,t) = 0$, then the signal enters the bus cluster x_y at bus $b_{x_y,z}$. If $last(x,t) = 1$, then the signal enters bus cluster x_y at bus $b_{x_y,(z+1 \pmod{n^\epsilon})}$. Since there are $\lfloor \frac{r_i}{n^{t\epsilon}} \rfloor$ input elements $a_{i'}$ that precede a_i in the list and which have $last(i',t) = 1$, the signal enters bus cluster i at bus $b_{i,j'}$, where $j' = \lfloor \frac{r_i}{n^{t\epsilon}} \rfloor \pmod{n^\epsilon}$. It is easy to see from the pseudo-code that $subrank(i,t+1) = j'$. Therefore, $rank(i,t+1) = rank(i,t) + subrank(i,t+1) \times n^{t\epsilon} = r_i \pmod{n^{t\epsilon}} + (\lfloor \frac{r_i}{n^{t\epsilon}} \rfloor \pmod{n^\epsilon})n^{t\epsilon} = r_i \pmod{n^{(t+1)\epsilon}}$. It is now clear from the pseudo-code that $last(i,t+1) = 1$ iff $rank(i,t+1) = n^{(t+1)\epsilon} - 1$. ∎

Theorem 1 *A CREW E-RMBM with $n^{1+\epsilon}$ processors and $n^{1+\epsilon}$ buses can rank a list of n elements in constant time, where $0 < \epsilon < 1$ is a constant.* ∎

Remarks: The E-RMBM algorithm uses $n^{1+\epsilon}$ processors; in contrast, the best constant time list ranking algorithm for the PARBS [4] uses $n^{2+\epsilon}$ processors. Both the E-RMBM and PARBS algorithms use $\Theta(n^{2+\epsilon})$ switches.

Applying our list ranking algorithm to the sorting algorithm proposed by Vaidyanathan [6] we have the following corollary.

Corollary 1 *A CREW E-RMBM with $n^{1+\epsilon}$ processors and $n^{1+\epsilon}$ buses can sort n $O(\log n)$-bit integers in constant time, where $0 < \epsilon < 1$ is a constant.* ∎

4 Relating CRCW E-RMBMs

As mentioned earlier, the E-RMBM uses the Common rule to resolve concurrent writes. In this section we give an efficient simulation of a "Priority" (and hence "Arbitrary") CRCW E-RMBM by a "Common" CRCW E-RMBM. Subsequently, we will use an Arbitrary CRCW E-RMBM, with the understanding that it can be simulated by a Common CRCW E-RMBM.

We now outline the simulation of each step of a Priority CRCW E-RMBM(P, B) Q on a Common CRCW E-RMBM$(P^{1+\epsilon}, BP^\epsilon)$ C, where $0 < \epsilon < 1$. A team $T(i)$ of P^ϵ processors in C will simulate each processor p_i of Q, and a collection $U(i)$ of P^ϵ buses in C will simulate each bus b_i of Q. We assume that for each i, each processor in $T(i)$ holds the index i of the corresponding processor p_i in Q. In one step a processor in Q may perform the following phases: internal computation, read, write, cut or fuse. Clearly, C can simulate all of the above phases, except resolution of concurrent writes. The processors of C cut and fuse buses and connect their ports to the buses exactly as in Q. Consequently, C contains P^ϵ identical copies of Q, including all fusing and segmenting of the buses. Let these copies of Q be $C_1, C_2, \cdots, C_{P^\epsilon}$.

To resolve concurrent writes, we proceed as follows. In a Priority CRCW E-RMBM, for each fused bus segment, the lowest indexed processor attempting to write on the bus segment succeeds in its write. The algorithm runs simultaneously and separately for each fused bus segment of C on which a processor is attempting to write.

For convenience, assume that P and P^ϵ are integer powers of 2. Each index $0 \le i < P$ can be expressed as the $\frac{1}{\epsilon}$-tuple $<\sigma_{i,\frac{1}{\epsilon}}, \sigma_{i,\frac{1}{\epsilon}-1}, \cdots, \sigma_{i,1}>$, where each $0 \le \sigma_{i,j} < P^\epsilon$ is a $\log P^\epsilon$-bit number; $\sigma_{i,\frac{1}{\epsilon}}$ and $\sigma_{i,1}$ are the most significant and least significant $\log P^\epsilon$-bit sections, respectively, of the binary representation of i.

For each fused bus segment, the teams in C that correspond to the processors attempting to write on the fused bus segment are the ones that perform the priority resolution. We outline the algorithm for one such fused bus segment. Initially, there are potentially P processors that could succeed in the write. In the first iteration, each processor i attempting to write indicates its presence by writing a 1 on the copy C_j of Q, where $j = \sigma_{i,1}$. Processor $p_{i,j}$ reads from C_j. Each team $T(i)$ now has a sequence of 0's and 1's; in fact, $p_{i,j}$ has a 1 iff there is at least one processor i' that is attempting to write and for which $j = \sigma_{i',1}$. Now each team determines (in constant time) the smallest j such that $p_{i,j}$ has a 1; let this value of j be $\sigma_{i,1}^*$. At this point we have determined that only processors i that have $\sigma_{i,1} = \sigma_{i,1}^*$, are eligible to write; that is, the search space has been reduced to at most $P^{1-\epsilon}$ processors. Proceeding in a similar fashion, C reduces the search space to at most $P^{1-t\epsilon}$ processors after $O(t)$ steps. At the end of $O(\frac{1}{\epsilon})$ steps, C selects a unique processor (for each fused bus segment) that is permitted to write.

The simulation of a Priority CRCW E-RMBM by a Collision CRCW E-RMBM is similar.

Theorem 2 *A Common (or Collision) CRCW E-RMBM$(P^{1+\epsilon}, BP^\epsilon)$ can simulate each step of a Priority (or Arbitrary) CRCW E-RMBM(P, B) in $O(\frac{1}{\epsilon})$ time, where $0 < \epsilon < 1$.* ∎

5 Graph Algorithms

Throughout this section, we assume that the undirected input graph $G = (V, E)$ has n vertices and m edges, and that G is represented as an $n \times n$ adjacency matrix.

Theorem 3 *For any $0 < \epsilon < 1$, a Common CRCW E-RMBM$(n^{2+\epsilon}, n^{1+\epsilon})$ can find the connected components of G in constant time.*

Proof outline: The idea is to configure the E-RMBM so that all processors that represent nodes in the same component of G are connected to the same fused bus segment. This can be accomplished in constant time. Next, each processor writes its index on the fused bus segment and

reads from the fused bus segment. Up to this point the algorithm uses only n^2 processors and n buses. If an Arbitrary CRCW E-RMBM is used, then this clearly gives the correct component numbers for each node of G. The rest follows from Theorem 2. ∎

A simple extension of the algorithm in Theorem 3 gives the following theorem.

Theorem 4 *For any $0 < \epsilon < 1$, a Common CRCW E-RMBM$(n^{2+\epsilon}, n^{1+\epsilon})$ can compute the transitive closure of G in constant time.* ∎

Remarks: The E-RMBM algorithms in Theorems 3 and 4 use $n^{2+\epsilon}$ processors. In contrast, the best known constant time PARBS algorithms for these problems run on an $n \times n \times n$ (or $n^2 \times n^2$) PARBS [7].

For brevity, we state the following results without proof.

Theorem 5 *For any $0 < \epsilon < 1$, a Common CRCW E-RMBM$(n^{3+\epsilon}, n^{2+\epsilon})$ can find all cut vertices of G in constant time.* ∎

Theorem 6 *For any $0 < \epsilon < 1$, a Common CRCW E-RMBM$(mn^{2+\epsilon}, \min(n^{2+\epsilon}, mn^{1+\epsilon}))$ can find all bridges of G in constant time.* ∎

References

[1] H. Li and M. Maresca, "Polymorphic-Torus Architecture for Computer Vision," *IEEE Trans. Pattern Analysis & Mach. Intelligence*, **11**, 1989, pp. 233–243.

[2] R. Miller, V. K. Prasanna-Kumar, D. Reisis, and Q. Stout, "Meshes with Reconfigurable Buses," *Proc. 5th MIT Conf. on Advanced Research in VLSI*, 1988, pp. 163–178.

[3] J. M. Moshell and J. Rothstein, "Bus Automata and Immediate Languages," *Info. and Control*, **40**, 1979, pp. 88–121.

[4] S. Olariu, J. L. Schwing, and J. Zhang, "Fundamental Algorithms on Reconfigurable Meshes," *Proc. 29th Annual Allerton Conf. on Comm., Control and Comput.*, 1991, pp. 811–820.

[5] R. K. Thiruchelvan, J. L. Trahan, and R. Vaidyanathan, "On the Power of Segmenting and Fusing Buses," To appear in *Proc. 7th Int'l Parallel Proc. Symp.*, 1993.

[6] R. Vaidyanathan, "Sorting on PRAMs with Reconfigurable Buses," *Info. Proc. Letters*, **42**, 1992, pp. 203–208.

[7] B. F. Wang, G. H. Chen, and F. Lin, "Constant Time Algorithms for the Transitive Closure and Some Related Graph Problems on Processor Arrays with Reconfigurable Bus Systems," *IEEE Trans. Parallel and Distr. Systems*, **1**, 1990, pp. 500–507.

ON THE PRACTICAL APPLICATION OF A QUANTITATIVE MODEL OF SYSTEM RECONFIGURATION DUE TO A FAULT

Gene Saghi, Howard Jay Siegel, and Jose A. B. Fortes

Parallel Processing Laboratory, School of Electrical Engineering
Purdue University, West Lafayette, IN 47907-1285 USA

Abstract -- *If a processor develops a permanent fault during the execution of a task on a dynamically reconfigurable parallel machine, three recovery options are task migration to a fault-free subdivision, task migration to another submachine, and task redistribution. Quantitative models of these three reconfiguration schemes are analyzed, together with the cost of making the wrong choice, to develop guidelines for selecting among these methods in a practical dynamic reconfiguration implementation. A multistage cube or hypercube inter-processor network is assumed. The PASM and nCUBE 2 parallel machines are used as vehicles for sudying the model parameters.*

1. INTRODUCTION

To provide reliable operation over extended periods of time, massively parallel processing systems must be capable of tolerating faults. A fault-tolerant system must be able to detect and locate faults, to reconfigure itself to "disconnect" and perhaps replace faulty components, to recover from possibly erroneous computations, and to restart operation from a correct state. When more than one reconfiguration option is available, the option that results in the earliest completion of the task is desirable. In [8], a quantitative model of dynamic reconfiguration was presented and it was observed that collecting precise values for all of these parameters is very difficult (if not impossible). The research here analyzes ranges of values that these parameters can assume to develop guidelines for making the best reconfiguration choice. Because there is no guarantee that these guidelines will produce the optimal reconfiguration strategy in all cases, the cost penalty of making the wrong choice is also considered. The analysis incorporates experimentally-derived parameters obtained on PASM [1, 13], an experimental SIMD/MIMD mixed-mode machine with a partitionable multistage cube communication network, and on nCUBE 2 [6], a commercially available MIMD machine with a partitionable hypercube communication network.

This research focuses on partitionable parallel processing systems where the set of processors can be partitioned to form multiple independent submachines. The execution of a parallel program on a submachine is defined as a task. It is possible to achieve fault tolerance in such a system by utilizing the reconfigurability of the system to effectively "disconnect" the faulty component. For example, parallel processing systems such as Intel Cube [3], nCUBE [2], IBM RP3 [7], and PASM [1, 13] incorporate partitionable interconnection networks and therefore have the ability to migrate a task from a faulty submachine to a fault-free submachine. The architecture assumed implements a physically distributed memory such that each processor is paired with local memory to form a processing element (PE). Accesses to memory locations located in a remote processor's memory require use of the interconnection network.

The system and fault models used to analyze fault-recovery options and the recovery options and associated costs are presented as background information in Section 2. An analysis of the range of costs for each option is examined in Section 3 to determine the relative weight of these costs in the reconfiguration decision. The penalty for making the wrong choice is considered in Section 4. Guidelines for choosing a reconfiguration strategy in the event of a fault are also presented in Section 4.

2. BACKGROUND INFORMATION

A mixed-mode machine can operate in either the SIMD or MIMD mode of parallelism and can switch modes at instruction level granularity [1]. A partitionable SIMD/MIMD machine can operate as one or more independent or cooperating submachines, where each submachine may operate as a mixed-mode machine [11]. The analyses here can be applied to MIMD, multiple-SIMD, or partitionable SIMD/MIMD parallel processing systems, utilizing a multistage cube or hypercube interconnection network.

At regular intervals during the execution of a task, the state of each PE, including register and allocated memory contents, is stored in a different PE within the same submachine. This state information is called checkpoint data and is used to restore a valid system state in the event of a fault. Further details about the system and fault models assumed in this research can be found in [8, 10] and are not included here due to space constraints.

When a permanent fault occurs in a submachine A, three possible reconfiguration/recovery options are as follows:

1) subdivide A into two equal-size system submachines, and use the one that is fault-free to complete the execution of the task,

This research was supported by the Office of Naval Research under grant number N00014-90-J-1483 and supported by the Innovative Science and Technology Office of the Strategic Defense Organization and administered through the Office of Naval Research under contract number N00014-88-K-0723.

2) migrate the task to another submachine that is fault-free,

3) redistribute the task programs and data among the fault-free PEs in A and complete the task using a modified algorithm that does not use the faulty PE.

These recovery options are discussed in detail in [8, 10].

3. CHOOSING AN OPTION

3.1. Overview

The time to reconfigure and complete a task for each reconfiguration option listed in Section 2 can be separated into three primary components: time to plan for the reconfiguration option (T_{Plan}), time to move the task data and code (T_{Trnsfr}), and time to complete the task execution ($T_{CmpExec}$). In this section, the relative impact these three components has on the overall reconfiguration cost is discussed to derive guidelines for choosing the best option. Experimentally determined ranges for these parameters on the PASM prototype and the nCUBE 2 are used in the analysis where applicable. Table I summarizes the most important notation used in the sections that follow.

Table I: Summary of notation used.

Notation	Meaning
FFS	fault-free subdivision reconfiguration option
TM	task migration reconfiguration option
TR	task redistribution reconfiguration option
T_{Plan}^{YYY}	time to plan for YYY^*
T_{Trnsfr}^{YYY}	time to move task data and code for YYY^*
$T_{CmpExec}^{YYY}$	time to complete task execution after YYY^*
T_{DA}	time to access disk
$\eta(x)$	estimated execution time for task on x PEs
τ	total time task executed prior to checkpoint
$F_{sm}(t)$	probability of submachine failure by time t
$R_{sm}(t)$	submachine reliability, $R_{sm}(t) = 1 - F_{sm}(t)$
$T_{Penalty}^{WC}$	worst case reconfig.-option-choice penalty

$^*YYY = FFS$, TM, or TR reconfiguration option.

In practical situations, the fault-free subdivision and task migration options may not be available. In some systems there may be a minimum size for a submachine. If the current submachine is of minimum size, it cannot be subdivided. It is also possible that no idle destination submachine exists to which to migrate a task. The task can be migrated to a submachine already being used to execute another task, and the two tasks can time-share the submachine. This is discussed in [10].

3.2. Range of T_{Plan} and T_{Trnsfr}

The results of an analysis of the range of values that T_{Plan} and T_{Trnsfr} can assume on PASM and nCUBE 2 for all three reconfiguration options considered is provided in Table II. The analysis details can be found in [10].

Table II: Approximate ranges, in microseconds, for T_{Plan} and T_{Trnsfr} for the FFS, TM, and TR reconfiguration options.

Option	T_{Plan} (PASM, nCUBE 2)	T_{Trnsfr}^* (PASM, nCUBE 2)
FFS	$(68 \rightarrow 80, 189 \rightarrow 329)$	$(0^{**}, 0^{***})$
TM	$(10 \rightarrow 100, 10 \rightarrow 100)$	$(> 480^{**}, > 320^{***})$
TR	$(1 \rightarrow 10, 1 \rightarrow 10)$	$(> 240^{**}, > 160^{***})$

* Upper bound determined by size of PE memories.
** Add T_{DA} for SIMD, MIMD, or mixed-mode tasks.
*** Add T_{DA} for MIMD tasks.

3.3. Range of $T_{CmpExec}$

Tasks can be divided into two categories based on execution time. These are tasks with data-independent execution times and tasks with data-dependent execution times. A task with a data-independent execution time does not depend on input data to make branching decisions. Thus, the number of times any branch in the task program code is taken can be determined by a compiler during program compilation, and a compiler can determine an expected execution time for the task. In contrast, a task with a data-dependent execution time has branch decisions that are based on data that is known only at run time. In this case, it is assumed that an estimated execution time for the task can be determined through the use of empirical studies (i.e., information about task execution time on various sets of data), an automatic complexity evaluator such as that presented in [5], or through analysis of the algorithm and data sets.

For all the reconfiguration options discussed, the number of PEs assigned to a task after the reconfiguration is equal to or less than the number of PEs originally assigned to the task. It is assumed that the average time for an inter-PE transfer does not increase when the task is executed on fewer PEs.

Once a task has been migrated from a submachine containing a faulty PE to a fault-free submachine of equal size, the time to complete the task execution will be the same as it would have been on the original fault-free submachine. Let $\eta(2^k)$ be the estimated execution time for a task on a submachine with 2^k PEs. It is assumed that the total amount of execution time spent on a task prior to a checkpoint is stored with that checkpoint. If a recovering task is to proceed from a checkpoint and the execution time stored with that checkpoint is τ, the expected amount of time required to complete task execution after migrating the task to another submachine is

$$T_{CmpExec}^{TM} = \eta(2^k) - \tau.$$

$T_{CmpExec}^{TM}$ assumes that the submachine size remains the same and that all the PEs in the submachine are fault-free. The expected range for $T_{CmpExec}^{TM}$ is

$$0 < T_{CmpExec}^{TM} \leq \eta(2^k).$$

Now, consider the completion time of a task that completes execution on a subdivision that is half the size of the original submachine. $T_{CmpExec}^{FFS}$ is bounded as follows:

$$T_{CmpExec}^{FFS} \leq 2(\eta(2^k) - \tau) = 2T_{CmpExec}^{TM}.$$

In addition, it is expected that $T_{CmpExec}^{FFS} > T_{CmpExec}^{TM}$ because the number of processors in a fault-free subdivision is assumed to be half the number that would be available if the task was migrated to another submachine. Although some tasks can execute faster on fewer PEs [4, 9, 12], it is assumed here that the original submachine size was selected for minimum execution time. That is, if a smaller submachine could be used to execute the task in the same or less time, the task would have been mapped to that smaller size submachine initially.

A more accurate remaining execution time estimate can be obtained if $\eta(2^{k-1})$ is known. Then, the estimated task execution time becomes a function of the submachine size and the estimate of the remaining execution time becomes

$$T_{CmpExec}^{FFS} = \frac{\eta(2^{k-1})}{\eta(2^k)} \left[\eta(2^k) - \tau\right] = \frac{\eta(2^{k-1})}{\eta(2^k)} T_{CmpExec}^{TM}.$$

Consistent with the assumptions given above, $\eta(2^k) < \eta(2^{k-1}) \leq 2\eta(2^k)$. Thus, using either the $\eta(2^{k-1})$ information, if it is known, or the inequalities stated in the previous paragraph, the expected range for $T_{CmpExec}^{FFS}$ is:

$$T_{CmpExec}^{TM} < T_{CmpExec}^{FFS} \leq 2T_{CmpExec}^{TM} \leq 2\eta(2^k).$$

An execution-time estimate for the task redistribution recovery option is more difficult than for the previous options. Consider a task executing on a submachine of size 2^k in MIMD mode. If a PE becomes faulty and its subtasks are distributed equally to the $2^k - 1$ fault-free PEs in the submachine, the remaining execution time is bounded as follows:

$$T_{CmpExec}^{TR} \leq \frac{2^k}{2^k - 1} \left[\eta(2^k) - \tau\right] = \frac{2^k}{2^k - 1} T_{CmpExec}^{TM}.$$

Consider the situation where the faulty PE's subtasks cannot be distributed equally among the fault-free PEs. In the worst case, all the faulty PE's subtasks would be assigned to a single PE and the remaining execution time could be twice that of the remaining execution time on a fault-free submachine.

In general, it is expected that $T_{CmpExec}^{TR} \leq T_{CmpExec}^{FFS}$ because the fault-free subdivision option can be thought of as a subset of the task redistribution option where the task is redistributed to half the PEs in the original submachine. Furthermore, it is expected that $T_{CmpExec}^{TR} > T_{CmpExec}^{TM}$ based on the earlier assumption that the original submachine size was selected for minimum

execution time. Therefore, the range on $T_{CmpExec}^{TR}$ is given by:

$$T_{CmpExec}^{TM} < T_{CmpExec}^{TR} \leq T_{CmpExec}^{FFS}.$$

In cases where an equal distribution of the task load among the fault-free PEs is possible, the upper bound of $T_{CmpExec}^{FFS}$ in the above inequality can be replaced by $\min((2^k/2^k - 1) T_{CmpExec}^{TM}, T_{CmpExec}^{FFS})$. By combining the results of the inequalities for remaining execution time determined in this subsection, the following ordering is established.

$$0 < T_{CmpExec}^{TM} < T_{CmpExec}^{TR} \leq T_{CmpExec}^{FFS} \leq 2\eta(2^k)$$

Although the above inequality indicates that task migration is the best reconfiguration option when $T_{CmpExec}$ is the dominant factor, it has already been shown that task migration is not the best option when considering T_{Plan} and/or T_{Trnsfr}. Therefore, no clear choice is apparent based on the analysis up to this point.

4. PENALTY FOR WRONG CHOICE

Thus far, a quantitative framework has been developed that attempts to relate various reconfiguration parameters. Some of the parameters can be predicted with good precision on real machines, while other parameters can only be coarsely bounded. The next step is to determine if a heuristic can be found that is based on the information available. In this section, a combination of probabilistic analysis and worst-case analysis is used to develop useful guidelines for choosing among reconfiguration options on real machines in practical situations.

Consider the relative magnitudes of $\eta(2^k)$, T_{Plan}, and T_{Trnsfr}. In general, for tasks with short execution times, it is better to restart the task when a PE becomes unusable rather than permanently and significantly increasing the execution time by including periodic checkpointing. Therefore, dynamic reconfiguration is generally not considered for tasks unless the estimated execution time for the task, $\eta(2^k)$, is orders of magnitude larger than T_{Trnsfr} and T_{Plan}.

One of the most common cumulative distribution functions assumed in reliability models is the exponential distribution, $F(t) = 1 - e^{-\lambda t}$ [14]. $F(t)$ represents the probability that a PE fault will occur between time 0 and time t, inclusive. The parameter λ describes the rate at which failures occur in time.

The reliability function, $R(t)$, is defined as $R(t) = 1 - F(t) = e^{-\lambda t}$. For a parallel system submachine of size 2^k PEs, where all the PEs must be operational for the submachine to be operational, the submachine reliability function, $R_{sm}(t)$, is the product of the individual PE reliability functions.

$$R_{sm}(t) = \prod_{i=1}^{2^k} R(t) = \prod_{i=1}^{2^k} e^{-\lambda t} = e^{-2^k \lambda t}$$

Thus, the submachine-failure probability distribution

function, $F_{sm}(t)$, for a submachine of size 2^k is given by:

$$F_{sm}(t) = 1 - e^{-2^k \lambda t}.$$

Consider the conditional probability that a failure occurs at or before time $.9\eta(2^k)$ given that a failure occurs at or before time $\eta(2^k)$.

$$Pr[t \le .9\eta(2^k) \mid t \le \eta(2^k)] =$$

$$\frac{Pr[t \le \eta(2^k) \mid t \le .9\eta(2^k)]Pr[t \le .9\eta(2^k)]}{Pr[t \le \eta(2^k)]} = \frac{1 - e^{-2^k \lambda .9\eta(2^k)}}{1 - e^{-2^k \lambda \eta(2^k)}}$$

This probability approaches 0.9 as $\eta(2^k)$ approaches zero from the positive direction, and it monotonically approaches 1 as $\eta(2^k)$ increases. Therefore, when $\eta(2^k)$ is 100 times greater than T_{Trnsfr}, there is a 0.9 or greater probability that a failure will occur by time $90T_{Trnsfr}$, given that a failure occurs by the time the program has completed. Thus, for this case, there is a high probability that τ, the time the failure occurs, will be less than or equal to $90T_{Trnsfr}$, and $T_{CmpExec} = \eta(2^k) - \tau >> T_{Trnsfr}$. For the case where $\eta(2^k)$ is more than 100 times greater than T_{Trnsfr}, there is an even greater probability that $\eta(2^k) - \tau >> T_{Trnsfr}$.

Here, the penalty of making the wrong choice of a reconfiguration option is examined. The worst-case penalty, $T_{Penalty}^{WC}$, is defined to be the worst-case difference between the expected completion time of a task after choosing a suboptimal reconfiguration option and the expected completion time of a task after choosing the optimal reconfiguration option. For example, if the task redistribution option was chosen, but the task migration option would have resulted in the earliest completion time for the task, the worst-case penalty would be:

$$T_{Penalty}^{WC} = \max (T_{Plan}^{TR} + T_{Trnsfr}^{TR} + T_{CmpExec}^{TR})$$
$$- \min(T_{Plan}^{TM} + T_{Trnsfr}^{TM} + T_{CmpExec}^{TM}),$$

where the maximum and minimum refer to the ranges for the parameters. Here, two cases are considered: 1) the reconfiguration choice was made assuming that the remaining execution time was much greater than the time to transfer the task code and data $((\eta(2^k) - \tau) >> T_{Trnsfr})$, and 2) the reconfiguration choice was made assuming that the remaining execution time was much less than the time to transfer the task code and data $((\eta(2^k) - \tau) << T_{Trnsfr})$.

First, consider the case where it was incorrectly assumed that $(\eta(2^k) - \tau) >> T_{Trnsfr}$. In this case, from the results of Subsection 3.3, the task migration option would have been chosen. If the fault-free subdivision option is the optimal one, the worst-case penalty would be:

$$T_{Penalty}^{WC} < \max(T_{Plan}^{TM} + T_{Trnsfr}^{TM}) - \min(T_{Plan}^{FFS} + T_{Trnsfr}^{FFS}),$$

because the best expected time to complete execution on a fault-free subdivision is greater than the expected time to complete execution after task migration. Furthermore, for machines like PASM and nCUBE 2, $T_{Plan}^{TM} - T_{Plan}^{FFS}$ is

negligible, and $T_{Trnsfr}^{TM} - T_{Trnsfr}^{FFS}$ is generally on the order of hundreds of milliseconds (see Table II). Thus, in this case, $T_{Penalty}^{WC}$ is on the order of T_{Trnsfr}^{TM} (recall $T_{Trnsfr}^{FFS} = 0$).

If instead the task redistribution option is the optimal choice for this example, the worst-case penalty would be:

$$T_{Penalty}^{WC} < \max(T_{Plan}^{TM} + T_{Trnsfr}^{TM}) - \min(T_{Plan}^{TR} + T_{Trnsfr}^{TR}).$$

Again, the best expected time to complete execution after task redistribution is greater than the expected time to complete execution after task migration. For PASM and nCUBE 2, $T_{Penalty}^{WC}$ is on the order of T_{Trnsfr}^{TM} (see Table II).

Now, consider the case where it was incorrectly assumed that $(\eta(2^k) - \tau) << T_{Trnsfr}$. In this situation, either the fault-free subdivision or task redistribution option would have been chosen. If the fault-free subdivision option was chosen when the task migration option would have been better (because in actuality $(\eta(2^k) - \tau) >> T_{Trnsfr}$), the penalty for making the wrong choice is given by:

$$T_{Penalty}^{WC} = \max(T_{Plan}^{FFS} + T_{Trnsfr}^{FFS} + T_{CmpExec}^{FFS})$$
$$- \min(T_{Plan}^{TM} + T_{Trnsfr}^{TM} + T_{CmpExec}^{TM}).$$

Substituting values from the analysis in Subsection 3.3, results in:

$$T_{Penalty}^{WC} = \max(T_{Plan}^{FFS} + T_{Trnsfr}^{FFS}) - \min(T_{Plan}^{TM} + T_{Trnsfr}^{TM})$$
$$+ (2T_{CmpExec}^{TM} - T_{CmpExec}^{TM})$$
$$= \max(T_{Plan}^{FFS} + T_{Trnsfr}^{FFS}) - \min(T_{Plan}^{TM} + T_{Trnsfr}^{TM}) + \eta(2^k) - \tau.$$

Recall the value of $\eta(2^k)$ is assumed to be much larger than T_{Trnsfr} when reconfiguration options are to be considered. Thus, in the worst case, the penalty for incorrectly assuming $(\eta(2^k) - \tau) << T_{Trnsfr}$ is much greater than incorrectly assuming $(\eta(2^k) - \tau) >> T_{Trnsfr}$. A similar analysis for the case where task redistribution was erroneously chosen over task migration results in the same potential for a large penalty.

To summarize this section, two conclusions are made: first, it is expected that there is a high probability that $T_{CmpExec}$ will be much greater than T_{Trnsfr} when a fault occurs, and second, that the worst-case penalty for incorrectly assuming this is true is far less than the worst-case penalty for incorrectly assuming the opposite. Therefore, a mathematical justification for choosing a reconfiguration option by considering only the time required to complete the task has been established. Combining this result with the results of Subsection 3.3, the choice of reconfiguration strategy becomes one of choosing to migrate the task if an idle submachine exists. If this option is not available, the next best option is task redistribution. Finally, if the task does not lend itself to redistribution, a fault-free subdivision can be used to complete the task.

The model parameter value ranges established in Section 3 are in some cases very coarse, e.g., $T_{CmpExec}$ for tasks with data-dependent (nondeterministic) execution times, and therefore do not provide the information needed to determine the best reconfiguration option. However, the analysis in this section has made it possible to establish a good set of reconfiguration guidelines.

5. CONCLUSIONS

The application of a quantitative model of system reconfiguration due to a PE fault was examined. The model parameters were categorized into one of three categories: time to plan for the reconfiguration option (T_{Plan}), time to move the task data and code (T_{Trnsfr}), and time to complete task execution after reconfiguration ($T_{CmpExec}$). The relative times for each reconfiguration option considered were examined for each category and the options were ranked when possible. Actual parameters collected on the PASM and nCUBE 2 parallel machine were used to support the analysis.

For the system architectures considered, T_{Plan} is generally much smaller than T_{Trnsfr}. Furthermore, when basing the reconfiguration decision only on T_{Trnsfr} (ignoring $T_{CmpExec}$), the fault-free subdivision or task redistribution options will result in the smallest total execution time for the task. The choice between the fault-free subdivision and task redistribution options will depend on the task being executed.

It was shown that for those tasks where dynamic reconfiguration should be considered, there is a high probability that the expected value of $T_{CmpExec}$ will be greater than T_{Trnsfr}. Thus, $T_{CmpExec}$ becomes the primary parameter to consider when choosing among reconfiguration options. When $T_{CmpExec}$ is the dominant factor, the task migration option results in the earliest task completion. Task redistribution is the next best option. However, in the worst case, task redistribution can require as much time as completing the task on a fault-free subdivision.

Task execution times used in the model may be just expected values when execution times are data dependent and therefore nondeterministic. Therefore, a worst case analysis was performed. An examination of the penalty for choosing the wrong reconfiguration option provides further justification for basing the reconfiguration decision on $T_{CmpExec}$. An analysis of the worst-case penalties reveals that the penalty for assuming $T_{CmpExec}$ to be much greater than T_{Plan} and T_{Trnsfr} is much less than the penalty for assuming otherwise. Thus, using a quantitative framework, it has been shown that task migration is the best dynamic recovery option when $T_{CmpExec}$ is expected to be much greater than T_{Plan} and T_{Trnsfr} for tasks with nondeterministic execution times.

Acknowledgments: The authors gratefully acknowledge discussions with James Armstrong and Dan Watson.

REFERENCES

[1] S. A. Fineberg, T. L. Casavant, and H. J. Siegel, "Experimental analysis of a mixed-mode parallel architecture using bitonic sequence sorting," *J. Parallel and Distributed Computing*, Vol. 11, Mar. 1991, pp. 239-251.

[2] J. P. Hayes, T. N. Mudge, Q. F. Stout, and S. Colley, "Architecture of a hypercube supercomputer," *1986 Int'l Conf. on Parallel Processing*, Aug. 1986, pp. 653-660.

[3] Intel Corporation, *A New Direction in Scientific Computing*, Order # 28009-001, Intel Corporation, 1985.

[4] R. Krishnamurti and E. Ma, "The processor partitioning problem in special-purpose partitionable systems," *1988 Int'l Conf. on Parallel Processing*, Vol. I, Aug. 1988, pp. 434-443.

[5] D. Le Métayer, "ACE: An Automatic Complexity Evaluator," *ACM Trans. on Programming Languages and Systems*, Vol. 10, Apr. 1988, pp. 248-266.

[6] nCUBE Corporation, *nCUBE 2 Processor Manual*, Order # 101636, nCUBE Corporation, Dec. 1990.

[7] G. F. Pfister et al., "The IBM Research Parallel Processor Prototype (RP3): introduction and architecture," *1985 Int'l Conf. on Parallel Processing*, Aug. 1985, pp. 764-771.

[8] G. Saghi, H. J. Siegel, and J. A. B. Fortes, "On the viability of a quantitative model of system reconfiguration due to a fault," *1992 Int'l Conf. on Parallel Processing*, Vol. I, Aug. 1992, pp. 233-242.

[9] G. Saghi, H. J. Siegel, and J. L. Gray, "Predicting performance and selecting modes of parallelism: a case study using cyclic reduction on three parallel machines," *J. Parallel and Distributed Computing*, to appear, Oct. 1993.

[10] G. Saghi, H. J. Siegel, and J. A. B. Fortes, *On a Quantitative Model of System Reconfiguration Due to a Fault*, Tech. Rep. in preparation, EE School, Purdue.

[11] H. J. Siegel, *Interconnection Networks for Large-Scale Parallel Processing: Theory and Case Studies*, Second Edition, McGraw-Hill, New York, NY, 1990.

[12] H. J. Siegel, J. B. Armstrong, and D. W. Watson, "Mapping computer-vision-related tasks onto reconfigurable parallel-processing systems," *Computer*, Vol. 25, Feb. 1992, pp. 54-63.

[13] H. J. Siegel, T. Schwederski, J. T. Kuehn, and N. J. Davis IV, "An overview of the PASM parallel processing system," in *Computer Architecture*, D. D. Gajski, V. M. Milutinovic, H. J. Siegel, and B. P. Furht, eds., IEEE Computer Society Press, Washington, DC, 1987, pp. 387-407.

[14] D. P. Siewiorek and R. S. Swarz, *The Theory and Practice of Reliable System Design*, Digital Equipment Corp., Bedford, MA, 1982.

Online Algorithms for Handling Skew
in Parallel Joins

Arun Swami   *Honesty C. Young*

IBM Almaden Research Center
650 Harry Road, San Jose, CA 95120-6099
{arun,young}@almaden.ibm.com

Abstract: When the work involved in a join is partitioned among multiple processors in the parallel join, the skew in the operand relations can result in significant imbalance in the work assigned to the different processors. This imbalance can cause significant degradation in the response time for the join operation.

In this paper, we present two new algorithms, the *pure online* **PO** and the *online/offline* algorithm **OO**, for handling skew in parallel join evaluation. Using the histogram data of the operand relations, the skew handling algorithms can achieve any permissible tolerance in the join work imbalance among the processors. Unlike previous work on handling skew in parallel joins, the algorithms are *online* in the sense that they process most (or all) of the histogram data as it is generated. We compare these algorithms and their variations experimentally and find that a particular variation of algorithm **OO** is usually the algorithm of choice.

1   Introduction

The join is an important but often expensive operation in relational database management systems. Significant improvements in response time may be attained by parallelizing the join. The work involved in the join is sensitive to the skew in the operand relations, and the correlation between the frequent values in the operand relations [6]. When the join work is partitioned among multiple processors in the parallel join, the skew in the operand relations can result in significant imbalance in the work assigned to the different processors. This imbalance can cause significant degradation in the response time for the join operation.

In the rest of this paper, we use the terms *load balancing* and *balancing the load* to denote *skew handling* and *handling the skew*. The term *load balancing* as used here should not be confused with balancing multi-user processor loads. We focus on load balancing in a shared nothing architecture where both memory and disks are local to processors. The processors communicate using messages over an interconnection network that supports point to point messages.

One may think of the join operation as consisting of multiple tasks. Each task corresponds to joining tuples from the two relations that have the same join column value. We may combine tasks corresponding to several join column values to form a larger task. If the join column values taken together form a range in the domain of the join column, we say that the task corresponds to

the *interval* of values. If the task corresponding to a join column value is large, it can be *split* by appropriately partitioning the tuples joined in this task. Such splitting is usually feasible but results in extra overhead. For the parallel join, splitting is achieved using a technique such as *fragment replicate* (FR) [2].

Let p denote the number of processors available to perform the join (*join* processors). For simplicity we assume that the operand relations for the join reside on processors (*storage* processors) that are distinct from the join processors. This assumption can easily be relaxed. The load balancing algorithms must assign the intervals so that each of the p processors performs approximately the same amount of work. We quantify the load imbalance permitted as follows. Let W_t denote the total join work including CPU and I/O, and let $W_p = W_t/p$ ($W_t = pW_p$) denote the join work per processor if load balancing is perfect. Let δ denote the permissible tolerance in the load imbalance. For the perfectly balanced case $\delta = 0$. Then we require that the load balancing algorithm assign the join work so that no processor is assigned work exceeding $(1 + \delta)W_p$. Note that this definition of load imbalance is effectively the same as the definition of skew in [6].

We first obtain histograms for the two operand relations. These histograms are input to the load balancing algorithms, which determine the intervals of join column values assigned to each join processor. Note that an interval may consist of a single distinct value (*single-valued interval*) or multiple distinct values (*multi-valued interval*). In the case of single-valued intervals, we often use the term *interval* to refer to the single value. We may also use the term interval to refer to the values comprising the interval or the join work corresponding to that interval. Each join processor is sent all or a subset of the tuples in its assigned intervals. This is done for each input relation.

For the parallel join, the load balancing algorithm determines how the histogram intervals are assigned to the join processors. If the number of histogram intervals is large, it may be expensive for the storage processors to use the load balanced distribution to determine the destination site for each tuple. Hence *combining* of intervals may be desired to achieve more efficient data distribution. Adjacent intervals assigned to the same processor can be combined.

An algorithm is said to be *offline* if it requires examination of all the tasks before any scheduling (distribution) of tasks is done. An *online* algorithm schedules tasks as they are presented to the algorithm. An algorithm is said to be *mixed* (online/offline) if it schedules most of the tasks as they are presented to the algorithm leaving a small number of them to be scheduled after all the tasks have been seen. We sometimes use the term *online algorithms* to refer to both online and mixed algorithms.

We describe two algorithms for balancing the work in a parallel join. The first algorithm, called the *pure online* algorithm or **PO**, processes *all* the histogram data *as it is generated*. It may split histogram intervals as necessary. The second algorithm called the *online/offline* algorithm or **OO** is a mixed algorithm. It consists of an *online* phase followed by an *offline* phase. The online phase of **OO** processes most of the histogram data as it is generated, and stores the rest to be processed by the offline phase. No splitting is done during the online phase. Splitting, if necessary, takes place in the offline phase after all the data has been seen.

2 Prior Work

Load balancing algorithms operating under different assumptions have been studied for a long time. The problem of distributing (or scheduling) independent tasks so that a task is restricted to run on a single processor, i.e., no splitting is allowed, has been shown to be NP-hard where the size of the problem is the number of tasks to be scheduled [3]. In [1], the problem is generalized to allow a task to be split up into k *equal* parts, $1 \leq k \leq p$. The problem continues to be NP-hard. A constraint they impose is that if a task is split, all the fragments of the task must begin executing at the same time. Wolf et al. [7] propose algorithms for the same problem except that they do not have the constraint that all the fragments of a task must begin executing at the same time.

All the algorithms mentioned above are offline algorithms. It is much harder to devise *online* or *mixed* algorithms. The algorithms described in this paper are either pure online algorithms or mixed online/offline algorithms that can meet any desired tolerance on the load imbalance, including the requirement that the load balancing be optimal.

In [4] Lu and Tan proposed a dynamic and load-balanced join algorithm with *task stealing* for the shared disk environment. When all join tasks are assigned to processors, an idle processor can *steal* subtasks from an overloaded processor to share its load using fragment and replicate join. The proposed algorithm cannot be extended directly to the shared nothing architecture we are considering.

3 Algorithm PO

In this section we describe the pure online algorithm **PO** that assigns each join task as it is seen. Let W_{seen}

denote the join work seen (and assigned) so far by algorithm **PO**. Initially W_{seen} is zero. Let I be the work in the next interval to be assigned. Let W_i denote the work on processor i before I is assigned, and W_i' the work on the same processor after I is assigned. In **PO**, we ensure that after assignment, the load imbalance does not exceed δ, i.e.,

$$W_i' \leq (1 + \delta) \frac{W_{seen} + I}{p}, \quad 1 \leq i \leq p \qquad (1)$$

First, **PO** tries to assign I to a single processor, i.e., without splitting, so that Equation 1 holds. If such an assignment is not possible, I needs to be split. It is easy to see that splitting each interval into p fragments will always work. However this may lead to excessive communication overhead. Hence **PO** computes the smallest number of splits of I required and assigns these fragments of I to the appropriate processors. Clearly when **PO** terminates after processing the complete histogram, the load balance criterion is satisfied since then $W_{seen} = W_t$ and $I = 0$. Thus all the intervals are processed as they are generated, and the algorithm is a pure online algorithm.

4 Algorithm OO: Phase I

In this section we describe the online phase of algorithm **OO**. To allow combining of intervals, **OO** processes the intervals in sorted order of join values. The online phase stores up to p single-valued intervals and their associated join work in a list called the *offline list*. When the online phase finishes, any intervals remaining in the offline list are processed by the offline phase. The other intervals have already been assigned by the online phase itself. Two statistics are of interest to the online phase. The first statistic, denoted by U_a, is the current average processor load.

$$U_a = \frac{\text{Work currently assigned to processors}}{p}$$

The second statistic, denoted by O_a, is the average work in the intervals in the offline list.

$$O_a = \frac{\text{Work currently in offline list}}{p}$$

Initially, both U_a and O_a are zero. As the intervals are assigned to the processors or put on the offline list, the statistics U_a and O_a are updated.

The online phase also maintains a list called the *combining list*. This list consists of adjacent intervals that will be combined together to form a multi-valued interval. This is done so that the *number* of intervals that need to be assigned is reduced. The idea is to achieve the load balancing criterion using as few intervals as possible. Initially the combining list is empty.

In the online phase, one of the processors is designated as the *current* processor. This is the processor to which the next task will be assigned. Initially, any processor can be chosen as the current processor. After the task is assigned to the current processor, the value of U_a is updated, and the least loaded processor is chosen as the current processor. As we shall see, by using the least loaded processor next, we tend to make the next multi-valued interval as large as possible. Note also that since the current processor is the least loaded processor, its load is never greater than U_a.

We now describe how the online phase processes a representative single-valued interval I. Let W_{curr} denote the load on the current processor, and let W_{comb} denote the work currently in the combining list (if there are no intervals currently in the combining list, W_{comb} is zero). As mentioned before, the number of intervals in the offline list is restricted to p.

1. Check the condition

$$(W_{curr} + W_{comb} + I) \leq (1 + \delta)(O_a + U'_a) \qquad (2)$$

where $U'_a = (U_a + (W_{comb} + I)/p)$ is the online average *if* the combining list including I were assigned to the current processor. If the condition in Equation 2 holds, I is added to the combining list. That is, we keep combining until the load on the current processor would exceed the permissible threshold given the intervals seen so far. Go to 5.

2. If the condition in Equation 2 does not hold, the interval already in the combining list (if any) is assigned. If $W_{comb} > 0$, U_a will change, and the current processor and W_{curr} may change.

3. Check the condition

$$(W_{curr} + I) \leq (1 + \delta)(O_a + U'_a) \qquad (3)$$

where $U'_a = (U_a + I/p)$ is the online average *if* I were assigned to the current processor. If the condition in Equation 3 holds, the combining list is initialized to contain I. Go to 5.

4. If the condition in Equation 3 does not hold, then assign I to the offline list. If the offline list is full, i.e., it already has p intervals, then the smallest interval in the offline list is identified, say I_m. I_m is assigned to the current processor, and is replaced on the offline list by I.

5. If more intervals in the input, go to 1.

Before step 4, I has not yet been assigned to the offline list. Let O'_a denote the offline average not including I_m before step 4. Then $O'_a = O_a - I_m/p$, and

$$(W_{curr} + I) > (1 + \delta)(O'_a + \frac{I_m}{p} + U_a + \frac{I}{p})$$

After I_m is assigned to the current processor, and I is put on the offline list, we want the condition in Equation 3 to hold, viz.,

$$(W_{curr} + I_m) \leq (1 + \delta)(O'_a + \frac{I}{p} + U_a + \frac{I_m}{p}) \qquad (4)$$

Since I_m is the smallest element in the offline list, we have that $I_m \leq O_a$, i.e., $I_m \leq (O'_a + I_m/p)$. Since the current processor is the least loaded processor, $W_{curr} \leq U_a$. Then we have

$$(W_{curr} + I_m) \leq (O'_a + \frac{I_m}{p} + U_a)$$

which clearly implies the condition in Equation 4.

In [5], we show that at the end of the online phase, the maximum load on any processor is $\leq (1+\delta)W_p$. We also show that we can assign the intervals on the offline list so that no processor's load exceeds the load balancing limit. The time complexity of the online phase is shown to be $O(d \log(p))$ in [5].

5 Algorithm OO: Phase II

We use the term *capacity* of a processor to denote the additional work that can be assigned to it without violating the load balancing requirement. In order to achieve perfect load balancing, the algorithm may need to split some single-valued intervals. Splitting intervals results in an increase in communication. We wish to minimize this extra communication. We present a polynomial time heuristic algorithm that tries to minimize the communication while achieving the desired load balancing.

The splitting of the work in an interval can be achieved by replicating one of the input relation intervals, and fragmenting the other input relation interval. This is called the *fragment replicate* (FR) algorithm [2]. To minimize the communication, the smaller of the two intervals is replicated and the other interval is fragmented. Let the two intervals being joined have m_i and M_i tuples. Without loss of generality, let $m_i \leq M_i$. We would like to minimize the number of splits of the intervals with the larger m_i. This is the motivation for the following heuristic algorithm that tries to minimize the communication. In the heuristic we always keep the processors sorted in descending order of remaining capacity.

1. Sort the intervals on the offline list in descending order on the join work. We process the intervals in this descending order.

2. Determine the least number of splits required for each interval (no splits may be required). Let s denote the number of pieces into which the interval must be split.

3. Use the smallest capacity s consecutive processors in the sorted order of processors such that the join work can be handled by these processors. By doing this, we try to preserve the larger capacity processors for minimizing later splits.

In [5] we show that the time complexity of the heuristic algorithm described above is $O(d_o \log d_o + d_o \log p)$. Then the time complexity of algorithm **OO** is $O(d \log p + p \log p) = O(d \log p)$.

6 Combining of Intervals

In this section we consider how we can increase combination of single-valued intervals into multi-valued intervals in order to reduce the number of intervals that have to be assigned. Each of the processors sending the data has to determine the destination site for each tuple. If the destination site is determined based on comparisons in an index, we wish to reduce the number of intervals to keep the number of levels in the index small.

Some combining of intervals is already done in the online part of **OO**. Combining of intervals can potentially be increased by removing the restriction on the size of the offline list. Note that this does not increase the space required by the online/offline algorithm, since when an interval is replaced on the offline list, it forms an interval by itself in the online assignment. Also, the bound on the load imbalance still holds as long as we continue to calculate the "average" of the offline list as the sum of the work in the offline list divided by p.

Let W_{off} denote the total work on the offline list. To understand the possible increase in combination, we examine Equation 2. Note that $W_{curr} \leq U_a$ and hence combining may be best increased by increasing O_a. Now $O_a = W_{off}/p$, and hence we need to increase W_{off}. By allowing the offline list to grow beyond p if necessary, we allow W_{off} and O_a to become as large as possible, allowing an increase in combining of intervals. In Section 7, we describe the results of experiments to compare this modified algorithm with the basic algorithm described in the earlier sections. In the full paper [5], we discuss additional variations.

7 Experiments

Different cost models can be used to evaluate the join work. The algorithms presented here can work with any set of cost equations for evaluating the join work, and are not dependent on any particular cost equations. In the experiments we describe, we used cost equations similar to those described in [7].

We implemented both the load balancing algorithms, the pure online algorithm (**PO**) and the online/offline algorithm (**OO**) . We performed experiments to measure the following for each pair of input histograms:

1. The number of intervals assigned to the processors. Note that splitting a frequent value does not increase the number of assigned intervals.

2. The total communication using FR when frequent values are split compared with the total work.

For each set of values for the parameters of the histograms, we performed replicate experiments, and obtained the mean over the replicates. The replications were generated by using random permutations of a given pair of histograms. In our experiments, we performed 1000 replications for each data point.

The allowed imbalance δ had the values (0.1, 0.5, 1). The number of processors had the values (2, 8, 32, 128). The input histograms had number of tuples $N = 10^6$, the number of distinct values $D = 10^4$, the distinct value distribution ranging from uniform to Zipf and the correlation between frequent values with parameter C taking on values in $(0.01D, 0.05D, 0.1D)$ In our experiments, both the operand relations have the same distinct values. Zipf-like distributions have often been observed when the distribution of column values is skewed. To model correlation between frequent values, we use the scheme described by [7].

In considering the number of intervals generated, we are not interested primarily in the absolute numbers. Since these intervals are going to be used in routing tuples based on comparison within a tree of these intervals, we are concerned with the increase in the number of levels in the search tree. Most of the overhead in performing the load balancing is due to the additional CPU time in navigating the increased number of levels in the search tree. Hence this increase, denoted by L in Equation 5 is used to characterize the performance of the load balancing algorithm with various algorithm and workload parameters. The minimum number of intervals is $\lceil p/(1+\delta) \rceil$, and hence the increase in the number of levels is given by

$$L = \log \frac{u}{\lceil p/(1+\delta) \rceil} \tag{5}$$

where u is the number of intervals generated by **OO**. Note that the values of L given below are statistics taken over a large number of replications (1000).

We first examine the number of intervals generated by **OO** with the following two variations

1. the offline list is bounded by p

2. the offline list is unbounded

We observed that there was no significant variation in L when experimenting with the two variations. As an example, we give the values of L for the two variations for the following set of values of the parameters

$$p = 8, \quad corr = 0.01D \;\; \delta = 0.1, \;\; \theta = 0.0 \text{ (Zipf)}$$

Variation 1:	Mean = 2.47	StdDev = 0.39
	Min = 0.81	Max = 3.44
Variation 2:	Mean = 2.50	StdDev = 0.41
	Min = 0.81	Max = 3.52

We observe that the differences in L are insignificant. Using the unbounded offline list is the simpler variation since the offline list does not have to be searched during the online part of **OO**. Notice that the sum of the space for the offline intervals and the online assignments is the same for the two variations. Hence, we recommend using the variation with the unbounded offline list.

We conducted further experiments for **OO** using the unbounded offline list. The experimental data and the results are given in the full paper [5]. We observe what happens when p is increased, when the correlation is decreased, when the imbalance permitted is increased, and when the distribution for the relations is changed to a uniform distribution. When one of the parameter values is changed the other parameters are fixed at their baseline values. Examining the complete data set reveals that the effects of these parameters is independent.

Experiments with the pure online algorithm (**PO**) showed that the number of intervals generated increased only slightly. Though we get many more splits using **PO**, splitting a frequent value does not increase the number of assigned intervals. The observed values of L are very close to the values of L for **OO**.

We observed that the FR communication cost is a negligible fraction of the total work in the join. Over all the experiments, the communication cost was less than 0.02% of the total work *irrespective* of the heuristic used to assign the offline intervals. Minimizing the number of splits is simpler since we have to only deal with the join work. Hence, if we are trying to minimize the communication we recommend trying to minimizing the number of splits. This is independent of the communication protocol.

For the pure online algorithm (**PO**), the communication cost was about 110% higher than the communication using the sophisticated heuristic in **OO**, and about 75% higher than using the simple heuristic in **OO**. This means that if the communication cost is a small fraction of the total cost we can do very well using algorithm **PO**.

In Section 3 we described the reasons for the increase in communication cost and number of intervals when **PO** is used instead of **OO**. The tradeoff is clear. If we expect the communication cost to be always low, algorithm **PO** can be used. Algorithm **OO** is somewhat more complex to implement, but is much more robust. If only one algorithm is to be implemented **OO** is the algorithm of choice using the unbounded offline list and the simple heuristic for splitting.

8 Summary

In this paper, we presented algorithms for balancing the work in a parallel join. Though the algorithms are described in the context of the join operation in parallel relational databases, the algorithms have wider applicability to balancing the work in any parallel computation when given the task sizes in some sequence. The algorithms can achieve any desired tolerance in load balance regardless of the skew in the task sizes.

The first algorithm, called the pure online algorithm (**PO**) processes *all* the histogram data as it arrives. The tradeoff is that the communication cost of the subsequent join is significantly higher than if (**OO**) were used

to perform the load balancing. Also, the number of intervals generated increases significantly, resulting in increased per tuple distribution costs.

The second algorithm, called the online/offline algorithm (**OO**) has two phases: an *online* phase and an *offline* phase. The online phase processes most of the histogram data as it arrives, generating and assigning intervals to processors. The offline phase complements the online phase and seeks to minimize the communication cost of frequent value splits. We proved that the combination of the online and offline phases can meet any desired tolerance in load balance.

The results of our experiments lead us to recommend using algorithm **OO** with the unbounded offline list. If a second load balancing algorithm can also be implemented, algorithm **PO** can be used in those cases where we expect the increased communication cost to be only a small fraction of the total join work.

Acknowledgements

Discussions with Ashish Gupta, Peter Haas, John McPherson, Robert Morris, A. L. N. Reddy, Eugene Shekita and Jim Stamos helped improve this paper.

References

[1] K. P. Belkhale and P. Banerjee. Approximate Algorithms for the Partitionable Independent Task Scheduling Problem. In *Int. Conf. on Parallel Processing*, pages I–72–I–75. Aug. 1990.

[2] R. Epstein, M. Stonebraker, and E. Wong. Distributed Query Processing in a Relational Database System. In *Proc. of ACM-SIGMOD Int. Conf. on Management of Data*, pages 169–180, 1978.

[3] R. L. Graham. Bounds on Multiprocessing Timing Anomalies. *SIAM Journal of Applied Mathematics*, 17(2):416–429, Mar. 1969.

[4] H. Lu and K.-L. Tan. Dynamic and Load Balanced Task Oriented Database Query Processing in Parallel Systems. In *Proc. of the 1992 Int. Conf. on Extending Database Technology*, Mar. 1992.

[5] A. Swami and H. C. Young. Online Algorithms for Handling Skew in Parallel Joins. Technical report, Sept. 1991. IBM Research Report RJ 8363 (76086).

[6] C. B. Walton, A. G. Dale, and R. M. Jenevein. A Taxonomy and Performance Model of Data Skew Effects in Parallel Joins. In *Proc. of the Seventeenth Int. Conf. on Very Large Data Bases*, pages 537–548, Barcelona, Spain, 1991. Morgan Kaufman.

[7] J. Wolf, D. Dias, and P. Yu. An Effective Algorithm for Parallelizing Sort Merge Joins in the Presence of Data Skew. In *Proc. of the 2nd Int. Symp. on Databases in Parallel and Distributed Systems*, pages 103–115. IEEE Computer Society, 1990.

A PARALLEL SCHEDULING METHOD FOR EFFICIENT QUERY PROCESSING

A. Hameurlain, F. Morvan

Université Paul Sabatier, Lab. IRIT

118 Route de Narbonne -31062 Toulouse - France

Tél. (33) 61 55 82 48 ; Fax : (33) 61 55 62 58 ; E-mail : hameur@irit.fr

Abstract -- *In this paper, we propose a method to determine parallel scheduling for SQL query operations. The parallelization strategy consists in determining a parallel scheduling based on serial methods. The criteria for operations scheduling is based on deadlines and consideration of the number of processors. Performance evaluation shows the efficiency of each type of parallelism (intra-operation and inter-operation) as a function of the number of processors.*

INTRODUCTION

Optimization of SQL query compilers may be achieved by more efficient use of parallelism. Usually, the optimization process for such compilers proceeds through the following three steps: logical optimization, physical optimization and parallelization. The parallelization phase (intra-operation and inter-operation) is the most critical step, notably the establishment of an inter-operation parallelization method in a database program. In the present paper, we will consider inter-operation parallelism only.

Previous work [6] considered execution strategies related to different formats representing a SQL query consisting of several joins. Such strategies, associated with left-deep trees, right-deep trees and bushy trees, are based on the use of hash-join algorithms [6]. The work demonstrates that strategies associated with right-deep trees are well suited to exploit the parallelism inherent in parallel database machines. Another approach [8] enlarges the search space to include an intermediate format lying between left-deep trees and right-deep trees. The execution strategy associated to such a so called zig-zag tree, improves response time, compared to the right-deep tree, in case of memory limitation.

The various parallelization methods offered in the literature consider SQL queries containing join operations only. In the present work, we propose a parallelization method for relational operators (Join, Union, Difference, Selection) and the fixed point operator. The method must respect posteriority constraints between operations as well as limitations in resources (number of processors). The process, integrated in an optimizer, must determine in an efficient way a parallel program such that query response time in optimal or close to optimal.

Our solution to the parallelization problem is based on serial methods [3]. Serial methods are based on priority mechanisms which allows quick construction of an approximate solution in presence of renewable resources (a processor is a renewable resource).

This paper is organized as follows: after this introduction, section 2 describes the inter-operation parallelization method based on serial methods. Next in section 3 performance evaluation shows the efficiency of each type of parallelism (intra-operation, independent inter-operation) as a function of the number of processors. Our conclusions and plans for future work are presented in section 4.

PARALLELIZATION METHOD

The intra-operations parallelization phase has led to associate a local response time LRT and a number of processors to each relational operator or fixed point operator [5]. We consider that the base relations are partitioned horizontally.

The question now arises as to how to schedule the operations of a graph associated with a user query for best results in term of global response time and use of processors. The scheduling process must respect posteriority constraints and limitation in processors. This process, integrated in an optimizer, must determine in an efficient way a schedule for optimal or close to optimal query response time.

A simple idea to resolve such a scheduling problem relies on serial methods. These methods are based on task (operation) priority classification and allow simple construction of approximate solutions in the presence of renewable resources (a processor is a renewable resource).

The principle of our method is to schedule the operations with time increments starting at time=0. At any time, resources are assigned to the operation with highest priority amongst the ready operations. The operations which verify the following conditions are considered as ready:

(1) All the direct predecessors of the operation are finished or the operation is a distribution operation.
(2) The operation does not require more processors than the amount available at time tp.

The priority rule we apply schedules the operation in the order of increasing **deadline**. The deadline is the time at which the operation must be finished so as to not increase the global response time.

The scheduling method has two steps:

E1: compute the deadlines for each operation.
E2: schedule the operations with the priority rules and conditions defined above.

E1: In the first step, to compute deadlines, we are led to compute the late start times LATE defined in a recursive way, as follows:
(* succ(i) gives the direct successor for operation i*)

$$LATE(i) = \begin{cases} TR(i)\text{-}LRT(i) & \text{if succ(i)}=\emptyset \\ LATE(succ(i))\text{-}LRT(i) & \text{else} \end{cases}$$

A first run through the query graph, from the distribution operation to the user query node, gives the response time TR. TR is computed as follows:

$$TR(i) = \begin{cases} LRT(i) & \text{if pred(i)}= \emptyset \\ LRT(i)+max(TR(j)) & \text{else with } j \in \text{pred(i)} \end{cases}$$

A second run, from the user query node to the distribution operations, gives LATE(i) for each operation i and the deadline (LATE(i)+LRT(i)) defined as follows:

$$deadline(i) = \begin{cases} TR(i) & \text{if succ(i)}=\emptyset \\ LATE(succ(i)) & \text{else} \end{cases}$$

E2: Prior to the development of the scheduling algorithm (below), we define the notations used:

I   : the set of query graph operations including distribution operations

S_N   : the set of non-processed operations,
O_E   : selected operation,
tp   : represents times,
P   : the set of operations ready at time tp,
E_O_E   : the set of operations being processed at time tp,
List   : the set of ordering operation,
Dmax   : number of processors not used at time tp,
slave_nb : total number of processors,
di   : number of processors minimizing the local response time for operation i.

Procedure scheduling
begin
    tp := 0; Dmax := slave_nb; S_N := I;
    While S_N<> $\emptyset$ do
    begin
        operation_ready (S_N, P, Dmax);
        if P <> $\emptyset$ then
        begin
            election (P, O_E);
            starting (O_E) := tp;
            add (List, O_E);
            S_N := S_N - {O_E};
            E_O_E := E_O_E $\cup$ {O_E};
            Dmax := Dmax - nb_processor (O_E);
        end
        else
            term_op (E_O_E, Dmax, tp);
    end
end.

Procedure election (P: operation, O_E: selected_op)
begin
    minima(P, S_P);
    % this function determines the operations set S_P such as deadline(i)) / i $\in$ P be minimum%
    if card(S_P) > 1 then
        minima2(S_P, O_E);
        % this function determines the selected operations O_E such as LATE(i) / i $\in$ S_P be minimum%
    else O_E := S_P;
end.

Procedure operation_ready (S_N, P: operation, Dmax: processors_nb);
    This procedure determines the ready operations set from the set of not treated operations S_N and the number of processors Dmax.

Procedure term_op (E_O_E: operation, Dmax: processors_nb, tp: times);
    This procedure determine the smaller instant tp such as at least one operation E_O_E is terminated. It updates the variables E_O_E and Dmax.

Applying the algorithm schedule to the data of the query graph (Fig 1 & Fig 2) yields a schedule represented by the Gantt diagram (Fig 3).

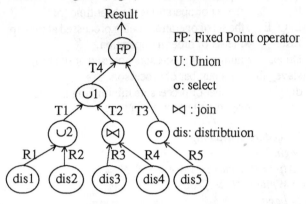

Figure 1: a Query Graph

FP: Fixed Point operator

U: Union

σ: select

⋈ : join

dis: distribtuion

	EC	∪1	∪2	⋈	σ	dis1	dis2	dis3	dis4	dis5
di	4	4	4	3	4	4	4	3	3	4
LRT(i)	13	7	5	7	6	5	4	3	4	7
TR(i)	31	18	10	11	13	5	4	3	4	7
LATE(i)	18	11	6	4	12	1	2	1	0	5
DL(i)	31	18	11	11	18	6	6	4	4	12

slave_nb=8   DL = deadline

Figure 2: Data of the Query Graph

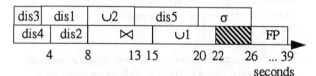

Figure 3: Scheduling representation with Gantt diagram

The advantage of this method lies in the fact that the processors are allocated to an operation as early as possible for best use of these processors. Furthermore, this method generates a time dependant operations schedule. For example, it is possible to express the operation sequences {join; U1} and {U2; dis5} are executed in parallel and U1 executed after U2. Conversely, with only the SEQ and PAR constructors for scheduling, one cannot guarantee that U1 will be executed after U2. The integration of the data communication operations which stall the following operations when no data is present, allows to express the scheduling in a manner which conforms to the execution plan of the method. In fact, each relational operator or fixed point operator is preceded by a data receiving operation and is followed by a data sending operation. For clarity and legibility of the execution plan, we did not integrate explicitly the data communication operations. The following program explains the above schedule.

```
Seq
  Par
    Seq
      dis4;
      dis2
    End_Seq
    Seq
      dis3;
      dis1
    End_Seq
  End_par;
  Par
    Seq
      Join;
      Union1
    End_Seq
    Seq
      Union2;
      dis5;
      select;
    End_Seq
  End_Par;
  Fixed_point
End_Seq
```

PERFORMANCE

In this section we first evaluate the response time (TR) of the query graph of figure 1, in order to analyse performance in the following cases:

(i) *full desclustering*: an operation is executed over all the machine's processors. The P1 type execution plan uses intra-operation parallelism only.

(ii) *partial desclutering*: an operation is executed over part of the machine's processors. The P2 type execution plan uses intra-operation parallelism and the independent inter-operation parallelism.

For performance evaluation we rely on the benchmarks described by [1]. The database is composed of six relations of 1, 000, 000 tuples. Each relation consists of thirteen 4-byte integer and three 52-byte string attributes. For flexibility and efficiency reasons, the architecture proposed for our Parallel Inference System PARIS is a on a distributed-memory architecture. This type of architecture has been validated by many database machines such as BUBBA [2] GAMMA [4]. Each node of the machine is composed of a processor, a local memory, and a disk. To measure the contribution of parallelism, the simulation parameters used are [6][7]:

The number of instructions per seconds for a processor = 4, 000, 000. The reading time of a 18 K disk page = 32 800 instructions. The writing time of a 18 K disk page = 61 500 instructions. The tuple transfer time = 5 μs. The time to send a message = 1 ms.

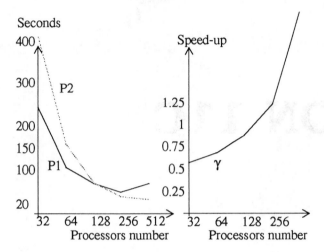

Figure 4: Response time for P1, P2   Figure 5: Speed-up values various parallelism types

In figure 4 the response time for different types of execution plan are presented as a function of the number of processors. Figure 5 shows speed-up values γ = TR(P1)/ TR(P2), as a function of the number of processors. Three statements can be made by observing the curves in figure 4 and 5:

1- The P1 type execution plan, which uses intra-operation parallelism only is more efficient for small processor numbers such as 32 to 130 processors for example.

2- The P1 execution plan suffers an increase in response time beyond 256 processors, due to larger data message communication times.

3- The P2 type execution plan using intra-operation and independent inter-operation parallelism has the best response time above 130 processors.

CONCLUSION

In this paper, we have presented an inter-operation parallelization method in a parallel deductive environment. This method determines a parallel execution plan, with the help of a query graph, which includes intra-operation parallelism, inter-operation independent parallelism and sequential execution mode. The originality of our approach comes from the fact it may be applied to all relational and specialized operators; it relies on a serial method with ap-

plied priority rules based on deadlines; it takes into account the number of processors of the target machine.

The inter-operation parallelization method described in this paper does not consider pipeline dependent parallelism. This type of parallelism consists in exchanging, as early as possible, small data sets between two communicating operations. We plan to evaluate pipeline efficiency in future work in order to integrate this type of parallelism in our overall strategy.

REFERENCES

[1]   D. Bitton et al., "Benchmarking Database Systems - a Systematic Approach", *Proc. of the 1983 VLDB conference*, (October, 1983), pp. 8-19.

[2]   H. Boral et al., "Prototyping Bubba, a Highly Parallel Database System", *IEEE TKDE*. Vol. 1, No. 1, (March, 1990), pp. 4-24.

[3]   P. Chretienne, "Scheduling and Parallelism", 20th spring school of LITP, Sables-d'Or-les Pins France, (may, 1992), pp. 297-312.

[4]   D.J. Dewitt, J. Gray, "Parallel Database Systems: The future of High Performance Database Systems", *Communication of the ACM*, Vol. 35, No. 6, (June, 1992), pp. 85-98

[5]   A. Hameurlain et al., "A Cost Evaluator for Parallel Deductive Database Systems", *Int. Conf. and* Exhibition on Parallel Computing and Transputer *Applications'92*, Barcelona, (Sept., 1992), pp. 1222-1228.

[6]   D. Schneider and D. Dewitt, "Tradeoffs in Processing Complex Join Queries via Hashing in Multiprocessor Database Machines", *Proc. of the 16th VLDB Conf.*, Brisbane, Australia (1990), pp. 469-480.

[7]   P. Valduriez, S. Khoshafian, "Parallel Evaluation of the Transitive Closure of a Database Relation", *Intl. Journal of Parallel Programming*, Vol. 17, No. 1, (Feb., 1988), pp. 19-42.

[8]   M. Ziane, et al. "Parallel Query Processing in DBS3", *2nd Int. Conf. on Parallel and Distribued Information Systems*, San Diego, (January, 1993).

SESSION 11C

RESOURCE ALLOCATION AND FAULT TOL-
ERANCE

AUTOMATED LEARNING OF WORKLOAD MEASURES
FOR LOAD BALANCING ON A DISTRIBUTED SYSTEM

Pankaj Mehra and Benjamin W. Wah[†]

Coordinated Science Laboratory

University of Illinois, Urbana-Champaign

1308 West Main Street

Urbana, IL 61801

{wah,mehra}@manip.crhc.uiuc.edu

ABSTRACT

Load-balancing systems use workload indices to dynamically schedule jobs. We present a novel method of automatically learning such indices. Our approach uses comparator neural networks, one per site, which learn to predict the relative speedup of an incoming job using only the resource-utilization patterns observed prior to the job's arrival. Our load indices combine information from the key resources of contention: CPU, disk, network, and memory. Our learning algorithm overcomes the lack of job-specific information by learning to compare the relative speedups of different sites with respect to the same job, rather than attempting to predict absolute speedups. We present conditions under which such learning is viable. Our results show that indices learnt using comparator networks correctly pick the best destination in most cases when incoming jobs are short; accuracy degrades as execution time increases.

1. INTRODUCTION

This papers addresses the computation of workload measures in distributed systems. Such measures are used by dynamic load-balancing policies to determine the site most suitable for executing an incoming task.

Our model of distributed systems comprises a network of interconnected sites. Each site has private resources such as CPU and memory; some have disks; and all share the network resource. These resources are *architecturally homogeneous*: they can service requests from programs at any of the sites. Resources such as network and secondary storage can be shared transparently among the processes at different sites; other resources, such as CPU and memory, can only be accessed by local processes. We do allow for *configurational heterogeneity*, i.e., sites having different processor speeds, memory space, or disk space.

We assume that the "background workload" at each site varies outside the control of the load-balancing software. Such load may be caused either by operating-system functions or by tasks that, for some reason, cannot be migrated. We assume no prior knowledge of the tasks to be scheduled, except

that they are drawn from a large population having stationary (but unknown) mean and variance of relative resource requirements. Local execution of an incoming task is the base case, relative to which all load-balancing decisions are evaluated. Since we assume independent tasks, our objective is to maximize the average speed-up, computed on a task-by-task basis, over local execution.

The unpredictable long-term variation of background workload rules out static strategies, and, coupled with our ignorance of task lengths, necessitates preemptive strategies which can "undo" the effects of poor initial placements. The distributed arrival pattern of tasks and the lack of constraining dependences between them entail that we prefer the less expensive and more robust decentralized strategies over the centralized ones. Therefore, our focus in this paper is on dynamic, decentralized, and preemptive load-balancing strategies.

Load-balancing strategies have two components: *workload measures*, which indicate each site's load; and *decision policies*, which determine both the conditions under which tasks are migrated and the destinations of incoming tasks. In this paper, we address automated learning of load measures; learning of policy parameters is described elsewhere [8].

Figure 1 shows the parameterized policy considered in this paper. The sender-side rules (SSRs) are evaluated at s, the site of arrival of a task. *Reference* can be either *0* or *MinLoad*; the other parameters — δ, θ_1, and θ_2 — take non-negative real values. A remote destination, r, is picked randomly from *Destinations*, a set of sites whose load indices fall within a small neighborhood of *Reference*. If *Destinations* is the empty set, or if the last SSR fails, then the task is executed locally at s; otherwise, site r is requested to receive the task. Upon receiving that

---

SENDER-SIDE RULES(s)

    Destinations = {site: Load(site)–Reference(s) < $\delta(s)$}

    Destination = Random(Possible_destinations)

    IF Load(s) – Reference(s) > $\theta_1(s)$ THEN Send

RECEIVER-SIDE RULES(r)

    IF Load(r) < $\theta_2(r)$ THEN Receive

Figure 1. The load-balancing policy studied

[†]Contact author: B. W. Wah. Research supported by National Aeronautics and Space Administration Grant NCC 2-481 and a grant from Sumitomo Electric Industries, Yokohama, Japan.

request, site r applies its receiver-side rule (RSR). If the RSR succeeds, the task is migrated; otherwise, the task is executed locally at s.

Our policy is a generalization of several well-known decentralized dynamic load-balancing policies [1, 7, 12]. It uses primitive workload measures (*Load*, denoting the load at each site) as well as an abstract measure (*MinLoad*, denoting the smallest *Load* value). Our focus in this paper is on methods for calculating *Load*.

A system supporting our load-balancing policy must support (i) measurement of low-level resource-utilization information; (ii) communication of *Load* values among sites; and (iii) determination of *MinLoad*.

The rest of this paper is organized as follows. In Section 2, we critique existing methods for computing load indices and motivate automated learning techniques. In Section 3, we describe how DWG, a workload generator, is useful for collecting the data necessary for learning. In Section 4, we describe our algorithm for learning load indices using neural networks. Section 5 contains empirical results on the quality of load indices learnt using our method. Section 6 concludes this paper.

2. LOAD INDICES

Load-balancing systems seek to maximize speed-up over local execution. This requires them to rank alternative destinations for each incoming task by their *expected* speed-ups over local execution. Usually, alternative destinations are ranked by their *Load* values, which are computed using a manually-specified formula as functions of current and recent utilization levels of various resources. Good load indices are difficult to design as they are sensitive to installation-specific characteristics of hardware devices as well as to the prevalent load patterns. This section reviews current approaches to load-index design, discusses their deficiencies, and presents our approach based on automated learning.

2.1. Existing Methods

Many existing methods for computing load indices use simplified queuing models of computer systems [3, 4, 12]. Almost all implemented systems use a function known as *UNIX-style load average* (hereafter, load average), which is an exponentially smoothed average of the total number of processes competing for CPU. Load average meaningfully compares loading situations across configurationally identical sites, but fails when the distributed system is heterogeneous. Assuming preemptive round-robin scheduling of processes within each site, let us consider the comparison between a site having a load average of 3 and another that is 5 times slower and has a load average of 0. An incoming task is likely to require 20% lesser time to complete at the fast site than the at the slow site, despite the former's high load average! A more fundamental problem with the traditional load-average function is that it completely ignores resources other than the CPU. Therefore, while it may be reasonable predict the performance of purely compute-bound tasks, its utility is questionable for tasks that use the other resources of contention: memory, disk, and network.

Typical workstation operating systems provide a number of different performance metrics. (For example, over 20 performance metrics are available in the SunOS kernel.) Although measuring the utilization levels of resources other than CPU may not require any hardware modifications, several of these metrics are unsuitable for inclusion in a load index because the overhead of estimating their values precludes frequent sampling. These include process-level metrics, which are sampled only once every 5 seconds. Even if we eliminate these, we are still left with a fairly large set of mutually dependent variables; for example, disk traffic is affected by the number of page swaps and process swaps. Other metrics, such as rate of data transfer, are fixed quantities for a given configuration, and affect only the (fixed) coefficients of a load index.

Ideally, workloads for load balancing should be characterized by a small set of performance metrics satisfying the following criteria: (i) low overhead of measurement, which implies that measurements can be performed frequently, yielding up-to-date information about load; (ii) ability to represent the loads on all the resources of contention; and (iii) measurable and controllable independently of each other. In the past, Zhou [12] has considered resource-queue lengths (the number of processes waiting for CPU, disk, and memory) in designing load indices. Although his approach of using multiple measurable metrics is similar to ours, he did not consider a systematic method for learning good workload indices. Further, he did not consider the design of a suitable workload generator for verifying his load index. In this paper we use the instantaneous utilization levels on the four basic resources — CPU, memory, disk, and network — that form a useful set of performance metrics satisfying all three criteria. We further present a synthetic workload generator that can be used for collecting workload patterns and for regenerating the workload collected in the presence of foreground jobs.

2.2. Automated Learning of Load Indices

Ideally, we would like to rank alternative destinations for an incoming job by their respective completion times. However, completion times can only be measured for completed jobs, whereas decisions needs to be made before jobs start. Therefore, we need to somehow *predict* completion times using only the information available before a job begins execution. Without knowing the resource requirements of jobs, we cannot predict absolute task-completion times.

Notice, however, that we only need to *compare* alternative destinations for the *same job*. Therefore, we only need to determine a relative (site-specific and configuration-specific) measure of completion time. It would, therefore, suffice to predict the *relative* completion times of a job at different sites. As a point of reference, we can use the completion time of the job on an idle file server [6]. Every site needs to predict the completion time of an incoming job relative to its completion time on the chosen idle file server, given only the loading conditions at job-arrival time. (Such prediction is only feasible for autocorrelated resource-utilization patterns [9].)

Predicted relative completion times can be used as load indices. If completion times of different sites for the same job vary widely, then the accuracy of prediction can be low; otherwise, if the difference between completion times are small relative to the reference, then more accurate prediction is needed. At the outset, it is unclear how much accuracy will suffice. Since the accuracy of convergent iterative learning procedures improves with training, we need to reformulate our problem so that we will know when adequate accuracy is attained.

Instead of predicting, for each site, the relative completion time of a task at that site, we now predict, *for each pair of sites*, the difference between their relative completion times. We express accuracy as the *percentage of correct comparisons*. We can stop refining our predictions when sufficient accuracy is achieved. We also need to ensure that the anti-symmetry of comparison — whenever A is better than B, B is worse than A — is preserved during learning.

3. DATA COLLECTION

No matter how many load indices we need to compare, the number of possible destinations for an incoming job is finite and equals the number of sites. Therefore, we can collect all the data necessary for learning load indices *off-line* and ahead of time. Such data take the form of before-after pairs: the first item of each pair is a trace of utilization of different resources; the second is the measured completion time of a job introduced at the end of the trace period. We can first measure realistic system-wide workloads, and then replay them repeatedly, each time introducing a job at precisely the same time into the experiment but at a different site of the network. If we repeat such experiments with several representative jobs, under several realistic loading conditions, then we can create a large database of realistic decision points.

Our experimentation environment must allow us to both *generate realistic loading conditions* and *repeat them as often as desired*. Representative test jobs must be run under such *synthetic workloads*, and the utilization of all resources recorded periodically. Both utilization levels and completion times must be measured with low overhead and high precision. With these in mind, we developed DWG [8, 10], a synthetic workload generator that helps us build our database of decision points.

DWG can measure and control the utilization levels of four key resources: CPU, memory, disk, and network. It supports a variety of data-collection operations: (i) precise measurement of resource-utilization patterns; (ii) precise generation of recorded patterns; (iii) initiation of foreground test jobs at precise times; and (iv) measurement of job-completion times. To accurately reproduce the behavior of the process population generating the measured load, we have implemented most of DWG inside the kernel [8] so that it has complete control over the utilization levels of local resources. DWG gives up a fraction of its resources allocated to the background load in response to the arrival of a foreground job, and reclaims these resources when the job terminates. The precise amount to give up is controlled by "doctoring" rules. Parameterization and tuning of these rules is described elsewhere [8].

We collected load patterns on a configurationally heterogeneous system consisting of (i) a diskless Sun 3/50 with 4 Mbytes of RAM; (ii) a diskful Sun 3/50 with 4 Mbytes; (iii) a diskful Sun 3/260 with 8 Mbytes; and (iv) a diskless Sun 3/60 with 24 Mbytes. The four workstations were connected by a single 10 Mb/s Ethernet.

We used a total of ten test jobs: three for sorting records of various sizes with various amounts of initial memory, two for uncompressing files of different sizes, and five from the Perfect Club suite of benchmarks (FLO52Q, TRFD, QCD, TRACK, and DYFESM) [2].

To study preemptive scheduling of jobs, we inserted checkpoints into each test job. Each checkpoint resembles a preemption point: the given job can be preempted and resumed at any of its checkpoints; consequently, we can treat the segments from the beginning of an instrumented job to each of its checkpoints as independent jobs. Although this technique creates a large database of training patterns, they could be less effective since different segments of the same test job could have similar resource requirements. This, however, does not affect the accuracy of our final result, as we are not interested in just the overall average speedup, but rather average speedups as a function of job length. (Speedups for different checkpoints of the same job are, therefore, never compared in the final result.)

Each job was instrumented to produce approximately 200 checkpoints during its execution. The Perfect Club benchmarks were modified so that they could complete within five minutes on an idle Sun 3/50 workstation. In order to avoid checkpoints too close to each other, our final database included the first checkpoint of each foreground job and, thereafter, the next checkpoint that took at least 5% longer than the one included before it. The final checkpoint of each job was always included. We ended up with a total of only 58 jobs, with about five checkpoints per test job; each of these jobs was run at each of the 4 sites and under 24 different background load patterns, for a total of 5,324 decision points.

Of the 24 background load patterns used in our experiments, 20 were created by running job-files created randomly from the pool of 10 test jobs described above. The remaining 4 load patterns were designed to create surprises for the load-balancing system. We started with an actual, heavy, system-wide load and patched together pieces of it and the idle load pattern, such that the resulting patterns would frequently contain loading conditions just the opposite of those prior to the job's arrival. Since the load-balancing system can only access the loading conditions before a job's arrival, its decisions would falter under these 'surprising' loads. Moreover, since real workloads often have unpredictable changes in workload, whose likelihood increases with the length of the load pattern, we packed more surprises near the end of the load pattern. We expect that, because of these load patterns, the performance (speed-up over local execution) attained by scheduling a job at the site with the least load index will tend to drop with the length of the job, approaching the performance of random choice for long jobs.

4. COMPARATOR NEURAL NETWORKS FOR LEARNING LOAD INDICES

We now describe our approach to off-line learning of load index functions. First, we formally state our learning task: learning to compare functions of multivariate time series. Then, we present the comparator-network architecture and its associated learning algorithms.

4.1. Design Goals

Different destinations (loading conditions) yield different completion times for an incoming job; our objective is to find the one with the minimum completion time. Comparison of job-completion times requires information about the utilization levels of local as well as remote resources. Raw utilization levels of resources are highly dynamic vector quantities, whose communication would incur too much overhead. It is desirable that comparison among sites be achieved solely via periodic and infrequent communication of scalar load indices, whose computation does not involve utilization levels of remote resources.

Completion times of jobs are unknown at their time of arrival. Given our lack of information about resource-intensities of different jobs, and unavailability of models relating resource-utilization patterns to job-completion times, we can predict completion times of jobs using functions statistically estimated from data. The data for estimation can be easily obtained using DWG; they contain information about (i) the utilization patterns of different resources prior to the arrival of a test job; and (ii) completion times of that job at each of the alternative destinations.

Our goal in load-index learning, therefore, is to estimate functions F_s, one per site, such that (i) the functions do not assume any prior information about the job being scheduled; (ii) they depend only on (smoothed, or otherwise processed) local resource-utilization values; and (iii) the ranking induced on alternative destinations by these functions is consistent with the ranking induced on those same destinations by the true completion times. Before the objective of learning is stated, we define the following symbols.

S: Set of S sites in the network.

F: Set of F foreground test jobs. 0 is the case with no foreground job.

T: Window size in time units during which workload is to be generated for the distributed computer system. (Note that the window of time covers all computers in the system, and that our unit of time equals 20 milliseconds, which is the period between successive interrupts of a real-time clock.)

B: Set containing B background load patterns, where a background load pattern is defined as a collection of background jobs, each characterized by a prescribed site and time of arrival. 0 is the case with no background load.

$l_{b,f(t),s}$: A vector containing the utilization levels at site s for CPU, memory, disk, and network at each time unit in the window for background load pattern b and foreground job f started at time t. $l_{b,f(t),s}$ is a matrix of 4 rows (representing

the four resources) and T columns. Note that $l_{b,0,s}$ represents the measured utilization when no foreground job is run, and that, whenever it is obvious, we will use f instead of $f(t)$.

$\hat{l}_{b,f(t),s}$: Vector of values derived from recent behavior of loads on different locally accessible resources at site s for background load pattern b before foreground job f is started at time t. $\hat{l}_{b,f(t),s}$, therefore, depends on t, the time that job f is started, but not on the behavior of job f. In subsequent discussions, we use f instead of $f(t)$ when t is constant.

$F_s^W\left[\hat{l}_{b,f,s}, f\right]$: Value of site s's load-index function, where W denotes the current values of the *weights* (or parameters) of F_s. We will omit the superscript W from our equations whenever it is obvious.

The objective of load-index learning, then, is

$$F_s\left[\hat{l}_{b,f,s_1}, f\right] - F_s\left[\hat{l}_{b,f,s_2}, f\right] = \frac{C\left[\hat{l}_{b,f,s_1}, f\right] - C\left[\hat{l}_{b,f,s_2}, f\right]}{C\left[\hat{l}_{0,f,s_{ref}}, f\right]} \tag{1}$$

for all foreground jobs f, for all background load patterns $\hat{l}_{b,f,s}$, and for all pairs (s_1, s_2) of sites. Note that $C\left[\hat{l}_{0,f,s_{ref}}, f\right]$ represents the completion time of job f at the idle reference site. Further, note that in Eq. (1), we assume that load indices are computed at time t for both sites s_1 and s_2. In practice, the above assumption is not correct, as it is difficult to synchronize the computation of indices at different sites. Moreover, there are delays in obtaining load indices from remote sites. These errors are not considered in the design of load-index function but rather in other parameters of the load balancing policy [8].

Let us examine whether the right-hand side of Eq. (1) depends on the job f being scheduled. Consider the expression $C\left[\hat{l}_{b,f,s}, f\right] \Big/ C\left[\hat{l}_{0,f,s}, f\right]$, which is commonly known as *stretch factor* [3]; its denominator is known as the *service time* of job f at site s. The latter depends only upon f and s, but not on the b. Further, the completion time $C\left[\hat{l}_{b,f,s}, f\right]$ can be expressed as a sum of service time and waiting time:

$$C\left[\hat{l}_{b,f,s}, f\right] = C\left[\hat{l}_{0,f,s}, f\right] + W\left[\hat{l}_{b,f,s}, f\right], \tag{2}$$

where $W\left[\hat{l}_{b,f,s}, f\right]$ is the time spent by job f waiting for resources. Under round-robin scheduling policies at the process level, waiting time grows linearly with service time. The rate of growth depends upon the load: when the load is high, even short jobs can take a long time to complete. Therefore, we can rewrite Eq. (2) as:

$$C\left[\hat{l}_{b,f,s}, f\right] = C\left[\hat{l}_{0,f,s}, f\right]\left[1 + G\left[\hat{l}_{b,f,s}\right]\right], \tag{3}$$

where $G\left[\hat{l}_{b,f,s}\right]$ is a site-specific function that increases monotonically with load.

Under the model of completion times described above, the stretch factor $C\left[\hat{l}_{b,f,s}, f\right] \Big/ C\left[\hat{l}_{0,f,s}, f\right]$ depends only upon the load $\hat{l}_{b,f,s}$ and site s, but not on the job f. If we further assume that the different sites of our distributed system come from the same architecture family and differ only in their raw speeds, then

$$C\left[\hat{l}_{0,f,s},f\right] = K(s)\, C\left[\hat{l}_{0,f,s_{ref}},f\right], \tag{4}$$

where $K(s)$ is some site-specific constant. Now, we can rewrite Eq. (1) as

$$F_s\left[\hat{l}_{b,f,s_1},f\right] - F_s\left[\hat{l}_{b,f,s_2},f\right] \tag{5}$$

$$= K(s_1)\left[1 + G\left[\hat{l}_{b,f,s_1}\right]\right] - K(s_2)\left[1 + G\left[\hat{l}_{b,f,s_2}\right]\right],$$

The right hand side does not depend upon the job f being scheduled. Therefore, under the assumptions described above, the objectives described by Eq. (1) are achievable without prior knowledge of the job being scheduled.

In short, the two assumptions required for the just-proven independence between the characteristics of a job f and the right-hand side of Eq. (1) are: (i) waiting time grows linearly with service time; and (ii) different sites come from the same architecture family. The first of these assumptions is a reasonable requirement of any fair scheduling policy, and is approximately true of UNIX-related operating systems that employ variants of round-robin scheduling. The second assumption is generally true of workstation-based computing environments, because even workstations from different vendors often employ the same microprocessor architecture.

How well can the objective described by Eq. (1) be satisfied depends upon several factors. Let $\mathbf{W}$ be the space of all parameter values for functions F_s, and let $\mathbf{F} = \{F^W : W \in \mathbf{W}\}$. The load-index function we shall learn must belong to the family of functions $\mathbf{F}$. Even when the best function from this family is used, the error between the left and right sides of Eq. (1) may not be zero for all data points; such errors contribute to the *approximation error*, which is defined independently of the training algorithm used for adjusting W. A given training algorithm may not be capable of finding the optimal weights W in reasonable amount of time. The error incurred due to suboptimality of weights W is called *estimation error*; it depends upon the particular training algorithm used. We need an approach that reduces both the approximation error as well as the estimation error.

To achieve our goal of obtaining *accurate comparison in most situations*, we would like both sides of Eq. (1) to have the same sign. We, therefore, seek to learn load-index functions that, when compared across sites, correctly predict (for most training patterns) the site with the lower relative completion time.

We partitioned our database of raw training patterns into a *raw training set* and a *raw test set*. We used the raw training set to tune the weights of the load-index functions, and the raw test set to evaluate the trained functions. Suppose that the set of raw training patterns is $\mathbf{R} = \{\tau(b, f, s_1, s_2) \mid b \in \mathbf{B}; f \in \mathbf{F}; s_1, s_2 \in \mathbf{S}; s_1 \neq s_2\}$. Each raw training pattern $\tau(b, f, s_1, s_2)$ is a 5-tuple $\left\{\hat{l}_{b,f,s_1}, C\left[\hat{l}_{b,f,s_1},f\right], \hat{l}_{b,f,s_2}, C\left[\hat{l}_{b,f,s_2},f\right], C\left[\hat{l}_{0,f,s_{ref}},f\right]\right\}$.

The objective function for load-index learning can be formally defined as follows.

Minimize $\sum_{\tau \in \mathbf{R}} E(\tau)$, where $\qquad\qquad$ (6)

$$E(\tau(b, f, s_1, s_2)) = \begin{cases} 0 & \text{if } \left[F_s\left[\hat{l}_{b,f,s_1},f\right] - F_s\left[\hat{l}_{b,f,s_2},f\right]\right] \\ & \times \left[C\left[\hat{l}_{b,f,s_1},f\right] - C\left[\hat{l}_{b,f,s_2},f\right]\right] > 0 \\ 1 & \text{otherwise} \end{cases}$$

4.2. The Comparator Network: Architecture and Learning Algorithms

Little is known about the problem of learning to compare functions. One exception is the work of Tesauro [11], who invented the comparator-neural-network architecture for learning to compare alternative moves for the game of backgammon. His approach does not directly carry over to the problem of comparing functions of time series. Our approach, described in this section, was motivated by Tesauro's work; however, in adopting his work to the index-learning problem, we have made significant departures from both his network configurations and training algorithms.

Figure 2 shows a schematic of our comparator neural network. It shows the details of the training algorithm, and the flow of information during a typical learning trial. Each learning trial involves one training pattern from the training set: first, resource-utilization information from a pair of randomly selected training patterns is presented at the inputs (to the left); then, the *actual outputs* of the two index functions are com-

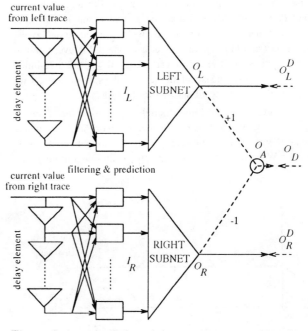

Figure 2. An episode in training a comparator network

puted; based on these outputs, the *desired outputs* for each of the index functions is computed as follows.

$$O_L^D = O_L - 2\eta(O_L - O_R - O^D) \tag{7a}$$

$$O_R^D = O_R - 2\eta(O_L - O_R - O^D) \tag{7b}$$

Finally, the two index functions are modified so that future presentations of similar inputs will generate outputs closer to their respective desired outputs.

Raw utilization patterns enter from the left of Figure 2; the delay elements create a window of recent values; the traces from each window are smoothed using low-pass filtering and an estimate of future resource utilization determined by extrapolating the smoothed trace. We use 5 different filters, with cutoff frequencies at, respectively, 1%, 5%, 10%, 25%, and 50% of maximum frequency. This yields 5 filtered traces per resource. Each of the filtered traces is projected a fixed interval into the future; the interval of projection equals the average completion time of test jobs at the reference site. Two different extrapolation techniques are used: one using linear fitting and the other using exponential fitting. The area under each extrapolated curve (see Figure 3) is used as an input to the load-index function. Since there are 4 different resources, 5 filters, and 2 extrapolation methods, our indices are functions of 40 variables. Vectors of these variables constitute the inputs I_L and I_R of the load-index functions (Figure 2).

We implemented load-index functions using feed-forward neural networks. Each network comprises three layers of units: an input layer, a hidden layer, and an output layer. The output layer has only one output unit. Links between units are uni-directional, and can connect either a unit in the input layer to a hidden unit or the output unit, or a unit in the hidden layer to the output unit. Associated with each link is a *weight* of that link. The weight on a link going from the i'th unit to the j'th unit of the network is denoted $w_{j,i}$. Every unit in the input layer is connected to every unit of the hidden and output layers, and every unit in the hidden layer is connected to every unit in the output layer. The set of parameters W for the load-index function at a site consists of all the $w_{j,i}$ values, where i and j are units in the feed-forward network for that site.

Given the actual and desired outputs for a feed-forward neural network, the "back-propagation algorithm" can determine the appropriate modifications to the weights of that network. We use the 'vanilla' back-propagation algorithm avail-

able in a public-domain simulation package [5].

The outputs of input-layer units are set at the beginning of each learning trial using filtered and extrapolated resource-utilization values, which are in turn derived from information contained in the training pattern chosen for that trial. The outputs of hidden-layer and output-layer units depend upon their *net inputs*. The net input of unit i (in the hidden or output layer) is given by $\sum_j w_{j,i} o_j$, where o_j denotes the output of unit j. The output of each unit of a feed-forward network is given by the *sigmoidal function* of its net input:

$$f(x) = \frac{1}{1 + e^{-x}}. \tag{8}$$

Let us denote by O_L the output of the left subnet; and, correspondingly, O_R, of the right subnet. In order to use the back-propagation learning procedure for training the subnets, we need to determine their desired outputs for every input. Let O_L^D denote the desired output of the left subnet; and, correspondingly, O_R^D, of the right subnet. Further, let us denote by O^A the actual output of the comparator network; that is,

$$O^A = O_L - O_R. \tag{9}$$

Given our objectives stated in Eq. (1), O^A corresponds to the left-hand side of that equation. Therefore, the desired output of the comparator network (denoted O^D) is given by the right-hand side of Eq. (1). That is,

$$O^D = O_L^D - O_R^D = \frac{C\left[\hat{l}_{b,f,s_1}, f\right] - C\left[\hat{l}_{b,f,s_2}, f\right]}{C\left[\hat{l}_{0,f,sref}, f\right]}. \tag{10}$$

The value of the objective function shown in Eq. (6) will be reduced if both O^A and O^D have the same sign. That can be achieved by driving their values closer together. Let us denote by E_{LMS} the sum (over all raw training patterns in the training set) of squared errors between the actual and the desired outputs of the comparator. That is,

$$E_{LMS} = \sum_\tau (O^A - O^D)^2. \tag{11}$$

We can minimize E_{LMS} by performing gradient descent; that is, by adjusting the outputs of the left and right subnets along their respective partial derivatives of error:

$$\Delta O_L = -\eta \frac{\partial E_{LMS}}{\partial O_L}, \qquad \Delta O_R = -\eta \frac{\partial E_{LMS}}{\partial O_R}, \tag{12}$$

where η is known as the *learning rate*. Hence, our training algorithm sets the desired outputs for the left and right subnets as shown in Eq's. (7a) and (7b).

One final detail needs to be worked out: ensuring the anti-symmetry of comparison. We resolve this problem by biasing the order of presentation of training patterns to the comparator. Raw training patterns are presented in pairs, one after another. If the first training pattern in the pair is $\tau_1(b, f, s_1, s_2)$, then the second training pattern must be $\tau_2(b, f, s_2, s_1)$. Thus, whenever index-functions are forced to predict that one completion time will be larger than another, they must (in the very next learning trial) predict that the latter will be *smaller* than the former.

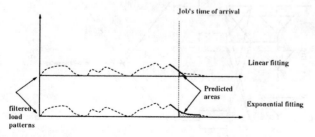

Figure 3. Trend extraction via curve fitting

5. EMPIRICAL RESULTS

This section presents our results on learning load indices for a system containing four sites. Each training pattern presented to the comparator network contains 40 projected resource-utilization values and 2 additional fields, $C\left[\hat{l}_{b,f,s}, f\right]$ and $C\left[\hat{l}_{0,f,s_{ref}}, f\right]$, where f identifies the foreground job, b identifies the background load pattern, and s the site at which f was executed.

5.1. Evaluation Methods

The training algorithm described in the last section was applied to 40×40×1 networks (*i.e.*, networks containing 40 hidden units each). In determining the number of hidden units, we used the popular rule of thumb that a network must contain approximately half as many weights as there are training patterns. The learning parameter η of the back-propagation algorithm was set to 0.001, and the momentum parameter set to 0.99.

We began by randomly assigning 10 percent of the jobs to the test set and the remaining 90 percent to the training set. The networks were trained using the training set, and their accuracy measured on the test set. Training and testing were applied alternately; during each *epoch* of training, the networks were trained on 1,000 randomly chosen comparisons from the training set, and tested on 100 randomly chosen comparisons from the test set. Each comparison involved two loading situations for the same job. Training was done in two stages. Stage 1 started with two identical networks (with random initial weights) for each site; these were trained to compare different loading conditions for the same job at the same site. In Stage 2, just the left networks from each site were further trained to compare different sites for the same job under the same system-wide load pattern. Unlike in Stage 1, the roles of the left and the right load-index functions (Figure 2) could not be reversed during Stage 2.

If we rank all available test-cases (all the different sites and loading patterns) of a job by their load indices, then we would like this ranking to have high correlation with the *true ranking* of these test-cases. The true ranking is, of course, the one induced by the measured completion times, $C\left[\hat{l}_{b,f,s}, f\right]$. One way to measure the correlation between two different rankings of the same data is to compute the *rank-correlation coefficient,* defined as:

$$r = 1 - \frac{6 \cdot \sum_i d_i^2}{n(n^2 - 1)},\qquad(13)$$

where d_i is the difference between the two different ranks of the i'th test-case, and n is the number of test-cases ranked. The value of r ranges between −1 and +1, where −1 indicates strong disagreement between rankings, 0 indicates no correlation, and +1 indicates strong agreement. This coefficient can be computed for each of the 58 jobs in the training set.

If there is *no relationship* between the two rankings, then the sampling distribution of r can be approximated with a normal distribution $N(0, \sigma)$, where σ, the standard deviation, equals $(n-1)^{-0.5}$. If we let $z = r/\sigma$, then z will be distributed according to the two-sided standard normal distribution $N(0,1)$. Using this fact, we can set up the following *null hypothesis*:

[H0] There is no relationship between the true ranking and the ranking induced by the load-index functions learnt using comparator networks.

We would like to *reject* H0 with high confidence. Given a confidence level α, we can look up the table of standard normal distribution to find an A such that $\Pr(\,|z|\leq A\,) \geq (1-\alpha)/2$. In particular, A takes on the values 2.58 for corresponding α value of 0.01 (99% confidence). Knowing A and n, we can determine *significance levels* for r. If, for instance, the absolute value of r exceeds its 99% significance level, then H0 can be rejected with 99% confidence. Our experimental results show that in all cases the null hypothesis can be rejected with 99% confidence [8].

5.2. Sensitivity of Load Index to Job Size

Load indices output by comparator networks at different sites are compared in order to determine the least-loaded site. Given that load balancing seeks maximum speed-up over local execution, we can evaluate the new load-index functions by the speed-up attained if each incoming job is scheduled at the least-loaded site. Figure 4 shows the results for our 4-processor system. Test jobs were introduced on a synthetic workload 2252 time units into the experiment, each time at a different site. The completion times of all the jobs and their checkpoints were recorded. Two policies were compared: (i) *opt*, which places each incoming job at the site with the optimum (least) completion time; (ii) *min*, which schedules incoming jobs at the site with the smallest load index. Assuming no overhead of remote execution, we calculated, for each test-case, the speed-up over local execution achieved using these policies. Figure 4 plots these speed-ups against the time of completion; it shows that while the site with the least load index behaves as well as the optimal site for short jobs, its performance is suboptimal for

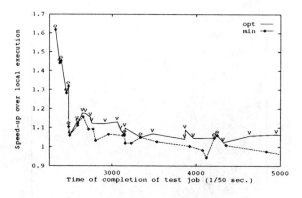

Figure 4. Performance under the worst-behaved load
Performance is computed for all test jobs under workload pattern 23; policies applied are *opt* and *min*.

long jobs. The letters on individual curves indicate the optimal site ('e' for elaine, and 'v' for 'vyasa'). The policy min runs all the jobs at elaine, the site with the least load index at the time of arrival; therefore, its performance is suboptimal only when a site other than elaine is optimal.

We evaluated min on all 10 jobs and all 24 load patterns. To relate job length with achieved speed-up, we created data-pairs $<l_i,e_i>$, where l_i is the length of the job and e_i, the corresponding speed-up. Figure 5 shows a contour plot of the probability of achieving certain speed-ups for jobs of certain length. The X-values represent speed-ups over local execution, and the Y-values, length of the job in seconds. Y-values range from 0.66 seconds to 439.94 seconds and, X-values, from 0.924524 and 4.92187. We used logarithmic scaling for the Y-axis. Nineteen contours, each connecting X-Y points having equal cumulative probability of speed-up, are shown; they divide up the space into twenty regions of equal (5%) probability. While speed-ups higher than 1.5 (more than 50% improvement over local execution) occur quite frequently for short jobs, they rarely occur for jobs more than 3 minutes long.

6. FINAL REMARKS

We have demonstrated automated learning of meaningful load-index functions from workload data. Using data collected by a synthetic workload generator, we trained neural networks to effectively compare alternative loading conditions for an incoming job. We collected workload and completion-time data on a network of 4 configurationally-distinct Sun 3 workstations, and performed off-line learning on a Sun Sparcstation 10, Model 20. The total down time was around 10 days; most of it was spent recording the 24 background loads and measuring completion times of 10 test jobs under these loads at each of the 4 sites. The training of comparator neural networks is fully automated; it requires no down time because it is performed off line (around 15 hours of CPU time).

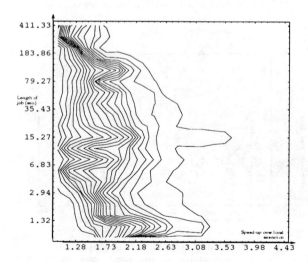

Figure 5. Policy: min; no overheads or delays.

REFERENCES

[1] K. Baumgartner and B. W. Wah, "GAMMON: A Load Balancing Strategy for a Local Computer System with a Multiaccess Network," *Trans. on Computers*, vol. 38, no. 8, pp. 1098-1109, IEEE, Aug. 1989.

[2] M. Berry et al., "The Perfect Club Benchmarks: Effective Performance Evaluation of Supercomputers," *International Journal of Supercomputing Applications*, vol. 3, no. 3, pp. 5-40, 1989.

[3] D. Ferrari, G. Serazzi, and A. Zeigner, *Measurement and Tuning of Computer Systems,* Prentice-Hall, Englewood Cliffs, NJ, 1983.

[4] D. Ferrari, "A Study of Load Indices for Load Balancing Schemes," pp. 91-99 in *Workload Characterization of Computer Systems and Computer Networks*, ed. G. Serazzi, Elsevier Science, Amsterdam, Netherlands, 1986.

[5] N. H. Goddard, K. J. Lynne, T. Mintz, and L. Bukys, Rochester Connectionist Simulator, Tech. Rep., Univ. of Rochester, Rochester, NY, Oct. 1989.

[6] K. Hwang, W. J. Croft, G. H. Goble, B. W. Wah, F. A. Briggs, W. R. Simmons, and C. L. Coates, "A UNIX-based Local Computer Network with Load Balancing," *Computer*, vol. 15, no. 4, pp. 55-66, IEEE, April 1982.

[7] M. L. Litzkow, M. Livny, and M. W. Mutka, "Condor - A Hunter of Idle Workstations," *Proc. 8th Int'l. Conf. Distributed Computer Systems*, pp. 104-111, IEEE, 1988.

[8] P. Mehra, *Automated Learning of Load Balancing Strategies for a Distributed Computer System*, Ph.D. Thesis, Dept. of Computer Science, Univ. of Illinois, Urbana, IL 61801, Dec. 1992.

[9] P. Mehra and B. W. Wah, "Adaptive Load-Balancing Strategies for Distributed Systems," *Proc. 2nd Int'l Conf. on Systems Integration*, pp. 666-675, IEEE Computer Society, Morristown, NJ, June 1992.

[10] P. Mehra and B. W. Wah, "Physical-Level Synthetic Workload Generation for Load-Balancing Experiments," *Proc. First Symposium on High Performance Distributed Computing*, pp. 208-217, IEEE, Syracuse, NY, Sept. 1992.

[11] G. Tesauro and T. J. Sejnowski, "A Parallel Network that Learns to Play Backgammon," *Artificial Intelligence*, vol. 39, pp. 357-390, Elsevier Science Pub., New York, 1989.

[12] S. Zhou, *Performance Studies of Dynamic Load Balancing in Distributed Systems*, Tech. Rep. UCB/CSD 87/376 (Ph.D. Dissertation), Computer Science Division, Univ. of California, Berkeley, CA, 1987.

Allocation of Parallel Programs With Time Variant Resource Requirements *

John D. Evans and Robert R. Kessler
Center for Software Science
University of Utah
Salt Lake City, Utah 84112
evans@cs.utah.edu and kessler@cs.utah.edu

Abstract: *This paper presents an innovative allocation strategy for parallel programs composed of communicating processes whose resource requirements vary during program execution. The primary difference between this and previous work is the incorporation into the allocation process of the estimation of variation in resource requirements over time. Experimental results are presented that compare this model with previous work and indicate that this method achieves superior allocation results without any increase in computational complexity. Results indicate a high correlation between the improvement achieved and the degree of process clustering in the parallel program.*

1. INTRODUCTION

The typical process information used in allocation reflects the computational or communication costs of each process. Parameters used may include the total amount of processing required by a process **process service load**, total communication load between process pairs, or the total degree to which service load demands of process pairs compete for processor time **interference**. In this paper the term **load** is used generically to refer to process service and communication loads. The **cost** of an allocation is viewed as a function of these load parameters and thus the allocation problem is formulated to find the allocation with the lowest **cost** [8, 1, 9]. Several researchers have concentrated on identifying significant process characteristics and generating parameters for minimization[2, 11].

The resource demands of a process can vary considerably during its lifetime due to a number of factors. These factors include dynamic memory allocation, synchronization between communicating process pairs caused by communication, precedence constraints between processes, or interaction with external devices. Another important factor is due to the typical byproduct of program structure which is clustering of processes into highly intercommunicating groups. When clustering occurs, it accentuates the degree of synchronization and precedence induced variation. In spite of the time dependent variation in process load, the load parameters used most often have been single values that represent the total load of each parameter type per process.

Several researchers have identified the dynamic nature of resource demands as a source of inefficiency in existing allocation approaches. Yan[3] notes that there may be poor correlation between a minimum load-cost allocation and a minimum run-time allocation. Chu, Lan and Hellerstein [11] have noted that

execution rates of modules and rates of intermodule communication in real-time monitoring systems are not uniformly distributed. They have observed that input monitor data largely determines these rates and have estimated peak module execution rates based on this. Hailperin[6] noted that in real-time systems, processor loading may exhibit complex and periodic patterns. Hailperin modeled loading as auto-regressive moving average random processes in order to forecast future load conditions. These approaches have been restricted to real-time systems. Lo[9] uses interference costs to characterize the degree of conflict that arises when two processes require the same resource simultaneously. Interference costs are modeled though, as the total sum of interference and do not capture the dynamic nature of interference. Dynamic allocation approaches may address the variation in resource demands at run-time by responding to fluctuations as they occur. The cost of dynamic response approaches, however, is often very high and these approaches only respond after problems have occured.

The degree of performance degradation due to resource demand fluctuations may be linked to the instantaneous loading on system resources. Performance suffers when there is contention among processes for system resources. No performance degradation may be seen, however, if processes never need resources at the same time, no matter what their total loads. If load parameters could be modeled in a time-sensitive manner, then the instantaneous loading of allocations could be estimated. This information may be useful in estimating peak-load periods, dynamic interprocess interference patterns or, potential idle periods when process migration becomes more attractive.

The resource utilization patterns of programs are allocation-dependent. The resource utilization pattern of one allocation may vary significantly from that of another allocation. A significant challenge is the estimation of resource requirements that reflects a reasonable cross section of efficient allocations. In the remainder of this paper we present an approach to the estimation of time variant resource requirements in an allocation independent manner. We show that many existing allocation methods can be adapted to use this approach and show significant performance improvement in so doing.

The remainder of this paper is organized as follows. Section 2 discusses the method of estimation of process load parameters. Section 3 studies the adaptation of existing strategies to use these parameters. Section 4 gives simulation performance results. Section 5 gives a summary and conclusions.

2. ESTIMATION METHOD

The representation of program resource load used

*This work is supported by The Hewlett Packard Corporation under the Mayfly project.

here is derived from precedence graphs. Precedence graphs are widely used in static **task** level allocation[7, 10, 4]. The program is represented as a directed acyclic graph. Graph nodes represent single threaded units of computations called **tasks**. Arcs represent data dependencies between tasks and are weighted to reflect the associated communication load parameters. Tasks may execute only when all data is present on input arcs and may produce output only when task execution is completed. Tasks are weighted to reflect associated load parameters. **Processes** are represented as chains of precedence related tasks. The arc weights, between tasks in the same process, reflect the cost of moving the process between processors.

The loadings and precedence constraints of the task precedence graph of a program, influence task relationships in several ways. The sequence of task execution within a program is only partially ordered by the precedence graph. Whenever intertask precedence constraints and communication costs allow more time for a task or chain of tasks to execute than is actually required by the tasks, **task shift** is possible. The load imposed on a processor by a task may be in conflict with another task only to the degree that the tasks compete for processor or system resources. If tasks can be shifted to allow execution at different times on the same processor, then no service-load conflict occurs. A high degree of task shift is not only possible, it may be necessary for effective program allocation. We refer to the length of potential task shift as the task's **range**.

In static task level allocation, heuristics related to the critical path heuristic have been found to perform very well[5]. These heuristics prioritize each task based on **exit path length** (epl) the longest computation path from the task to the end of the program. Each task is thus prioritized according to how critical it is to the total program run-time. The longest exit path length of a precedence graph is the **critical path length** (cpl). Tasks along the critical path are the most tightly constrained and have the least potential task shift.

The precedence relationships of the task graph determine which pairs of tasks can, and which can never, execute concurrently. Tasks in the transitive closure of a task node in the precedence graph, can never execute concurrently with it. Their loads are not considered to be in conflict.

We use an algorithm for estimating task interactions based on relaxation methods. The procedure RELAX_GRAPH (see Figure 1) produces an abstract program allocation that incorporates the concepts of task precedence, tightness of constraints (task criticality) and task range. The algorithm determines the available range for precedence related sequences of tasks. The sequence is averaged (or relaxed) simultaneously by dividing the available range for the entire sequence proportionally among all the tasks in the sequence. The algorithm repeatedly finds the path (or sequence) of unrelaxed tasks with the least potential shift and relaxes it. This process is repeated until all tasks have been relaxed. The tasks start time and end times are set to the beginning and ending of its relaxed range. The tasks load is then divided by the length of its relaxed range producing a task **load rate**.

```
Procedure: RELAX_GRAPH( G: task graph )
    SETQ, PATHQ, DELAYQ :priority queue
    T,C,P: task
    DONE :boolean
1 determine exit path length epl for each task
    determine the critical path
    mark critical path tasks and insert in SETQ
        (epl priority)
    set start and end times of critical path tasks
    for each non-critical task T: T_start(T)=0,
        T_end(T)=0, delay(T)=∞
    DONE = false
2 while not DONE
    2a loop SETQ entries { T = POP(SETQ) }
        for each unmarked child C of T
            T_start(C) = max( T_start(C), T_end(T))
            insert C in PATHQ (T_start priority)
        for each unmarked parent P of T
            if( delay(P) > T_start(T) - T_end(P))
                delay(P) = T_start(T) - T_end(P)
            remove P from DELAYQ if present
            insert P in DELAYQ (min. delay pri.)
    2b loop PATHQ entries { T = POP(PATHQ) }
        T_end(T) = T_start(T) + service_time(T)
        for each child C of T
            if C is marked
                delay(T) = min(delay(T), T_start(C) -
                    T_end(T))
                delay_time(T) = T_start(C)
                remove T form DELAYQ if present
                insert T in DELAYQ (min. delay pri.)
            else
                if( T_end(T) > T_start(C) )
                    T_start(C) = T_end(T)
                    backpath(C) = T
                    remove C from PATHQ if present
                    insert C in PATHQ (T_start priority)
    2c if DELAYQ is empty then DONE = true
        else
            T = POP(DELAYQ)
            RELAX_TIGHTEST_PATH( T, SETQ )
```

Figure 1: Path Relaxation Algorithm

Figure 1 outlines the path relaxation algorithm. SETQ is used to store the most recently relaxed path. DELAYQ stores the down stream ends of unrelaxed paths. PATHQ is used to propagate path lengths through the unrelaxed portions of the graph.

STEP 1: The critical path tasks are determined as in [5]. Task times are set to critical path position and tasks are inserted in SETQ.

STEP 2: Repeat until all paths are relaxed.

LOOP 2a: Process tasks in the most recently relaxed path. Outgoing branches (unmarked children) are updated with new start times and inserted in PATHQ. Incoming branches (unmarked parents) are ends of paths. The available delay is determined and these tasks are inserted in DELAYQ.

LOOP2b: Propagate constraints along unrelaxed paths and set task end times. Tasks with marked children are path ends, their delay times are up-

Procedure: RELAX_TIGHTEST_PATH(T, SETQ)
    weight = 0
    new_T_end = delay_time(T)
    TMP = T
    while(TMP is unmarked)
        weight += load(TMP)
        TMP = backnode(TMP)
    load_rate = weight / (weight + delay(T))
    TMP = T
    while(TMP is unmarked)
        mark TMP
        T_end(TMP) = new_T_end
        new_T_end -= load(TMP) / load
        load(TMP) = load_rate
        T_start(TMP) = new_T_end
        insert TMP in SETQ (epl priority)
        TMP = backnode(TMP)

Figure 2: Relaxation Algorithm

dated and they are inserted in DELAYQ. Unmarked children have their start times updated and are inserted in PATHQ. Backpaths are set to point to the most tightly constrained parent task.

STEP 2c: Select the most tightly constrained path in DELAYQ and relax this path.

Figure 2 outlines RELAX_TIGHTEST_PATH: The first loop traverses the path backward following **back-path** links and determines the total weight for the sequence. The second loop traverses the path and updates each tasks times and load rates. The path tasks are marked and inserted in SETQ.

Example: Figure 3(a) shows a three process task graph (similar to Gantt chart form) with task loads labeled. Tasks within each process execute in sequence (interprocess arcs may be omitted in this representation). The height of each task indicates the task's load rate. The length of each task indicates the length of task execution at the given load rate. The critical path is indicated by shading in Figure 3(a) (tasks a, h, i and g). These tasks are initially set in step 1. The path in the remaining graph with the least potential shift is (b, d) with shift of 30. Path (b, c, f) has shift of 40 and path (b, d, e, f) has 60. Figure 3(b) shows the graph after path (b, d) has been relaxed. Two more iterations are required for path (c, f) and path (e). Figure 3(c) shows the final graph state.

Discussion: The graph produced by RELAX_GRAPH will be used for allocation. Task loads are considered to overlap if the relaxed tasks overlap. Intertask conflict is estimated to be proportional to task load rates in the overlapping interval. The relaxation estimation approach models several heuristic goals:

1. It distributes work evenly through program execution by distributing available range through entire sequences of tasks.
2. Highly critical tasks maintain high load rates. The relationships between highly critical tasks maintain relatively fixed relationships. This is due to the relatively high constraints upon these tasks.

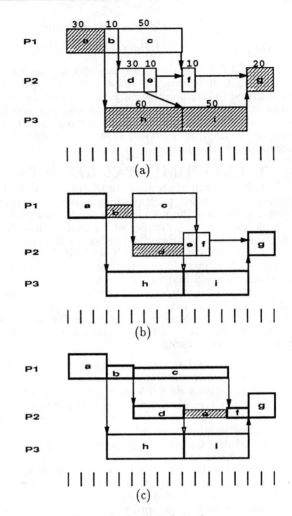

Figure 3: Path Relaxation

3. Less critical tasks have broader ranges, lower load rates and therefore contribute less to the total instantaneous load rate at any point.
4. Precedence constraints are never violated. Tasks constrained to execute in sequence, do not show up as concurrent in the load estimation.

3. ADAPTATION APPROACH

Many existing strategies address the allocation problem as the minimization of an objective function of the load parameters. We define parameter functions $Par_i(X)$ to return parameter i for process X. Many objective functions can then be specified as:

$$OF = f(Par_1, Par_2, ..., Par_n).$$

We propose the minimization of parameters over time intervals. For this purpose, the length of the relaxed graph is divided into a fixed number of discrete intervals. For each process a load value vector is computed with values for each parameter for each interval. Each task in the relaxed graph contributes its load parameters proportionally to the intervals that it overlaps. We define modified parameter functions $Par_{ij}(X)$ to return parameter i for process X in interval j. Two corresponding objective functions then are:

$$OF' = \sum_{k=0}^{maxint} f(Par_{1k}, Par_{2k}, ...Par_{nk})$$

$$OF'' = \max_{k=0}^{maxint}(f(Par_{1k}, Par_{2k}, ...Par_{nk}))$$

OF' and OF'' subsume OF as a special case where the number of intervals $maxint$ is 1. The interval values need to be computed only once for each process, prior to allocation. The allocation cost complexity is therefore unchanged. In the following section we discuss the minimum number of intervals required for acceptable performance.

4. EXPERIMENTAL RESULTS

· Extensive simulation tests were conducted using randomly generated precedence graphs. The experiments compared the performance of allocation algorithms with varying number of intervals. Objective functions of the type described in Section 3 were used. Single valued load parameter balancing is the case where $maxint$ is 1. The optimization algorithm was local search using first-improvement.

Properties of the precedence graphs were correlated against algorithm performance. The properties considered reflect the degree of clustering of the processes. In order to measure different graph properties, several graph types were used.

1. **Unclustered Processes:** The number of processes used was randomly selected. Pairs of tasks are iteratively added to the graph by randomly selecting a sender/receiver pair of processes and adding a task to each. Tasks and arcs are assigned weights randomly.

2. **Partially Clustered Processes:** Approach 1 is modified as follows: Processes are initially assigned to clusters. Sender/receiver pairs are selected using probability value C1 that sender and receiver are selected from the same cluster and probability C2 **damping** that the sending cluster will become inactive if receiver is in a different cluster. The ratio $\frac{1}{1-C1}$ is the average number of tasks in a period of cluster activity. Senders are selected only from active clusters. Inactive clusters become active when selected as a receiver.

3. **Fully Clustered Processes:** Approach 1 is used to form an initial graph. Each process is subsequently expanded into a cluster of processes. Each process task is expanded into a subgraph of its process cluster using approach 1. Clusters formed this way are fully synchronized by intercluster communication.

For each test case graph, processor configurations were varied from 2 to 10 processors making 9 configurations. FCFS task scheduling was used throughout. The average number of tasks per graph was 1200 and the number of processes ranges from 6 to 84.

Experiments 1 and 2 compared the performance of the allocation algorithm with varying number of intervals on unclustered graphs and on fully clustered graphs respectively. Task run-time is the only load parameter **L**. The objective function is:

$$OF1 = \sum_{k=0}^{maxint} \max_{Pe_a} \left\{ \sum_{proc_i \in Pe_a} L(proc_i, k) \right\}$$

For each configuration, allocations were made for each number of intervals **i**. For number of intervals n

Table 1. EXPERIMENTS 1 and 2

# Intervals	Unclustered (400 cases) % Improvement	Clustered (300 cases) % Improvement
2	0.74	9.08
4	1.67	15.24
8	2.57	17.63
16	3.19	18.70
32	3.44	19.38
64	3.52	19.78
128	3.50	19.93
256	3.52	19.99
512	3.49	20.01

the function $100*(1 - \frac{Time(n)}{Time(1)})$ was computed indicating the improvement seen using n intervals over the run-time using a single interval. The average improvements are shown in Table 1. Minimal improvement is seen on unclustered graphs, although there is always some improvement[1]. On fully clustered graphs a very significant improvement is seen. On both experiments, increasing the number of intervals increases performance. Increases above 32 intervals, however, show minimal difference in performance improvement.

Table 2. EXPERIMENT 3

C1 = 0.95		1 vs. 32 ints
CASES	C2	% Improvement
100	0.30	9.3
100	0.50	9.4
100	0.70	9.6
100	0.80	11.2
100	0.85	12.2

Table 3. EXPERIEMENT 4

C2 = 0.85			1 vs. 32 ints
CASES	C1	Period Size	% Improvement
100	0.55	2.22	3.96
100	0.65	2.85	4.94
100	0.75	4.00	4.29
100	0.85	6.67	6.72
100	0.90	10.00	7.53
100	0.92	12.50	9.52
100	0.94	16.70	11.97
100	0.96	25.00	13.17
100	0.98	50.00	18.54

Experiments 3 and 4 used the same objective function with partially clustered graphs consisting of 4 clusters of 12 to 20 processes. In experiment 3, C1 = 0.95 and C2 was varied from 0.3 to 0.85. The results in Table 2 indicate that variation in damping has little impact below 0.7, but corresponds to significant differences above that level. In experiment 4, C1 was varied from 0.55 to 0.98 while C2 was held at 0.85. The results in Table 3 indicate a dramatic correspondence between cluster activity periods and the impact of interval load balancing. Graphs with higher degree of clustering are indicated by a higher C2 and longer periods of cluster activity C1. Experiments 1 through

---

[1]Short graphs (not shown) under 500 nodes have less homogeneous work distribution and produce more significant improvement.

4 indicate that when clustering is high, dramatic improvement is seen using interval load balancing.

Table 4. EXPERIMENT 5

CASES	R	1 int	32 ints
		Avg. % Above Best Time	
180	0.25 - 0.30	231.2	172.8
180	0.35 - 0.40	230.1	60.1
180	0.45 - 0.50	225.9	16.0
180	0.55 - 0.60	201.0	5.4
180	0.65 - 0.70	111.3	3.1
180	0.75 - 0.80	66.6	3.2
180	0.85 - 0.90	45.7	3.2
180	0.95 - 1.00	12.2	3.4

In experiment 5, we considered the impact of communication costs. Partially clustered graphs were used: $C2 = 0.85$ and $C1 = 0.98$. For the simulated multiprocessor system, communication delays were dependent on instantaneous communication loading. Processors continue processing while messages are in transit. Average message size is equal to the average task load. The delay for message transfer $Delay = D * K1 * \frac{K2-M}{K2}$. Where the D = size of the message, M = number of messages in the system, $K2 = 10.0$ and $K1 = 1.75$ indicating moderately low impact from communication costs. The objective function OF2 minimizes the sum of maximum processor loading plus total communication load:

$$OF2 = \sum_{k=0}^{maxint}\{(R * Lfn) + ((1 - R) * Cfn)\}$$

Where:
$$Cfn = \sum_{p_i,p_j \in Procs}(Prd(p_i, p_j) * C(p_i, p_j, k))$$
$$Lfn = \max_{Pe_a}(\sum_{p_i \in Pe_a} L(p_i, k))$$

$Prd(P_i, P_j)$ returns 1 if the tasks are assigned different processors, and 0 otherwise.

$C(P_i, P_j, k)$ returns the communication load between P_i and P_j in interval k.

The relationship between computation and communication load is nonlinear due to the dependence on instantaneous loading making the optimal coefficient **R** graph dependent. **R** was varied from 0.3 to 1.0 for testing. Table 4 indicates the average percentage increase above the best time achieved using any **R** setting: $100 * (\frac{Time-BEST}{BEST})$. The results indicate that the single-valued load balance is relatively unresponsive to communication cost and performs best when the coefficient is near 1.0. In contrast, interval load balancing showed a significant improvement and a broad range of response indicating low susceptibility to variation in the **R** coefficient.

5. SUMMARY AND CONCLUSIONS

This paper presents an innovative allocation strategy for parallel programs composed of communicating processes whose resource requirements vary during program execution. The primary difference between this and previous work is the incorporation of the estimation of variation in resource requirements into the allocation process. A model for the estimation of time variant resource requirements was introduced. We showed that this load estimation model is easily adapted to many existing load balancing approaches. Experimental results were presented that

compare this model with single-valued load balancing and indicate that this method achieves superior allocation results. The results also indicated a significant tolerance for inaccuracy in the objective functions when compared with single-valued load balancing. The results indicate a high correlation between the improvement achieved and the degree of process clustering in the parallel program.

REFERENCES

[1] E. HADDAD. "Partitioned Load Allocation for Minimum Parallel Processing Execution Time". In *Proceedings of the International Conference on Parallel Processing* (1989), pp. II–192–II–199.

[2] E. WILLIAMS. "Assigning Processes to Processors in Distributed Systems". In *Proceedings of the International Conference on Parallel Processing* (1983), pp. 404–406.

[3] J. C. YAN. "New 'Post-Game Analysis' Heuristics for Mapping Parallel Compuataions to Hypercubes". In *Proc. of the Intl. Conference on Parallel Processing* (1991), pp. II–236–II–242.

[4] J. EVANS, AND R. KESSLER. *A Communication-Ordered Task Graph Allocation Algorithm*. Tech. Rep. UUCS-92-026, Computer Science Department, University of Utah, 1992. Submitted to: IEEE Transactions on Parallel and Distributed Systems.

[5] L. HO, AND K. IRANI. "An Algorithm For Processor Allocation in a Dataflow Multiprocessing Environment". In *Proc. of the Intl. Conference on Parallel Processing* (1983), pp. 338–340.

[6] M. HAILPERIN. "Load Balancing for Massively-Parallel Soft-Real-Time Systems". In *2nd Symposium on the Frontiers of Massively Parallel Computations* (1988), pp. 159–163.

[7] R. L. GRAHAM. "Bounds on Multiprocessing Timing Anomalies". *SIAM Journal on Applied Mathematics 17*, 2 (March 1969), 416–429.

[8] V. CHAUDHARY, AND J. AGGARWAL. "Generalized Mapping of Parallel Algorithms Onto Parallel Architectures". In *Proceedings of the International Conference on Parallel Processing* (1990), pp. II–137–II–141.

[9] V. LO. "Heuristic Algorithms for Task Assignment in Distributed Systems". In *Proceedings of the International Conference on Parallel Processing* (1984), pp. 30–39.

[10] V. SARKAR, AND J. HENNESSY. "Compile-time Partitioning and Scheduling of Parallel Programs". In *Proc. of the SIGPLAN '86 Symposium on Compiler Const.* (1986), ACM, pp. 17–26.

[11] W. CHU, M. LAN, AND J. HELLERSTEIN. "Estimation of Intermodule Communication (IMC) and Its Applications in Distributed Processing Systems". *IEEE Transactions on Computers C-33*, 8 (August 1984), 691–699.

Impact of Data Placement on Parallel I/O Systems*

J. B. Sinclair, J. Tang, P.J. Varman, B. R. Iyer [†]

Abstract

The I/O performance of several concurrent external merge jobs sharing a parallel I/O system is studied. The placement of the runs is found to have a significant impact on performance. Placements leading to serialization among jobs were identified and analyzed, and solutions discussed. Contrary to the behavior of a single merge, increasing the buffer size in this situation may actually degrade performance.

1 Introduction

This paper studies the impact of data placement on the performance of a parallel I/O system. The workload consists of a number of independent and concurrent external-merge [3] jobs. The problem of deciding on which disks the data of different jobs should be placed arises frequently in high-performance database management systems executing multiple, concurrent sort queries. The results show that external merge sorting programs can achieve significant speedup through better data layout schemes, and demonstrate the importance and relevance of providing this capability to application processes. We investigate how the placement of the input and output runs of each merge job affects the disk utilizations, and the job completion times.

The reader is referred to [1, 10, 5, 8, 7, 6, 11] for external merging methods for a single job, using multiple disks. Here we study the interaction of several concurrent merge jobs in a parallel I/O system.

2 Run Placement Policies

In this section, we compare the performance of data layout policies for multiple, independent merge jobs. We observe that the dynamic interaction of identical, symmetrically placed jobs can sometimes lead to a race among them, resulting in significant disk serialization. A model of such racing behavior is analyzed in Section 3.

2.1 Simulation Model

The workload consists of J identical merge jobs. The input of a job consists of several sorted runs, and its output is a single sorted run. We assume (in keeping with current high-performance database systems like DB2, SQL/DS, etc.) that each run is placed entirely on a single disk without striping.

*Partially supported by a grant from IBM Corporation and by NSF and DARPA Grant CCR 9006300.

[†] The first three authors are with the ECE Department, Rice University, Houston, TX 77251. B. Iyer is with IBM, DBTI, P.O. Box 49023, San Jose, CA, 95161.

The system model consists of CPUs, a set of independent disks and a disk cache. The merge is simulated using the random-block-depletion model [4]. The block to be depleted is chosen with uniform probability from all non-empty runs. The CPU depletes this block and requests the next one from the same run. If this block is in the cache the request is immediately satisfied; else the CPU is blocked until the I/O for the block is completed. Reads are done in units of several physically contiguous blocks (N), to amortize disk latency. Anticipatory prefetching is employed. A read of N blocks is initiated when the first block of the *previously fetched* N blocks is processed. For each depleted block, a write block is generated; when M blocks have been buffered, they are written to disk in one I/O operation. If a free cache block is unavailable when requested, a job waits until one is released. To focus on the I/O performance, we assume infinite-speed CPUs here. A linear seek model with seek time of 0.04 ms/cylinder, average rotational latency of 8.33 ms, and block transfer time of 1.024 ms/block are used. A job has 20 runs of 1000 blocks each. The read and write blocking factors are $N = 12$ and $M = 40$, respectively. Our study was carried out using the YACSIM simulation package [2].

2.2 Performance of Placement Policies

The four data layout policies studied are shown in Figure 2.1 for the case of $J = 2$ and $D = 5$. The input runs of a job are indicated by Read and its output run by Write. Disks containing only input (output) runs are called *read (write) disks*; disks with both input and output runs are called *shared disks*. No disk will hold more than one output run. To utilize all disks efficiently the I/O load should be divided as equally as possible among the D disks. Since every output run is placed on a separate disk, we need only consider the placement of the input runs.

- **Case 1**: *Dedicated Write Disk for Each Job*. Disk k, $0 \le k \le J-1$, is used exclusively for the output run of job k; the input runs of each job are spread evenly among the remaining $D - J$ read disks.

- **Case 2**: *Intra-job Separate Read and Write Disks* Job k, $0 \le k \le J - 1$, uses disk k for its output run, and its input runs are spread evenly among the remaining $D - 1$ disks.

- **Case 3**: *Intra-job Shared Read and Write Disks* This allocation is obtained by permuting the allocation of Case 2, such that the input runs of job k, $0 \le k \le J-1$, are moved from disk $(k-1) \bmod J$ to disk k.

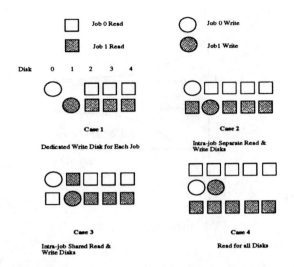

Figure 2.1: Run placement policies, D = 5 and J = 2.

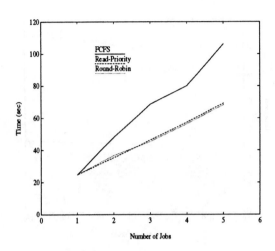

Figure 2.3: Completion time with different disk scheduling policies

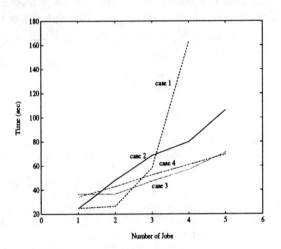

Figure 2.2: Completion time of all jobs

- **Case 4:** *Read from All Disks* A job uses all D disks for input and one disk for its output run.

Case 1 is motivated by the fact that a single job is write bound; hence an output disk is not loaded further. Figure 2.2 shows that as J increases, and the system becomes input-bound, the performance degrades steeply. This allocation is suitable if it is known in advance that there will be only a small ($\leq D/2$) number of jobs. Case 2 has the advantage that a job's allocation is independent of other jobs, and can therefore be easily implemented in the case of staggered job arrivals. Note that the increase in the time for increasing J *cannot* be accounted for by just the increased load on a disk. Despite symmetry in the run placement jobs progress at different rates, with a significant amount of serialization among all jobs. The race conditions occur because the data placement strategy allows a large number of writes from a single job to be queued together, effectively blocking other jobs which

attempt to read from the same disk. We present solutions for the race are in Section 2.3, and study an analytic model in Section 3. In Case 3, the input runs of the jobs are placed so that every write disk also contains input runs from the *same* job. This data placement scheme prevents a job from racing ahead, since the effect of its queued writes would slow down its own reads as well. Figure 2.2 shows the significant performance improvement of Case 3 over that of Case 2. A disadvantage of Case 3, however, is that the number of jobs must be known prior to laying out data on the disks. In Case 4 all input runs are distributed evenly on the set of D disks. Because each disk has some number of input runs from each job, no race can develop. The performance is consistent, albeit suboptimal, over the entire range.

2.3 Race Condition and Solutions

While appropriate run placement can sometimes be used to eliminate the race condition (Case 3), global constraints (staggered arrivals) may make the method unsuitable. We discuss disk-scheduling and buffer management policies to control this behavior.

One mechanism is to use either a Round Robin or Read Priority disk scheduling policy. These policies allow reads to get service even though there may be a large backlog of pending writes at that queue. Figure 2.3 shows that both policies are very effective in improving the performance. Simulation data also suggests that the performance is quite insensitive to the buffer size.

A second mechanism for avoiding races is to partition the cache and give each job a fixed, limited amount of buffer space. When a job fills up its allocated space it automatically must wait until some of its writes are serviced. Figure 2.4 shows something quite unexpected; that is, increasing buffer size by even a small amount may degrade performance significantly. This phenomenon is contrary to our expectation that more buffer space improves performance.

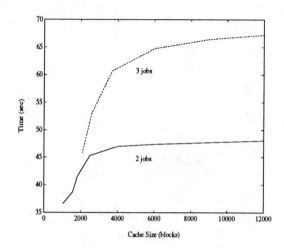

Figure 2.4: Completion time versus Buffer Size.

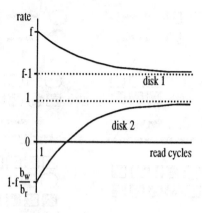

Figure 3.5: $f > 2$

3 Analysis of Race Conditions

The observed race condition occurs even though the run placement strategy balances the read and write demands equally among all of the disks. In this section, we present simple analytic models of concurrent merge tasks that exhibit race conditions. We show that the race conditions occur because the data placement strategy allows a large number of writes from a single task to be queued together, effectively blocking other tasks which attempt to read from the same disk. The analysis permits us to predict potential races based on I/O system parameters, including the number of blocks read or written at a time, and the average access time per block for a read and a write.

We first analyze a simple model of a two-disk, two-job system which may exhibit a race condition, depending on how many blocks are read or written on each disk access, as well as other factors. The race condition causes the two jobs to be effectively serialized, with one finishing far behind the other.

Consider a system consisting of two disks, d_0 and d_1, and two jobs, j_0 and j_1. j_0 reads from d_0 and writes to d_1, while j_1 reads from d_1 and writes to d_0. Scheduling at each disk is FCFS. Jobs block on reads but not on writes. To simplify the analysis, we assume times for reads and writes that are functions only of the number of blocks read or written at one time. Let

$$
\begin{aligned}
b_r &= \text{read blocking factor} \\
b_w &= \text{write blocking factor} \\
r &= \text{read time per block} \\
w &= \text{write time per block} \\
f &= \tfrac{w}{r}
\end{aligned}
$$

Assume that at time $t = 0$ each disk queue contains one read job, d_0 has no queued write demand, and d_1 has N_0 total write service demand. Let T_1 be the time required for d_1 to service the demand queued at $t = 0$. Then

$$T_1 = N_0 + rb_r$$

Let N_1 be the total write service demand that arrives at d_1 during the interval $(0, T_1)$. During $(0, T_1)$, d_1 services only one read request. As soon as j_1 has this read request serviced, it sends $\frac{b_r}{b_w}$ write requests to d_0. The rate at which d_0 completes reads is $\frac{[T_1 - wb_r]}{T_1} \frac{1}{rb_r}$. The total write service demand N_1

$$
\begin{aligned}
N_1 &= T_1 \frac{[T_1 - wb_r]}{T_1} \frac{1}{rb_r} \frac{b_r}{b_w} wb_w \\
&= N_0 f + wb_r(1 - f)
\end{aligned}
$$

For the system to be stable, $N_1 \le N_0$, or $N_0(1 - f) \ge wb_r(1 - f)$. Thus, the conditions for stability are $f \le 1$ and $wb_r < N_0$ or $f > 1$ and $wb_r > N_0$. wb_r, the critical threshold for N_0, is the amount of write demand generated for each read completed.

This race condition arises because j_0, by writing to d_1, interferes with the only mechanism that can slow j_0, that is, writes by j_1 to d_0. This suggests that the race condition might not exist in a system with more jobs and disks, so that j_0's writes to a disk do not interfere directly with the job which writes to d_0. However analysis shows that, while adding another disk and another job does change the system behavior, a race condition may still exist, but it will manifest itself in an markedly different way.

Consider a system with three disks, d_0, d_1, and d_2, and three jobs, j_0, j_1, and j_2. j_0 reads from d_0 and writes to d_1, j_1 reads from d_1 and writes to d_2, and j_2 reads from d_2 and writes to d_0. As before, scheduling is FCFS, and jobs block only on reads. Assume that the queues for d_0 and d_1 contain no writes and one read each, and d_2's queue contains N_0 total write demand (from j_1) and one read (from j_2).

A *read cycle* is the interval during which the initial demand in d_0's queue is serviced. In [9] we show that at the end of the read period, d_0's queue still contains a single read (perhaps partially completed), d_2's queue contains a single read (just beginning service), and d_1's queue contains a read plus a number of writes corresponding to a total write demand of N_1. This is similar to the state in which the read period began, but with each disk ready to play a different role in a

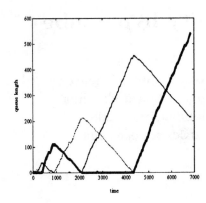

Figure 3.6: Simulation results for 3 disks, 3 jobs, f = 3

new read period just beginning.

The graph in Fig. 3.5 illustrates one (of three) possible scenarios for the sequence of values $\{N_0, N_1, N_2, \ldots\}$, the total write demand in the backlogged queue at the beginning of successive read periods. The curve labeled disk 1 (disk 2) shows the rate at which the service demand changes at the disk with increasing (decreasing) queued service demand during a read period. For $f > 2$ (Fig. 3.5), the two rate curves do not intersect. Consequently, regardless of N_0, N_1 will be greater than N_0, and when N_0 is large enough, $N_1 \approx (f-1)N_0 > N_0$. For $2 > f > 1$ the rate curves intersect and hence the system has a steady-state behavior in which the lengths of the read periods are all approximately the same. For $f < 1 - f\frac{b_w}{b_r}$, the two rate curves never intersect, and the rate at which N_0 is depleted is always greater than the rate at which the backlog increases on d_1. Consequently, queues are always quite short.

Simulation of a similar model with uniformly distributed rotational latencies gives results which agree closely with this deterministic analysis. Fig. 3.6 illustrate this for $f = 3$. The y-axis is the number of writes in a disk queue. The solid, dotted, and bold lines correspond to d_0, d_1, and d_2, respectively. Fig. 3.6 corresponds to the situation shown in Fig. 3.5. Successive peaks eventually grow by a factor of 2 as predicted.

During a read period one job receives virtually all the service capacity of its read and write disks, one job uses the remaining disk almost 100%, and the remaining job is essentially blocked. In the next read period, this last job receives nearly 100 % of the service from its read and write disks, and so forth. The jobs alternate taking the lead in the number of reads completed, and the job that finishes first is determined by how this sequence first started and the total number of reads to be performed.

4   Summary

When several concurrent external merge jobs share a parallel I/O system, the placement of the runs on the disks has a significant impact on the I/O performance.

The difference in the time for reads and writes (due to differing blocking factors or input parallelism) may result in a race among the jobs. resulting in disk serialization and a corresponding performance penalty. Contrary to the behavior of a single merge, increasing the buffer size in this situation may actually degrade performance. Analytical models for the race condition are presented. Solutions based on run placement, buffer control, and disk scheduling policies are shown to be effective.

References

[1] A. Aggarwal and J. S. Vitter. The Input/Output Complexity of Sorting and Related Problems. *Comm. ACM*, 31(9):1116–1127, 1988.

[2] J. R. Jump. *YACSIM Reference Manual, Version 1.1*. Electrical and Computer Engineering, Rice University), April, 1992.

[3] D. Knuth. *The Art of Computer Programming. Volume 3: Sorting and Searching*. Addison-Wesley, 1973.

[4] S. C. Kwan and J. L. Baer. The I/O Performance of Multiway Mergesort and Tag Sort. *IEEE Transactions on Computers*, 34(4):383–387, 1985.

[5] M. H. Nodine and J. S. Vitter. Large-Scale Sorting in Parallel Memories. In *Proc. 1991 ACM Symposium on Parallel Algorithms and Architectures*, pages 29–39, 1991.

[6] M. H. Nodine and J. S. Vitter. Optimal Deterministic Sorting in Large-Scale Parallel Memories. In *Manuscript(To appear in ACM SPAA 92)*, 1992.

[7] V. S. Pai, A. A. Schäffer, and P. J. Varman. Markov Analysis of Multiple-Disk Prefetching Strategies for External MergeSort. In *Proc. Int. Conf. on Parallel Processing*, 1992.

[8] V. S. Pai and P. J. Varman. Prefetching with Multiple Disks for External Mergesort: Simulation and Analysis. In *Proc. 8th Intl. Conference on Data Engineering*, pages 273–282, 1992.

[9] J. Sinclair, J. Tang, P. Varman, and B.Iyer. Performance Study of Concurrent I/O. Technical Report TR 9209, Rice University, ECE, July 1992.

[10] J. S. Vitter and E. H. A. Shriver. Optimal Disk I/O with Parallel Block Transfer. In *Proc. 1990 ACM Symposium on Theory of Computing*, pages 159–169, 1990.

[11] L. Q. Zheng and P.-A Larson. Speeding Up External Mergesort. Technical Report TR CS-92-40, University of Waterloo, Computer Science, August 1992.

Prefix computation on a faulty hypercube *

C. S. Raghavendra
School of EECS
Washington State University
Pullman, WA 99164
raghu@eecs.wsu.edu

M. A. Sridhar
Dept. of Computer Science
University of South Carolina
Columbia, SC 29208
sridhar@cs.scarolina.edu

S. Harikumar
School of EECS
Washington State University
Pullman, WA 99164
hsivaram@eecs.wsu.edu

Abstract

The fundamental question addressed in this paper is that of computing the parallel prefix operation. In particular, we study the problem of performing such an operation in an n-dimensional SIMD hypercube, Q_n, with upto $n-1$ node faults. In an SIMD hypercube, during a communication step, nodes can exchange information with their neighbors only across a specific dimension. We exhibit an $n + 5 \log n$ algorithm for this problem. The development of the algorithm is based on the existence of two so-called *free* dimensions in such a faulty hypercube [6].

1 Introduction

The hypercube is a regular and symmetric structure that has proved very popular for parallel processing applications. An important area of research is to obtain efficient algorithms for parallel processing applications that operate gracefully even when some of the nodes of the cube fail. Such algorithms are very useful in mission-critical environments including medicine and space exploration.

Fault tolerance in machines such as hypercubes is important in order to achieve sustained high performance computing. Therefore, it is necessary to compute important primitive functions even in the presence of faults. The hypercube network is quite robust [1, 8]; in fact, at least n faults are needed to disconnect Q_n into 2 components. The symmetry and robustness of hypercube can be exploited to compute many functions efficiently even when there are about n faults in the systems. Several researchers have developed algorithms for solving a wide range of different problems in the presence of processor and/or communication link faults [1, 3, 7, 2]. Many of these papers, however, exhibit results that are either restricted to

handling a small number of faults, or else incur significant performance degradation.

This paper addresses the particular problem of computing the parallel prefix [5]. The parallel prefix computation problem is a fundamental one, and arises very frequently in many parallel processing algorithms [4]. The computing model we consider is an n-dimensional hypercube Q_n containing upto $n-1$ node faults. The algorithms operate under the assumption that as far as the non-faulty nodes are concerned, the value in a faulty node is invalid or zero; but faulty nodes cannot be used for any kind of communication. In this model, we develop two algorithms for performing parallel prefix computation on an n-dimensional SIMD cube. The first algorithm uses $3n-2$ time steps, and the second uses $n + 5 \log n - 4$ time steps. Both algorithms make use of the existence of so-called *free* dimensions, the properties of which have been established elsewhere [6].

2 Background and notation

We will use the term *k-cube* to mean a k-dimensional hypercube. We will use the term *k-subcube* to mean a k-dimensional subcube of some larger cube.

Every subset $S \subseteq \{1, 2, \ldots, n\}$ of the dimensions of Q_n induces a collection of $2^{n-|S|}$ $|S|$-subcubes, each obtained by varying the bits of the dimensions in S over all possible values while fixing the rest of the bits at an arbitrary value. But when given a particular node $x \in Q_n$ and a particular subset $S \subseteq \{1, 2, \ldots, n\}$ of dimensions, there is a unique $|S|$-subcube induced by S containing x. We will denote this subcube by $x \langle S \rangle$. For example, given a node $x \in Q_5$, we denote by $x \langle 2, 4, 5 \rangle$ the 3-subcube induced by the dimensions $2, 4$, and 5 containing x.

It is well known that the hypercube Q_n can be viewed as the Cartesian product of the single edge (Q_1) with itself, taken n times. We will occasionally

*This research is supported in part by the NSF Grants No. MIP-9296043 and MIP-9103086.

find it convenient to take a viewpoint related to this, i.e., that Q_n can be thought of as the Cartesian product $Q_k \oplus Q_{n-k}$, for any k in the range $1 \le k \le n-1$. This essentially means viewing Q_n as being an $(n-k)$-dimensional cube, each of whose nodes is really a "supernode" that contains a k-dimensional cube.

Given a specific set D of dimensions of Q_n, i.e., $D \subseteq \{1, 2, \ldots, n\}$, we denote by Q_n/D the $(n - |D|)$-dimensional cube that uses the dimensions not in D, whose supernodes use the dimensions in D. We will say that the dimensions in D are the *internal* dimensions (used within the supernodes), and those not in D are the *external* ones.

Here we point out a simple fact that we subsequently use.

Lemma 1. In Q_n with at most $n-1$ faults, given any set H of $n - \log n$ of the dimensions of Q_n, there exists a $(n - \log n)$-dimensional subcube of Q_n whose edges use the dimensions in H that is entirely fault-free.

Proof. There are $2^{\log n} = n$ such subcubes, and since we have only $n - 1$ faults, one of the subcubes must be fault-free. □

It is straightforward to generalize this result to larger fault set sizes; the above result suffices for our purposes.

We will also need the notion of *free* dimensions. Free dimensions were first introduced by [6], and detailed results about them can be found in that paper.

Definition. A dimension d of an n-dimensional cube Q_n containing faults is said to be *free* if no edge along dimension d has both its end-points faulty.

Theorem [6]. In any n-dimensional hypercube containing at most $n-1$ node faults, there exist at least two free dimensions. Furthermore, every 2-subcube induced by these two dimensions contains at most one faulty node.

The problem of prefix computation is the following. Given a hypercube Q_n in which each node i contains a certain initial value v_i, and given a binary operation defined on these values, each node needs to compute the value

$$\sum_{0 \le j \le i} v_i.$$

This problem is of fundamental importance in parallel processing; numerous algorithms exist in the literature that make extensive use of prefix computation.

3 Prefix computation in a faulty hypercube

In this section, we develop the algorithms claimed earlier. First consider the case where there are no faulty nodes. A simple algorithm that accomplishes this task is the following.

Algorithm A.

for $i \leftarrow 1$ to n do

(1) **for** every node $v = b_n b_{n-1} \ldots b_1$ **do**

(1a) $t \leftarrow v^{(i)}.sum$, where $v^{(i)}$ is the neighbor of v along dimension i
(1b) **if** $b_i = 1$ **then** $v.value \leftarrow v.value + t$
(1c) $v.sum \leftarrow v.sum + t$

end

end

This algorithm executes one iteration corresponding to each dimension of the cube. The invariant maintained by this algorithm is that after the iteration on dimension i, every node v with label $b_n b_{n-1} \ldots b_2 b_1$ has $v.sum$ equal to the sum of the values of all the nodes in the subcube denoted by $b_n b_{n-1} \ldots b_{i+2} b_{i+1} * * \cdots *$. It is easy to see that the algorithm uses n parallel time steps for execution.

Algorithm B. Now consider the case where we have upto $n - 1$ faulty nodes in the cube. The final values obtained by faulty nodes are of no significance; however, during the execution of the prefix computation, other non-faulty nodes may rely on faulty nodes for the correctness of intermediate values. Therefore, we will need to assign the role of each faulty node to some other non-faulty node in the network to ensure correctness of the final result.

One simple assignment would be to use a free dimension d of the network and to require, for each faulty node w, the neighbor $w^{(d)}$ of w along dimension d to perform its own computation as well as the computation of w. With this arrangement, it becomes necessary to re-route data intended for w to $w^{(d)}$ during the intermediate steps of the algorithm. This can be achieved, for example, by modifying step 1a of the above algorithm.

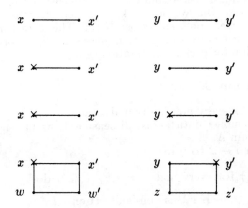

Figure 1: Four possible fault scenarios

After any given iteration of the above algorithm on dimension i, the only nodes that may have invalid values stored are the ones whose neighbors along dimension i are faulty. Let x be such an "invalid" node, and y its faulty neighbor along dimension i.

Let d_1 and d_2 represent the two free dimensions of the hypercube. Note that after execution of the iteration, the 2-subcube $x\langle d_1, d_2\rangle$ contains at most one fault, as does the 2-subcube $y\langle d_1, d_2\rangle$. Suppose x' and y' are, respectively, the neighbors of x and y along dimension d_1. Now there are four possible scenarios, shown in Figure 1. Three of these have at least one invalid node.

In the first case, there are no invalid nodes; so a single exchange across dimension i preserves correctness of data. In the second case, x is faulty, and its role is taken by x'; so y's data should be routed to x'. This can be done via y' using two extra steps along dimensions d_1 and i.

In the third case, both x and y are faulty, and their roles are taken by x' and y' respectively. The correction requires just a single exchange along dimension i.

The last case involves a data transfer using an additional free dimension, i.e., dimension d_2. Here, x and y' are faulty, so that data must be routed between x' and y. This can be done using the sequence of dimensions d_2, d_1, i, d_2. Since there are two pieces of data to be routed between these two nodes, this correction could take as many as eight time steps.

In the worst case, the total execution time will be $n + 8n = 9n$ steps.

For Algorithms C and D, we will assume the restriction $d_1 = 1$ and $d_2 = 2$. After presenting these two algorithms, we will show how to remove this restriction and deal with the general case.

With this restriction, can do significantly better than Algorithm B above. We can exploit the fact that the valid sum values of the nodes in a given 2-subcube C at the end of iteration i of Algorithm A will be identical to the sum values in the nodes in C's image along dimension i. Furthermore, any such subcube C has at most one fault, and therefore at most one invalid node; thus we can render this node valid by copying from one of the valid nodes within C, along dimension 1 or dimension 2. But we do not know which of these two is the correct one, so we use both. This uses three data transfers for each step, giving us a $(3n - 2)$-step algorithm for prefix computation:

Algorithm C.

Mark every non-faulty node 'valid.'
Exchange data across dimension 1, as in Algorithm A.
for $i \leftarrow 2$ **to** n **do**

(1) **for** every node $v = b_n b_{n-1} \ldots b_1$ **do**

    if $v^{(i)}$ is non-faulty **then**

        $t \leftarrow v^{(i)}.sum$, where $v^{(i)}$ is the neighbor of v along dimension i

        $v.sum \leftarrow v.sum + t$
        Mark v valid

    endif
    {At this point, if v is invalid, then one of its neighbors across dimensions 1 or 2 will have a valid sum. }
    if $v^{(i)}$ is faulty **then**

        $t \leftarrow v^{(1)}.sum$;

        $t \leftarrow v^{(2)}.sum$;

        { If either of these two is a faulty node, the copied value is discarded. If they are both non-faulty, they will both have the same sum value.}

    endif
    if $b_i = 1$ **then** $v.value \leftarrow v.value + t$

  end

end

Algorithm D. A significant improvement in total time by using the following technique. Imagine an algorithm that executes in three phases. The first phase is merely an execution of the first $\log n$ iterations of Algorithm C. At the end of this phase, every node y remembers the $value$ it has computed so far, in $x.value1$. After this phase, of Algorithm C, every node x is part of the $(\log n)$-subcube all of whose nodes have completed computation within C_x. Since $Q_n/\{\log n + 1, \log n + 2, \ldots, n\}$ contains $n = 2^{\log n}$ supernodes, at least one supernode must be fault-free. Each such supernode contains $2^{n-\log n}$ nodes.

The second phase comprises of executing iterations $\log n + 1$ through n of Algorithm A augmented with a marking step: during the phase, if a node exchanges data with a faulty or invalid node, it marks itself invalid. Any node that was never involved in such an exchange during the second phase qualifies to be part of a fault-free supernode. After these iterations, every node x of a fault-free supernode C will have attained the correct sum value for itself; that is, the value of $x.sum$ is the sum of the initial values of *all* nodes in C with address values smaller than x. Also, $x.value$ is the correct final value for x.

Now if y is any node in the subcube $x\langle 1, 2, \ldots, \log n\rangle$, the correct sum value for y would be obtained by adding $x.sum$ to $y.value1$. Therefore, the third phase is simply a broadcast of $x.sum$ to all the non-faulty nodes within $x\langle 1, 2, \ldots, \log n\rangle$; this can be accomplished by merely using the data transfer steps of Algorithm C. Note that in making this claim, we make use of the fact that the two free dimensions are dimensions 1 and 2, so that it is possible to complete this broadcasting phase in $3 \log n - 2$ time steps.

It is easy to see that the total time taken by all three phases is $(3 \log n - 2) + (n - \log n) + (3 \log n - 2) = n + 5 \log n - 4$.

We now sketch briefly how to relax the restriction that the free dimensions must be dimensions 1 and 2.

If we are given free dimensions d_1 and d_2, we can view the set of all dimensions as being re-ordered so that the dimensions 1 and 2 are exchanged with dimensions d_1 and d_2. This means that the processor numbers are reassigned within 4-subcubes induced by the dimensions $1, 2, d_1$ and d_2. Therefore, we can simply use one of the above algorithms for the prefix computation, and remap the final results to the correct processor numbers using a constant number of additional steps. Thus the time bounds attained by our algorithms remain essentially the same, to within an additive constant.

4 Conclusions

We have considered the problem of computing the parallel prefix on a faulty hypercube with $n-1$ faults, and have shown that it is possible to perform this operation with very little slowdown relative to the non-faulty case.

Further work involves applying these techniques to design algorithms for other commonly-used global operations, including selection and sorting, on faulty hypercubes. Also of interest are the matrix and permutation problems, as well as the question of tolerating greater numbers of faults. These issues are currently being investigated.

5 References

[1] B. Becker and H. U. Simon, "How robust is the n-cube?," *Information and Computation* 77 (1988), pp. 162–178.

[2] J. Bruck, R. Cypher, and D. Soroker, "Running algorithms efficiently on faulty hypercubes," *Computer Architecture News* 19:1 (1991), pp. 89–96.

[3] S. M. Hedetniemi, S. T. Hedetniemi, and A. L. Liestman, "A survey of gossipping and broadcasting in communication networks," *Networks* 18 (1988), pp. 319–349.

[4] C. P. Kruskal, L. Rudolph, and M. Snir, "The power of parallel prefix," *IEEE Trans. on Computers* 34 (1985).

[5] R. E. Ladner and M. J. Fischer, "Parallel prefix computation," *J. ACM* 27 (October 1980), pp. 831–838.

[6] C. S. Raghavendra, P. J. Yang, and S. B. Tien, "Free dimensions – an effective approach to achieving falut tolerance in hypercubes," *Proc. 22nd Internaitonal symposium on fault tolerant computing*, 1992.

[7] P. Ramanathan and K. G. Shin, "Reliable broadcasting in hypercube multicomputers," *IEEE Trans. on Computers* 32:12 (December 1988), pp. 1654–1657.

[8] Y. Saad and M. H. Schultz, "Topological properties of hypercubes," Research Report 389, Dept of Computer Science, Yale University, June 1985.

TASK BASED RELIABILITY FOR LARGE SYSTEMS:
A HIERARCHICAL MODELING APPROACH

Teresa A. Dahlberg, Member, IEEE   Dharma P. Agrawal, Fellow, IEEE
Department of Electrical & Computer Engineering
Box 7911, N.C. State University
Raleigh, NC 27695
email: dpa@ncsu.edu

Abstract -- *An efficient algorithm to compute Task Based Reliability of large multiprocessor systems or Distributed Computer Systems, using hierarchical modeling approach, is introduced. Heterogenous clusters with irregular topologies are handled.*

I. INTRODUCTION

Increased demand for computer systems with high processing power has led to multiprocessor architectures, which extend limited physical technology by hardware replication of functional units. Since inter-processor communications are necessary to coordinate concurrent activities, the reliability of the interconnection network of a multiprocessor system largely affects its performance.

Reliability calculation is inherently complex, due to state space explosion. This has prompted researchers to devise algorithms based on hierarchical models [2], which is seen to drastically reduce algorithm complexity by breaking the problem into smaller pieces. A 2-level hierarchical model naturally models systems which are inherently hierarchical, such as cluster-based multiprocessing systems, and approximately models large Distributed Communications Systems (DCSs), treating Local Area Networks (LAN) as clusters and the Wide Area Network (WAN) as the global model [2].

However, if local communications is exploited to maximize use of faster local networks, common reliability parameters do not adequately measure the degree to which a cluster-based multiprocessing system or DCS can perform its intended function. One is not solely interested in the connectivity of any source-destination pair, as measured by terminal reliability [1], but rather, in the connectivity of resources within each cluster and the global connectivity of clusters. Similarly, network reliability [4]-[5] is only useful if connection of all system resources is essential for success. Of more use is Task Based Reliability (TBR) [6], which measures the ability of the system to satisfy resource requirements for execution of a function, or "task".

This paper introduces a decomposition technique to efficiently compute TBR for large systems such as cluster-based multiprocessor systems and DCSs, which may have heterogeneous clusters and irregular topologies at the same or different level of the hierarchy. Section II describes hierarchical modeling. Section III details the algorithm. Examples are shown in section IV. Section V concludes the paper.

II. HIERARCHICAL MODELING

Terminology from graph theory is used to model the system. The algorithm operates on the System Graph (SG) having V nodes and E links (edges). Each node of SG represents a processor of a multiprocessor system or a communications' site of a large network. Links of SG represent communications between nodes. Each node of SG is assigned to a cluster, and modeled by a Cluster Model graph, at level 1 of the hierarchy. A link of SG that is incident with two nodes from the same cluster is called a cluster link. Level 2 of the hierarchy is the Global Model graph. Each node of the global graph represents a level 1 cluster. An SG link that is incident with two nodes from different clusters is called a global link.

Global Task Requirement, $K_{min} <= K <= K_{max}$, is the number of connected clusters required.
Cluster Task Requirement, $k_{min} <= k <= k_{max}$, is the number of connected nodes/cluster required.
Subtree is a particular configuration of working and faulty links, such that the working links form a tree.
Gateway Node is a cluster node that could be incident to global links.
Cluster Type defines the topology, gateways, and link reliability values of a cluster. Clusters of the same type need only be evaluated once.
Reliability Value (rval) is a numerical value indicating the probability of occurrence of a subtree.

The rval of edge e is x_e for a local link or g_e for a global link. E.g., rval of subtree x_1x_2' is $x_1(1-x_2)$ and is "the probability that link x_1 is up and x_2 is down".

Gateway Vector (gv) is assigned to a cluster subtree. Each bit of the vector represents a gateway of the cluster and is set if the corresponding gateway node is included in the subtree.

Gateway Vector List (gvlist) is a list of gateway vectors assigned to a global subtree, one gv for each cluster that is a node of the global subtree. A bit is set in a gv if its corresponding cluster gateway node is incident to the global subtree.

Assumptions
- Self loops are not present.
- Links are full duplex and good or faulty.
- Link failures are independent.
- Nodes are always be operational (but faulty nodes could be handled).
- Local communications [2] is assumed. Otherwise, results are a lower bound.

III. TBR EVALUATION

The algorithm efficiently computes the probability that K clusters of the global graph are connected, and each of these clusters has at least k nodes connected which include a path to the gateways incident to the global links being used.

During evaluation of each cluster graph, reliability terms of all connected subtrees on k nodes are found and made disjoint in one step, over all values of k, and for all gateway combinations. Results are stored in lookup tables. A similar procedure is followed for the global graph for all values of K. As each reliability term of the global model is found, it is multiplied by appropriate cluster reliability values found in the lookup tables and added to a partial result. An appropriate cluster rval represents a cluster subtree on at least k nodes that includes, or has added, a path to gateways used by the particular global subtree under consideration that includes this cluster as a macro node.

Expansion step
The important feature of the algorithm is the expansion steps, in which cluster and global model tree structures are generated. The goal is to expand a tree structure (not to be confused with a subtree of the graph) such that each vertex of the structure represents a unique subtree of the graph being evaluated, along with reliability value and gateway information pertaining to that subtree. Gateway information is a gateway vector in cluster tree structures and a gateway vector list for the global tree structure. The tree structures are arranged in levels, beginning with level 2, such that vertices at level j represent all possible j node disjoint subtrees of the graph. Subtrees are made disjoint as they are generated using Exhaustive Conservative Policy [3]. The method used to generate each tree structure vertex is equivalent to graph contraction/fusion and to a 1(0)-SUB operation used for decomposition of the connection matrix [3]. Incidence lists are used for more efficient implementation since the connection matrix is typically sparse.

Once the tree structures are expanded, all of the information exists to find the probability that K clusters are connected with k connected nodes/cluster for values of K and k ranging from 2 to the maximum level of expansion.

Procedures
PROCESS: This recursive, depth-first-search procedure creates level i+1 tree structure vertices such that each subtree includes the subtree represented by its "parent" plus exactly one additional good link, and is disjoint from its "sibling" subtrees.

TRAVERSE calculates gateway vector lookup table values by traversing the expanded tree structure for each gv value, starting at level k, for each value of k. The gv at each vertex is compared to the needed gv value. If the vertex gateway vector is a superset of the needed value, the rval at that vertex is added to a running sum and the procedure moves to examine the next "sibling" vertex. Otherwise, all "children" of the vertex are examined for a match. The sum represents the probability that at least k nodes of the graph are connected and include or have added a path to the gateways specified by the bits set in the gv value.

Algorithm description
Cluster Model Evaluation:
Input Cluster type graph incidence list.
Expand cluster tree structure.
Traverse tree structure to calculate gateway vectors.
Repeat for each cluster type.

Global Model Evaluation:
Input Global graph incidence list.
Expand global tree structure using gvlist to combine cluster reliability values with global reliability value, and add to a running sum in Result Matrix.

IV. EXAMPLE SYSTEMS

Example 1 illustrates the algorithm for the network shown in Fig. 1. The network is divided into 4 clusters of the same type with 3 gateways defined for the cluster type. Fig. 2 shows the tree structure generated during the cluster expansion step, and table 1 lists the corresponding gateway vector values calculated by TRAVERSE. Similarly, Table 6 lists the first 3 vertices of the tree structure generated during Global Model expansion. Note that this tree structure is never stored. Table 2 displays results for $k_{min}=K_{min}=2$ and $k_{max}=K_{max}=4$.

Example 2 evaluates TBR for the hypercube shown in Fig. 3 with results listed in Table 4.

The capability of the algorithm to consider gateways makes it excellent for evaluating TBR for an extended hypercube. Example 3 evaluates an EH(2,1) shown in Fig. 4 with results listed in Table 5. A run time comparison for all three examples is listed in Table 6.

V. CONCLUSION

An efficient algorithm has been presented to compute Task Based Reliability of large multiprocessor systems or DCSs using a hierarchical model. Time saving techniques incorporated into the algorithm, include one-step generation of disjoint multi-terminal connections and use of incidence lists rather than connection matrix. Breaking the problem into smaller subproblems using the hierarchical model provides the most drastic savings in run time and space. The capability to handle heterogeneous, irregular cluster topologies with multiple gateways and the calculation of results over a range of values in one pass of the algorithm provide network designers with a tool to evaluate the effect of cluster size and gateway selection on the network's ability to handle task based problems.

REFERENCES

[1] S. Rai, and D.P. Agrawal, *Advances in Distributed System Reliability, Tutorial Text*, IEEE Computer Society Press, 1990, 333 pp.

[2] N. Sharma and D.P. Agrawal, "Hierarchical Reliability Evaluation of Large Networks," *Proc. Third IEEE Symp. on Parallel and Dist. Proc.*, Dec. 1991, pp. 444-451.

[3] A. Kumar and D.P. Agrawal, et. el., "Impact of Network Failures on the Performance Degradation of a Class of Cluster-Based Multiprocessors," *IEEE Tran. Reliability*, April 1991, pp. 39-44.

[4] I.O. Mahgoub, "Reliability of Cluster-Based Multiprocessor Systems", *Proc. 12th International Conf. Distr. Computing Systems, IEEE*, June 1992, pp. 204-209.

[5] D. Mandaltsis and J.M. Kontoleon, "A Decomposition Technique For The Overall Reliability Evaluation of Large Computer Communication Networks", *Microelectronics Reliability*, Vol. 27, No. 2, 1987, pp. 299-312.

[6] C.R. Das and J. Kim, "A Unified Task-Based Dependability Model for Hypercube Computers", *IEEE Trans. Parallel and Dist. Systems*, Vol. 3, No. 3, May 1992, pp. 312-324.

ILLUSTRATIONS

Figure 1: Example 1, System Graph

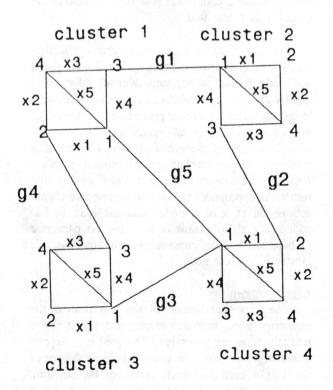

TABLE 1: Ex.1, Gateway Vector Lookup Table
p = 0.9 for all links

k=	2	3	4
001	.998190	.997110	.976860
010	.990000	.988200	.976860
011	.988290	.987390	.976860
100	.988380	.988200	.976860
101	.987480	.987390	.967860
110	.978480	.978480	.976860
111	.977670	.977670	.976860

TABLE 2: Ex. 1, Result Matrix

K\k	2	3	4
2	.985934	.984615	.954246
3	.965198	.963228	.930235
4	.938854	.936675	.889532

TABLE 3: Ex. 2, Result Matrix
p = 0.9 for all links

K\k	2	3	4
2	.978264	.944690	.898045

TABLE 4: Ex. 3, Result Matrix
p = 0.9 for all links

K\k	2	3	4
2	.999700	.999628	.998848
3	.979806	.979700	.978553
4	.947321	.947185	.945706

TABLE 5: Run Time Comparison
Number of PROCESS calls for examples 1,2,3.

Ex.	Hier-archical	Nonhier-archical
1	42	65,536
2	16	2,146
3	62	7,806

TABLE 6: Ex. 1, Global Tree Structure
(3 vertices)

SUBTREE	GVLIST	T/N
[1] g_1	100	1/1
	001	1/2
[2] g_1g_2	100	1/1
	101	1/2
	010	1/4
[3] $g_1g_2g_3$	100	1/1
	101	1/2
	001	1/3
	011	1/4

T=cluster type, N=node number

Figure 2: Ex.1, Cluster Tree Structure

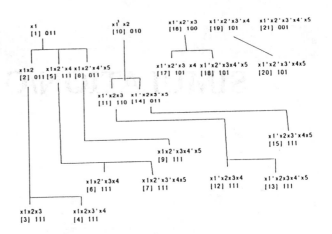

Figure 3: Example 2, 3-cube

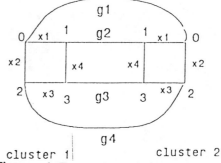

Figure 4: Example 3, EH(2,1

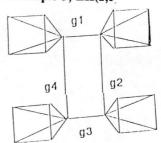

SESSION 12C

SIMULATION/OPTIMIZATION

Performance of a Globally-Clocked Parallel Simulator

Gregory D. Peterson and Roger D. Chamberlain
Computer and Communications Research Center
Washington University, St. Louis, MO 63130
greg@wuccrc.wustl.edu   *roger@wuccrc.wustl.edu*

Abstract *A performance model for a globally-clocked, discrete-event queueing network simulator is developed and validated against measured results. The use of architectural enhancements for improving the performance of the algorithm is investigated. Both scaled and fixed problem sizes are investigated, with a maximum measured scaled speedup of 47 on 64 processors. The model is very accurate, predicting runtime within 5% of measured results.*

INTRODUCTION

Most recent work on parallel simulation algorithms has concentrated on time synchronization protocols for locally-clocked distributed simulation algorithms [9, 12, 15]. However, there are classes of problems for which the conceptually simpler globally-clocked simulation algorithm has adequate performance, eliminating the need to use the more complex, locally-clocked protocols. In addition, the performance of globally-clocked simulation is more predictable, avoiding the sometimes erratic behavior of other techniques. (Lin and Lazowska [13] coined the term "S phenomenon" to describe the observation that speedup curves for an optimistic local clock algorithm often have several local minima and maxima. This observation was made over a large set of different simulation applications [7, 10, 11, 20].)

This paper presents a performance model of a globally-clocked queueing network simulator. The queueing networks simulated are symmetric, closed networks of the type used by Nicol [16] and Fujimoto [8] for the evaluation of locally-clocked protocols. The model is validated by comparison with a parallel simulator, *qnet*, that has been implemented on an NCUBE multiprocessor [17].

The performance of the simulations is quite good, with scaled speedups of 47 on 64 processors. (Scaled speedups increase the problem size in proportion to the

number of processors; fixed speedups do not change the problem size with the number of processors). This corresponds to a simulation that requires over ten hours on a uniprocessor completing in under fifteen minutes. Fixed problem size speedups are considerably lower, reaching approximately 5 on 64 processors. However, the performance of the fixed size problems doubles when proposed synchronization hardware is included in the parallel architecture.

The accuracy of the model is very high, with predicted execution times consistently within 5% of measured results on a variety of network topologies and job populations. This enables us to confidently use the model to investigate issues such as alternative input parameters, underlying technology performance, and architectural alternatives.

The model is applied to investigate the performance implications of an architectural enhancement to the standard hypercube topology. We ask, and answer, the question "How will performance be impacted by the addition of synchronization hardware?" This is an important issue in globally-clocked simulations, where the number of required barrier synchronization operations is quite large.

The paper is organized as follows. The next section describes the parallel simulator, *qnet*, and details the queueing systems to be simulated. Section 3 presents the performance model development. Section 4 validates the model against measured results. Section 5 uses the model to predict the performance implications of an alternative architecture, and the final section presents conclusions and continuing work.

SIMULATOR

Qnet is an event-driven parallel discrete-event queueing network simulator written in C for execution on an NCUBE hypercube multiprocessor. A variety of capabilities are built into the simulator that allow the analysis of different classes of networks. The user defines the size of the network, the interconnection of the queues and servers, the service discipline for each

---
[1] This work partially supported by the Department of Education and National Science Foundation grant CCR-9021041.

server, initial job populations, branching probabilities for outputs, and sinks and sources for open queueing networks. Additionally, the user can specify the length of the simulation, the number of processors in the multicomputer, the partitioning of the network, the mapping of queues and servers to particular processors, and the data collection to be performed during the course of the simulation. The queueing networks are limited to a first-come, first-served queueing discipline.

The simulation algorithm is presented in Figure 1. The program is first initialized and then enters its main loop. In this loop, the current events are processed, event messages are transmitted, event messages are received, and acknowledgments are sent, received, and processed. Once all the acknowledgments are processed, the processor enters a barrier synchronization. This barrier synchronization includes the determination of the next global time. The loop is then repeated using this new simulated time. Following the last iteration, data pertaining to the queueing network behavior and the performance of the simulator are collected and the simulation is complete.

Initialization
Read network configuration information
Perform complete exchange barrier synchronization
While time left in simulation
 Process events for current simulation time,
   send event msgs for non-local events
 While waiting for acknowledgments
  Process incoming event messages
   Send acknowledgments
  Process incoming acknowledgments
 End while
 Calculate next local time
 Perform complete exchange barrier sync
 If there are incoming event messages
  Process incoming event messages
   Send acknowledgments
 End if
 Simulation time ← next global time
End while
Collect trace data

Figure 1: Parallel Algorithm

Each event message sent from one processor to another is answered by an acknowledgment. This is to prevent the possibility of a message being in transit *after* the barrier synchronization.

The queueing network topologies studied include the ring, the torus, and the hypercube. The bidirectional ring consists of a ring of stations where each station has two neighbors. The torus is a mesh with the edges "wrapped around" so that each station has four neighbors. The hypercube is a standard binary n-cube. In all three topologies, a job is equally likely to be sent to each neighboring station following job completion. Because the performance of *qnet* with each queueing network topology is similar, we only include the results of experiments with the torus topology here. Detailed results from the other topologies are presented in [17].

One of the principal issues involved in the use of parallel architectures is the method that is used to decompose a problem into parts that can be operated on by the individual processors. Since the primary focus of this work is to understand the capabilities of globally-clocked simulation, an ideal partitioning and mapping method is used that results in minimizing the amount of communication with the best load balance. The partitioning and mapping results in all processors simulating an equal number of stations. The amount of communication is also minimized and communication is limited to neighboring processors [17]. In general, finding the optimal partitioning and mapping is difficult [19] and beyond the scope of this paper.

MODEL DEVELOPMENT

In a globally-clocked simulator, each processor is executing the same program with local data pertaining to the portion of the queueing network that it is responsible for simulating. An event in the simulation is defined as the arrival of a job at a station. At the time of arrival, the amount of time spent waiting in the queue and receiving service are calculated, with the job being output to the next appropriate station. The use of a nonpreemptive first-in, first-out queueing discipline allows this method of calculating waiting and service times. Since there is only one job class, events at the same simulation time can be safely processed in any order. The flow of jobs through the network is accomplished by passing event messages between processors. Finally, after all events have been processed and all messages have been received that correspond to job arrivals at nonlocal stations, all of the processors participate in a barrier synchronization and calculate the next simulation time by finding the time of the earliest future event.

Figure 2 illustrates a sample iteration of the simulator with two processors. Processor two (P2) completes its event processing after processor one (P1). Some time during its event processing, P1 sends an event message to P2. Upon completion of its local event processing, P2 processes its incoming event messages. When it has completed the processing of the incoming

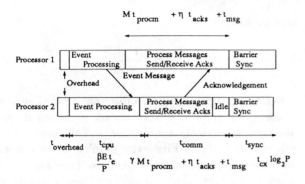

Figure 2: Sample Iteration of Simulator

event message from P1, P2 sends an acknowledgment back to P1 and enters the barrier synchronization. Let us assume that the last acknowledgment sent in the current iteration originates from P2. This acknowledgment is sent to P1, which must update the number of acknowledgments that are anticipated, and check to see that there are no more acknowledgments outstanding (there will be none). P1 will then proceed to the barrier synchronization. Note that P2 spends some idle time waiting for P1 to enter the barrier synchronization.

One can model the runtime of the simulation over B busy ticks (simulation time units when there is an event) on P processors [5] as:

$$R_P = B[t_{cpu} + t_{comm} + t_{sync} + t_{overhead}]$$

For this equation, t_{cpu} is the time spent processing events for the most heavily loaded processor. Similarly, t_{comm} is the time spent processing event messages. The t_{sync} term corresponds to the amount of time required to perform a barrier synchronization using a complete exchange algorithm. Finally, the simulator has overhead that is incurred in each iteration for calculating the next simulation time, cleaning data structures, and for initializing variables. The time spent on these activities is represented by $t_{overhead}$.

Each iteration, a barrier synchronization is performed using the complete exchange algorithm [5, 22]. The number of steps in the complete exchange algorithm is logarithmic with respect to the number of processors in the hypercube. Assuming that the time to perform each step is t_{cx}, the time for synchronizing each iteration can be expressed as:

$$t_{sync} = t_{cx} \log_2 P$$

Event Processing

We next consider the role of event processing. In the serial case, the time spent processing events can be expressed as the product of the number of events

and the average time spent processing an event. Assuming that all processors have an equal portion of the events to process, the time spent processing events in the parallel case is equal to the serial event processing time divided by the number of processors. Unfortunately, this assumption is not true in general because some processors will have more events to process than others. A factor β is defined as the ratio of the number of events at the most heavily loaded processor to the number of events at a processor with an average load. This results in the time spent processing events for the most heavily loaded processor:

$$t_{cpu} = \frac{\beta E t_e}{P}$$

For this equation, E is the mean number of events that are processed in each iteration. Each event takes t_e time to be processed. P is the number of processors, and β is the imbalance factor. Note that the ideal value of β is 1, which corresponds to a perfectly balanced system. As the load imbalance worsens, β increases.

The simulator is restricted to queueing networks of first-in, first-out queues with nonpreemptive queueing disciplines. Here, the queueing networks are further restricted to closed symmetric networks of identical stations with a single job class and all have a torus topology where each station has an identical exponential service distribution. Note that this is not a limitation of the simulator (*qnet* allows each station to have a unique distribution and supports a number of service distributions), rather we make this assumption to make the analysis more tractable. For scaled problems, we have 1024 stations located on each processor. In the case of fixed problems, we have a total of 1024 stations. We vary the number of jobs initially located at each station from one to five.

An analysis of the characteristics of a closed symmetric queueing network results in simple equations describing the first-order statistics [21]. If the servers all have identical service distributions, the utilization of each server is:

$$U = \frac{J}{J + Q - 1}$$

where J is the total number of jobs and Q is the total number of stations in the network. Note that if there are Q stations that are busy a fraction of time equal to U, then one expects that at any time, the number of stations that are busy is equal to UQ. In addition, if each server has a mean service time of S, then one expects that every S time units that the server is busy, another event is generated by a job departing the station. This implies that the expected number of events processed at any time is:

$$E = \frac{UQ}{S} = \frac{QJ}{S(J + Q - 1)}$$

Because of the stochastic nature of the service requirements and output distributions from servers, each processor has a number of events to process each simulation step that may be different from the expected value. Two methods are used to determine the load imbalance factor β. First, experiments are run to measure the value of β, thus finding it empirically. This method results in the best accuracy possible, but essentially removes the possibility of using the model to predict performance in different circumstances. The alternative is to describe β analytically. We now present a method to approximate β through analytic means.

The load on each processor is considered to be a random variable sampled from some distribution. Given P processors, the most heavily loaded processor has the maximum value of P samples. To find β, we take the expectation of this maximum value. In order to simplify the model, the event distributions at each simulation time point are assumed to be independent and identically distributed. Although the distributions will not be truly independent, there is empirical evidence that for many simulation applications the error introduced by this assumption is minimal. We present evidence below for queueing networks, and Agarwal and Chakradhar [1] observe a similar situation for logic simulation.

Assuming the existence of a distribution with probability distribution function $F_X(x)$ from which P random variables are drawn, the maximum of the P random variables is a random variable Z with a probability distribution function:

$$F_Z(z) = (F_X(z))^P \tag{1}$$

The probability density function of the maximum can be found by differentiating and yields:

$$f_Z(z) = P(F_X(z))^{P-1} \left(\frac{d}{dz}\right) F_X(z)$$

Finding the expectation of the maximum is accomplished by integrating the product of the maximum and its density:

$$E_P[Z] = \int_{-\infty}^{\infty} \alpha f_Z(\alpha) d\alpha \tag{2}$$

The value that is assigned to β is the load imbalance of the system. This is determined by finding the ratio of the load at the most heavily loaded processor to the average load.

$$\beta = \frac{E_P[Z]}{E/P} \tag{3}$$

Before attempting to find a distribution that approximates the number of events at each processor in each iteration, some of the requirements of the distribution will be considered. First, the number of events will be discrete-valued and finite because there are a finite number of servers that can produce events in each iteration. This implies that the distribution must not have "tails" that are asymptotic. If one assumes each server has an equal probability in each iteration of producing an event, then each processor would have an equal number of Bernoulli trials using this probability, yielding a distribution that approaches a Gaussian distribution as the number of servers approaches infinity. It is likely that the distribution would then have a "bell" curve and will have a finite range of possible values. Experimental results confirm that the number of events is distributed consistently with the assumptions above. See Figure 3 for a histogram of the number of events per iteration measured in a typical simulation.

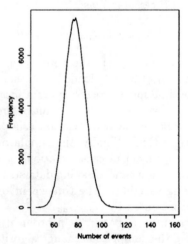

Figure 3: Measured Number of Events per Iteration

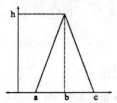

Figure 4: Triangle Distribution (Events per Iteration)

As a means of reducing the mathematical complexity, a triangle distribution is assumed. The triangle distribution has a mean equal to the expected number of events at a processor each simulation time and is symmetric about this point (labeled b). See Figure 4. The area of the triangle must equal one to allow it to be a probability density. The height of the triangle therefore is:

$$h = \frac{2}{c - a}$$

where a and c are the two vertices of the triangle on the x-axis. The probability density of the random variable can be written as:

$$f_X(x) = \begin{cases} 4\frac{(x-a)}{(a-c)^2} & \text{if } a \leq x \leq b \\ 4\frac{(c-x)}{(a-c)^2} & \text{if } b \leq x \leq c \\ 0 & \text{otherwise} \end{cases}$$

And the probability distribution function is:

$$F_X(x) = \begin{cases} 0 & \text{if } x \leq a \\ 2\frac{(x-a)^2}{(a-c)^2} & \text{if } a \leq x \leq b \\ 1 - 2\frac{(c-x)^2}{(a-c)^2} & \text{if } b \leq x \leq c \\ 1 & \text{if } c \leq x \end{cases}$$

To solve for the probability distribution function for the maximum of P samples from the distribution above, the earlier solution presented in Equation (1) is applied. Differentiating to find the probability density function for the maximum results in:

$$f_Z(z) = \begin{cases} 2P\left(\frac{2}{(a-c)^2}\right)^P (z-a)^{2P-1} & \text{if } a \leq z \leq b \\ 4P\frac{(c-z)}{(c-a)^2}\left(1 - 2\frac{(c-x)^2}{(a-c)^2}\right)^{P-1} & \text{if } b \leq z \leq c \\ 0 & \text{otherwise} \end{cases}$$

Finding the expectation of the maximum is performed as in Equation (2).

$$E_P[Z] = \int_a^b 2P\alpha\left(\frac{2}{(a-c)^2}\right)^P (\alpha - a)^{2P-1} d\alpha +$$
$$\int_b^c 4P\alpha\frac{(c-z)}{(c-a)^2}\left(1 - 2\frac{(c-x)^2}{(a-c)^2}\right)^{P-1} d\alpha$$

Solving the first integral by parts and the second by using a binomial expansion results in the following:

$$E_P[Z] = \frac{a + Pa + Pc}{(2P+1)2^P} -$$
$$P\sum_{i=0}^{P-1} \binom{P-1}{i}\left[-\frac{1}{2}\right]^{i+1}\frac{ai + a + ci + 2c}{(i+1)(2i+3)}$$

This is then substituted into Equation (3) to yield a value for β.

Message Processing

The cost for communications is the amount of time spent processing event messages and ensuring that all messages have been received through the use of acknowledgments. Each event message that is received is processed by inserting an event into the event queue for the appropriate simulation time. The reception of these messages is followed by sending an acknowledgment message to the processor where the event message originated. The amount of time required to process an individual message is denoted by t_{procm}. The final acknowledgment sent will take t_{msg} to arrive at its destination. Finally, an amount of time is spent checking for acknowledgments and processing them.

In an iteration of the simulator, incoming event messages are processed immediately following the completion of event processing for the local servers. Once the event messages have been processed, the acknowledgments are received. The processor may not proceed to the barrier synchronization until all the acknowledgments have been received. The order in which these actions are completed indicates that the time spent with communications can be expressed as:

$$t_{comm} = \gamma M t_{procm} + t_{msg} + \eta t_{acks}$$

Each iteration, one of the processors will receive the largest number of event messages. The number of messages that arrive at this processor is denoted by M. Because the most heavily loaded processor (in terms of event processing) might not receive the most messages, γ is included to account for the proportion of the communications contribution that is performed concurrently with the event processing at the most heavily loaded processor. In other words, the time spent processing messages that "overlaps" the time spent processing events is ignored. Note that $\gamma = 1$ indicates that the processor with the most messages does not start its message processing until after the most heavily loaded processor processes all its events (there is no overlap). If $\gamma = 0$, then all message processing is complete before the most heavily loaded processor completes event processing.

To find the value of M, one must examine the topology of the queueing network and the partitioning that was utilized to break the network into pieces for each processor. The expected maximum number of messages received over all processors is the expected communications load at the bottleneck processor. If there are N_{edge} stations that each have a probability P_{off} of sending an event message to another processor, then the number of event messages sent from each processor is a binomially distributed random variable. The probability of a service completion at one of the N_{edge} servers is U/S. Thus each processor has a number of events that are sampled from N_{edge} trials with a probability of sending a message of UP_{off}/S. The following binomial distribution describes the number of event messages on each processor in each iteration:

$$F_X(x) = \sum_{i=0}^{x} \binom{N_{edge}}{i} \left(1 - \frac{UP_{off}}{S}\right)^{N_{edge}-i} \left(\frac{UP_{off}}{S}\right)^i$$

Because we are finding the number of event messages at the busiest processor, we solve for the maximum over the P random variables sampled. The probability distribution function for the maximum of P samples from this distribution is found using Equation (1).

To find the probability that the maximum is z, one simply takes $F_Z(z) - F_Z(z-1)$. Now, the expectation of the maximum is the sum of the probability of each weighted by the event count.

$$M = \sum_{i=1}^{N_{edge}} i(F_Z(i) - F_Z(i-1))$$

The term t_{acks} is used to represent the time spent processing acknowledgment messages. After all event messages that have been received are processed, a check is made for acknowledgments and the count of outstanding acknowledgments is decremented accordingly. Once the received acknowledgments have been processed, the processor checks to see if additional acknowledgments are expected. If all of them have been received, the processor is finished with the current iteration and proceeds to the barrier synchronization. Otherwise, these steps are repeated, beginning with a check for additional incoming event messages.

To consider the possibility that the final acknowledgment will arrive while the receiving processor is looking for event messages or checking count variables, we introduce the term η. The term η indicates the expected number of times that the bottleneck processor will have to perform scans of the count variables and check for event messages after all event message computations are complete. η is at least one (reflecting the necessity of checking for incoming event messages and acknowledgments at least once) but less than two. Given that a message is sent and an acknowledgment needed, we assume that, on average, an acknowledgment will arrive halfway through the acknowledgment and event message processing loop. Assuming that P_m is the probability of an event message being generated, then η can be expressed as follows:

$$\eta = 1 + \frac{P_m}{2}$$

In order to determine P_m, some knowledge of the queueing network topology, mapping, and partitioning is required. Thus P_m must be found independently for each simulation. Assuming that a service completion occurs every U/S simulation time steps at any station, the likelihood that there is at least one event generated is approximately:

$$P_m \approx 1 - \left(1 - \frac{P_{off}U}{S}\right)^{PN_{edge}}$$

We now take the pieces and use them in the model. Table 1 shows the parameters of the model and their definitions. Figure 2 shows the time that is spent performing a sample iteration with graphic illustration of the values calculated by the model, allowing insight into the runtime spent on various components of the simulator.

Table 1: Parameters for Performance Model

R_P	runtime with P processors
P	number of processors
B	number of busy simulation time points
β	load imbalance factor
E	number of events per iteration
U	utilization of servers
Q	number of stations
J	number of jobs
S	mean service time
M	max # event msgs at proc per iteration
N_{edge}	number of boundary stations (with nonlocal outputs)
P_{off}	prob boundary station outputs to nonlocal station
γ	comm overlap with event proc factor
η	scanning ack count factor
P_m	prob of event msg in an iteration
t_e	time to process an event
t_{procm}	time to process an event message
t_{msg}	time to send a message
t_{acks}	time to process acknowledgments
$t_{overhead}$	time to complete overhead tasks
t_{cx}	time to perform pairwise exchange

MODEL EVALUATION

To evaluate the accuracy of the performance model just developed, a comparison between the model's predicted values and measured values is presented. Discrepancies in the model's performance are discussed to understand the limitations of the model.

A number of factors are parameters of the queueing network being simulated. For example, B represents the number of busy ticks in the simulation. E, the expected number of events that occur in each iteration is a function of the queueing network. Specifically, E is a function of Q, J, and S, the number of queues, number of jobs, and average service time, respectively. Finally, the probability that a message is sent for any iteration is dependent on the topology and the utilization of the network. These values are all found by examining the network.

Other parts of the model are related to the machine architecture and implementation. For example, t_e represents the average time to process an event. The time to process a message is denoted by t_{procm}, and the time to process acknowledgments is t_{acks}. The costs associated with looping, calculating the next time, and setting variables each iteration are incorporated through $t_{overhead}$. These four components are related to the algorithm that is used for the simulation. Costs that are more directly related to the machine architecture are t_{msg} and t_{cx} which are primarily functions of the speed of the interconnection network between processors. Another architectural contribution is P, the number of processors used in the simulation.

The other factors, β and γ, are more difficult to find. A method for determining β was described. Empirical results have shown γ to be approximately equal to unity. Consequently, we assume $\gamma = 1$. The difficulty in finding the parameters of the triangle distribution remains to be addressed.

The triangle distribution that is used to determine β (see Figure 4) is centered around the mean number of events at a processor each iteration. Having already found that the mean number of events on any processor each simulation step is E/P, this is the value assigned to b. In addition, the values of a and c must be found that will produce an accurate model of the number of events at each processor each simulation time. For the following discussion, the values for a and c are found by examining the measurements of the number of events at each processor (collected in histograms). These histograms contain the distribution of the number of events that occur in each iteration for each processor. Using these values for the parameters of the triangle distribution, the estimated maximum number of events is calculated and compared to empirical results in order to ascertain the accuracy of the β model. The measured values of β are collected by comparing the number of events in each iteration at the heaviest loaded processor (the processor with the largest number of events to process in the current iteration) and the mean number of events in each iteration.

Table 2: Calculated β Values

Jobs per Processor	Processors					
	2	4	8	16	32	64
1024	1.10	1.18	1.24	1.29	1.33	1.35
2048	1.08	1.14	1.19	1.23	1.26	1.28
3072	1.07	1.14	1.18	1.22	1.25	1.27
4096	1.07	1.13	1.18	1.21	1.24	1.26
5120	1.07	1.13	1.17	1.21	1.23	1.25

Table 3: Measured β Values

Jobs per Processor	Processors					
	2	4	8	16	32	64
1024	1.08	1.14	1.20	1.25	1.29	1.33
2048	1.07	1.12	1.17	1.21	1.25	1.28
3072	1.06	1.12	1.16	1.20	1.23	1.27
4096	1.06	1.11	1.15	1.19	1.22	1.24
5120	1.06	1.11	1.15	1.19	1.22	1.25

Tables 2 and 3 present calculated β values and measured β values for the torus topology. Note that the measured values are within 5% of the calculated values. Increasing the number of processors (and the size of the problem proportionally) results in increased load imbalance. The mean number of events in each iteration at a processor is approximately equal for all the processor populations, but the maximum number of events increases. In addition, as the number of jobs in the network increases, the utilization of each server will also increase. As more servers are busy, the variance in the distribution of the number of events will increase accordingly. Note that this does not imply that the load imbalance becomes more pronounced, since the number of jobs in the network increases and the utilization approaches one. To the contrary, the number of events and the variance in this number will increase, but the proportion of the load at the heaviest loaded processor to the average load will decrease. In other words, the load imbalance decreases with the increase in the utilization of each server.

Table 4 contains information detailing the measured values for t_e, t_{procm}, t_{msg}, t_{acks}, $t_{overhead}$, and t_{cx}. The overhead for the serial case is smaller because there is no need to initialize acknowledgment count variables and similar activities.

The queueing networks that are considered here have a torus topology with 1024 stations per processor. The multicomputer has P processors on which simulations

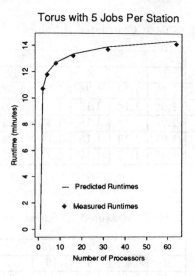

Figure 5: Measured and Predicted Runtimes

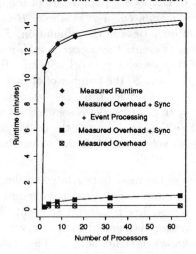

Figure 6: Performance of Model Components

can be run, thus the networks contain $1024P$ stations. The job population is varied from one to five jobs initially located at each station. As stated earlier, all the servers have the same service distribution which is an exponential distribution with a mean service time of 10 simulation time steps. All these simulations are executed for a total of 10,000 simulation time steps.

Table 4: Performance Model Parameters

Parameter	Time Cost (Microseconds)
t_e	742
t_{procm}	320
t_{msg}	512
t_{acks}	320
$t_{overhead}$	1280 (960 for serial case)
t_{cx}	922

The model predicts runtimes that are quite accurate, as can be seen in Figure 5. Table 5 shows the percentage error for each processor population. To further investigate how well the model predicts the performance of the simulator, we compare the predicted and measured values of the time spent on overhead, synchronization, event processing, and communications. These are shown in Figure 6. The model is quite accurate in predicting the time spent in performing the overhead and synchronization. The largest error is associated with the time for processing events. Most of the difference between the predicted runtimes and the measured runtimes can be attributed to the error in predicting the value of β.

Table 5: Performance Model Error

Jobs per Station	Percent Error						
	Processors						
	1	2	4	8	16	32	64
1	0.5	2.4	2.3	3.1	3.8	4.0	3.5
2	0.5	1.0	0.4	1.1	1.4	1.5	1.1
3	0.3	1.2	0.6	1.3	1.8	2.0	1.7
4	0.0	0.8	0.3	1.2	1.3	1.6	1.3
5	0.1	1.1	0.3	1.1	1.5	1.7	1.5

Note that these results assume symmetric, closed queueing networks. For non-symmetric networks, different techniques will be required to determine some of the parameters. The number of events per iteration and the load imbalance will be more difficult to characterize, implying additional difficulty in determining values for β and E. In addition, the communications may be difficult to characterize, resulting in more complicated expressions for M and P_m. Finally, the overlap of event processing and message processing could be impacted, resulting in different values for γ.

Even though some of the model parameters will have to be determined differently, the model is not limited to predicting the performance of symmetric, closed queueing networks, but is applicable to a range of applications. Earlier versions of the model have already been used to investigate logic simulation on hypercube architectures [4, 5] and special purpose hardware for logic simulation [22], where some of the more difficult to predict parameters (e.g., E, β) were determined empirically using traces from uniprocessor simulations [6, 23].

In addition, Agrawal and Chakradhar [1] have proposed techniques for analytically predicting the computational load for the logic simulation application that will fit well into the model presented here.

MODEL APPLICATION

The globally-clocked synchronization method is used to coordinate the clocks of all of the processors after each simulation time step. Because this operation must be performed repeatedly, examining the performance implications of synchronization is important in evaluating the overall system performance. In this section, we use the model to evaluate the time required for the synchronization and its implications on performance, followed by considering the improvement in performance that can be anticipated through the use of faster synchronization mechanisms. Instead of increasing the size of the problem with the number of processors, we use a fixed problem size for this analysis since synchronization has a larger impact with fixed problem sizes, occasionally exceeding 50% of the simulation execution time.

To improve the performance of discrete-event simulation and several other important iterative algorithms[3], the use of architectural enhancements has been proposed. Of interest here is the impact that such an improvement would have on simulators such as *qnet*.

The architectural enhancement considered for a hypercube parallel processor is a synchronization network consisting of an additional mechanism for performing combining operations on subsets of the processors in the hypercube[2]. Typical operations include logical operations and minimum and maximum calculations. The synchronization network would consist of combinational logic that is asynchronously interfaced with the processors. The network will allow the completion of such operations as a barrier synchronization much faster than through the use of the message passing capabilities of the hypercube. Conservative estimates are that a global minimum operation would complete in less than 1% of the time it would take through the use of message passing as modeled earlier[2, 3]. This architectural enhancement is similar in concept to the control network of Thinking Machine's CM-5.

To investigate this issue, the model is modified to reflect the addition of a synchronization network. The time spent performing barrier synchronizations is reduced to 0.5% of the time used earlier (i.e., the value of t_{sync} is decreased by a factor of 200). Figure 7 shows the fixed speedup both with the synchronization network and without it for a torus topology with 1024 stations and five jobs initially placed at each station.

As can be seen, the decrease in runtime exceeds 50% for 64 processors, resulting in more than double the speedup. For fixed problem size executions, the amount of time spent synchronizing is a large portion of the overhead of *qnet*, and this effect increases with larger processor populations. With the addition of the synchronization network, the performance is improved dramatically. Note that the load imbalance is the principal reason why the speedup is not improved even more dramatically. At 64 processors, β is 4.5 for this simulation. An accurate performance model is quite helpful in evaluating the quantitative performance improvements associated with synchronization hardware. This allows for informed decision making when considering whether or not to include synchronization hardware in parallel computer designs.

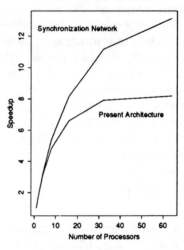

Figure 7: Speedup for Torus

CONCLUSIONS

In this paper, we develop a performance model that accurately predicts the runtime of a globally-clocked, discrete-event simulator. This model enables us to better understand the simulation algorithm and the impact on the runtime of the tasks involved. In addition, the model is used to predict the impact of an architectural enhancement on simulation performance. Workloads are described that are efficiently simulated using a globally-clocked algorithm.

Although the model has been presented for the synchronous simulation of symmetric, closed queueing networks, we are confident that it is applicable to the general class of globally-clocked simulations. The model has been previously used to model logic simulation performance on both general purpose multiprocessors and special purpose architectures. Agrawal and Chakradhar [1] have proposed a method for generally characterizing the value of β by using Bernoulli trials for estimating the number-of-events probability distribution. Agrawal and Chakradhar's work used logic simulation as the application vehicle. We have also applied the model to a wider set of synchronization algorithms [17, 18], including a synchronous conservative algorithm and the speculative algorithm of Mehl [14].

The results are encouraging, but many questions remain. The performance of globally-clocked simulation and the model with more general workloads must be investigated. For example, what is the impact of alternate (non-symmetric) topologies, open queueing networks, different job service requirements, and finer time resolution? Nonetheless, having a better understanding of the algorithm allows us to investigate improvements. Some preliminary work has been completed, but we must still address additional improvements that are possible, such as exploiting the lookahead capability in the simulation or utilizing speculative computation within a globally-clocked framework [17]. Finally, the impact of partitioning the simulated networks into subnetworks and mapping these subnetworks to the processors needs to be investigated. Quantifying the impact of the partitioning and mapping and the development of algorithms for performing this task are necessary for a general parallel simulation capability to be practical.

REFERENCES

[1] V.D. Agrawal and S.T. Chakradhar, "Performance Analysis of Synchronized Iterative Algorithms on Multiprocessor Systems," *IEEE Trans. Parallel and Dist. Sys.*, Nov. 1992.

[2] R.D. Chamberlain, *Multiprocessor Synchronization Network: Preliminary Design Description.* Tech. Report WUCCRC-90-12, Washington Univ., Nov. 1990.

[3] R.D. Chamberlain, *Gaussian Elimination on a Hypercube Architecture Augmented with a Synchronization Network.* Tech. Report WUCCRC-91-12, Washington Univ., Apr. 1991.

[4] R.D. Chamberlain and M.A. Franklin, "Discrete-Event Simulation on Hypercube Architectures," *Proc. of IEEE Int. Conf. on Computer-Aided Design*, 1988.

[5] R.D. Chamberlain and M.A. Franklin, "Hierarchical Discrete-Event Simulation on Hypercube Architectures," *IEEE Micro*, Aug. 1990.

[6] R.D. Chamberlain and M.A. Franklin, "Collecting Data About Logic Simulation," *IEEE Trans. CAD*, July 1986.

[7] M. Ebling et al., "An Ant Foraging Model Implemented on the Time Warp Operating System," *Proc. SCS Multiconf.*, Mar. 1989.

[8] R.M. Fujimoto, "Performance of Time Warp Under Synthetic Workloads," *Proc. SCS Multiconf.*, Jan. 1990.

[9] R.M. Fujimoto, "Parallel Discrete Event Simulation," *CACM*, Oct. 1990.

[10] P. Hontalas et al., "Performance of the Colliding Pucks Simulation on the Time Warp Operating System (Part 1: Asynchronous Behavior & Sectoring)," *Proc. SCS Multiconf.*, Mar. 1989.

[11] D. Jefferson et al., "Distributed Simulation and the Time Warp Operating System," *Proc. 11th ACM Symp. on Operating Sys. Principles*, Nov. 1987.

[12] D. Jefferson, "Virtual Time," *ACM Trans. Prog. Lang. Sys.*, July 1985.

[13] Y.B. Lin and E.D. Lazowska, "Processor Scheduling for Time Warp Parallel Simulation," *Proc. SCS Multiconf.*, Jan. 1991.

[14] H. Mehl, "Speed-Up of Conservative Distributed Discrete-Event Simulation Methods by Speculative Computing," *Proc. SCS Multiconf.*, Jan. 1991.

[15] J. Misra, "Distributed Discrete-Event Simulation," *Computing Surveys*, Mar. 1986.

[16] D.M. Nicol, "Parallel Discrete-Event Simulation of FCFS Stochastic Queueing Networks," *SIGPLAN Not.*, Sep. 1988.

[17] G.D. Peterson, *Qnet: A Globally-Clocked Discrete-Event Queueing Network Simulator*, MS Thesis, Dept. Elec. Eng., Washington Univ., May 1992.

[18] G.D. Peterson and R.D. Chamberlain, "Exploiting Lookahead in Synchronous Parallel Simulation," *1993 Winter Simulation Conf.*, (submitted).

[19] A.N. Tantawi and D. Towsley, "Optimal static load balancing in distributed computer systems," *JACM*, Apr. 1985.

[20] F. Wieland et al., "Distributed Combat Simulation and Time Warp: The Model and Its Performance," *Proc. SCS Multiconf.*, Mar. 1989.

[21] K. Wong, *Solution of Symmetric, Product Form Closed Queueing Networks.* Tech. Report WUCCRC-91-10, Washington Univ., Jan. 1991.

[22] K. Wong and M.A. Franklin, "Performance Analysis and Design of a Logic Simulation Machine," *Proc. 14th Int. Symp. Comp. Arch.*, June 1987.

[23] K.F. Wong et al., "Statistics on Logic Simulation," *IEEE Proc. of 23rd Design Automation Conf.*, June 1986.

Fast enumeration of solutions for data dependence analysis and data locality optimization [*][†]

C. Eisenbeis[‡] O. Temam[§] H. Wijshoff[¶]

Abstract Most of the sophisticated optimization tools dealing with data dependence analysis, parallelization and data locality exploitation cannot provide satisfactory solutions to a number of problems because they are, in general, unable to precisely handle some complex systems of linear equations coming from dependence equations between array subscripts. For such cases, these algorithms either provide rough estimates or resort to unefficient and therefore costly enumeration strategies. In this paper, we present an efficient technique, named *Fast Determination*, for dealing with the enumeration problem which degrades the behavior of sophisticated algorithms Incorporating *Fast Determination* would enhance the performance and accuracy of existing algorithms, and widen their scope of application.

Keywords: data dependence analysis, data locality, parallelization, integer linear programming, linear Diophantine equations, enumeration.

1 Introduction

Many problems related to parallelization, data dependence analysis and data locality exploitation are equivalent to either checking whether a system of equations has solutions, or estimating the number of solutions, or characterizing the solutions. On a first step, in each domain (data dependence analysis, parallelization, data locality exploitation) heuristics have been developed in order to cope with most classic, frequent and urgent problems, and to provide unperfect but satisfactory solutions. For instance, in data locality exploitation, techniques relied on rough *estimates* [4] of the amount of reuse for deciding which data should be kept in upper levels of memories. In parallelization, a loop would not be concurrentized as far as dependencies *might* exist [5]. In data dependence analysis, simple tests provided quick but *unprecise* answers on the existence of solutions [1].

Then, on a second step, theories were extrapolated from these heuristics and experiments. Such theories lead to the design of more global and more efficient algorithms. Thus, the *Data Locality Optimization Algorithm* [11] and the *Window Theory* [2] now provide powerful frameworks for dealing with data locality. The *Window Theory* is based on the characterization of the solutions of sets of integer linear equations which actually correspond to dependency equations between array references. Parallelizing compilers now integrate some simple algorithms in data dependence analysis [5, 1] resulting in a greater number of loop nests parallelized per code. Finally, based on integer linear programming techniques, substantial improvements, such as the Omega Test [9] occured in data dependence analysis techniques [9]. Most achievements were obtained by improving the way systems of linear equations are handled. Number theory and linear algebra have been extensively used and helped clearly representing and solving linear equations problems. Now, though most techniques have gained maturity, they are still being revisited because, prior to real implementation, such techniques must be improved so as to exhibit the following properties:

- Provide answers to problems for almost any parameter range (instead of most usual and simple ones).

- Provide steady performance over a wide range of parameters.

- Provide very accurate or even exact answers.

In the techniques cited above, most of these improvements are not yet achieved because some systems of equations are still relatively poorly handled. For instance, in data locality exploitation, the *Window Theory* requires to estimate the number of reuses per

---

[*]Part of this work was done while the two authors O. Temam and H. Wijshoff were employed by the Center for Supercomputing Research and Development, University of Illinois at Urbana–Champaign, USA.

[†]This work was partly funded by the DGXIII BRA Esprit III European Project APPARC

[‡]INRIA, Domaine de Voluceau, 78153 Le Chesnay CEDEX, France

[§]IRISA/INRIA, Campus de Beaulieu, 35042 Rennes CEDEX, France *and* HPC Division, Department of Computer Science, University of Leiden, Niels Bohrweg 1, 2333 CA Leiden, the Netherlands

[¶]HPC Division, Department of Computer Science, University of Leiden, Niels Bohrweg 1, 2333 CA Leiden, the Netherlands

element, which sums up to counting the number of solutions of a linear equation with finite boundaries. For complex linear subscripts, or non-rectangular loops, this computation is difficult and therefore only cheap and unprecise estimates are provided, thereby degrading the efficiency of the technique. In data dependence analysis, a sophisticated test such as the Omega Test [9] exhibits unstable performances (see section 2.1) because it still lacks the ability to efficiently scan a solution space in order to check for the presence of solutions. And performing this enumeration step is sometimes compulsory for the test to be exact. Regarding parallelization, though sophisticated data dependence analysis is being implemented in compilers, the suspected presence of a dependence still condemns a loop nest to serial execution. However, in many cases, dependences do not exist between *all* iterations. Therefore, after extracting some faulty iterations, it would be possible, in many cases, to parallelize the remaining ones.

So, many flaws of existing high performance software optimization techniques lay in unsufficiently precise characterization of solutions: there is a need for *precise and fast enumeration* in order to check for the presence of a solution and contingently count solutions. In this paper, we have addressed this issue. We propose a technique named *Fast Determination* for exactly and efficiently enumerating the number of solutions. Some examples of application are provided. First, we show that such a technique can act as a valuable complement to data dependence analysis tests such as *Omega Test*. Second, it is also shown how data locality optimization algorithms such as the *Window Theory* might benefit from *Fast Determination*. Performance and complexity of the technique are discussed.

In section 2, it is shown why some current optimization techniques need efficient enumeration for better and steadier performances. In section 3, it is explained how to represent in a simple form, and then to reduce the complexity of the original problem. Then the principles of *Fast Determination* itself are presented.

2 Applications requiring efficient enumeration

To give an insight at the range of applications that could benefit from *Fast Determination*, we detail two applications in data dependence analysis (section 2.1) and exploitation of data locality (section 2.2).

2.1 Enumeration in data dependence analysis: enhancing the Omega Test

In addition to basic tests like Banerjee test [1], a number of dependence tests has appeared [9, 7]

that aim at solving exactly dependence equations and moreover, at providing informations on dependence distances and dependence vectors. However, because these tests are in general based on elimination methods and projections, they may reach a point that no conclusion can be given on whether a system has solutions. In such cases, these tests resort to enumeration techniques, i.e they basically scan the solution space defined by the system constraints. Because of that particular flaw, such tests cannot provide stable and regular performance. Whenver such a case is encountered the performance of integer programming tests, which otherwise is high in general, may be extremely poor. Because performance stability is a prerequisite to any real implementation and commercial use of such analysis techniques, it is critical to provide tools for performing fast solution space scanning, i.e fast enumeration of solutions.

Our method is particularly well suited to be a complement to existing integer programming tests because it is focused on the efficiency of solutions enumeration. It is not a competitor for such tests as Omega because it does not provide symbolic information on solutions, it is less efficient at handling large systems of equations, and finally it does not provide informations on dependence distance and vectors, at least not at a cheap cost. On the other hand, it is an exact test in all cases, and its performance, though in average lower than Omega test, is much more stable because its strong asset is the capacity to efficiently scan the solution space. Therefore, it is a very proper tool for performing efficient enumeration when Omega or other tests would fail.

In the remaining of this section, three different examples are used to illustrate the relative behavior of Omega Test and *Fast Determination*.

Note: Because finding a "*representative* set of problems" is extremely difficult if not impossible considering the diversity of possible systems of equations, the purpose of the examples below is to *illustrate* the main behavior trends of *Omega Test* and *Fast Determination* rather more than experimentally *validate* them.

Complex systems of equations Omega Test performs better than *Fast Determination* when complex systems of equations must be handled (cf figure 1). For the following system, Omega Test can handle directly the set of equations; using elimination it can state that there are solutions.

$$0 \le j_1 \le N$$
$$0 \le j_2 \le N$$
$$2\, j_1 + 3\, j_2 = 15$$
$$7\, j_1 + 9\, j_2 = 10$$

On the other hand, *Fast Determination* collapses the two equations with 2 unknowns into one equation with 4 unknowns:

$$0 \le j_1 \le N$$
$$0 \le j_2 \le N$$
$$0 \le j_1' \le N$$
$$0 \le j_2' \le N$$
$$2\, j_1 + 3\, j_2 - 7\, j_1' - 9\, j_2' = 5$$

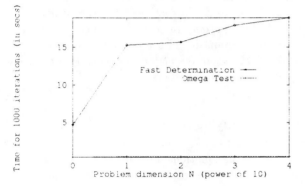

Figure 1: *Complex systems of equations.*

Because the number of unknowns in a single equation increases significantly, the performance is degraded. The performance of *Fast Determination* increases linearly with the solution soace size but it increases exponentially with the number of unknowns. So, in this case, performance of Omega Test is better than *Fast Determination*.

Simple systems of equations For simple systems of equations, the performance of Omega Test and *Fast Determination* are of the same order magnitude. For 4 unknowns and less, both tests have in general similar performances. Above 4 unknowns the complexity of *Fast Determination* is increasing and consequently Omega Test may perform better (cf figure 2).

Since the systems in figure 2 have only a single equation and several inequations, *Fast Determination* needs not collapse equations and therefore does not loose efficiency. Though in some cases *Fast Determination* behaves worse than Omega Test. it must be noted that Omega Test only guesses whether there is a solution while *Fast Determination* actually *computes* the exact number of solutions.

$$0 \le j_i \le 100$$
$$j_1 + 2j_2 = 20 \qquad (1)$$
$$2j_1 + 3j_2 + 12j_3 = 30 \qquad (2)$$
$$4j_1 + 6j_2 + 8j_3 + 10j_4 = 50 \qquad (3)$$

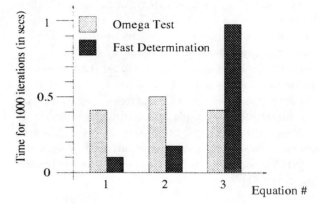

Figure 2: *Simple systems of equations.*

Systems requiring enumeration Finally, Omega Test seems to particularly lack efficiency when the number of solutions is high and/or the domain to be scanned is very large (cf figure 3). For the system of figure 3, the efficiency of Omega Test degrades severely when solution space increases, while *Fast Determination* performance remains stable (for greater increments of N, it would actually increase slowly). In this case, Omega Test has to perform enumeration for which it does not have any efficient technique.

$$0 \le j_i \le N$$
$$10j_1 + 162j_2 + 1299j_3 = 10000$$

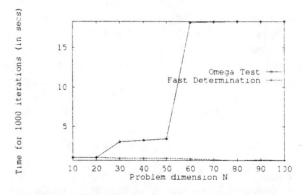

Figure 3: *Systems requiring enumeration.*

2.2 Enumeration in data locality exploitation: enhancing the *Window* strategy

The *Window Theory* is a quantitative algorithm for data locality optimization. The technique provides a framework for dealing with *windows*, i.e sets of array elements which are alive at a given moment [2].

Most other data locality optimization algorithms rely on similar methods, though the notion of *window* is not formally defined [11]. One of the main assets of the *window algorithm* is to provide the compiler with the *window* size along with a strategy for deciding whether a window should actually be kept in local memories, since many such windows generally compete for upper-level memory space. The window size is directly related to the average distance between two reuses of an element, and the decision to keep it is related to the average number of reuses per element. When array subscripts are simple, estimating the number of references per item and the average reuse distance is straightforward. It is not true anymore when subscripts are complex. In that case, only rough estimates are provided.

Let us consider the array reference $A(a_1 j_1 + a_2 j_2)$ (where $\gcd(a_1, a_2) = 1$ and j_1, j_2 are loop indexes). First, the number N_A of distinct elements of array A that are touched must be computed. Then the average number of references per element is given by $\frac{N_1 \times N_2}{N_A}$, where $0 \leq j_i \leq N_i$. A rough estimate for N_A is the difference between the smallest and highest array subscript values, assuming that all elements in-between have been touched, i.e $a_1 N_1 + a_2 N_2 - a_1 0 + a_2 0$. The number of values of a for which the following system has solutions indicates the number of distinct elements of A that have been referenced.

$$a \in [0, a_1 N_1 + a_2 N_2]$$
$$0 \leq j_i \leq N_i$$
$$a_1 j_1 + a_2 j_2 = a$$

Fast Determination can be used so as to quickly enumerate the number of values of variables a that are solutions, thus providing the *exact* value of N_A. Therefore, a precise average number of references per element can be computed, thereby considerably refining the decision to keep the window in local memory.

Computing the average reuse distance can be done using similar techniques. First, the enumeration must be done with parameter a as the outermost variable (enumeration is done in lexicographic order) so that distances correspond to references to the *same* array element. Then, the variables are ordered in the way the loop nest is structured, so that the distances correspond to *consecutive* references of a given array element. Then, the final modification is to record the distance between two references instead of the number of references. These precautions were not necessary for computing N_A because the order in which elements are referenced was not important. So, using this technique the *sum of distances* between consecutive references to elements of A, D_A, is *exactly* evaluated. Then $\frac{D_A}{N_1 \times N_2}$

gives a precise evaluation of the average reuse distance, thereby refining the window size computation.

While all these computations would be very costly with classic enumeration techniques, *Fast Determination* can perform them very quickly.

3 Efficient enumeration of the solutions

Most straightforward enumeration techniques lack efficiently because there is no attempt made to simplify the representation and reduce the complexity of the original problem. In section 3.1, a methodology for simplifying the original probleme representation is proposed. In section 3.2, two techniques for respectively reducing variable bounds and coefficient values are also proposed. In section 3.3.1, the principles of *Fast Determination* are presented. Let us now give the exact formulation of the problem considered.

Count the solutions of the following system of one equation and n inequations:
$$M_i \leq j_i \leq N_i, \quad (M_1, N_1), \dots, (M_n, N_n) \in \mathbf{Z} \times \mathbf{Z}$$
$$a_1 j_1 + \dots + a_n j_n = C, \quad a_1, \dots, a_n, C \in \mathbf{Z}$$

3.1 Simplifying the original problem

In this section, it is first shown how to collapse a set of multiple equations into one single equation (section 3.1.1). Normalizations for one such equation are proposed (section 3.1.2). Then, it is shown how to express a problem where variables are bounded using several problems having unbounded variables (section 3.1.3). In [10], a similar method is presented for dealing with non-constant boundaries.

3.1.1 Collapsing a set of equations

In some cases, the system to be solved is composed of several equations. In all cases, *linearization* and *equations rewriting* techniques can be used to collapse a set of equations into one single equation. Let us consider for instance the following problem:
$$A(b_1 j_1 + b_2 j_2 + b, c_1 j_1 + c_2 j_2 + c) = \dots$$
$$\dots = A(d_1 j_1 + d_2 j_2 + d, e_1 j_1 + e_2 j_2 + e)$$
The number of dependences between the two statements must be computed. The corresponding system of equations is the following:

$$b_1 j_1 + b_2 j_2 = b$$
$$d_1 j_1 + d_2 j_2 = d$$
$$c_1 j_1 + c_2 j_2 = c$$
$$e_1 j_1 + e_2 j_2 = c$$

If array subscripts are linearized before equations are written, collapsing the system is simplified because less equations appear. Let us call N the leading dimension of array A, then the system obtained after linearization is the following:

$$(b_1 j_1 + b_2 j_2 + b) + N(c_1 j_1 + c_2 j_2 + c) = 0$$
$$(d_1 j_1 + d_2 j_2 + d) + N(e_1 j_1 + e_2 j_2 + \epsilon) = 0$$

Then, using simple equations rewriting techniques, the two equations can be collapsed into the following single equation:

$$(b_1 j_1 + b_2 j_2 + b) + N(c_1 j_1 + c_2 j_2 + c)$$
$$= (d_1 j_1' + d_2 j_2' + d) + N(e_1 j_1' + e_2 j_2' + \epsilon)$$

where j_i' has the same bounds as j_i. These techniques can be employed for any set of equations. After system collapsing the complexity is neither reduced nor increased.

3.1.2 Normalizations

In this section, in order to fit the hypothesis of *Fast Determination*, the problem is normalized to the following form:

$$0 \le j_i \le N_i, \ N_1, \ldots, N_n \in \mathbf{N}$$
$$a_1 j_1 + \ldots + a_n j_n = C, \ a_1, \ldots, a_n, C \in \mathbf{N}$$

It must be noted that any original problem can be transformed to such a system (actually, C may not be made positive, but, in that case the system has no solution). The simplifications proposed are all straightforward. Let us shortly list them:

- *Make all $a_i \ge 0$.* If $a_i < 0$, write $a_i j_i = |a_i| j_i'$ with $j_i' = -j_i$, i.e., $j_i' \in [-N_i, -M_i]$. Therefore, through a modification on the bounding intervals of the j_i, it is now possible to assume that $a_i \ge 0, \forall i \in \{1, \ldots, n\}$.

- *Make all $M_i = 0$.* If $M_i \ne 0$, write $j_i' = j_i - M_i$, i.e., $j_i' \in \left[0, N_i'\right]$, with $N_i' = N_i - M_i$. The equation is now equivalent to $a_1 j_1 + \ldots + a_i j_i' + \ldots + a_n j_n = C'$ with $C' = C - a_i M_i$. Therefore, through a modification on C, we can now assume that $M_i = 0, \forall i \in \{1, \ldots, n\}$.

- *Divide the equation by the gcd of the coefficients.* Let $d_{1 \ldots n}$ denote the gcd of $a_1, \ldots, a_n$ $(\bigwedge_{i=1}^{n} a_i)$. If $d_{1 \ldots n}$ does not divide C then there is no solution to the equation. Otherwise, write $C' = \frac{C}{d_{1 \ldots n}}$ and $a_i' = \frac{a_i}{d_{1 \ldots n}}, \forall i 1 \le i \le n$. Therefore, through a modification on C, it is now possible to assume that $\bigwedge_{i=1}^{n} a_i = 1$. It must be noted that the gcd of several integers can be computed very quickly using Euclid's algorithm [6].

3.1.3 Unbounding the variables

A more complex modification of the problem is to unbound variables j_i, that is, to have $j_i \ge 0$ instead of $j_i \in [0, N_i]$. The reason for unbounding these variables is twofold. First, the bounds on the variables make the problem much more difficult to deal with. Essentially,

because of the greater number of unknowns and the fact that the solution space is finite, the complexity of finding solutions increases. Second, most number theory results on linear Diophantine equations do not take into account such bounds, and therefore cannot be applied to the problem considered.

Definition: Let $\Delta_{N_1, \ldots, N_n}(a_1, \ldots, a_n; C)$ be the number of solutions of the simplified problem of section 3.1.2, and $\Delta_\infty(a_1, \ldots, a_n; C)$ be the number of solutions of the same problem without bounds on variables j_i (i.e $0 \le j_i$). $\Delta_{N_1, \ldots, N_n}(a_1, \ldots, a_n; C)$ can be expressed as a function of $\Delta_\infty(a_1, \ldots, a_n; C)$.

Let us first recall that any rational fraction $\frac{P(x)}{Q(x)}$ where $P(x), Q(x)$ are polynomials, can be developed into an infinite serie: $\sum_{i=0}^{\infty} c_i x^i$. Then, a classic result in number theory [8] is that $\Delta_\infty(a_1, \ldots, a_n; C)$ is the coefficient of x^C in the series development of $\frac{1}{(1-x^{a_1}) \ldots (1-x^{a_n})}$, and that $\Delta_{N_1, \ldots, N_n}(a_1, \ldots, a_n; C)$ is the coefficient of x^C in the series development of $\frac{(1-x^{(N_1+1)a_1}) \ldots (1-x^{(N_n+1)a_n})}{(1-x^{a_1}) \ldots (1-x^{a_n})}$. In figure 4, it is shown how to decompose the bounding problem into a combination of unbounding problems using this property for $n = 2$.

Similarly, $(1 - x^{(N_1+1)a_1}) \ldots (1 - x^{(N_n+1)a_n})$ can be written, for any n, as $\sum_{i=0}^{2^n} \gamma_i x^{\delta_i}$, that is, a polynomial of 2^n terms. Therefore, $\Delta_{N_1, \ldots, N_n}(a_1, \ldots, a_n; C)$ can be expressed as $\sum_{i=0}^{2^n} \gamma_i \Delta_\infty(a_1, \ldots, a_n; C - \delta_i)$, an expression with 2^n different terms of $\Delta_\infty(a_1, \ldots, a_n; C)$.

Since the values of n considered are small enough, 2^n is not too large, rendering the computations of $\Delta_\infty(a_1, \ldots, a_n; C)$ feasible. So, it is possible to restrict our effort to determine the expression of $\Delta_\infty(a_1, \ldots, a_n; C)$.

Therefore, the new problem to be considered is the following:

$$0 \le j_i \qquad a_1, \ldots, a_n, C \in \mathbf{N}$$
$$a_1 j_1 + \ldots + a_n j_n = C$$

3.2 Reducing problem complexity

In this section, a techniques for reducing variables bounds is presented. In [10] a technique for also reducing coefficients values is proposed.

Direct reduction of problem variables In section 3.1.3, the number of solutions of a bounded equation is given as a linear combination of the number of solutions of unbounded equations. In this paragraph, a geometric interpretation of this transformation is provided and techniques for improving its efficiency are derived.

Let us consider case $n = 3$. Finding the number of solutions of $a_1 j_1 + a_2 j_2 + a_3 j_3 = C$ is equivalent

$$\frac{(1-x^{(N_1+1)a_1})(1-x^{(N_2+1)a_2})}{(1-x^{a_1})(1-x^{a_2})} = \frac{1}{(1-x^{a_1})(1-x^{a_2})} - \frac{x^{(N_1+1)a_1}}{(1-x^{a_1})(1-x^{a_2})} - \frac{x^{(N_2+1)a_2}}{(1-x^{a_1})(1-x^{a_2})} + \frac{x^{(N_1+1)a_1+(N_2+1)a_2}}{(1-x^{a_1})(1-x^{a_2})}$$

$$\Rightarrow \Delta_{N_1,N_2}(a_1,a_2;C) = \Delta_\infty(a_1,a_2;C) - \Delta_\infty(a_1,a_2;C-(N_1+1)a_1) - \Delta_\infty(a_1,a_2;C-(N_2+1)a_2) + \Delta_\infty(a_1,a_2;C-(N_1+1)a_1-(N_2+1)a_2)$$

Figure 4: *Relation between bounded and unbounded problem for $n = 2$.*

to computing the number of integer points within a triangle bounded by the axes, i.e defined by extreme points $j_i^{max} = \frac{C}{a_i}$ (cf figure 5). If the variables are bounded ($j_i \leq N_i$), then the problem is now equivalent to finding the number of integer points within the intersection of previous surface and the 3-cube $(0,0,0),(N_1,N_2,N_3)$ (coordinates of diagonally opposite corner points). Two cases may then occur:

- $j_i^{max} \geq N_i$ and the upper bound N_i is minimum.

- $j_i^{max} < N_i$, then the bound N_i can be taken equal to $\lfloor j_i^{max} \rfloor$.

Similarly, lower bounds of variables j_i are $j_i^{min} = \frac{C-\sum_{k\neq i} a_k N_k}{a_i}$. Again, two cases may occur:

- $j_i^{min} \leq 0$ and the lower bound 0 is maximum.

- $j_i^{min} > 0$, then j_i^{min} should be the new lower bound of j_i. This can be achieved by changing variable j_i into $j_i' = j_i - \lceil j_i^{min} \rceil$. The new upper bound of j_i' is then $N_i' = N_i - \lceil j_i^{min} \rceil$, and the right-hand side term C is now equal to $C' = C - a_i \lceil j_i^{min} \rceil$.

In addition to reducing the value of bounds and C, these transformations have an even more profitable impact on the computations complexity, which is explained in the following paragraphs.

The number of solutions of the unbounded problem is always greater or equal to that of the bounded problem. Actually the solutions of the bounded problem is a subset of the solutions of the unbounded problem. In the geometric interpretation, this assertion appears clearly because the solutions of the bounded problem are located within the intersection of an 3-cube and the surface enclosing the solutions of the unbounded problem (cf figure 5). Therefore, it is possible to relate the bounded and unbounded case by removing points which are solutions of the unbounded problem and not solutions of the bounded problem.

Choosing the best origin The n-cube has two opposite corner points of respective coordinates $O(0,\ldots,0), o'(N_1,\ldots,N_n)$. The bounded problem

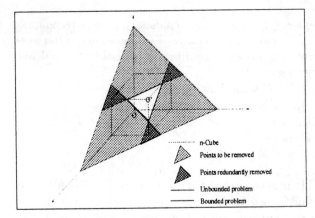

Figure 5: *Geometric interpretation of unbounded and bounded problems.*

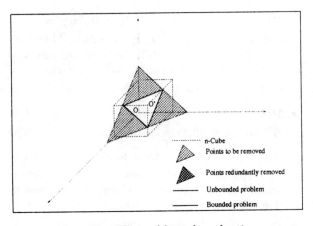

Figure 6: *Effect of bounds reduction.*

can be solved by considering either of the two points as the origin of j_i-axes. This simple transformation may again have an important impact on the length of computations. Indeed, depending on the position of the origin, the number of elements to be removed can vary. Figure 7 illustrates a case where by changing the origin the number of terms to be computed varies from 3 to 1. Figure 7 corresponds to figure 6 but with origins switched.

A criterium is now needed for determining when the origin must be changed. Basically, a first order term $\Delta_\infty(a_1,\ldots,a_n;C-a_iN_i)$ has to be computed if $N_i < \frac{C}{a_i}$, i.e if a corner point is outside the n-cube.

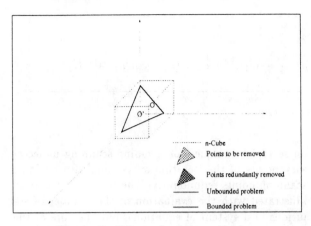

Figure 7: *Effect of origin switching.*

So, for each origin, the number of corner points that are located outside the n-cube is evaluated, and the origin for which the number of inner corner points is greater is selected for computations.

Switching th origin again corresponds to a simple transformation: $j_i' = N_i - j_i$, and consequently $C' = \sum_{i=1}^{n} a_i N_i - C$.

3.3 *Fast Determination*

The problem now considered is the unbounded problem of section 3.1.3, with all previous simplifications already performed (problem reductions of section 3.2 are not compulsory).

3.3.1 **Principles of** *Fast Determination*

In the next paragraphs, the computation of $\Delta_\infty(a_1, \ldots, a_n; C)$ is described for $n \in \{2, 3, 4\}$. The method can be directly generalized to greater values of n.

Case $n = 2$: The problem can be written as follows:

$$j_1 \geq 0, j_2 \geq 0$$
$$a_1, a_2, C \in \mathbf{N}$$
$$a_1 \wedge a_2 = 1$$
$$a_1 j_1 + a_2 j_2 = C$$

where hypothesis $a_1 \wedge a_2 = 1$ comes from the *gcd* simplification of section 3.1.2. Under these constraints, it is possible to apply Bezout's theorem to the equation. This theorem states that there exist two integers u_1, u_2 such that $a_1 u_1 + a_2 u_2 = 1$, and that all solutions of the equation have the following expression:

$$0 \leq j_1 = C u_1 - a_2 \lambda$$
$$0 \leq j_2 = C u_2 + a_1 \lambda$$

where λ is an integer parameter. The two constraints $j_1 \geq 0, j_2 \geq 0$ can be transformed into two constraints on λ (recall that $a_1 > 0$ and $a_2 > 0$):

$$\lambda \leq \frac{C u_1}{a_2} \text{ and } \lambda \geq \frac{-C u_2}{a_1}$$

Since λ is an integer, the number of solutions of the problem i.e., the number of possible values of λ, is given by the following expression:

$$\Delta_\infty(a_1, a_2; C) = \lfloor \frac{C u_1}{a_2} \rfloor - \lceil \frac{-C u_2}{a_1} \rceil + 1$$

The computation of the Bezout numbers u_1, u_2 using Euclid's algorithm is very efficient and fast. Besides, it is mentioned in section 3.1.2 that it is necessary to compute the *gcd* of a_1, a_2 in order to perform one of the simplifications, and that Euclid's algorithm can be used for that purpose also. Actually, it is possible to compute u_1, u_2 and $d_{12} = gcd(a_1, a_2)$ at the same time, during one single execution of Euclid's algorithm [6], which renders the computation of these three numbers very cheap.

Case $n = 3$: For $n = 3$, the equation can be rewritten as $a_1 j_1 + a_2 j_2 = C - a_3 j_3$. Consequently,

$$\Delta_\infty(a_1, a_2, a_3; C) = \sum_{all\ possible\ j_3} \Delta_\infty(a_1, a_2; C - a_3 j_3)$$

Let us find the possible values of j_3. If j_3 is a solution then d_{12} necessarily divides $(C - a_3 j_3)$, i.e.,

$$C - a_3 j_3 = 0 \bmod d_{12}$$
$$\Leftrightarrow C - a_3 j_3 \equiv 0 \, (d_{12})$$
$$\Leftrightarrow j_3 \equiv a_3^{-1} C \, (d_{12}), \text{ since } a_3 \wedge d_{12} = 1$$

Define $j_3^0 = a_3^{-1} C \bmod d_{12}$, then the possible values of j_3 are of the form $j_3 = j_3^0 + \lambda d_{12}$ (where λ is an integer parameter) [1]. There are two constraints on j_3 ($j_3 \geq 0$, $C - a_3 j_3 \geq 0$) which give the interval within which λ varies, i.e., $\lambda \in \left[0, \lfloor \frac{C - a_3 j_3^0}{a_3 d_{12}} \rfloor\right]$. Therefore,

$$\Delta_\infty(a_1, a_2, a_3; C) = \sum_{\lambda=0}^{\lfloor \frac{C - a_3 j_3^0}{a_3 d_{12}} \rfloor} \Delta_\infty(a_1, a_2; (C - a_3 j_3^0) - \lambda a_3 d_{12})$$

---

[1] The *integer inverse of numbers* $a_3^{-1} (d_{12})$ can be computed very quickly using Euclid's algorithm. For instance, $a_3^{-1} (d_{12})$ is an integer such that $a_3^{-1} a_3 \equiv 1 (d_{12})$, i.e., such that $\exists u \in \mathbf{N} : a_3^{-1} a_3 = 1 + u d_{12}$, which can be rewritten as $a_3^{-1} a_3 - u d_{12} = 1$. Therefore, finding $a_3^{-1} (d_{12})$ is strictly equivalent to computing the Bezout numbers of equation $a_3^{-1} j_1 - d_{12} j_2 = 1$, which justifies the use of Euclid's algorithm.

$$\Delta_\infty(a_1, a_2, a_3, a_4; C) = \sum_{\lambda=0}^{\lfloor \frac{C - d_{12}\mu^0}{d_{12}d_{34}} \rfloor} \Delta_\infty(a_1, a_2; C - \mu^0 d_{12} - \lambda d_{12}d_{34}) \times \Delta_\infty(a_1, a_2; \mu^0 d_{12} + \lambda d_{12}d_{34})$$

Figure 8: *Expression of the number of solutions in case $n = 4$.*

Case $n = 4$: For $n = 4$, the equation can be rewritten as $a_1 j_1 + a_2 j_2 = C - a_3 j_3 - a_4 j_4$.

For any value of (j_1, j_2) and (j_3, j_4), there are two integer parameters μ and ν such that

$$a_1 j_1 + a_2 j_2 = \mu d_{12} \tag{1}$$
$$a_3 j_3 + a_4 j_4 = \nu d_{34} \tag{2}$$

For given values of μ and ν, there are respectively $\Delta_\infty(a_1, a_2; \mu d_{12})$ pairs (j_1, j_2) solutions of (1), and $\Delta_\infty(a_3, a_4; \nu d_{34})$ pairs (j_3, j_4) solutions of (2). Since $\mu d_{12} = C - \nu d_{34}$, for each value of μ there exists at most one single value for ν. Now, for each pair (j_1, j_2), solution of $a_1 j_1 + a_2 j_2 = \mu d_{12}$ and each pair (j_3, j_4), solution of $a_3 j_3 + a_4 j_4 = C - \mu d_{12}$, the 4-tuplet (j_1, j_2, j_3, j_4) constitutes a solution to the equation. Therefore, for a given value of μ, there are $\Delta_\infty(a_1, a_2; \mu d_{12}) \times \Delta_\infty(a_3, a_4; C - \mu d_{12})$ solutions to the equation. Let us now give the expression of all possible values of μ. If μ is a solution then d_{34} divides $(C - \mu d_{12})$ (since $C - \mu d_{12} = \nu d_{34}$)

$$\Leftrightarrow C - \mu d_{12} \equiv 0 \ (d_{34})$$
$$\Leftrightarrow \mu \equiv d_{12}^{-1} C \ (d_{34}), \text{ since } d_{34} \wedge d_{12} = 1$$

Let $\mu^0 = d_{12}^{-1} C \mod d_{34}$, then the possible values of μ are of the form $\mu = \mu^0 + \lambda d_{34}$. There are two constraints on μ ($\mu \geq 0$, $C - d_{12}\mu \geq 0$), which give the interval within which λ varies. That is, $\lambda \in \left[0, \lfloor \frac{C - d_{12}\mu^0}{d_{12}d_{34}} \rfloor \right]$.

Therefore, the number of solutions is given by the expression of figure 3.3.1.

The method described in previous paragraphs can be straightforwardly extended to any value of n, except that instead of a formula or a single sum, several nested sums will have to be computed to obtain $\Delta_\infty(a_1, \ldots, a_n; C)$. So, the complexity of the computations will grow exponentially with n because the degree of nesting of the sums is equal to $\frac{n-2}{2}$.

4 Conclusion

We have presented a technique named *Fast Determination* for efficient enumeration of solutions of a system of equations and inequations. The technique can be employed for any system of equations and provides exact answers in all cases. Its efficiency is based on several techniques for avoiding scanning unnecessary points of the iteration space, and also avoiding scanning points redundantly. The technique has been illustrated with the evaluation of the number of solutions of a system of equations, but the core of the method, i.e the enumeration process, can be modified so as to serve other purposes, like evaluating distances between solutions...

The enumeration problem is a major obstacle to steady performance, accuracy and scope of application of many sophisticated algorithms dealing with data dependence analysis and data locality exploitation. *Fast Determination* can be inserted in such algorithms to perform efficient enumeration when it is necessary. The cost of *Fast Determination* is very moderate compared to that of standard enumeration strategies.

References

[1] U. Banerjee: *Dependence analysis for supercomputing*, Kluwer Academic Publisher, Norwell, Massachusetts, 1988.

[2] C. Eisenbeis, W. Jalby, D. Windheiser, F. Bodin: *A strategy for array management in local memory*, Advances in Languages and Compilers for Parallel Processing, MIT Press, 1991.

[3] C. Eisenbeis, O. Temam, H. Wijshoff: *On Efficiently Characterizing Solutions of Linear Diophantine Equations*, IRISA/INRIA Report 633, Jan. 1991.

[4] K. Gallivan, W. Jalby, U. Meier: *The use of BLAS3 in linear algebra on a parallel processor with a hierarchical memory*, SIAM Journal on Scientific and Statistical Computing, Nov. 1986.

[5] D. Gannnon and al.: *SIGMA II: A Tool Kit for Building Parallelizing Compiler and Performance Analysis Systems*, University of Indiana technical report, 1992.

[6] R. L. Graham, D. E. Knuth, O. Patashnik: *Concrete mathematics*, Addison-Wesley, 1989.

[7] Y. Jegou: *An algorithm for symbolic analysis of dependence equations*, IRISA/INRIA technical report, Dec. 1992.

[8] G. Polya, G. Szego: *Lessons and exercises of analysis*, Springer-Verlag, 1964.

[9] W. Pugh: *The Omega Test: a fast and practical integer programming algorithm for dependence analysis*, to appear in Communications of the ACM.

[10] C. Eisenbeis, O. Temam, H. Wijshoff: *Characterizing the solutions of linear Diophantine equations*, IRISA/INRIA Technical report, January 1992.

[11] M. E. Wolf and M. Lam: *A Data Locality Optimizing Algorithm*, Proc. of PLDI.

Square Meshes Are Not Optimal For Convex Hull Computation *

(Extended Abstract)

D. Bhagavathi[†]   H. Gurla[†]   S. Olariu[†]   R. Lin[‡]   J. L. Schwing[†]   J. Zhang[§]

Summary

Recently it has been noticed that for semigroup computations and for selection, rectangular meshes with multiple broadcasting yield faster algorithms than their square counterparts. The contribution of this paper is to provide yet another example of a fundamental problem for which this phenomenon occurs. We show that the problem of computing the convex hull of a set of n sorted points in the plane can be solved in $O(n^{\frac{1}{8}}\log^{\frac{3}{4}}n)$ time on a rectangular mesh with multiple broadcasting of size $n^{\frac{3}{8}}\log^{\frac{1}{4}}n \times n^{\frac{5}{8}}/\log^{\frac{1}{4}}n$. The fastest previously-known algorithms on a square mesh of size $\sqrt{n} \times \sqrt{n}$ run in $O(n^{\frac{1}{6}})$ time in case the n points are pixels in a binary image, and in $O(n^{\frac{1}{6}}\log^{\frac{2}{3}}n)$ time for sorted points in the plane.

1 Introduction

Recently, an architecture referred to as mesh with multiple broadcasting (MMB, for short), has been proposed. The MMB is obtained by adding one bus to every row and to every column of a mesh connected computer (see Figure 1). It has been shown that semigroup operations can be performed faster if rectangular MMB's are used instead of square ones [2]. A similar phenomenon occurs in selection [3, 5].

Our contribution is to exhibit an algorithm that finds the convex hull of a sorted set of n points in the plane in $O(n^{\frac{1}{8}}\log^{\frac{3}{4}}n)$ time on an MMB of size $n^{\frac{3}{8}}\log^{\frac{1}{4}}n \times n^{\frac{5}{8}}/\log^{\frac{1}{4}}n$. The fastest previously-known algorithms solve the problem in $O(n^{\frac{1}{6}})$ time in the special case where the n points are pixels in a binary image, and in $O(n^{\frac{1}{6}}\log^{\frac{2}{3}}n)$ time for n sorted points in the plane, both on an MMB of size $\sqrt{n} \times \sqrt{n}$.

*Work supported by NASA under grant NCC1-99 and by the NSF grant CCR-8909996

[†]Dept. of Computer Science, Old Dominion University, Norfolk, VA 23529

[‡]Dept. of Computer Science, SUNY, Geneseo, NY 14454

[§]Dept. of Mathematics and Computer Science, Elizabeth City State University, Elizabeth City, NC 27909

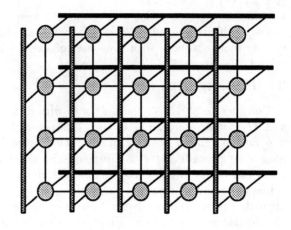

Figure 1: A 4 × 5 mesh with multiple broadcasting

2 Preliminaries

Let S be a set of points in the plane and let $P = (p_1, p_2, \ldots, p_k)$, be the upper hull of S. Let the points of P be stored in one row of an MMB.

Lemma 2.0. The supporting line from a point to an upper hull stored in a row of an MMB can be determined in $O(1)$ time. $\square$

A *sample* of P is an ordered subset of P. Consider an arbitrary sample $A = (p_1 = a_0, a_1, \ldots, a_s = p_k)$ of P. A partitions P into s pockets $A_1, A_2, \ldots, A_s$, such that A_i involves the points in P lying between a_{i-1} and a_i. Let the points of P be stored in an MMB with points belonging to sample A in one row. Let σ be the largest size of a pocket in P and let $t(\sigma)$ be the time needed to move the points of the largest pocket to one row.

Lemma 2.1. The supporting line to an upper hull from a point can be determined on an MMB in time $t(\sigma)$. $\square$

Lemma 2.2. The supporting line of two separable upper hulls U_1 and U_2 stored in one row of an MMB takes $O(\log\min\{|U_1|, |U_2|\})$ time. $\square$

Consider the problem of finding the common tangent of two upper hulls $P = (p_1, p_2, \ldots, p_k)$ and

$Q = (q_1, q_2, \ldots, q_l)$. Consider samples $A = (p_1 = a_0, a_1, \ldots, a_s = p_k)$ and $B = (q_1 = b_0, b_1, \ldots, b_t = q_l)$ of P and Q, respectively. The two samples determine pockets $A_1, A_2, \ldots, A_s$ and $B_1, B_2, \ldots, B_t$ in P and Q. Let the supporting line of A and B be achieved by a_i and b_j, and let the supporting line of P and Q be achieved by p_u and q_v. The following result has been established in [1].

Proposition 2.3. [1] At least one of the following statements is true: (a) $p_u \in A_i$; (b) $p_u \in A_{i+1}$; (c) $q_v \in B_j$; (d) $q_v \in B_{j+1}$. $\square$

Lemma 2.4. Computing the common tangent of two separable upper hulls P and Q stored in column-major order in a mesh with row buses of size $y \times 2z$ takes $O(y + \log z)$. $\square$

3   The Algorithm

Consider an MMB $\mathcal{R}$ of size $M \times N$ with $M \geq N$ and $MN = n$. The input is an arbitrary set S of n points in general position in the plane sorted by increasing x coordinate. It is assumed to be stored in block row-major order. Our goal is to determine the values of M and N such that the running time of the algorithm is *minimized*, over all possible choices of rectangular meshes containing n processing elements.

Proposition 3.1. [6] The convex hull of a set on n points in the plane can be computed in $O(\sqrt{n})$ time on a mesh-connected computer of size $\sqrt{n} \times \sqrt{n}$. $\square$

Corollary 3.2. The convex hull of a set on n points in the plane can be computed in $O(\max\{a, b\})$ time on a mesh-connected computer of size $a \times b$, with $a * b = n$.

We start out by setting our overall target running time to $O(x)$, with x to be determined later, along with M and N. Throughout the algorithm we view the original mesh as consisting as a set of submeshes R_j $(1 \leq j \leq \frac{M}{y})$ of size $y \times N$ each, with the value of y ($y \leq x$) to be specified later. We further view each R_j as consisting of a set of submeshes $R_{j,k}$ $(1 \leq j \leq \frac{M}{y}; 1 \leq k \leq \frac{N}{x})$ of size $y \times x$. Our algorithm is partitioned into three distinct stages whose description follows.

Stage 1. {Preprocessing} View the mesh $\mathcal{R}$ as consisting of the submeshes $R_{j,k}$ $(1 \leq j \leq \frac{M}{y}, 1 \leq k \leq \frac{N}{x})$ described above. In each $R_{j,k}$ compute the upper hull using the optimal algorithm in [6]. By virtue of Corollary 3.2 this takes $O(x)$ time. In every $R_{j,k}$, in addition to computing the upper hull we also choose a sample, that is, a subset of the points on the corresponding upper hull. The sample is chosen to contain the first hull point in every column of $R_{j,k}$ in topdown order. In addition to this the following information is computed in Stage 1.

(H) for every hull point, its rank on the upper hull, along with the identity and coordinates of its left and right neighbors on the upper hull computed so far;
(S) for every sample point, its rank within the sample, along with the identity and coordinates of its left and right neighbors (if any) in the sample.

Note that the information specified in (H) and (S) can be computed in time $O(x)$ by using local communications within every $R_{j,k}$.

Stage 2. {Horizontal Stage} This stage involves computing the upper hull of the points in every R_j $(1 \leq j \leq \frac{M}{y})$, while maintaining the conditions (H) and (S) invariant. It involves repeatedly finding the common tangent of two neighboring upper hulls and merging them, until only one upper hull remains. The definition of the submeshes R_j guarantees that no pocket contains more than y points.

At the beginning of the i-th step of Stage 2, a generic submesh R_j contains the upper hulls U_1, U_2, $\ldots, U_{\frac{N}{2^{i-1}x}}$, with U_1 being the upper hull of the points in $R_{j,1}$, $R_{j,2}$, $\ldots$, $R_{j,2^{i-1}}$, U_2 standing for the upper hull of the points in $R_{j,2^{i-1}+1}$, $\ldots$, $R_{j,2^i}$. In this step, we merge each of the $\frac{N}{2^i x}$ consecutive pairs of upper hulls. We now show how U_1 and U_2 are merged into the upper hull of the points in $R_{j,1}$, $R_{j,2}$, $\ldots$, $R_{j,2^i}$. We also show that the invariants (H) and (S), assumed to hold at the beginning of the i-th step, continue to hold at the end of the step.

We restrict the broadcasting involved in the computation of the common tangent of U_1 and U_2 to the bus in the first row of R_j. Since all that is required to satisfy this condition is to move the sample and the pockets to the one row, Lemma 3.4 guarantees that the whole computation runs in $O(y + \log 2^{i-1}x)$ time. Once the common tangent is known, we need to eliminate from U_1 and U_2 the points that no longer belong to the new upper hull. Note that, this is done while preserving the invariants (H) and (S).

Assume, without loss of generality, that some points u in U_1 and v in U_2 are the touching points of the common tangent. To correctly update the upper hull of the union of U_1 and U_2, we need to eliminate all the points that are no longer on the upper hull. Firstly, we mandate the processor holding the point u to send up to the first row of R_j a packet containing the coordinates of u and v, along with the rank of u in U_1 and the rank of v in U_2. The corresponding processor in the first row of R_j will broadcast the packet along the row bus. Every processor in the first row of R_j belonging to $R_{j,1}$, $\ldots$, $R_{j,2^i}$ will transmit the packet southbound in its own column, using local communications. Every processor in $R_{j,1}$, $R_{j,2}$, $\ldots$, $R_{j,2^i}$ storing a point in U_1 or U_2 decides whether the point it stores remains in the upper hull or not. Therefore, the update can be correctly performed in $O(y)$ time. To preserve (II), every point on the upper hull of the union of U_1 and U_2 must compute its rank in the new hull and to identify its left and right neighbors. Clearly, every point in the new upper hull keeps its own neighbors except for u and v, which become each other's neighbors. To see that every point on the new convex hull is in a position to correctly update its rank, note that all points on the hull to the left of u keep their own rank; all points to the right of v update their ranks by first subtracting 1 plus the rank of v in U_2 from their own rank, and then by adding the rank of u in U_1 to the result. Thus, the invariant (H) is preserved.

To see that invariant (S) is also preserved, note that every sample point to the left of u is still a sample point in the new hull; similarly, every sample point to the right of v is a sample point in the new hull. The only sample points that may change are those in the same column as v. Therefore, all sample points in the new upper hull can be correctly identified. In addition, all of them keep their old neighbors in the sample set, except for two sample points: one is the sample point in the column containing u and the other is v. In another broadcast these sample points can find their neighbors. Similarly the rank of every sample point within the new sample set can be computed. Therefore, the invariant (S) is also preserved.

Lemma 3.3. The common tangent of U_1 and U_2 can be computed in $O(y + \log 2^{i-1}x)$ time in the first row of R_j. Furthermore, the invariants (H) and (S) are preserved. $\square$

The reason behind computing the common tangent of U_1 and U_2 in the first row of R_j is to dedicate the first row bus to the first pair of upper hulls, the second bus to the second pair of upper hulls, and so on. Thus, the computation involving the first y pairs of hulls can be performed in parallel. By virtue of Lemma 4.3, the common tangents in each group of y pairs of upper hulls in R_j can be computed in $O(y + \log 2^{i-1}x)$ time. Since there are $\frac{N}{2^i xy}$ such groups, synchronizing the computation in all the groups guarantees that the i-th step of Stage 2 takes $O(y + \frac{N \log 2^{i-1}x}{2^i xy})$ time.

It is easy to verify that the running time of Stage 2 is in $O(y \log n + \frac{N \log n}{xy})$ [4]. Since we want the overall running time to be restricted to $O(x)$ we write

$$(y + \frac{N}{xy}) \log n \leq x. \tag{1}$$

Stage 3. {Vertical Stage} Recall that at the end of

Stage 2, every submesh R_j ($1 \leq j \leq \frac{M}{y}$) contains the upper hull of the points stored by processors in R_j. The task specific to Stage 3 involves repeatedly merging pairs of two neighboring groups of R_j's as described below.

At the beginning of the i-th step of Stage 3, the upper hulls in adjacent pairs each involving 2^{i-1} consecutive R_j's are being merged. We only show how the pair of upper hulls of points in $R_1, R_2, \ldots, R_{2^{i-1}}$ and $R_{2^{i-1}+1}, R_{2^{i-1}+2}, \ldots, R_{2^i}$ is updated into a new upper hull of the points in $R_1, R_2, \ldots, R_{2^i}$. We refer to the submeshes $R_1, R_2, \ldots, R_{2^{i-1}}$ as group G_1 and to $R_{2^{i-1}+1}, R_{2^{i-1}+2}, \ldots, R_{2^i}$ as group G_2. In Stage 3 we no longer need sampling. We shall also prove that in the process the invariant (H), assumed to hold at the beginning of the i-th step, continues to hold at the end of the step.

Let U_1 and U_2 be the upper hulls of the points stored in G_1 and G_2, respectively. As a first step, we compute the common tangent of U_1 and U_2; once this common tangent is available, the two upper hulls will be updated into the new upper hull of all the points in $R_1, R_2, \ldots, R_{2^i}$. In stage 3, the horizontal buses within every group will be used to broadcast information, making it unnecessary to move data to a prescribed row. Let $U_1 = u_1, u_2, \ldots, u_p$ and $U_2 = v_1, v_2, \ldots, v_q$, with all the points in U_1 to the left of U_2. We assume that (H) holds, that is, the points in U_1 and U_2 know their rank within their own upper hull, as well as the coordinates of their left and right neighbors (if any) in the corresponding upper hull.

To begin, the processor storing the point $u_{\frac{p}{2}}$ broadcasts on the bus in its own row a packet consisting of the coordinates of $u_{\frac{p}{2}}$ and its rank in U_1. In turn, the corresponding processor in the first column of the mesh will broadcast the packet along the bus in the first column. Every processor in the first column of the mesh belonging to $R_{2^{i-1}+1}, R_{2^{i-1}+2}, \ldots, R_{2^i}$ read the bus and then broadcast the packet horizontally on the bus in their own row. Note that as a result of this data movement, the processors in the group G_2 have enough information to detect whether the points they store achieve the supporting line to U_2 from $u_{\frac{p}{2}}$. Using the previous data movement in reverse, the unique processor that detects this condition broadcasts a packet consisting of the coordinates and rank of the point it stores back to the processor holding $u_{\frac{p}{2}}$.

By checking its neighbors on U_1, this processor detects whether the supporting line to U_2 from $u_{\frac{p}{2}}$ is supporting for U_1. In case it is, we are done. Otherwise, the convexity of U_1 guarantees that half of the points in U_1 can be eliminated from further consideration. This process continues for at most $\lceil \log 2^{i-1} yN \rceil$ iterations. Consequently, the task of computing the common tangent of U_1 and U_2 runs in $O(\log 2^{i-1} yN)$ time. Now, eliminate from U_1 and U_2 the points that no longer belong to the new upper hull. We now show that, all this is done while preserving the invariant (II).

Assume, without loss of generality, that some points u in U_1 and v in U_2 are the touching points of the common tangent. To correctly update the upper hull of the union of U_1 and U_2, we need to eliminate all the points that are no longer on the upper hull. As a first step, the processor holding the point u broadcasts on its own row a packet containing the coordinates of u and v, along with the rank of u in U_1 and the rank of v in U_2. The corresponding processor in the first column will broadcast the packet along the column bus. Every processor in the first column belonging to R_1, ..., R_{2^i} will broadcast the packet horizontally on the bus in their own row. Every processor in $R_{j,1}$, $R_{j,2}$, ..., $R_{j,2^i}$ storing a point in U_1 or U_2 decides whether the point is stores should remain in the upper hull or not. The task of eliminating points that no longer belong to the new upper hull can be performed in $O(1)$ time. To preserve (II), every point on the upper hull of the union of U_1 and U_2 must be able to compute its rank on the new hull and also identify its left and right neighbors.

Clearly, every point in the new upper hull keeps its own neighbors except for u and v, which become each other's neighbors. To see that every points on the new convex hull is in a position to correctly update its rank, note that all points on the hull to the left of u keep their own rank; all points to the right of v update their ranks by first subtracting 1 plus the rank of v in U_2 from their own rank and by adding the rank of u in U_1. All the required information was made available in the packet previously broadcast. Thus, the invariant (II) is preserved. To summarize our discussion we state the following result.

Lemma 3.4. The common tangent of U_1 and U_2 can be computed in $O(\log 2^{i-1} yN)$ time using vertical broadcasting in the first column of the mesh only. Furthermore, the invariant (II) is preserved. $\square$

Notice that we have computed the common tangent of U_1 and U_2 restricting vertical broadcasting to the first column of the mesh. The intention was to assign the first column bus to the first pair of upper hulls, the second bus to the second pair of upper hulls, and so on. The computation involving the first N pairs of hulls can be performed in parallel. Therefore, by virtue of Lemma 4.4 the common tangents in each of the first N upper hulls in R_j can

be computed in $O(\log 2^{i-1} yN)$ time. Since there are $M/2^i yN$ such pairs, the i-th step of Stage 3 takes $O(M \log 2^{i-1} yN / 2^i yN)$ time. It is easy to verify that the overall running time of Stage 3 is $O(M \log n / yN)$ [4]. As we want to restrict this to $O(x)$, we write

$$\frac{M \log n}{yN} \leq x. \qquad (2)$$

It is a straightforward, albeit slightly tedious, to verify that the values of x, y, M, and N that simultaneously satisfy constraints (1), and (2) so as to minimize the value of x are:
$x = n^{\frac{1}{8}} \log^{\frac{3}{4}} n$; $y = n^{\frac{1}{8}} / \log^{\frac{1}{4}} n$; $M = n^{\frac{5}{8}} / \log^{\frac{1}{4}} n$; $N = n^{\frac{3}{8}} \log^{\frac{1}{4}} n$.

Theorem 3.5. The problem of computing the convex hull of a set of n points in the plane sorted by increasing x coordinate can be solved in $O(n^{\frac{1}{8}} \log^{\frac{3}{4}} n)$ time on an MMB of size $n^{\frac{3}{8}} \log^{\frac{1}{4}} n \times n^{\frac{5}{8}} / \log^{\frac{1}{4}} n$. $\square$

References

[1] M. J. Atallah and M. T. Goodrich, Parallel algorithms for some functions of two convex polygons, *Algorithmica* 3, (1988) 535–548.

[2] A. Bar-Noy and D. Peleg, Square meshes are not always optimal, *IEEE Transaction on Computers*, C-40, 1991, 196–204.

[3] D. Bhagavathi, P. J. Looges, S. Olariu, J. L. Schwing, and J. Zhang, A fast selection algorithm on meshes with multiple broadcasting, *Proc. International Conference on Parallel Processing*, 1992, St-Charles, Illinois, III, 10–17.

[4] Bhagavathi *et al.*, Square meshes are not optimal for convex hull computation, *TR-92-31*, December 1992, Dept. of Computer Science, Old Dominion University, USA.

[5] Y. C. Chen, W. T. Chen, G. H. Chen and J. P. Sheu, Designing efficient parallel algorithms on mesh connected computers with multiple broadcasting, *IEEE Transactions on Parallel and Distributed Systems*, vol. 1, no. 2, 1990.

[6] C. S. Jeong and D. T. Lee, Parallel Convex Hull Algorithms in 2- and 3-dimensions on Mesh-Connected Computers, *Proc. Internat. Conference on Parallel Processing for Computer Vision and Display*, Leeds, UK, 1988.

[7] F. P. Preparata and M. I. Shamos, Computational Geometry – An Introduction, Springer-Verlag, 1988.

Embedding Large Mesh of Trees and Related Networks in the Hypercube with Load Balancing

Kemal Efe,
The Center for Advanced Computer Studies,
University of Southwestern Louisiana, Lafayette, LA 70504

Abstract

The ability to embed arbitrarily large graphs in smaller graphs has important applications in mapping problems which require more processors than is available in a parallel architecture. We address this problem with the main focus on balancing processor loads. We show that maximum level of system utilization can be obtained when each host node emulates an equal number of "busy" guest nodes for each step of computation. While the embedding methods used are applicable to a variety of guest graphs, the main focus of the paper is on meshes of trees due to their importance as parallel architectures. Methods are also developed for embedding arbitrarily large complete binary trees and grids.

1 Introduction

Mesh of trees are powerful architectures for parallel computation. Therefore, it is highly desirable to develop efficient embeddings of mesh of trees on the available architectures. A r-dimensional $N \times N \times \cdots \times N$ mesh of trees is obtained from r-dimensional $N \times N \times \cdots \times N$ grid by replacing the linear connections at each dimension for N-leaf complete binary trees. The resulting network contains $(r + 1)N^r - rN^{r-1}$ nodes.

Earlier, Leighton showed that [6] the r-dimensional $N \times N \times \cdots \times N$ meshes of trees can be embedded in $(2N)^r$-node hypercubes with dilation cost 2. The same result was independently discovered in [8]. Other researchers [2], showed that two dimensional $N \times N$ mesh of trees (containing $3N^2 - 2N$ nodes) are subgraphs of $4N^2$ node hypercubes, eliminating the dilation of 2.

We are mainly interested in embedding large guest graphs in smaller hypercubes. Many parallel algorithms assume that there are as many processors as needed by the algorithm. To maintain this virtual picture of "unlimited number of processors" to a programmer of the system, there needs to be a way to automatically map large computations to the available number of processors. This raises a need for many-to-one mappings from the nodes of the guest to the nodes of the host.

1.1 Definitions and Motivations

Expansion of embedding plays a key role in the efficiency of emulating one architecture by another. Expansion of an embedding is defined as the ratio N_h/N_g, where N_h is the number of nodes in the host graph, and N_g is the number of nodes in the guest graph. We consider an embedding optimal if this ratio is unity. If the ratio is $O(1)$, then we say that the embedding is "processor preserving," since then the host uses the same number of nodes (within a constant factor) as the guest to do the same computation. If the mapping is many-to-one, we require that the mapping be "work preserving."

The concept of work preserving emulations was originally introduced in [5]. The *work* of a computation is defined as the product of running time and the number of processors used. If the size of the host is smaller by a factor of k, we require for the running time to increase not more than by a factor of $O(k)$. As a corollary of theorem 3 in Section 3, we tie together the two concepts of processor preserving emulations (for one-to-one mapping) and work preserving emulations (for many-to-one mapping), by showing that if the one-to-one embedding is not processor preserving, then the corresponding many-to-one embedding cannot be work preserving.

Given these restrictions, the embedding methods in [2, 6, 8] are processor preserving for r-dimensional mesh of trees when r is fixed. For r-dimensional mesh of trees, the embedding uses $O((r+1)N^r)$ nodes while the corresponding hypercube in this embedding contains $(2N)^r$ nodes. By dividing the two values we find that the expansion is $O(2^r/(r+1))$, which can be considered as $O(1)$ when r is fixed. Nevertheless, this expansion is too large for practical purposes.

The following example illustrates the importance of expansion: Suppose we wish to use a matrix multiplication algorithm by embedding the mesh of trees in the hypercube. To multiply a 4×4 matrix with a 4×1 vector, we need a 40-node mesh of trees. The smallest hypercube with enough number of nodes contains 64 nodes. Then, 24 nodes out of 64 would never be used by the algorithm. Each of the remaining 40 nodes would be used only for a small fraction of the computation time, due to the "normal" nature of the computation.

The concept of *normal* algorithms was originally introduced by Ullman [7] to refer to the type of computations which use only a particular subset of processors at a time. For instance, normal algorithms running on the binary tree architecture, e.g. all semigroup computations, use only one level of the tree nodes at a time. For mesh of trees, the concept of normal algorithms is defined similarly. For example, all of the graph problems presented in [6] for mesh of trees have this property; i.e. they use only one level of trees at a time.

We show in this paper that the "normal" nature of parallel computations is not necessarily a bad property. We can tailor the embedding methods to particularly take advantage of this normal behavior, and increase the processor utilization significantly. As a result, the proposed embeddings are even better than work preserving embeddings.

1.2 Results

We first show that r-dimensional $N \times N \times \cdots \times N$ mesh of trees are subgraphs of $(2N)^r$-node hypercubes. This result represents no improvement over that of [6] in terms of expansion, but reduces the dilation from 2 to 1. It also represents an improvement over the method in [2] by generalizing the embedding to r-dimensions. This embedding will help understand some of the difficulties of some straightforward methods to achieve load balanced embeddings.

Next, we show that arbitrarily large complete binary trees can be embedded in a fixed size hypercube with uniform load, dilation 2, and all hypercube nodes are utilized. This result is then used to show that r-dimensional $N \times N \times \cdots \times N$ mesh of trees can be embedded in N^r-node hypercube with the uniform load of $r + 1$, and there are no unused hypercube nodes. This embedding also has the remarkable feature that out of the $r + 1$ mesh of trees nodes mapped to a hypercube node, exactly one is a leaf, and exactly one is an internal tree node from dimension i for all $i \in \{1 \cdots r\}$. This property of the embedding is highly desirable for normal algorithms since different dimension trees tend to be used at different times. Moreover, the expansion cost of $2^r/(r+1)$ in the above embedding is eliminated. For example, embedding of the 4×4 mesh of trees requires a 16-node hypercube only.

This result is further extended for arbitrary size mesh of trees and fixed size hypercubes with special attention for the needs of normal algorithms. This is a natural extension of the embedding method above. This feature of the embedding ensures that every hypercube node is assigned the same amount of "busy" load, leading to maximum processor utilization. The dilation of embedding is 2 and congestion is proportional to the load.

These results are formally presented as follows:

Theorem 1 *The r-dimensional $N \times N \times \cdots \times N$ mesh of trees is a subgraph of the $(2N)^r$-node hypercube.*

Theorem 2 *The r-dimensional $N \times N \times \cdots \times N$ mesh of trees can be embedded in N^r-node hypercube with dilation 2, and uniform load of $r+1$ for each hypercube node.*

From this theorem and Theorem 3 in Section 3, we have the following result:

Corollary 1 *The r-dimensional meshes of trees with $n + k$ levels in each tree can be embedded with dilation 2, and uniform load of $(r+1)2^k$ in $(r \times n)$-dimensional hypercube.*

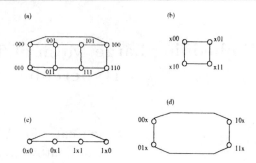

Figure 1: Contracting a 3-dimensional hypercube. b: along dimension 2, c: along dimension 1, and, d: and along dimension 0.

2 Embedding with Large Expansion

To avoid duplication, we assume that the reader is familiar with [2].

Proof of Theorem 1 (Sketch): The basic idea is to start with an embedding of the two-dimensional mesh of trees as in [2], and then (a) Expand the hypercube containing the two-dimensional mesh of trees by adding another dimension, (b) Connect each leaf to its adjacent neighbor in that new dimension, and (c) Attach tails as in [2]. (d) Using these tails in a similar way as in the construction of two dimensional mesh of trees, we can expand the $N \times N$ mesh of trees to $N \times N \times N$. Details are provided in [3].

Similarly, r-dimensional mesh of trees can be obtained from the $(r - 1)$-dimensional mesh of trees by the construction method above. The initial 2-dimensional mesh of trees requires a $(2N)^2$-node hypercube since it contains $3N^2 - 2N$ nodes, and therefore the expansion is minimal for two dimensions. Going from i-dimensional mesh of trees to $i + 1$ dimensions, we multiply the size of required hypercube by $2N$ (2 for step 1, and N for step 4), which yields the result of $(2N)^r$ for r dimensions. ∎

From an academic point of view this result is interesting, but the expansion is quite large and makes it practically impossible to utilize in real applications. In the remainder of this paper, we focus on embedding mesh of trees in smaller hypercubes.

3 Work Preserving Embeddings

Earlier, Gupta [4] showed that large hypercubes can be embedded in smaller hypercubes by dimension contractions. In this method, $(n + k)$-dimensional hypercube is shrunk down to n dimensions by replacing k bits in address labels by don't care symbols. Figure 1 shows how a 3-dimensional hypercube can be shrunk into a 2-dimensional hypercube.

A straightforward method of embedding a large guest graph to a smaller hypercube can be achieved by embedding the guest graph to a large enough hypercube, and then shrinking the hypercube to a smaller size. While this method works well for certain guest graphs, it may not produce the best possible embedding in general. The theorem below states the conditions for obtaining work preserving embeddings. Proof is provided in [3].

Theorem 3 *We state the theorem in two parts:*

1. *To construct an embedding of an arbitrary guest graph G containing N_G nodes in n-dimensional (smaller) hypercube with dilation d (where $d = O(1)$) and uniform load of $L = N_G/2^n$ by the method of dimension contraction, it is necessary and sufficient to find an embedding of G with dilation d and uniform load of ℓ (where $\ell < L$) embedding on some $(n + k)$-dimensional hypercube with optimal expansion.*

2. *If every dilation d and uniform load ℓ embedding of G requires a hypercube of size e times the minimal expansion, the difference between the maximum and the minimum processor loads for embedding in n-dimensional (smaller) hypercube is at most $e(N_G/2^n)$.*

The result below immediately follows:

Corollary 2 *If $O(1)$ dilation embedding of G requires more than $O(1)$ expansion in a larger hypercube, the dimension contraction method does not yield a work preserving embedding for the minimal size hypercube.*

Proof: If G has close to (and not more than) 2^n nodes and requires k dimensions more that the minimal size, where $k = O(f(n))$, then at least $2^n(2^k - 1)$ processors are never used for one-to-one mapping. Since the edges of G must span all of the $n + k$ dimensions, each contraction step must increase the load of some host nodes, which are already assigned some guest nodes. After k steps of contraction these loads may be at most multiplied by 2^k, which leads to load levels varying between $\ell = 0$ and $\ell = 2^k$.

To complete the proof, we can show that if G is connected, at the end of k contraction steps (a) at least k host nodes must have no guest nodes assigned to them, and, (b) at least one host node must have $O(k)$ guest nodes assigned to it. Since every contraction step must increase the load of some host node already containing a guest node, claim (a) follows since assigning two or more guest nodes to a host node leaves one or more host node vacant. Claim (b) follows from the observation that for every subset of k dimensions out of the $n + k$ dimensions spanned, there must be a k-dimensional subcube which contains at least $O(k)$ guest nodes assigned to its nodes. Then, contracting these k dimensions will bring all of those $O(k)$ guest nodes into a single host node. If for some subset of k dimensions there were no subcube which contains at least $O(k)$ guest nodes, then dilation of embedding would be larger than $O(1)$, which leads to a contradiction. ∎

The above corollary only focuses on the first k steps of contractions. By a similar reasoning, it is easy to see that further steps of contraction may preserve this load imbalance, or even amplify it, except for some special cases. While the worst case load distribution need not happen for every guest graph, the result is significant in that, it points out how bad the situation can get.

For the embedding of a complete binary tree, if we consider the second root as used, then the following result immediately follows:

Corollary 3 *Double rooted complete binary trees with 2^n nodes can be embedded in smaller hypercubes with unit dilation and uniform load.*

Proof: The result follows since the N-node double rooted complete binary tree is a subgraph of N-node hypercube. ∎

Corollary 4 *For any $2^{r_1} \times 2^{r_2} \times \cdots \times 2^{r_k}$ grid, there is a uniform load and dilation-1 embedding in a smaller hypercube.*

Proof: Follows from part-1 of the theorem since the grid is a subgraph of 2^r-node hypercube where $r = r_1 + r_2 + \cdots + r_k$. Figure 2 shows the case for a linear array. ∎

The significance of the above theorem is mainly due to its focus on its selection of the initial large hypercube at a size which allows the minimal load at a level ℓ greater than one. This allows us to realize at least that *even if a graph does not have optimal expansion embedding with unit load, it may have optimal expansion embedding with a higher level of uniform load.* For all the results presented here, we determine the initial load ℓ at its minimum possible level. It will turn out, for example, that uniform load-ℓ and unit dilation embedding of the 3-dimensional $128 \times 128 \times 128$ mesh of trees is not possible with $\ell < 509$, requiring a 14-dimensional hypercube. Under the uniform load requirement, there is no embedding for a hypercube larger than this. On the other hand, if we allow dilation-2 embedding, $\ell = 4$ (or $\ell = r + 1$ for r-dimensional mesh of trees) is possible.

While these positive results are encouraging, it is natural to ask if there are graphs which cannot be embedded with uniform load and $O(1)$ dilation in a small enough hypercube. The corollary below provides an example class of graphs for which dilation may remain at its maximum possible level throughout the steps of contractions.

Corollary 5 *For any network of N nodes with $O(N)$ edges and minimum bisection width at least $O(N)$, there is no $O(1)$-dilation embedding in a smaller hypercube.*

Proof: Follows from Theorem 3, and the result that embedding of such a graph in N-node hypercube requires a dilation of $O(Log N)$ [6]. ∎

In general, edge congestion is another factor which may affect the efficiency of embeddings. We adress the edge congestion problem in [3], and show that for the embedding methods based on dimension contractions, edge congestion cannot make an emulation not work preserving, if the embedding is work preserving from the load and dilation points of view. Hence we are not concerned about congestion in the rest of this paper.

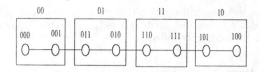

Figure 2: A linear array embedding based on gray codes preserves the initial edge congestion when hypercube is contracted along the lowest dimension. This scheme can be extended to higher dimensional arrays.

4   Load Balanced Embeddings

Load balanced embeddings try to maximize processor utilization beyond that of work preserving embeddings. For an embedding to be load balanced, each hypercube node must emulate an equal number of "busy" guest nodes for each step throughout the computation.

This requirement can be better explained with an example. Suppose we mapped a 127-node complete binary tree to a 16-node hypercube with load-8 everywhere (with the exception of one hypercube node which would receive 7 binary tree nodes). If a normal algorithm is being computed, say for summation of numbers initially stored two values per leaf, all the 64 leaves of the binary tree will be active during the first step. If the leaves are not equally distributed between hypercube nodes, some nodes may be emulating 8 busy leaves while others are emulating 0 or more leaves during the first step. This work imbalance, if repeated at every step of emulation, may cause the emulation to operate much slower than necessary.

This is illustrated in Figure 3 for a smaller example, where 8-node double rooted binary tree is first mapped to the 8-node hypercube and then contracted into 4 nodes. If a normal algorithm is executed on this embedding, in the case of Figure 3.b for instance, processor 11x (where x is a don't care symbol) will operate twice as slow as the other nodes since 11x contains two nodes from the same level. In general, this type of load imbalance depends on the algorithm being run, but if we are able to identify a common pattern which is valid for a large number of computations, it becomes worthwhile to investigate mappings that are particularly efficient for that type of computation. Normal algorithms running on binary trees and mesh of trees have the special level oriented characteristic. Due to widespread validity of this computation pattern, the embedding methods we present in the rest of this paper are designed with this special focus.

4.1   Embedding Large Binary Trees

The main reason we are interested in embedding large binary trees in this paper is because the results are directly applicable for the case of mesh of trees. Nevertheless, the method is important in its own right.

Note that 8-node double-rooted tree contains 4 leaves and 4 internal nodes. If we find a way to map the leaves to the internal nodes, then every processor will be busy while they are emulating the leaves. Once the work of the leaves are done and the results are sent to parents, the rest of the steps can be carried out as

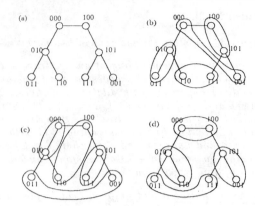

Figure 3: A 8-node double-rooted complete binary tree is embedded in 8-node hypercube, and then b: the hypercube is contracted along dimension 0, c: along dimension 1, and d: along dimension 2.

normal binary tree computations. Such a method for unit dilation embedding of double rooted trees is provided in [3]. Here we show a method for dilation 2 embedding since it is simpler to describe.

We start with a one-to-one embedding method for binary trees, originally due to [1]. This method is based on assigning labels $0 \cdots 2^n - 1$ to the nodes of complete binary tree tracing the nodes in inorder sequence. Figure 4 shows the assignment of labels $0000 \cdots 1111$ where it is assumed that the tree is a left subtree of an auxiliary root node with label 1111. The labels assigned this way on tree nodes indicate which hypercube nodes they are assigned to. In this embedding each right child is connected to its parent through the left child with a dilation of 2.

The following theorem gives the details of the load balanced embedding method [3].

Theorem 4 *For the embedding of a complete binary tree in a cube by inorder labeling, the load of each processor can be doubled by contracting the cube along the lowest dimension. In the resultant smaller cube each leaf will be at distance 0 or 1 from its parent. The internal nodes will be at the same distance from their parent as before contraction.*

To embed a large binary tree to a smaller hypercube, we can start with a unit load embedding of the required size and repeatedly apply the above theorem, shrinking the hypercube along the next lowest dimension. Figure 4 shows this process starting with a 15 node binary tree.

4.2   Embedding Large Mesh of Trees

In this section we finally address the proof of Theorem 2 and Corollary 1. If we use the unit load embedding as in Theorem 1 and then apply contractions, the resulting embedding will not satisfy the uniform load requirement. In fact, since r-dimensional mesh of trees require $(r + 1)N^r - rN^{r-1}$ nodes and this number is not a power of 2, there is no unit load and unit dilation embedding with optimal expansion.

It turns out, however, that if we release the requirement of unit dilation, simple embedding methods exist

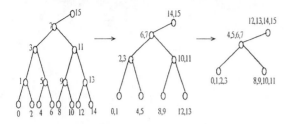

Figure 4: Inorder labeling of complete binary tree lends itself easily for uniform load embedding by contracting the hypercube along the lowest dimension.

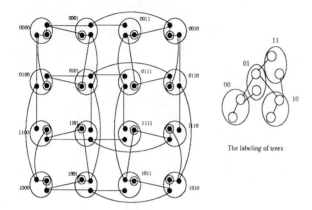

Figure 5: Embedding of 4 × 4 mesh of trees in 16-node hypercube. Each hypercube node emulates exactly one leaf (indicated by double circles), one internal tree node from a row tree, and one internal node from a column tree. The labeling of trees shown applies for both row trees and column trees. For clarity, only the mesh of trees connections are shown.

for any desired sizes of the mesh of trees on hypercubes. As a staring point we provide a method of dilation 2 and initial load $\ell = r + 1$ embedding which *almost* satisfies the uniform load requirement. This method is designed to meet the needs of normal algorithms, so that each hypercube node is equally busy throughout the computation.

Proof of Theorem 2:

From Theorem 4 we know that $(2N-1)$-node complete binary tree can be embedded in N-node hypercube with dilation 2 and uniform load of 2. Note that N^r-node hypercube contains $r LogN$ dimensions. Divide the address bits into r groups of $LogN$ bits each. Since each group defines a N-node subcube, we can use the method for the $2N$-node binary tree embedding for each group. As a result, every node of the hypercube will be assigned one leaf plus one internal tree node for each of the r dimensions. Figure 5 shows the embedding of 4 × 4 mesh of trees in 16-node hypercube. ■

Proof of Corollary 1:

If inorder labeling method is used for embedding binary trees, then we can initially map the mesh of

trees to a large enough hypercube as in Figure 5, and then by Theorem 3 we can contract it down to any desired size by selecting contractions at "appropriate" dimensions, i.e. the lowest dimension at each of the r groups. From Theorem 4, we know that these contractions will preserve the level property of the embedding needed by normal algorithms. ■

5   Summary and Conclusions

Efficient methods for emulating arbitrarily large graphs on fixed size hypercubes are presented with special focus on meshes of trees. Similar methods for grid embedding are sugested briefly. Along with the other main results of this paper, we believe Theorem 3 is a valuable tool when one considers embedding other networks in smaller hypercubes. In general, to prove that a given network can be embedded in a smaller hypercube with uniform load, one needs to prove that there exists a uniform load ℓ embedding of the network in a larger hypercube with optimal expansion. Algorithm specific information may be useful to obtain emulations which are better than work preserving. Some of the other networks which can be embedded this way are outlined in [3]. We expect these results to be particularly useful for mapping large computations onto large grain massively parallel computers, such as the NCube.

References

[1] S.N. Bhatt and I.C.F. Ipsen (1985), "How to embed trees in hypercubes," Rep. No. YALE/DCS/RR-43, Department of Computer Science, Yale University.

[2] K. Efe "Embedding Mesh of Trees in the Hypercube," *Journal of Parallel and Distributed Computing*, March 1991, pp. 222-230.

[3] K. Efe, "Embedding Large Mesh of Trees and related Networks in the Hypercube With Load Balancing," Tech. Rep. TR-93-8-5, Center for Advanced Computer Studies, University of Southwestern Louisiana.

[4] A.K. Gupta, A.J. Boals, N.A. Shervani, and S.E. Hambrusch, "A Lower Bound on Embedding Large Hypercubes into Small Hypercubes," *Congressus Numerantium*, Vol. 78, December 1990, pp. 141-151.

[5] R. Koch, T. Leighton, B. Maggs, S. rao, and A. Rosenberg, "Work-Preserving Emulations of Fixed-connection Networks," *proc. 21st Annual ACM Symposium on Theory of Computing*, 1989, pp. 227-240.

[6] F.T. Leighton, "Introduction to Parallel Algorithms and Architectures: Arrays, Trees, and Hypercubes," Morgan Kaufman, 1992.

[7] J. D. Ullman, "Computational Aspects of VLSI," *Computer Science Press*, 1984.

[8] M. Zubair, and S.N. Gupta, "Embeddings on a Boolean Cube," *BIT*, Vol 29, 1989.

TABLE OF CONTENTS - FULL PROCEEDINGS

Volume I = Architecture
Volume II = Software
Volume III = Algorithms & Applications

A3

A6

DATE DUE

MAY 12 1995   MAY 24 1995

GAYLORD

PRINTED IN U.S.A.